Environmental Bioavailability Handbook

Bioavailability processes are essential for assessing environmental risk, environmental health, and decision-making in environmental management. This practical guide on environmental bioavailability explains its importance and includes specific methods for testing the bioavailability of pollutants in water (freshwater and seawater), soils, and sediments. Detailed methods and techniques are given to measure and assess bioavailability using biotic methods, abiotic methods, and modeling. Case studies help environmental researchers and practitioners understand the applications of bioavailability in field and laboratory.

This book contributes to a deeper understanding of the role of environmental bioavailability in contemporary environmental science in the following ways:

- Provides information on the importance of environmental bioavailability of pollutants.
- Explains, in a "how-to" manner, the use of environmental bioavailability in the context of Environmental Risk Assessment and environmental management decision-making.
- Includes areas of applicability, optimum conditions, and pros and cons for bioavailability methods provided.
- Summarizes state-of-the-art in modeling environmental bioavailability and discusses current best practices.
- Includes case studies illustrating the assessment of bioavailability using different approaches.

This handbook is intended for professionals, researchers, academics, environmental practitioners, and policymakers working in the areas of Environmental Risk Assessment, environmental health, environmental management, sustainability, and related fields. It is also an excellent resource for students studying in any of these fields and looking for a foundational text on environmental bioavailability.

Environmental Assessment and Management

Series Editors: Glenn W. Suter II, U.S. Environmental Protection Agency, Cincinnati, OH, USA and Mona Wells, Ronin Institute, USA

Published Titles

Ecological Causal Assessment
Edited by Susan B. Norton, Susan M. Cormier, and Glenn W. Suter II

Multi-Criteria Decision Analysis: Case Studies in Engineering and the Environment, Second Edition
Igor Linkov, Emily Moberg, Benjamin D. Trump, Boris Yatsalo, and Jeffrey M. Keisler

Management of Water Resources Using Electromechanical Methods
Edited by Guoshuai Liu, Yong Jiang, and Changyong Zhang

Environmental Bioavailability Handbook: Methods and Assessments
Xiaokai Zhang and Mona Wells

For more about this series, please visit https://www.routledge.com/Environmental-Assessment-and-Management/book-series/CRCENVASSANDMAN

Environmental Bioavailability Handbook

Methods and Assessments

Xiaokai Zhang and Mona Wells
With the collaboration of Boling Li

CRC Press
Taylor & Francis Group
Boca Raton London New York

CRC Press is an imprint of the
Taylor & Francis Group, an **informa** business

Designed cover image: Shutterstock

First edition published 2026
by CRC Press
2385 NW Executive Center Drive, Suite 320, Boca Raton FL 33431

and by CRC Press
4 Park Square, Milton Park, Abingdon, Oxon, OX14 4RN

CRC Press is an imprint of Taylor & Francis Group, LLC

ISBN: 9781032404721 (hbk)
ISBN: 9781032409016 (pbk)
ISBN: 9781003355229 (ebk)

DOI: 10.1201/9781003355229

Typeset in Times
by codeMantra

Contents

Preface

In the context of today's increasingly serious global environmental problems, the environmental behavior of pollutants and their ecological risks have become a focus of attention for both the scientific community and the public. Traditional approaches toward the analysis of the total concentration of pollutants, however, often fail to accurately reflect their actual ecotoxicity. Only those amounts of pollutants that can be absorbed by organisms and produce effects – i.e., environmental bioavailability – truly determine the degree of threat a pollutant poses to the ecosystem. A deepening scientific understanding of the importance of environmental bioavailability has led to a paradigm shift in the field of environmental science from "total content analysis" to "bioavailability assessment". It might be said that within this paradigm shift daring exceeds engineering – i.e., much research is conducted, but a relatively modest amount of this has translated into standardized practice.

Research on environmental bioavailability spans a wide range of disciplines, including environmental chemistry, ecotoxicology, analytical chemistry, and modeling. The key challenge uniting studies of environmental bioavailability in all these fields is how to accurately measure and assess the "availability to biota" of pollutants in complex environmental media. Although a large number of studies have been conducted to address this issue, the relevant methods are often scattered across different academic literature or technical reports, and systematic integration is lacking. Yet the value of environmental bioavailability research lies not only in academic innovation but also in its practical guidance for environmental management. Environmental quality standards in many countries and regions are still based on total concentration, which is, at best, a blunt and inadequate tool for Environmental Risk Assessment.

This book has been compiled precisely to address the gap between theory and practice, adducing, in an approachable way, the state of scientific progress in this field. The publication of this book provides a methodological reference for the study of pollutant bioavailability, providing systematized methodological guidance to support researchers, decision-makers, and environmental managers with a set of tools that closely integrates theory and practice. Through the organization of five major sections – from basic concepts, biological methods (true bioavailability), chemical methods (chemoavailability), bioavailability modeling, and case studies – for the first time, the two types of mainstream assessment methods (biomonitoring and chemical substitution) appear under the same framework for comparative analysis, supplemented by modeling tools and examples, to help readers understand the applicable scenarios and limitations of different methods to assess environmental bioavailability.

While results from research are discussed at times, the contents of this book are primarily applied; we hope that overall the approach will also help bridge the divide between researchers and practitioners in future. As a first attempt to systematize a standard of practice in the field of environmental bioavailability, there are inevitably shortcomings in this book, but we hope that it can become a guiding light for scholars, engineers, and decision-makers to illuminate the way forward to achieve more fit-for-purpose outcomes in Environmental Risk Assessment, Environmental Impact Assessment, and the environmental management of pollutants that are increasingly widespread and pernicious.

Xiaokai Zhang

About the Authors

Xiaokai Zhang completed his PhD at the University of Liverpool, focusing on topics involving bioavailability in environmental risk and impact assessment, the use of environmental biotechnology to measure bioavailability, and risk-based environmental management interventions. Currently, he is an associate professor at Jiangnan University in Wuxi, Jiangsu Province, China, where he continues to work on topics in bioavailability-based risk and environmental management, as well as the environmental behavior of biochar and remediation of contaminated soils using biochar. His research in his areas of expertise has been published in several book chapters and in notable environmental science journals including *Environmental Science & Technology*, *Water Research*, and *Environmental Pollution*, with some works rated as highly cited papers.

Mona Wells is an interdisciplinary environmental scientist specializing in Environmental Impact Assessment, Environmental Risk Assessment, and environmental management. Her roles have included posts in different locales internationally as an academic and as an environmental practitioner on projects spanning a broad portfolio including technical project work, research, commercialization, program development, policy advisement, science communication, public consultation, pedagogical development, and continuous improvement.

Section 1

Background

1 Bioavailability and Influencing Factors

1.1 WHAT IS ENVIRONMENTAL BIOAVAILABILITY?

As a result of human activities, an unparalleled number and amount of pollutants have been released into the environment in the form of heavy metals (inorganic pollutants), organic pollutants, and ***emerging pollutants***, i.e., pollutants that are not often controlled or monitored in the environment, including many synthetic or natural substances, such as pharmaceuticals and personal care products, hormones, flame retardants, and per-/polyfluorinated substances (Gavrilescu et al., 2015; Krafft and Riess, 2015; Mas et al., 2010; Vardhan et al., 2019).[1] These pollutants put unprecedented pressure on the environment and pose a serious threat to ecosystems and human health. Therefore, it is of great importance to develop new approaches to effective ***Environmental Risk Assessment,*** a formal process whereby environmental risk associated with pollutant ***exposure*** is quantified and assessed.

In recent years, it has been increasingly recognized that the total amount of a pollutant is not equal to its availability to biota or ***bioavailability***, which is a prerequisite for toxic effects. Environmental Risk Assessment based on the bioavailability of pollutants is gradually being recognized as more fit-for-purpose than assessment based on total concentration, the latter of which contains biounavailable, ergo non-hazardous, pollutants. But how do we characterize environmental bioavailability? There are many different definitions of bioavailability, which vary from discipline to discipline. In the fields of toxicology and pharmacology, the term "bioavailability" was first used to describe how readily a xenobiotic (chemical not naturally produced or expected to be present in an organism) enters the human body following intravenous or oral administration (FDA, 1977). Definitions of environmental bioavailability are quite different from this and have been gradually evolving over time, as summarized by representative examples in Table 1.1. For the purposes of this book, we use the term bioavailability in a manner consistent with the definitions in Table 1.1, i.e., bioavailable pollutant is that readily available to be taken up into an organism from the ***medium*** the organism inhabits at a given time (Semple et al., 2004).

When we talk about bioavailability, there are a few key processes that need to be clarified. ***Sorption*** is a phenomenon that occurs universally in nature and a term that describes the association of a chemical species with a solid phase, i.e., in environmental chemistry, the accumulation of pollutants from solution onto or into surfaces of suspended particulates of soil/sediment (Landrot and Sparks, 2023), though the term can reference organisms as a type of "solid surface" as well. Desorption is the inverse process of sorption. The major mechanisms of association include ***absorption*** (characterizes the process of a solid material/organism taking in or soaking up gases, liquids, or dissolved substances into the interior volume), ***adsorption*** (the ability of solid surfaces to attract and hold gases, liquid components, or dissolved substances to their surfaces), surface precipitation, and polymerization (Dąbrowski, 2001). Sorption is one of the most important mechanisms controlling the rate at which pollutants move in soil and subsurface systems, hence bioavailability. It affects a number of other pathways such as volatilization, hydrolysis, oxidation, biodegradation, and so on.

Sorption relates to two other important concepts, ***partitioning*** and ***distribution***. Partitioning refers to how pollutant concentration partitions, i.e., is divided, between two distinct, immiscible phases—such as organic vs. aqueous, gas vs. liquid, or solid vs. aqueous (Leo et al., 1971). Common

[1] Note: Throughout this book, except for some commonly used Quality Assurance/Quality Control terms, key terminology appears in bold italic on first introduction. See Chapter 2 for a full compilation of terms used in the book and associated definitions.

DOI: 10.1201/9781003355229-2

TABLE 1.1
Representative Examples of Definitions Used with Reference to Environmental Bioavailability in Recent Decades

Definition	Source
If a chemical element may easily transition into a free-ion species or is present in plant roots during a time crucial to plant growth and development, and if its absorption by the roots impacts the plant's life cycle, then it is considered bioavailable to plants.	Sposito (1989)
The degree to which a chemical in a potential source is free for uptake.	Hamelink et al. (1994)
A physiologically driven desorption process.	Peijnenburg et al. (1997)
A chemical's accessibility for absorption and potential harm.	Alexander (2000)
The extent to which a contaminant is available for biological conversion, which is a function of the biological system, physicochemical properties of the contaminant (pollutant), and environmental factors.	Juhasz and Naidu (2000)
Bioavailable: For chemicals, the state of being potentially available for biological uptake by an aquatic organism when that organism is processing or encountering a given environmental medium (e.g., the chemicals that can be extracted by the gills from water as it passes through the respiratory cavity or the chemicals that are absorbed by internal membranes as the organism moves through or ingests sediment).	US EPA (2000)
What is readily available to pass through into an organism from the medium it resides in at any particular time is referred to as a bioavailable substance.	Semple et al. (2004)

examples of partitioning mentioned in this book include the ***n-octanol–water partition coefficient***, which describes how pollutants divide between *n*-octanol (a non-polar solvent) and water, and the ***soil/sediment organic carbon–water partition coefficient***, which explains how pollutants divide between water and solid phases, as influenced by organic carbon content (Hanson et al., 2019). While partitioning describes the division of a pollutant between two immiscible phases at chemical equilibrium, the term distribution is more commonly used in environmental contexts to capture how pollutants behave within complex, heterogeneous solid-phase systems, namely soil and sediment. Though closely related, distribution accounts for factors beyond what phases are present, e.g., mineral composition. Partitioning and distribution can be related, for instance, the soil/sediment organic carbon–water partition coefficient is derived from the soil/sediment–water distribution coefficient and the soil/sediment organic carbon content (Zhu et al., 2004). Because the organic carbon component in soil/sediment acts as a distinct organic phase, analogous to *n*-octanol in octanol–water systems, the division of pollutant across soil/sediment with organic carbon present and water is referred to as a partitioning process.

There are a number of other terms in use relating to bioavailability, many of which will be introduced in this book. In particular, bioavailability is often confused with ***bioaccessibility***; however, these two terms have very different meanings (see Figure 1.1). Bioaccessible pollutant is that which may not be bioavailable at any given point in time, but which may *become* bioavailable, i.e., that which is potentially bioavailable in view of, e.g., changing environmental conditions or the manner in which an organism interacts with the medium containing the bioaccessible pollutant. As reported by Wells et al. (2005), as an example, pollutant bioaccessibility may occur by a mechanism of mediated time-dependent release from environmental materials (notably soils and sediments).

In the natural environment, the biogeochemical processes occurring in different environmental media (e.g., natural waters, soil/sediments, different biota) are complex, interrelated, and may strongly affect the bioavailability of pollutants. The effect of a pollutant on an organism as a result of exposure depends on the interaction between the pollutant and the ***matrix*** within which the pollutant occurs. Matrix refers to the chemical components of an environmental sample excluding the analyte (pollutant). In other words, the matrix is the material in which the analyte or pollutant is suspended in environmental samples. For different environmental media, important physicochemical

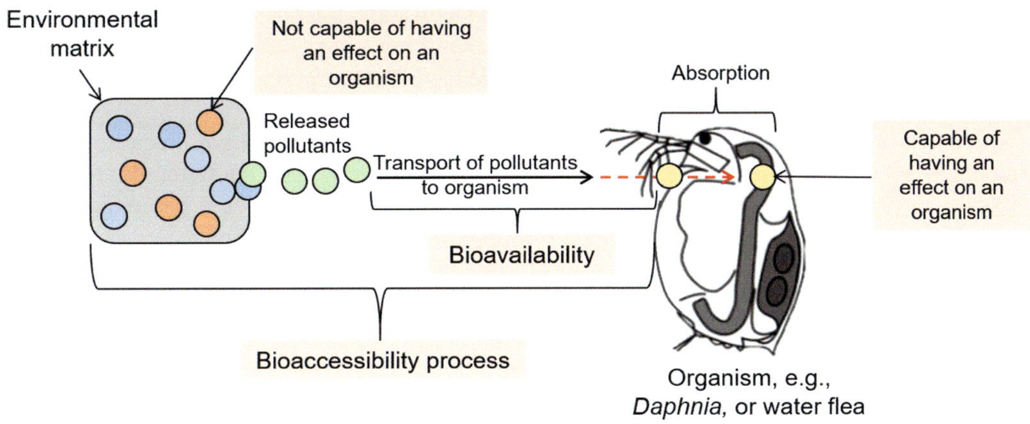

FIGURE 1.1 Conceptual diagram illustrating the process of how bioaccessible pollutants become bioavailable. In the example given, some pollutants may become bioavailable (light green color), i.e., they are bioaccessible (light blue color), through bioaccessibility processes affecting the release of pollutants from environmental matrices. Subsequently, bioavailable pollutants are subject to absorption/uptake by organisms (light yellow color), e.g., such as *Daphnia* sp. (a type of water flea). Some pollutants are neither bioavailable nor bioaccessible; these pollutants could be described as effectively permanently biounavailable (orange color).

factors that affect the bioavailability of a pollutant often relate to the matrix (e.g., natural organic matter, and pH), or other environmental factors (temperature, light) (Naidu et al., 2008; NRC, 2003). These physicochemical factors affect the bioavailability of pollutants by influencing the chemical ***speciation*** of pollutants, discussed in the next section. In addition to physicochemical factors, biological factors, notably chemical reactions associated with metabolism, absorption and excretion, binding and sequestration, and other processes, can affect speciation and the bioavailability of pollutants (Cerniglia, 1984; Sucharitha et al., 2022). Studying the speciation of pollutants is essential to understand their bioavailability. The next two sections discuss speciation and some important specific factors that affect pollutant bioavailability.

1.2 SPECIATION

According to the International Union of Pure and Applied Chemists (IUPAC), a chemical species, in the sense that we use it here, is a "*specific form of an element defined as to … electronic or oxidation state, and/or complex or molecular structure*", with the "complex or molecular structure" being most applicable for environmental bioavailability. Later in this book, when modeling speciation is discussed, the concept of ***components*** and ***species*** will be introduced. So, for instance, while the total concentration of iron (Fe) may represent one chemical component in a system, there may be many different species of Fe such as Fe-humic acid, $Fe(OH)_3$/ferric oxyhydroxide, and many others (Hoffmann, 2005; Namieśnik and Rabajczyk, 2010). Generally, highly soluble and reactive species are more bioavailable as they can easily enter organisms through direct uptake or diffusion across cell membranes (de Paiva Magalhães et al., 2015; Traina and Laperche, 1999). For example, de Paiva Magalhães et al. (2015) showed that the more soluble metals are more bioavailable. Traina and Laperche (1999) found that reactive species of lead (Pb) are more bioavailable than those less reactive. It has been demonstrated that high pH can lead metals to be transformed into less bioavailable species, however, at low pH metals tend to be more bioavailable (Olaniran et al., 2013). Temperature is another factor that may affect pollutant speciation and bioavailability. It has been found that, at high temperatures, pyrene adsorbed by ***ligands*** in the environment is released to form dissolved pyrene species, thus increasing its bioavailability (Zhang et al., 2020a). Other effects are, for instance, non-heavy metal cations in solution compete with heavy metal ions for binding sites of natural ligands, affecting the bioavailability of heavy metals (and ligands) (Song et al., 2014). There are large differences in the properties of heavy metal and organic pollutants, however, translating into differences in the manner in which environmental factors affect pollutant bioavailability, which is discussed in greater detail below.

Speciation, and how it relates to bioavailability, has been the subject of both measurement and modeling (e.g., Tipping et al., 2002; Wells et al., 2013; Zhang et al., 2020b). Often the focus is on speciation in the aqueous phase; however, additional considerations that come into environmental play include gas-phase and gas-aqueous phase reactions, aqueous-solid sorption reactions, and redox reactions. Some pollutants may exist in gaseous forms, such as volatile organic compounds or atmospheric pollutants such as sulfur dioxide (SO_2) and nitrogen oxides (NO_x), which can be absorbed by organisms directly from the atmosphere or via dissolved aqueous species. Gas and aqueous-phase reactions, aqueous solubility of gases, and pollutant vapor pressure are additional important factors controlling the speciation of gaseous pollutants (Brusseau and Chorover, 2019; NRC, 2014).

The speciation of pollutants that participate in sorption processes can be particularly complex. Sorption reactions may be reversible or irreversible, having an impact on speciation and bioavailability. During sorption, factors including properties of the solid phase, temperature, and parameters reflected in sorption isotherms (the relationship between the amount of pollutant in the sorbed state and the amount in the aqueous phase, or vapor phase in the case of vapor-phase sorption[2]) determine and/or characterize pollutant speciation and thereby the mechanisms of sorption and strength of pollutant association with the solid-phase (Gunay Gurer et al., 2021; Nethaji et al., 2013).

Redox systems, explained more in the next section, are always present, even ubiquitous, in environmental media. Redox is an acronym for reduction-oxidation, denoting a chemical reaction in which there is simultaneous occurrence of both an oxidation (loss of electrons from a molecule, an atom, or an ion) and a reduction (gain of electrons) reaction. Redox-linked factors that particularly affect speciation include pH, ***Eh* (*redox potential*)**[3], ***ionic strength***[4] (a measure of the total concentration of ions in a solution), and temperature (Lu et al., 2022; Ma et al., 2021; Peng et al., 2022). An excellent example of how gas-phase/gas-aqueous phase, sorption, and redox reactions affect speciation and bioavailability is the example of mercury (Hg) biogeochemical cycles and speciation (Gojkovic et al., 2022; Lee and Fisher, 2019; Li et al., 2024). Mercury oxidation/reduction and methylation/demethylation processes in particular are influenced/mediated by light (photooxidation), microbes, and organic carbon cycling dynamics, among other factors key to the cycling of Hg, Hg bioavailability, and the consequent danger of Hg to ecosystem and human health (Figure 1.2, Ci et al., 2023; Shah et al., 2021).

As shown in Figure 1.2, oxidation-reduction processes in the atmosphere and surface waters that change the volatile Hg(0) species ("elemental" Hg, the least toxic form) into soluble Hg(II), and vice versa, are a primarily reaction in controlling the Hg cycle. Various forms of atmospheric Hg have brief atmospheric residence periods (minutes to weeks) and are therefore effectively eliminated from the atmosphere by wet and dry deposition into adjacent landscapes (Mason and

---

[2] More formally, a mathematical relationship between a substance sorbed to solid particles, Q (often measured in mol or mass substance per dry mass solid), and the remaining solute concentration of the substance, C (measured in, e.g., mol or mass substance per volume solution), such that Q is parameterized as a function of C, i.e, $Q = f(C)$.

[3] pH: quantitative assessment of a liquid solution's basicity or acidity. Eh: When an electrode made of a redox couple (such as SO_4^{2-}/H_2S) is measured in a galvanic cell against a normal hydrogen electrode, the resultant is known as the redox potential (Eh). It measures an environment's propensity to oxidize or reduce substrates.

[4] Ionic strength, I, reflects the concentration of electrically charged species in a solution as quantified by the expression

$$I = \frac{1}{2}\sum\nolimits_n C_i Z_i^2,$$

where C_i is the molar concentration of ion i, the charge of ion i is Z_i, and the product of C_i and Z_i^2 is summed over the total number of all ions in solution, n. The coefficient of $\frac{1}{2}$ reflects adjustment for both anions and cations being included in the calculation.

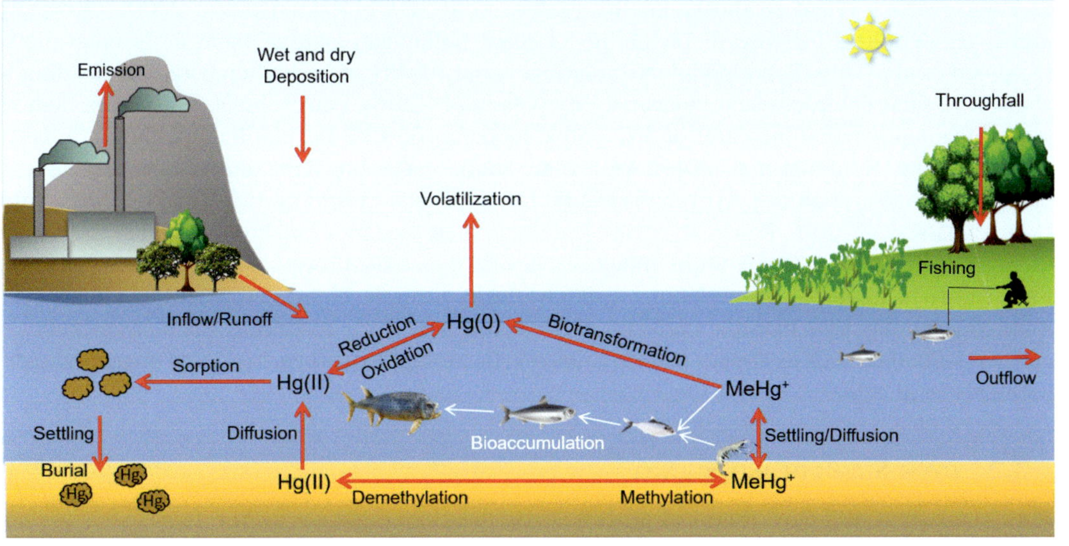

FIGURE 1.2 A schematic showing the cycling of Hg and Hg speciation in aquatic systems. Details and abbreviations are as given in the text. The transfer of $MeHg^+$ between sediment and water column ("Settling/Diffusion", to the right in the figure) occurs in the same manner as shown in greater detail for Hg (II) on the left side in the figure.

Sheu, 2002), ultimately collecting in surface water. As a result of the oxidation of Hg(0), reactive Hg(II) is formed. In surface water, Hg(II) may be reduced back to Hg(0) by biotic (Poulain et al., 2007) or photochemical (Amyot et al., 1994) mechanisms, after which it may be revolatilized, or sorbed to particles as Hg(II) and deposited with aquatic sediments. After burial, under anoxic conditions, bacteria convert some Hg(II) to methylmercury ($MeHg^+$); a demethylation reaction may also happen and convert $MeHg^+$ back to Hg(II) through both biotic and abiotic processes. Organomercury species are generally more toxic than inorganics forms, with $MeHg^+$ being particularly toxic to biota; $MeHg^+$ is typically the most common species of organic Hg. In some circumstances, $MeHg^+$ can undergo biotic transformation back to the elemental Hg(0) form.

Mercury species in the water column are subject to ***bioaccumulation*** by fish (a phenomenon whereby a pollutant accumulates in tissues of organisms as a result of all environmental exposure routes, including water, sediment or suspended particulates, food sources).[5] Per Figure 1.2, fish consumption exposes humans to bioaccumulated $MeHg^+$, i.e., higher bioavailable concentrations, bringing full circle our consideration of how Hg cycling is an excellent example illustrating the importance of understanding speciation in predicting pollutant risk.

1.3 FACTORS AFFECTING BIOAVAILABILITY

1.3.1 Temperature

Environmental temperature affects the solubility of pollutants and the chemical reactions that occur between pollutants and other constituents in the sample matrix and environmental media, thereby

[5] Another term that is used in aquatic toxicology is ***bioconcentration***, which describes the build-up of a pollutant in or on an organism from water as the source of the pollutant. ***Biomagnification*** is yet another related term that signifies an increase in concentration of a pollutant in organisms at successively higher levels in a food chain. Bioconcentration, bioaccumulation, and biomagnification are all applicable to Hg, and biomagnification is particularly problematic because the concentration of $MeHg^+$ increases with each increase in trophic level, increasing the chances of toxic effects from $MeHg^+$ at each successive trophic level.

exerting a strong effect on bioavailability (Kibria et al., 2021; Ragnarsdottir, 2000). Generally, higher temperatures increase the solubility of some pollutants, making them more bioavailable (Koproch et al., 2019). Some pollutants volatilize more readily at higher temperatures, leading to changes in their environmental bioavailability (Ma et al., 2016). For example, it has been demonstrated that temperature increases of 3°C lead to increases of 9.3% in volatilization fluxes for hexachlorobenzene (Komprda et al., 2013), while other studies have shown that increases in temperature decrease the adsorption of polycyclic aromatic hydrocarbons by soil particles and organic matter (Lamichhane et al., 2016; Ren et al., 2018), leading to increased bioavailability to plants and microorganisms. In addition, temperature influences various metabolic processes of organisms, affecting their ability to interact with pollutants as bioavailable. Overall, higher temperatures (notably climate change) may have multiple implications for speciation, likely increasing the bioavailability of certain pollutants, such as copper (Cu), Hg, polycyclic aromatic hydrocarbons, and some pesticides (Schiedek et al., 2007).

1.3.2 Ionic Strength and Salinity

Ionic strength affects most chemical processes and therefore plays a critical role in speciation and bioavailability. Interactions between organic compounds and dissolved metal species within a single solution can be intricate. For organic compounds, organic molecules compete with other solutes in aqueous solutions to become "hydrated". The equilibrium between the dissolved and undissolved forms of a less soluble organic compound shifts in favor of the undissolved form when a second more soluble component is present (Hutacharoen et al., 2017). This effect is expressed generally by the term ***desalting***, a phenomenon whereby a decrease in the solubility of organic compounds results from an increase in the ionic strength of the solution due to dissolved salts. The desalting phenomenon reduces the bioavailability of organic compounds.

The concentration of free ions in solution is decreased with increasing ionic strength, mainly via ***ion pairing*** (ion pair being a term in physics and chemistry used to describe a duplex of charged particles, often charged atoms or molecules, one positive and one negative) and ***complexation*** (complex, a term usually used in chemistry to describe molecules or ensembles formed by coordination bonding through one or more atoms, ions, or molecular ligands donating electrons to a metal center). As a result, the thermodynamic driving force behind the precipitation of solids from the aqueous phase drops, i.e., the formation of ion pairs and complexes shifts equilibria, describing the formation of these species forward, which results in a backward shift in precipitation reactions. In this way, high ionic strength enhances the solubility and ***saturation index*** of salts (George et al., 2023).

In environmental aqueous media in particular, calcium and magnesium ions (Ca^{2+} and Mg^{2+}, respectively, e.g., from carbonates[6]) compete with other divalent metal ions (such as heavy metal ions of cadmium, Cd^{2+}, zinc, Zn^{2+}, Pb^{2+}, and cobalt, Co^{2+}) for binding sites in organisms (Kozlova et al., 2009), acting as blockers of heavy metal entry into cells and reducing heavy metal bioavailability. In most cases, the effect of Ca^{2+} on the bioavailability of heavy metals is greater than that of Mg^{2+}, because Ca^{2+} competes more strongly. Analogously, for fish, for example, Ca^{2+} reduces the permeability of the gills to other ions by binding to the gill surfaces, making the gills positively charged, repelling other cations, and reducing the likelihood that heavy metals will enter fish through the gill surface (McWilliams and Potts, 1978). Therefore, higher concentrations of the Ca^{2+} species may act as a protective factor to reduce the bioavailability of some environmental heavy metal pollutants (De Schamphelaere and Janssen, 2004).

The ionic strength of soil/sediment pore waters also affects pollutant bioavailability in these media. As ionic strength increases, the concentration of cations and anions in soil/sediment pore water increases, which can influence the solubility and mobility of pollutants. As an example, in

[6] Major contributors to ionic strength in many aqueous environments.

soils/sediments, cations such as Ca^{2+} and Mg^{2+} compete with heavy metals for binding sites on soil/sediment particles, and hence affect heavy metal bioavailability, while desalting should make organics more likely to sorb (Acosta et al., 2011; Wang et al., 2020).

For the transition from freshwater to brackish and seawater, ionic strength increases with ***salinity***, and increases in both are reflected in increases in solution electrical conductivity. Salinity is defined as a measure of the total amount of dissolved salts in a solution and, as such, affects the bioavailability of pollutants in the same manner as ionic strength (dissolved salts originating from compounds composed of ions). As an example of a study specific to the effect of salinity on bioavailability, one study showed that higher salt concentrations were conducive to the increased formation of heavy metal complexes with anions (e.g., cadmium chloride complexes $CdCl^{+}$ and $CdCl^{3-}$), and these forms are biounavailable or less bioavailable (de Paiva Magalhães et al., 2015). For organics, studies have shown that under high salinity conditions, organics are more likely to be desorbed from sediments into water and thus increase their bioavailability, which is due to the decrease of pollutant solubility (Topaz, 2023). Additional studies have indicated that the increase in pesticide bioavailability caused by salinity may be associated with elevated salt levels, encouraging the generation of more harmful metabolites (Hutton et al., 2021; Xing et al., 2022).

1.3.3 pH

pH is possibly the most important variable affecting heavy metal speciation and bioavailability in the environment (US EPA, 2007a). It determines the degree of hydrolysis, polymerization, aggregation, and precipitation of heavy metals, as well as proton competition for available ligands (de Paiva Magalhães et al., 2015). Typically, low pH induces dissociation of heavy metals from complexes/natural organic matter, which increases bioavailable heavy metal free-ion species, and thus heavy metal toxicity, whereas at alkaline pH, heavy metals precipitate as oxides and hydroxides (with some exceptions, for instance, for oxides consisting of the alkali metal hydroxides Al^{3+} and Ba^{2+}, which become less bioavailable and thus less toxic) (Bourg and Loch, 1995). In certain cases, however, the bioavailability of metals is lower at low pH. For example, for *Ceriodaphnia dubia* (a type of water flea), *Pimephales promelas* (fathead minnow), and *Hyalella azteca* (a type of scud), the bioavailability of Cd, nickel (Ni), and Zn, hence toxicity, was lower at pH = 6.3 than at 8.3 (Schubauer-Berigan et al., 1993). This is because at low pH the higher relative concentration of H^{+} competes with Cd^{2+}, Ni^{2+}, and Zn^{2+} for biological ligand binding sites, thereby reducing the uptake of heavy metals. At higher pH, the competition is weakened as H^{+} decreases, increasing uptake of these metals by organisms (Niyogi et al., 2008).

The pH of aqueous solutions also has a significant influence on organic pollutants. Organic pollutants that have ionic forms typically have solubility that is several orders of magnitude greater than those that do not ionize. For instance, quetiapine, a pharmaceutical pollutant, dissociates to its cationic form in acidic conditions and is 1,000 times more soluble at pH 2 than at pH 8. Propylparaben (a common pollutant from personal care products), in contrast, has an anionic form at high pH and is 25 times more soluble at pH 9.5 than at pH 7.5 (Shoghi et al., 2013). These types of dissociation reactions have a strong influence on speciation and hence bioavailability of organic pollutants.

1.3.4 Redox Potential

The total reducing or oxidizing capability of a system is expressed in terms of its redox potential, Eh. It has been demonstrated that under positive Eh (oxidizing conditions), metals usually exist in free ions, which are more bioavailable. However, under negative Eh (reducing conditions) conditions, metals more easily form precipitates with media components (Olaniran et al., 2013). In field conditions in particular, ***oxidation-reduction potential*** (ORP) is measured. ORP is a measurement, usually expressed in millivolts, that indicates the tendency of a chemical substance to either oxidize or reduce another chemical substance. In theory, field-measured ORP values can be translated to

Eh, however, in practice the former is more subject to artifacts, and so field-measured ORP values should only be considered as a rough qualitative indicator of Eh. It is widely recognized that the ORP in soil/sediment is also an important parameter controlling the mobility of heavy metals (De Jonge et al., 2012; Kelderman and Osman, 2007; Peng et al., 2009). ORP may be described as a measure of electron availability, and thus affects the valence state, stability, and consequent bioavailability of heavy metals in soil/sediment. Because of the relationship between Eh and ORP, Eh will be used henceforward for simplicity.

Depending on oxygen (O_2) content, the redox zones of soil/sediment can often be vertically stratified into three categories: oxidic (zone of O_2 reduction), suboxic (zone of nitrate, NO_3^-, reduction; manganese, Mn, reduction; Fe reduction), and anoxic (zone of sulfate, SO_4^{2-}, reduction and methanogenesis, i.e., CH_4 production) (Fiedler et al., 2007). Microbial activity and the availability of a carbon supply for the microbes – which use organic materials as electron donors (oxidation) during respiration – are the primary drivers of redox processes in soils/sediments (Valenzuela et al., 2017). The ideal electron acceptor for reduction reactions is molecular oxygen, as long as there is an adequate supply of oxygen. Microbial respiration rangers from ***aerobic*** (alive, functioning, or being only when oxygen is present) to facultative (able to switch from aerobic to anaerobic) and ultimately to ***anaerobic*** (alive, functioning, or taking place without access to free oxygen). Other electron acceptors (NO_3^-, Mn, Fe, and SO_4^{2-}) are employed after molecular oxygen is exhausted (Fiedler et al., 2007). Microbial activity in soils, sediments, and aquatic systems determines Eh, and Eh determines the composition of microbial consortia. For instance, molecular O_2 and the reactions in which it participates are a strong determinant of Eh, and aerobic microorganisms predominate in and are adapted to Ehs of well-oxygenated habitats, whereas anaerobic microbes require an environment with low oxygen to live.

Shifting Eh from redox reactions tends to be relatively slow; however, microbial activity in soil/sediment may substantively enhance redox reaction rates. Under anaerobic conditions, Eh is associated with the formation of precipitates of heavy metals (notably heavy metal sulfides in the SO_4^{2-} reduction zone), thereby reducing bioavailability. For example, of the total amount of Zn added to anaerobic cultures enriched with acetic acid, in which Zn is normally soluble, only 1% dissolved in the aqueous phase (Majumdar et al., 1999). As redox zone/conditions transition from anaerobic to aerobic due to microbial activity within the system (or, e.g., flood impacts or other causes), the bioavailability of heavy metals changes, generally increasing, albeit with a strong dependence on the redox behaviors of organic matter and how this affects the release of adsorbed/complexed heavy metals (Strom et al., 2011).

Eh also plays a crucial role in determining the bioavailability of organic compounds because the electron transfer processes that take place during the degradation of organic pollutants are influenced by Eh. Higher Eh generally favors oxidation reactions, while lower Eh favors reduction reactions, thus Eh has an effect on the availability of electron donors (such as organic matter) or acceptors (importantly, O_2) that aid in the degradation of organic pollutants (Brusseau and Chorover, 2019). The different Ehs that are associated with make-up and activity of different microbial communities in turn affect if, and how quickly, organic pollutants degrade (Crawford et al., 2000; Oktyabrskii and Smirnova, 2012). Because there may be many reactions in sequence in order for any given organic pollutant to completely degrade or mineralize (i.e., degrade to carbon dioxide, CO_2), the entire set of transformation pathways leading to mineralization of organic pollutants is directly impacted by Eh. Different kinds of reactions, including hydrolysis, reduction, oxidation, and methylation, occur or not as a result of varying Eh (Crawford et al., 2000; Sarkar et al., 2021). The metabolites produced as a result of these reactions represent species that may have a strong impact on the bioavailability and environmental impact of organic pollutants inasmuch any given metabolite may be much more, or much less, toxic than the original pollutant precursor. Redox reactions may lead to significant acidification or alkalization of aquatic systems (Bourg and Loch, 1995). These changes will also strongly affect the bioavailability of pollutants in aquatic systems.

1.3.5 Alkalinity

Alkalinity is simply defined as the capacity of water to resist acidification based on the relative concentrations of weak acids and bases in solution; it is measured as the molar amount of bases in a solution that can be converted to uncharged species by a strong acid (typically expressed in milliequivalents base per liter or meq/L). Alkalinity in natural waters typically arises primarily from carbonate and bicarbonate ions, with phosphates, borates, silicates, ammonia, and even conjugate bases of organic acids also contributing to alkalinity, albeit usually to a lesser extent. Higher aqueous alkalinity reduces the bioavailability of metal ions either through ion pair formation between heavy metals and species contributing to alkalinity (Santore et al., 2001) or through the formation of insoluble precipitates (e.g., metal carbonates and phosphates) that reduce their bioavailable concentration (de Paiva Magalhães et al., 2015). The role of alkalinity in affecting bioavailability dominates discussions of heavy metal bioavailability, and is not generally known to be a strong determinant of the bioavailability of organic pollutants, other than the contribution of alkalinity to ionic strength/desalting.

1.3.6 Organic Matter and Ligands

Natural organic matter in the environment affects the bioavailability of organic pollutants through various mechanisms: for instance, it can affect the bioavailability of organic pollutants by partitioning organic pollutants or accelerating the degradation of organic pollutants by providing a carbon source for microorganisms. The many sources of particulate and dissolved organic matter in the environment likewise greatly influence its characteristics. Increasingly organic matter in the environment is of anthropogenic origin (e.g., complex reagents contained in detergents, compost, sewage sludge, food waste digestate) in addition to the naturally occurring organic matter (e.g., humic and fulvic acids) (Reeve, 2002). Given the abundance of oxygen-containing functional groups – ligands including carboxy-, phenoxy-, hydroxy-, and carbonyl-substituents – natural organic matter may have both polar and non-polar domains. It appears that the sources of organic matter in terrestrial environments (e.g., from decomposition of leaves) produce residues that are somewhat more polar than the equivalent residues that are generated from within natural water bodies (so-called autochthonous organic matter, e.g., algae, Liu et al., 2020). These characteristics of varying polarity give natural organic matter a potent ability to adsorb and partition both polar and less polar organic pollutants (Ahmed et al., 2015).

The sorption of organic pollutants onto surfaces containing organic matter has been demonstrated in numerous studies to considerably lower pollutant bioavailability and mobility in the environment (Parsons, 2020). Many studies have demonstrated that the presence of dissolved organic matter inhibits the bioaccumulation of organic compounds that are hydrophobic (Haitzer et al., 1998; Parsons 2020). For instance, *Daphnia magna*'s accumulation of benzo[a]pyrene, pyrene, and 3,3′,4,4′-tetrachlorobiphenyl was decreased in the presence of dissolved organic matter from various sources (Akkanen and Kukkonen, 2001). Organic pollutants sequestered into organic matter may be involved in various processes that lead to flocculation and settling out of the water column, which may reduce their concentration in the water column (Gießl, 1999).

Organic matter can also affect the biodegradation of organic pollutants by microorganisms. Microorganisms use organic matter as a source of energy and nutrients, and the presence of organic matter can stimulate microbial activity, leading to the degradation of organic pollutants in the environment. This process is known as co-metabolism, where microorganisms degrade organic matter and associated organic pollutants at the same time. It has been reported that a polycyclic aromatic hydrocarbon-degrading bacterium biodegraded aqueous-phase phenanthrene and pyrene at rates enhanced by a factor of 1.32 and 1.26, respectively, in the presence of dissolved organic matter at concentrations of 43.14 and 42.15 mg/L compared to in the absence of organic matter (Haftka et al., 2008).

For heavy metal pollutants, complexation is the primary mechanism by which organic matter affects bioavailability (Kumpiene et al., 2008). Complexation occurs with ligands including hydroxyl, carboxylate, cyanide, and amino functional groups within organic matter, as well as ligands such as inorganic carbonates, phosphates, cyanides, and chlorides, depending on whether the organic matter is of natural or anthropogenic origin. Chelating agents can also strongly affect the metal bioavailability. Research has shown that model complexing agents (i.e., mimicking environmental complexation) ethylenediaminetetraacetic acid (EDTA) and nitrilotriacetic acid reduce the bioavailability/effects of heavy metals on fish (Klinck et al., 2005), bivalve mollusks (Funes et al., 2006), bioluminescent bioreporters (Rodea-Palomares et al., 2009; Zhang et al., 2017b), crustaceans (Borgmann et al., 2004), and unicellular algae (Vignati et al., 2010). Heavy metal complexes with higher-molecular weight organic matter in particular are often less bioavailable or are biounavailable because they are unable to cross biological membranes (Richards et al., 1999). The relative stability of aquatic humic substances with heavy metal strongly influences the complexation and transport mechanisms, hence bioavailability, of heavy metals in the environment (Zhang et al., 2020b).

How heavy metal complexes affect bioavailability is not consistently dependable. While most models still count heavy-metal free ion as being the primary if not sole bioavailable species (Groenenberg and Lofts, 2014), i.e., species capable of exerting toxic effects, exceptions are known to occur. For some aquatic organisms, heavy metal chelated into a lipophilic organic ligand is able to passively diffuse through cell membranes; for instance, bioavailability (as demonstrated via toxicity testing) of Cd in the presence of some natural organics increased the toxicity of Cd to rainbow trout (*Oncorhynchus mykiss*) compared to a Cd-only control (Schwartz et al., 2004). For *Vibrio fischeri*, the toxicity of Pb increased in the presence of some concentrations of humic acid, but not others. Humic acid caused a relatively high increase in toxicity at Pb concentrations below 1 mg/L, but for Pb concentrations above 1 mg/L, toxicity was almost constant (Tsiridis et al., 2006). Zhang et al. (2017b), however, found that the presence of humic acid strongly reduced Pb bioavailability to the bioreporter zntA. Modeling is generally able to predict heavy metal bioavailability and bioaccumulation on the basis of free-ion speciation alone; however, the speciation of heavy metals in the presence of organic matter can be complex and depends on many factors, two important ones for further research being the complexity and non-uniformity of natural organic matter and what this entails for exposure. In natural environments, as a fraction of natural organic matter, metallophores,[7] such as siderophores and chalkophores produced by algae or bacteria, can complex with metals, reducing metal bioavailability (Zhang et al., 2020b, 2020c).

Another process with important implications concerning how complexation with metals affects bioavailability is that functional groups in aquatic humic substances may complex with "dangling bonds" on particle surfaces and thereby form a link by which heavy metals are distributed between soluble and solid phases (Pédrot et al., 2010). While complexation is a very important reaction process between heavy metals and organic matter, adsorption and cation exchange reactions are also prominent (Reuter and Perdue, 1977).

1.3.7 Biological Factors

Organisms in the natural environment often have an effect on the bioavailability of pollutants in environmental media, per the examples already given above. Microbial activity, various enzymatic processes, and plant–root interactions are examples of ways in which organisms influence the fate and transport of pollutants in soil and water systems (McLaughlin et al., 1998; Zhao et al., 2022).

[7] A metallophore is a molecule that binds metal ions with exceptionally high affinity and strength, forming very stable complexes. The high binding strength is a key characteristic that distinguishes metallophores from other molecules with lower affinity for metals. Siderophores and chalkophores are, respectively, metallophores with extremely high affinity for binding with Fe and Cu.

Microbial enzymatic processes such as oxidation, reduction, and methylation, for instance, metabolize and break down various organic pollutants, thus reducing their bioavailability (Maier and Gentry, 2015; Mbachu et al., 2020; Megharaj et al., 2011). Microorganisms in the environment may affect heavy metal bioavailability as well, e.g., through extracellular secretions that complex heavy metals (Chen et al., 2015).

In soil systems, plant–root interactions notably play an important role in managing pollutant bioavailability. Plants release exudates that alter the physicochemical properties of the plant root ***rhizosphere***, i.e., the region surrounding a plant root subject to chemical, biological, and physical influences generated by root growth and activity and also a region where microorganisms transform organic and inorganic species in biochemical interaction with the plant root. Processes occurring in the rhizosphere exert a great influence on the mobility and bioavailability of pollutants. Research shows that the formation of complexes between root exudates and free-ion metal species influences the mobility and bioavailability of metals (Chen et al., 2017). Furthermore, plants can take up both organic and inorganic pollutants from soil and water, bioaccumulating such pollutants in plant tissues and decreasing the overall bioavailability of bioaccumulated pollutants to other organisms in the ecosystem (Jeevanantham et al., 2019; Zhang et al., 2017a). Certain plant species hyperbioaccumulate pollutants (Ghori et al., 2016).

1.4 BIOAVAILABILITY: A KEY FOCUS FOR THE FUTURE

In the past, regulations focused on the total concentration of pollutants in the environment without considering pollutant bioavailability. The modern focus on bioavailability is to provide insights into the pathways and mechanisms through which pollutants enter and affect ecosystems (NRC, 2003), because it is increasingly recognized that bioavailability plays a crucial role in determining the actual exposure and potential harm caused by pollutants to the ecosystem (Toutain and Bousquet-Mélou, 2004; Traina and Laperche, 1999). In recent years, government policies and regulations are evolving to incorporate bioavailability as a key factor in evaluating the risks posed by pollutants (Janssen et al., 2003; Maddaloni, 2004). The European Union's Registration, Evaluation, Authorization, and Restriction of Chemicals (REACH) regulation has begun applying bioavailability theories to enhance the evaluation of chemical hazards (ECHA 2023). Similarly, the US Environmental Protection Agency (US EPA) is actively attempting to incorporate bioavailability into its recommendations after recognizing its significance in risk assessment (US EPA, 2007b). By incorporating bioavailability into policies and regulations, decision-makers can develop more accurate and effective risk assessment and management strategies. Given a fixed total concentration of a given pollutant, low bioavailability scenario is considered less of a concern than high bioavailability scenario. This approach of assessing bioavailability rather than focusing on total concentration allows for more targeted allocation of resources and prioritization of pollutants based on their actual risks (Kim et al., 2015). It is foreseeable that bioavailability will play an increasingly important role in Environmental Risk Assessment in the future.

1.5 TRUE BIOAVAILABILITY, CHEMOAVAILABILITY, AND MODELED BIOAVAILABILITY

While bioavailability is often confused or conflated with toxicity, the same is true of chemoavailability. "True bioavailability" refers to a measurement that involves an organism; hence, results are a true reflection of bioavailability. A number of chemical methods have been developed, whereby a chemical proxy that is intended to mimic true bioavailability is used in testing. These abiotic or non-biological methods for estimating bioavailability have been referred to as ***chemoavailability***

(Wells, 2012). Put another way, if bioavailability is defined as a pollutant's availability to biota, then chemoavailability is the pollutant's availability to a chemical proxy. Well-known examples of chemoavailability proxies include the membranes used in DGT (Diffusive Gradients in Thin-Films) (Zhang et al., 2014), chelate extractions (Cao et al., 2008), permeable liquid membranes (Marr et al., 1990), anodic stripping voltammetry (Abollino et al., 2019), and Donnan dialysis (Chen et al., 2023). The putative "chemoavailable" proportion of pollutant extracted or sequestered from a sample matrix by these methods is measured and then subjected to instrumental analysis in these chemoavailability procedures.

We distinguish between true bioavailability and chemoavailability here as, strictly speaking, chemoavailability does not reflect the bioavailability of pollutants, as will be discussed in some chapters and case studies herein and is otherwise well known in the literature. It has been demonstrated that some organisms may sense the bioavailability of chemical substances in their surroundings and, in some situations, react in a way that modifies the substances' bioavailability, e.g., by sequestering a toxic pollutant or use of a pollutant as a carbon source/for nutritive purposes (Wells, 2012; Zhang et al., 2024). In general, however, while chemoavailability methods may be a crude or inadequate, substitute for bioavailability, they may be in many cases superior to the even more crude/inadequate measurement of total concentration. In addition, it usually takes much more time and effort to determine the true bioavailability of pollutants, and chemoavailability methods often have the advantage of convenience. An exception to this generalization is the use of bioreporters (discussed herein) to determine bioavailability, which makes this form of biotechnology an exciting avenue to pursue because it combines the authenticity of a true bioavailability measurement while being even more convenient than chemoavailability measurements in common use.

Within the last two decades, modeling true bioavailability and chemoavailability has increasingly been the topic of research focus. If chemoavailability methods are convenient, modeling, when models are validated, has the potential to be an even more facile method for estimating bioavailability, vastly reducing the need for costly and time-consuming laboratory analyses. Approaches such as modeling and the use of bioreporter biotechnology streamline Environmental Risk Assessment and management processes and reduce overall costs with efficacy.

1.6 USE OF THIS BOOK

While the critical role of bioavailability in environmental decision-making has been increasingly recognized, there is as yet no standard resource for environmental practitioners to refer to for guidance and methodology in estimating bioavailability. This book is a first effort to provide both guidance and methods for field and laboratory techniques that are routine or in emergent use to measure bioavailability. The book is structured in five sections. This chapter has provided orientation and guidance concerning key concepts in environmental bioavailability and factors affecting bioavailability, and other chapters in this section provide needed background to the book or to several chapters in the book. The next three sections are methods-oriented, with the first two of these sections consisting of different methods or techniques to measure bioavailability and chemoavailability. Section 2 will deal with biotic methods, or true bioavailability, and Section 3 will deal with abiotic methods, i.e., chemoavailability. Section 4 will cover modeling of bioavailability, with a broad overview of models in common use, and a method for one of the most widely accepted methods in use. Table 1.2 summarizes the methods discussed in Sections 2–4. Finally, Section 5 will provide illustrative case studies on environmental bioavailability that cover true bioavailability, chemoavailability, and modeling bioavailability.

TABLE 1.2
Summary of Methods Covered in This Book

Basis of Method	Suitable Environmental Media	Pollutant
Bivalve mollusks, TB	Freshwater to seawater	Heavy metals and organic pollutants
Worms, TB	Submerged soils and sediments	Heavy metals and organic pollutants
Macrophytes, TB	Freshwater, seawater, submerged soils and sediments	Heavy metals and organic pollutants
Whole-cell bioreporter, TB	Freshwater, seawater, soils, and sediments	Heavy metals and organic pollutants
Diffusive Gradients in Thin-Films, CA	Freshwater, seawater, soils, and sediments	Heavy metals and organic pollutants
Permeation Liquid Membrane technology, CA	Freshwater to seawater	Cd, Cu, Ni, Pb, and Zn
Chelating resin, CA	Freshwater to seawater	Ag, As, Cd, Co, Cr(III), Cu, Ni, Pb, Zn
Tenax extraction, CA	Soils and sediments	Organic pollutants
Speciation, model	Freshwater	Heavy metals

Abbreviations: CA, chemoavailability; TB, true bioavailability.

REFERENCES

Abollino, O., Giacomino, A., Malandrino, M., 2019. In: Worsfold, P., Poole, C., Townshend, A., Miró, M. (eds.) *Encyclopedia of Analytical Science*, 3rd Edition, pp. 238–257. Academic Press, Cambridge.

Acosta, J.A., Jansen, B., Kalbitz, K., Faz, A., Martínez-Martínez, S., 2011. Salinity increases mobility of heavy metals in soils. *Chemosphere* 85(8), 1318–1324. https://doi.org/10.1016/j.chemosphere.2011.07.046

Ahmed, A.A., Thiele-Bruhn, S., Aziz, S.G., Hilal, R.H., Elroby, S.A., Al-Youbi, A.O., Leinweber, P., Kühn, O., 2015. Interaction of polar and nonpolar organic pollutants with soil organic matter: Sorption experiments and molecular dynamics simulation. *Sci. Total Environ.* 508, 276–287. https://doi.org/10.1016/j.scitotenv.2014.11.087

Akkanen, J., Kukkonen, J.V., 2001. Effects of water hardness and dissolved organic material on bioavailability of selected organic chemicals. *Environ. Toxicol. Chem.* 20(10), 2303–2308. https://doi.org/10.1002/etc.5620201024

Alexander, M., 2000. Aging, bioavailability, and overestimation of risk from environmental pollutants. *Environ. Sci. Technol.* 34(20), 4259–4265. https://doi.org/10.1021/es001069

Amyot, M., McQueen, D.J., Mierle, G., Lean, D.R., 1994. Sunlight-induced formation of dissolved gaseous mercury in lake waters. *Environ. Sci. Technol.* 28(13), 2366–2371. https://doi.org/10.1021/es00062a022

Borgmann, U., Norwood, W.P., Dixon, D.G., 2004. Re-evaluation of metal bioaccumulation and chronic toxicity in *Hyalella azteca* using saturation curves and the biotic ligand model. *Environ. Pollut.* 131(3), 469–484. https://doi.org/10.1016/j.envpol.2004.02.010

Bourg, A.C.M., Loch, J.P.G., 1995. Mobilization of heavy metals as affected by pH and redox conditions. In: Salomons, W., Stigliani, W.M. (eds.) *Biogeodynamics of Pollutants in Soils and Sediments: Risk Assessment of Delayed and Non-Linear Responses*, pp. 87–102. Springer, Berlin, Heidelberg.

Brusseau, M.L., Chorover, J., 2019. Chemical processes affecting contaminant transport and fate. In: Brusseau, M.L., Pepper, I.L., Gerba, C.P. (eds.) *Environmental and Pollution Science*, 3rd Edition, pp. 113–130. Academic Press, Cambridge.

Cao, A., Cappai, G., Carucci, A., Lai, T., 2008. Heavy metal bioavailability and chelate mobilization efficiency in an assisted phytoextraction process. *Environ. Geochem. Health* 30(2), 115–119. https://doi.org/10.1007/s10653-008-9136-2

Cerniglia, C.E., 1984. Microbial metabolism of polycyclic aromatic hydrocarbons. *Adv. Appl. Microbiol.* 30, 31–71. https://doi.org/10.1080/10406639408014720

Chen, A., Shang, C., Zeng, G., Chen, G., Shao, J., Zhang, J., Huang, H., 2015. Extracellular secretions of *Phanerochaete chrysosporium* on Cd toxicity. *Int. Biodeter. Biodegr.* 105, 73–79. https://doi.org/10.1016/j.ibiod.2015.08.017

Chen, H., Rose, M., Fleming, M., Souizi, S., Shashvatt, U., Blaney, L., 2023. Recent advances in Donnan dialysis processes for water/wastewater treatment and resource recovery: A critical review. *Chem. Eng. J.* 455, 140522. https://doi.org/10.1016/j.cej.2022.140522

Chen, Y.T., Wang, Y., Yeh, K.C., 2017. Role of root exudates in metal acquisition and tolerance. *Curr. Opin. Plant Biol.* 39, 66–72. https://doi.org/10.1016/j.pbi.2017.06.004

Ci, Z., Tang, X., Shen, W., Chen, B., 2023. Mercury (Hg) in the "skin" of the ocean: Dissolved gaseous Hg, total Hg, and Hg redox chemistry in sea surface microlayer and implication for air–Sea Hg exchange. *Environ. Sci. Technol.* 57(40), 15037–15046. https://doi.org/10.1021/acs.est.3c05170

Crawford, J.J., Traina, S.J., Tuovinen, O.H., 2000. Bacterial degradation of atrazine in redox potential gradients in fixed-film sand columns. *Soil Sci. Soc. Amer. J.* 64(2), 624–634. https://doi.org/10.2136/sssaj2000.642624x

Dąbrowski, A., 2001. Adsorption–from theory to practice. *Adv. Colloid Interfac.* 93(1), 135–224. https://doi.org/10.1016/S0001-8686(00)00082-8

De Jonge, M., Teuchies, J., Meire, P., Blust, R., Bervoets, L., 2012. The impact of increased oxygen conditions on metal-contaminated sediments part II: Effects on metal accumulation and toxicity in aquatic invertebrates. *Water Res.* 46(10), 3387–3397. https://doi.org/10.1016/j.watres.2012.01.052

de Paiva Magalhães, D., da Costa Marques, M.R., Baptista, D.F., Buss, D.F., 2015. Metal bioavailability and toxicity in freshwaters. *Environ. Chem. Lett.* 13(1), 69–87. https://doi.org/10.1007/s10311-015-0491-9

De Schamphelaere, K.A.C., Janssen, C.R., 2004. Development and field validation of a biotic ligand model predicting chronic copper toxicity to *Daphnia magna*. *Environ. Toxicol. Chem.* 23(6), 1365–1375. https://doi.org/10.1897/02-626

ECHA. 2023. Guidance on Information Requirements and Chemical Safety Assessment. European Chemicals Agency

FDA. 1977. FDA Issued Final Regulations in Part 320(21CFR320) Establishing Definitions and Requirements of rBA and BE Studies. Food and Drug Administration.

Fiedler, S., Vepraskas, M.J., Richardson, J.L., 2007. Soil redox potential: Importance, field measurements, and observations. In: Sparks, D.L. (ed.) *Advances in Agronomy*, pp. 1–54. Academic Press, Cambridge.

Funes V., Alhama J., Navas JI., López-Barea J., Peinado J., 2006. Ecotoxicological effects of metal pollution in two mollusc species from the Spanish South Atlantic littoral. *Environ. Pollut.* 139, 214–223. https://doi.org/10.1016/j.envpol.2005.05.016

Gavrilescu, M., Demnerová, K., Aamand, J., Agathos, S., Fava, F., 2015. Emerging pollutants in the environment: Present and future challenges in biomonitoring, ecological risks and bioremediation. *New Biotechnol.* 32(1), 147–156. https://doi.org/10.1016/j.nbt.2014.01.001

George, S., Mattei, M.R., Frunzo, L., Esposito, G., van Hullebusch, E.D., Fermoso, F.G., 2023. Dynamic modelling the effects of ionic strength and ion complexation on trace metal speciation during anaerobic digestion. *J. Environ. Manage.* 343, 118144. https://doi.org/10.1016/j.jenvman.2023.118144

Ghori, Z., Iftikhar, H., Bhatti, M.F., Nasar um, M., Sharma, I., Kazi, A.G., Ahmad, P., 2016. Phytoextraction: The use of plants to remove heavy metals from soil. In: Ahmad, P. (ed.) *Plant Metal Interaction: Emerging Remediation Techniques*, pp. 385–409. Elsevier, Amsterdam.

Gießl, H., 1999. DOM-Flocculation: A suitable approach for separating free and DOM-bound herbicides?. *Environ. Sci. Pollut. Res.* 6, 77–82. https://doi.org/10.1007/BF02987552

Gojkovic, Z., Skrobonja, A., Funk, C., Garbayo, I., Vílchez, C., 2022. The role of microalgae in the biogeochemical cycling of methylmercury (MeHg) in aquatic environments. *Phycology* 2, 344–362. https://doi.org/10.3390/phycology2030019

Groenenberg, J.E., Lofts, S., 2014 The use of assemblage models to describe trace element partitioning, speciation, and fate: A review. *Environ. Toxicol. Chem.* 33(10), 2181–2196. https://doi.org/10.1002/etc.2642

Gunay Gurer, A., Aktas, K., Ozkaleli Akcetin, M., Erdem Unsar, A., Asilturk, M., 2021. Adsorption isotherms, thermodynamics, and kinetic modeling of methylene blue onto novel carbonaceous adsorbent derived from bitter orange peels. *Water Air Soil Pollut.* 232(4), 138. https://doi.org/10.1007/s11270-021-05090-7

Haftka, J.J., Parsons, J.R., Govers, H.A., Ortega-Calvo, J.J., 2008. Enhanced kinetics of solid-phase microextraction and biodegradation of polycyclic aromatic hydrocarbons in the presence of dissolved organic matter. *Environ. Toxicol. Chem.* 27(7), 1526–1532. https://doi.org/10.1897/07-544.1

Haitzer, M., Höss, S., Traunspurger, W., Steinberg, C., 1998. Effects of dissolved organic matter (DOM) on the bioconcentration of organic chemicals in aquatic organisms—A review. *Chemosphere* 37(7), 1335–1362. https://doi.org/10.1016/S0045-6535(98)00117-9

Hamelink, J., Landrum, P.F., Bergman, H., Benson, W.H., 1994. *Bioavailability: Physical, Chemical, and Biological Interactions*. CRC Press, Karabas.

Hanson, K.B., Hoff, D.J., Lahren, T.J., Mount, D.R., Squillace, A.J., Burkhard, L.P. 2019. Estimating n-octanol-water partition coefficients for neutral highly hydrophobic chemicals using measured n-butanol-water partition coefficients. *Chemosphere* 218, 616–623. https://doi.org/10.1016/j.chemosphere.2018.11.141

Hoffmann, P., 2005. Speciation of iron. In: Cornelis, R., Caruso, J., Crews, H., Heumann, K. (eds.) *Handbook of Elemental Speciation II – Species in the Environment, Food, Medicine and Occupational Health*, pp. 200–217. John Wiley & Sons, Ltd., New Jersey.

Hutacharoen, P., Dufal, S., Papaioannou, V., Shanker, R.M., Adjiman, C.S., Jackson, G., Galindo, A., 2017. Predicting the solvation of organic compounds in aqueous environments: From alkanes and alcohols to pharmaceuticals. *Ind. Eng. Chem. Res.* 56(38), 10856–10876. https://doi.org/10.1021/acs.iecr.7b00899

Hutton, S.J., St Romain, S.J., Pedersen, E.I., Siddiqui, S., Chappell, P.E., White, J.W., Armbrust, K.L., Brander, S.M., 2021. Salinity alters oxicity of commonly used pesticides in a model euryhaline fish species (*Menidia beryllina*). *Toxics* 9(5). https://doi.org/10.3390/toxics9050114

Janssen, C.R., Heijerick, D.G., Schamphelaere, K.A.C.D., Allen, H.E., 2003. Environmental risk assessment of metals: Tools for incorporating bioavailability. *Environ. Int.* 28(8), 793–800. https://doi.org/10.1016/S0160-4120(02)00126-5

Jeevanantham, S., Saravanan, A., Hemavathy, R., Kumar, P.S., Yaashikaa, P., Yuvaraj, D., 2019. Removal of toxic pollutants from water environment by phytoremediation: A survey on application and future prospects. *Environ. Technol. Innov.* 13, 264–276. https://doi.org/10.1016/j.eti.2018.12.007

Juhasz, A.L., Naidu, R., 2000. Bioremediation of high molecular weight polycyclic aromatic hydrocarbons: A review of the microbial degradation of benzoapyrene. *Int. Biodeter. Biodegr.* 45(1–2), 57–88. https://doi.org/10.1016/S0964-8305(00)00052-4

Kelderman, P., Osman, A.A., 2007. Effect of redox potential on heavy metal binding forms in polluted canal sediments in Delft (The Netherlands). *Water Res.* 41(18), 4251–4261. https://doi.org/10.1016/j.watres.2007.05.058

Kibria, G., Nugegoda, D., Rose, G., Haroon, A.K.Y., 2021. Climate change impacts on pollutants mobilization and interactive effects of climate change and pollutants on toxicity and bioaccumulation of pollutants in estuarine and marine biota and linkage to seafood security. *Mar. Pollut. Bull.* 167, 112364. https://doi.org/10.1016/j.marpolbul.2021.112364

Kim, R.Y., Yoon, J.K., Kim, T.S., Yang, J.E., Owens, G., Kim, K.R., 2015. Bioavailability of heavy metals in soils: Definitions and practical implementation—A critical review. *Environ. Geochem. Health* 37(6), 1041–1061. https://doi.org/10.1007/s10653-015-9695-y

Klinck, J., Dunbar, M., Brown, S., Nichols, J., Winter, A., Hughes, C., Playle, R.C., 2005. Influence of water chemistry and natural organic matter on active and passive uptake of inorganic mercury by gills of rainbow trout (*Oncorhynchus mykiss*). *Aquat. Toxicol.* 72(1), 161–175. https://doi.org/10.1016/j.aquatox.2004.11.013

Komprda, J., Komprdová, K., Sáňka, M., Možný, M., Nizzetto, L., 2013. Influence of climate and land use change on spatially resolved volatilization of persistent organic pollutants (POPs) from background soils. *Environ. Sci. Technol.* 47(13), 7052–7059. https://doi.org/10.1021/es3048784

Koproch, N., Dahmke, A., Köber, R., 2019. The aqueous solubility of common organic groundwater contaminants as a function of temperature between 5 and 70°C. *Chemosphere* 217, 166–175. https://doi.org/10.1016/j.chemosphere.2018.10.153

Kozlova, T., Wood, C.M., McGeer, J.C., 2009. The effect of water chemistry on the acute toxicity of nickel to the cladoceran *Daphnia pulex* and the development of a biotic ligand model. *Aquat. Toxicol.* 91(3), 221–228. https://doi.org/10.1016/j.aquatox.2008.11.005

Krafft, M.P., Riess, J.G., 2015. Per-and polyfluorinated substances (PFASs): Environmental challenges. *Curr. Opin. Colloid Interface Sci.* 20(3), 192–212. https://doi.org/10.1016/j.cocis.2015.07.004

Kumpiene, J., Lagerkvist, A., Maurice, C., 2008. Stabilization of As, Cr, Cu, Pb and Zn in soil using amendments–a review. *Waste Manage.* 28(1), 215–225. https://doi.org/10.1016/j.wasman.2006.12.012

Lamichhane, S., Krishna, K.B., Sarukkalige, R., 2016. Polycyclic aromatic hydrocarbons (PAHs) removal by sorption: A review. *Chemosphere* 148, 336–353. https://doi.org/10.1016/j.chemosphere.2016.01.036

Landrot, G., Sparks, D.L., 2023. Sorption–desorption kinetics. In: Goss, M.J., Oliver, M. (eds.) *Encyclopedia of Soils in the Environment: Volume 2–Chemistry*, 2nd Edition. Academic Press, Cambridge.

Lee, C.S., Fisher, N.S., 2019. Microbial generation of elemental mercury from dissolved methylmercury in seawater. *Limnol. Oceanogr.* 64, 679–693. https://doi.org/10.1002/lno.11068

Leo, A., Hansch, C., Elkins, D., 1971. Partition coefficients and their uses. *Chem. Rev.* 71(6), 525–616. https://doi.org/10.1021/cr60274a001

Li, S., Li, Z., Wu, M., Zhou, Y., Tang, W., Zhong, H., 2024. Mercury transformations in algae, plants, and animals: The occurrence, mechanisms, and gaps. *Sci. Total Environ.* 911, 168690. https://doi.org/10.1016/j.scitotenv.2023.168690

Liu, S., He, Z., Tang, Z., Liu, L., Hou, J., Li, T., Zhang, Y., Shi, Q., Giesy, J.P., Wu, F., 2020. Linking the molecular composition of autochthonous dissolved organic matter to source identification for freshwater lake ecosystems by combination of optical spectroscopy and FT-ICR-MS analysis. *Sci. Total Environ.* 703, 134764. https://doi.org/10.1016/j.scitotenv.2019.134764

Lu, H.L., Li, K.W., Nkoh, J.N., He, X., Xu, R.K., Qian, W., Shi, R.Y., Hong, Z.N., 2022. Effects of pH variations caused by redox reactions and pH buffering capacity on Cd(II) speciation in paddy soils during submerging/draining alternation. *Ecotox. Environ. Saf.* 234, 113409. https://doi.org/10.1016/j.ecoenv.2022.113409

Ma, J., Hung, H., Macdonald, R.W., 2016. The influence of global climate change on the environmental fate of persistent organic pollutants: A review with emphasis on the Northern Hemisphere and the Arctic as a receptor. *Glob. Planet. Change* 146, 89–108. https://doi.org/10.1016/j.gloplacha.2016.09.011

Ma, Z., Liu, W., Yang, W., Li, W., Han, B., 2021. Temperature effects on redox potentials and implications to semiconductor photocatalysis. *Fuel* 286, 119490. https://doi.org/10.1016/j.fuel.2020.119490

Maddaloni, M.A., 2004. Bioavailability of soil-borne chemicals: A regulatory perspective. *Hum. Ecol. Risk Assess.* 10(4), 657–663. https://doi.org/10.1080/10807030490484246

Maier, R.M., Gentry, T.J., 2015. *Environmental Microbiology*, pp. 377–413, Elsevier, Amsterdam.

Majumdar, P., Bandyopadhyay, S., Bhattacharya, S.K., 1999. Effects of heavy metals on the biodegradation of organic compounds. In: Hinchee, R.E. (ed.) *Applied Biotechnology for Site Remediation*, pp. 354–359. Lewis Publishers, Boca Raton.

Marr, R., Bart, H.J., Draxler, J., 1990. Liquid membrane permeation. *Chem. Eng. Process.* 27(1), 59–64. https://doi.org/10.1016/0255-2701(90)85007-Q

Mas, S., de Juan, A., Tauler, R., Olivieri, A.C., Escandar, G.M., 2010. Application of chemometric methods to environmental analysis of organic pollutants: A review. *Talanta* 80(3), 1052–1067. https://doi.org/10.1016/j.talanta.2009.09.044

Mason, R.P., Sheu, G.R., 2002. Role of the ocean in the global mercury cycle. *Glob. Biogeochem. Cycles.* 16(4), 40–41. https://doi.org/10.1029/2001GB001440

Mbachu, A., Chukwura, E., Mbachu, N., 2020. Role of microorganisms in the degradation of organic pollutants: A review. *Energy Environ. Eng.* 7(1), 1–11. https://doi.org/10.13189/eee.2020.070101

McLaughlin, M.J., Smolders, E., Merckx, R., 1998. Soil-root interface: Physicochemical processes. *Soil Chem. Ecosyst. Health* 52, 233–277. https://doi.org/10.2136/sssaspecpub52.c9

McWilliams, P.G., Potts, W.T.W., 1978. The effects of pH and calcium concentrations on gill potentials in the Brown Trout, *Salmo trutta*. *J. Comp. Physiol.* 126(3), 277–286. https://doi.org/10.1007/bf00688938

Megharaj, M., Ramakrishnan, B., Venkateswarlu, K., Sethunathan, N., Naidu, R., 2011. Bioremediation approaches for organic pollutants: A critical perspective. *Environ. Int.* 37(8), 1362–1375. https://doi.org/10.1016/j.envint.2011.06.003

Naidu, R., Semple, K.T., Megharaj, M., Juhasz, A.L., Bolan, N.S., Gupta, S.K., Clothier, B.E., Schulin, R., 2008. Chemical bioavailability in terrestrial environments. In: Hartemink, A.E., McBratney, A.B., Naidu, R. (eds.) *Developments in Soil Science*, pp. 39–51. Elsevier, Amsterdam.

Namieśnik, J. Rabajczyk, A., 2010. The speciation and physico-chemical forms of metals in surface waters and sediments. *Chem. Spec. Bioavailab.* 22(1), 1–24. https://doi.org/10.3184/095422910X12632119406391

Nethaji, S., Sivasamy, A. and Mandal, A.B., 2013. Adsorption isotherms, kinetics and mechanism for the adsorption of cationic and anionic dyes onto carbonaceous particles prepared from *Juglans regia* shell biomass. *Int. J. Environ. Sci. Technol.* 10(2), 231–242. https://doi.org/10.1007/s13762-012-0112-0

Niyogi, S., Kent, R. Wood, C.M., 2008. Effects of water chemistry variables on gill binding and acute toxicity of cadmium in rainbow trout (*Oncorhynchus mykiss*): A biotic ligand model (BLM) approach. *Comp. Biochem. Phys. C* 148(4), 305–314. https://doi.org/10.1016/j.cbpc.2008.05.015

NRC. 2003. *Bioavailability of Contaminants in Soils and Sediments: Processes, Tools, and Applications*. The National Academies Press, Washington, DC.

NRC. 2014. *A Framework to Guide Selection of Chemical Alternatives*. The National Academies Press, Washington, DC.

Oktyabrskii, O.N., Smirnova, G.V., 2012. Redox potential changes in bacterial cultures under stress conditions. *Microbiology* 81(2), 131–142. https://doi.org/10.1134/s0026261712020099

Olaniran, A.O., Balgobind, A., Pillay, B., 2013. Bioavailability of heavy metals in soil: Impact on microbial biodegradation of organic compounds and possible improvement strategies. *Int. J. Mol. Sci.* 14(5), 10197–10228. https://doi.org/10.3390/ijms140510197

Parsons, J.R., 2020. In: Ortega-Calvo, J.J., Parsons, J.R. (eds.) *Bioavailability of Organic Chemicals in Soil and Sediment*, pp. 113–128. Springer, Berlin, Heidelberg.

Pédrot, M., Dia, A., Davranche, M., 2010. Dynamic structure of humic substances: Rare earth elements as a fingerprint. *J. Colloid Interf. Sci.* 345(2), 206–213. https://doi.org/10.1016/j.jcis.2010.01.069

Peijnenburg, W., Posthuma, L., Eijsackers, H., Allen, H., 1997. A conceptual framework for implementation of bioavailability of metals for environmental management purposes. *Ecotoxicol. Environ. Saf.* 37(2), 163–172. https://doi.org/10.1006/eesa.1997.1539

Peng, J.F., Song, Y.H., Yuan, P., Cui, X.Y., Qiu, G.l., 2009. The remediation of heavy metals contaminated sediment. *J. Hazard. Mater.* 161(2), 633–640. https://doi.org/10.1016/j.jhazmat.2008.04.061

Peng, X.X., Gai, S., Cheng, K., Yang, F., 2022. Roles of humic substances redox activity on environmental remediation. *J. Hazard. Mater.* 435, 129070. https://doi.org/10.1016/j.jhazmat.2022.129070

Poulain, A.J., Chadhain, S.M.N., Ariya, P.A., Amyot, M., Garcia, E., Campbell, P.G.C., Zylstra, G.J., Barkay, T., 2007. Potential for mercury reduction by microbes in the high arctic. *Appl. Environ. Microbiol.* 73(7), 2230–2238. https://doi.org/10.1128/AEM.02701-06

Ragnarsdottir, K.V., 2000. Environmental fate and toxicology of organophosphate pesticides. *J. Geol. Soc.* 157(4), 859–876. https://doi.org/10.1144/jgs.157.4.859

Reeve, R., 2002. *Introduction to Environmental Analysis*. Wiley, Hoboken.

Ren, X., Zeng, G., Tang, L., Wang, J., Wan, J., Liu, Y., Yu, J., Yi, H., Ye, S., Deng, R., 2018. Sorption, transport and biodegradation–An insight into bioavailability of persistent organic pollutants in soil. *Sci. Total Environ.* 610, 1154–1163. https://doi.org/10.1016/j.scitotenv.2017.08.089

Reuter, J.H., Perdue, E.M. 1977. Importance of heavy metal-organic matter interactions in natural waters. *Geochim. Cosmochim. Acta* 41(2), 325–334. https://doi.org/10.1016/0016-7037(77)90240-X

Richards, J.G., Burnison, B.K., Playle, R.C., 1999. Natural and commercial dissolved organic matter protects against the physiological effects of a combined cadmium and copper exposure on rainbow trout (*Oncorhynchus mykiss*). *Can. J. Fish. Aquat. Sci.* 56(3), 407–418. https://doi.org/10.1139/f98-187

Rodea-Palomares, I., González-García, C., Leganés, F., Fernández-Piñas, F., 2009. Effect of pH, EDTA, and anions on heavy metal toxicity toward a bioluminescent cyanobacterial bioreporter. *Arch. Environ. Con. Toxicol.* 57(3), 477–487. https://doi.org/10.1007/s00244-008-9280-9

Santore, R.C., Di Toro, D.M., Paquin, P.R., Allen, H.E., Meyer, J.S., 2001. Biotic ligand model of the acute toxicity of metals. 2. Application to acute copper toxicity in freshwater fish and daphnia. *Environ. Toxicol. Chem.* 20(10), 2397–2402. https://doi.org/10.1002/etc.5620201035

Sarkar, B., Mukhopadhyay, R., Ramanayaka, S., Bolan, N., Ok, Y.S., 2021. The role of soils in the disposition, sequestration and decontamination of environmental contaminants. *Philos. T. Royal Soc. B* 376(1834), 20200177. https://doi.org/10.1098/rstb.2020.0177

Schiedek, D., Sundelin, B., Readman, J.W., Macdonald, R.W., 2007. Interactions between climate change and contaminants. *Mar. Pollut. Bull.* 54(12), 1845–1856. https://doi.org/10.1016/j.marpolbul.2007.09.020

Schubauer-Berigan, M.K., Dierkes, J.R., Monson, P.D., Ankley, G.T., 1993. pH-Dependent toxicity of Cd, Cu, Ni, Pb and Zn to *Ceriodaphnia dubia*, *Pimephales promelas*, *Hyalella azteca* and *Lumbriculus variegatus*. *Environ. Toxicol. Chem.* 12(7), 1261–1266. https://doi.org/10.1002/etc.5620120715

Schwartz, M.L., Curtis, P.J., Playle, R.C., 2004 Influence of natural organic matter source on acute copper, lead, and cadmium toxicity to rainbow trout (*Oncorhynchus mykiss*). *Environ. Toxicol. Chem.* 23(12), 2889–2899. https://doi.org/10.1897/03-561.1

Semple, K.T., Doick, K.J., Jones, K.C., Burauel, P., Craven, A., Harms, H., 2004. Defining bioavailability and bioaccessibility of contaminated soil and sediment is complicated. *Environ. Sci. Technol.* 38(12), 228a–231a. https://doi.org/10.1021/es040548w

Shah, V., Jacob, D.J., Thackray, C.P., Wang, X., Sunderland, E.M., Dibble, T.S., Saiz-Lopez, A., Černušák, I., Kellö, V., Castro, P.J., Wu, R., Wang, C., 2021. Improved mechanistic model of the atmospheric redox chemistry of mercury. *Environ. Sci. Technol.* 55(21), 14445–14456. https://doi.org/10.1021/acs.est.1c03160

Shoghi, E., Fuguet, E., Bosch, E., Ràfols, C., 2013. Solubility–pH profiles of some acidic, basic and amphoteric drugs. *Eur. J. Pharm. Sci.* 48(1–2), 291–300. https://doi.org/10.1016/j.ejps.2012.10.028

Song, N., Zhong, X., Li, B., Li, J., Wei, D., Ma, Y., 2014. Development of a multi-species biotic ligand model predicting the toxicity of trivalent chromium to barley root elongation in solution culture. *PLoS One* 9(8), e105174. https://doi.org/10.1371/journal.pone.0105174

Sposito, G., 1989. *The Chemistry of Soils*. Oxford University Press, Oxford.

Strom, D., Simpson, S.L., Batley, G.E., Jolley, D.F., 2011. The influence of sediment particle size and organic carbon on toxicity of copper to benthic invertebrates in oxic/suboxic surface sediments. *Environ. Toxicol. Chem.* 30(7), 1599–1610. https://doi.org/10.1002/etc.531

Sucharitha, P., Ramesh Reddy, K., Satyanarayana, S.V., Garg, T., 2022. Absorption, distribution, metabolism, excretion, and toxicity assessment of drugs using computational tools. In: Parihar, A., Khan, R., Kumar, A., Kaushik, A.K., Gohel, H. (eds.) *Computational Approaches for Novel Therapeutic and Diagnostic Designing to Mitigate SARS-CoV-2 Infection*, pp. 335–355. Academic Press, Cambridge.

Tipping, E., Rey-Castro, C., Bryan, S.E., Hamilton-Taylor, J., 2002. Al (III) and Fe (III) binding by humic substances in freshwaters, and implications for trace metal speciation. *Geochim. Cosmochim. Acta* 66(18), 3211–3224. https://doi.org/10.1016/S0016-7037(02)00930-4

Topaz, T., 2023. Attenuation of organic pollutants and the effects of salinity and seasonality in a Mediterranean micro-estuary. *Sci. Total Environ.* 856, 158919. https://doi.org/10.1016/j.scitotenv.2022.158919

Toutain, P.L., Bousquet-Mélou, A., 2004. Bioavailability and its assessment. *J. Vet. Pharmacol. Ther.* 27(6), 455–466. https://doi.org/10.1111/j.1365-2885.2004.00604.x

Traina, S.J., Laperche, V., 1999. Contaminant bioavailability in soils, sediments, and aquatic environments. *Proc. Natl. Acad. Sci. USA* 96(7), 3365–3371. https://doi.org/10.1073/pnas.96.7.3365

Tsiridis, V., Petala, M., Samaras, P., Hadjispyrou, S., Sakellaropoulos, G., Kungolos, A., 2006. Interactive toxic effects of heavy metals and humic acids on *Vibrio fischeri*. *Ecotoxicol. Environ. Saf.* 63(1), 158–167. https://doi.org/10.1016/j.ecoenv.2005.04.005

US EPA. 2000. Bioaccumulation testing and interpretation for the purpose of sediment quality assessment: Status and needs. EPA-823-R-00-001. U.S. Environmental Protection Agency, Washington, DC.

US EPA. 2007a. Guidance for evaluating the oral bioavailability of metals in soils for use in human health risk assessment. OSWER 9285.7-80, May 2007. U.S. Environmental Protection Agency, Washington, DC.

US EPA. 2007b. Framework for metals risk assessment. Office of the Science Advisor, Risk Assessment Forum. EPA/120/R-07/001. U.S. Environmental Protection Agency, Washington, DC.

Valenzuela, E.I., Prieto-Davó, A., López-Lozano, N.E., Hernández-Eligio, A., Vega-Alvarado, L., Juárez, K., García-González, A.S., López, M.G., Cervantes, F.J., 2017. Anaerobic methane oxidation driven by microbial reduction of natural organic matter in a tropical wetland. *Appl. Environ. Microbiol.* 83(11), e00645-17. https://doi.org/10.1128/AEM.00645-17

Vardhan, K.H., Kumar, P.S., Panda, R.C., 2019. A review on heavy metal pollution, toxicity and remedial measures: Current trends and future perspectives. *J. Mol. Liq.* 290, 111197. https://doi.org/10.1016/j.molliq.2019.111197

Vignati, D.A.L., Dominik, J., Beye, M.L., Pettine, M., Ferrari, B.J.D., 2010. Chromium(VI) is more toxic than chromium(III) to freshwater algae: A paradigm to revise? *Ecotoxicol. Environ. Saf.* 73(5), 743–749. https://doi.org/10.1016/j.ecoenv.2010.01.011

Wang, F., Zhang, M., Sha, W., Wang, Y., Hao, H., Dou, Y., Li, Y., 2020. Sorption behavior and mechanisms of organic contaminants to nano and microplastics. *Molecules* 25(8). https://doi.org/10.3390/molecules25081827

Wells, M., 2012. Polycyclic aromatic hydrocarbon (PAH) sensitive bacterial biosensors in environmental health. In: Hunter, R.J., Preedy, V.R. (eds.) *Biosensors and Environmental Health*, pp. 170–182. CRC Environmental Health Series, Enfield, New Hampshire.

Wells, M., Buck, K.N., Sander, S.G., 2013. New approach to analysis of voltammetric ligand titration data improves understanding of metal speciation in natural waters. *Limnol. Oceanog. Methods* 11(9), 450–465. https://doi.org/10.4319/lom.2013.11.450

Wells, M., Wick, L.Y., Harms, H., 2005. Model polymer release system study of PAH bioaccessibility: The relationship between "rapid" release and bioaccessibility. *Environ. Sci. Technol.* 39(4), 1055–1063. https://doi.org/10.1021/es035067b

Xing, S.Y., Li, Z.H., Li, P., You, H., 2022. A mini-review of the toxicity of pollutants to fish under different salinities. *Bull. Environ. Con. Toxicol.* 108(6), 1001–1005. https://doi.org/10.1007/s00128-022-03528-0

Zhang, C., Ding, S., Xu, D., Tang, Y., Wong, M.H., 2014. Bioavailability assessment of phosphorus and metals in soils and sediments: A review of diffusive gradients in thin films (DGT). *Environ. Monit. Assess.* 186(11), 7367–7378. https://doi.org/10.1007/s10661-014-3933-0

Zhang, C., Yao, F., Liu, Y.W., Chang, H.Q., Li, Z.J., Xue, J.M., 2017a. Uptake and translocation of organic pollutants in plants: A review. *J. Integr. Agr.* 16(8), 1659–1668. https://doi.org/10.1016/S2095-3119(16)61590-3

Zhang, Q., Wang, H., Xia, X., Bi, S., Lin, H., Chen, J., 2020a. Elevated temperature enhances the bioavailability of pyrene to *Daphnia magna* in the presence of dissolved organic matter: Implications for the effect of climate warming. *Environ. Pollut.* 266, 115349. https://doi.org/10.1016/j.envpol.2020.115349

Zhang, X., Li, B., Deng, J., Qin, B., Wells, M., Tefsen, B., 2020b. Advances in freshwater risk assessment: Improved accuracy of dissolved organic matter-metal speciation prediction and rapid biological validation. *Ecotoxicol. Environ. Saf.* 202, 110848. https://doi.org/10.1016/j.ecoenv.2020.110848

Zhang, X., Li, B., Deng, J., Qin, B., Wells, M., Tefsen, B., 2020c. Quantitative high-throughput approach to chalkophore screening in freshwaters. *Sci. Total Environ.* 735, 139476. https://doi.org/10.1016/j.scitotenv.2020.139476

Zhang, X., Qin, B., Deng, J., Wells, M., 2017b. Whole-cell bioreporters and risk assessment of environmental pollution: A proof-of-concept study using lead. *Environ. Pollut.* 229, 902–910. https://doi.org/10.1016/j.envpol.2017.07.068

Zhang, X., Zhu, Y., Elçin, E., He, L., Li, B., Jiang, M., Yang, X., Yan, X.P., Zhao, X., Wang, Z., Wang, F., Shaheen, S.M., Rinklebe, J., Wells, M., 2024. Whole-cell bioreporter application for rapid evaluation of hazardous metal bioavailability and toxicity in bioprocess. *J. Hazard. Mater.* 461, 132556. https://doi.org/10.1016/j.jhazmat.2023.132556

Zhao, S., Wang, J., Feng, S., Xiao, Z., Chen, C., 2022. Effects of ecohydrological interfaces on migrations and transformations of pollutants: A critical review. *Sci. Total Environ.* 804, 150140. https://doi.org/10.1016/j.jhazmat.2023.132556

Zhu, L., Chen, B., Tao, S. 2004. Sorption behavior of polycyclic aromatic hydrocarbons in soil–water system containing nonionic surfactant. *Environ. Eng. Sci.* 21, 263–272. https://doi.org/10.1089/109287504773087410

2 Terms and Definitions

2.1 COMMENTS

This chapter provides a curated list of key terms needed for understanding the concepts discussed throughout this book. The definitions adhere to customary usage within the field of environmental science while being specifically tailored to the context of this book to avoid ambiguities. Some definitions are also crafted to be accessible and useful to general readers. To aid in cross-referencing, terms given in this chapter appear in ***bold italic*** on their first use in other chapters. Readers may refer back to this chapter for definitions whenever this formatting is encountered.

Definitions for key terms used throughout this book are given below. Terms that relate primarily to ***Quality Assurance*** (QA) and ***Quality Control*** (QC) also appear in Chapter 4. The QA/QC terms are extensive and complex and have therefore been placed with the discussion on QA/QC in order for that material to be more convenient and apprehensible to readers who are not familiar with QA/QC. Definitions given in this chapter reflect common usage in all cases; however, some terms have usage that varies widely, hence the need for specificity in the list below.

2.2 TERMS AND DEFINITIONS

Absorption – is a physicochemical process in which one substance (typically a gas, liquid, or solute) is taken up into the bulk phase of another material — usually a solid or liquid — and distributed throughout its internal structure. This uptake occurs via molecular diffusion, capillary action, or chemical interaction, resulting in the penetrant becoming part of the absorbing medium's internal volume rather than merely adhering to its surface.

Acceptance Criteria – preestablished QC criteria that must be met in order for data to be accepted as meeting standards for precision, accuracy (bias), representativeness, completeness, comparability, and sensitivity (PARCCS), and therefore being fit-for-purpose for use in data analysis and subsequent conclusions about the data. PARCCS standards are often given in Standard Methods.

Acclimation – the process in which an individual organism (for the purposes of this book, a test organism) adjusts to a change in its environment, such as a change in temperature, pH, or water chemical composition, allowing it to maintain fitness.

Accuracy – the degree of agreement between an observed value and an accepted reference or "true" value. Inaccuracy results from a combination of a random error and systematic error (bias – an error in accuracy that is systematic and causes a persistent positive or negative deviation from the true value). Accuracy is determined as absolute or relative, respectively, according to

$$\text{Accuracy} = \bar{x} - \mu$$

and

$$\%\text{Accuracy} = \frac{\bar{x} - \mu}{\mu} \times 100\%,$$

where $\bar{x}$ is the mean value of a measurement on a sample and μ is the true value. The standard approach to evaluation of accuracy is via the use of Certified or Standard Reference Materials.

Activation – to make something reactive or more reactive. In gene expression, activation is a form of positive control, and with reference to bioreporters means activating a bioreporter transcription factor response under the induction of target pollutant(s).

 DOI: 10.1201/9781003355229-3

Activity (chemical) – may be thought of as the effective concentration of a species in solution. The activity of a species is related to its concentration, C, by an activity coefficient, γ, such that $a = \gamma C$. The activity coefficient accounts for deviations from ideality due to ionic interactions and is therefore a function of ionic strength.

Activity coefficient – often denoted as γ, is a correction factor that accounts for deviations from ideal chemical behavior, adjusting concentration to reflect effective chemical potential. Activity represents the thermodynamic "effective concentration" of a species, which differs from measured concentration according to molecular interactions and environmental conditions, which are influenced by factors such as ionic strength, charge, and interactions with other dissolved species.

Adsorption – the ability of solid surfaces to attract and hold gases, liquid components, or dissolved substances to their surfaces.

Aerobic – requiring oxygen to survive, function, or sustain biological or chemical processes. Organisms classified as aerobic depend on oxygen for cellular respiration.

Alkalinity – the quantitative capacity of a natural water to neutralize acidic inputs, primarily governed by the concentrations of weak acid conjugate bases such as bicarbonate (HCO_3^-), carbonate $\left(CO_3^{2-}\right)$, and to a greater or lesser extent silicates (SiO_4^{4-}, $HSiO_3$), ammonia/ammonium (NH_3, NH_4^+), borates BO_3^{3-}, phosphates (HPO_4^{2-}, PO_4^{3-}), and organic bases (e.g., amino acids, humic substances).

Anaerobic – surviving, functioning, or sustaining biological or chemical processes without access to free oxygen. Organisms classified as anaerobic rely on electron acceptors other than molecular oxygen for cellular respiration.

Analytical batch – a group of samples that are processed together using the same method, the same lots of reagents, and at the same time in continuous, sequential time periods. Batches always contain one or more laboratory QA/QC samples that are not counted as part of the batch number. Samples in each batch should be of similar composition and all should be subject to the same treatment. Samples collected from the same area would normally be grouped together for batching purposes, and samples in the same batch are analyzed together on a single instrument and at the same time. In some cases, a batch may contain similar samples, though from different sites/locales. Field QA/QC samples are often site-specific, and a batch having samples from more than one site/locale would require more QA/QC samples to accommodate needs.

Aquatic microcosm – refers to small-scale, controlled environments replicating some aspects of natural aquatic ecosystems.

Autovalidation – the process of validating a model using one set of data that is used to both parameterize/calibrate the model and then subsequently test the model for predictive accuracy. While autovalidation is often used, best practice is to instead or additionally use one set of data for model parameterization/calibration and an independent set of data to test for accuracy of a model's predictions. Autovalidation ensures internal consistency, however, independent validation provides a more robust assessment of model accuracy.

Back extraction – as used in liquid–liquid extraction, a loaded solvent results when using an organic solvent to extract solutes from a sample, and back extraction is the process of returning the extracted solutes from the loaded organic phase to an aqueous phase using a counter-extractant.

Binding gel – in DGT is a specialized hydrogel layer that contains immobilized ligands or chelating agents, e.g., Chelex-100 for metals, designed to irreversibly trap solutes (target pollutants) after they diffuse through an overlying diffusive gel.

Bioaccessibility – refers to the portion of a pollutant that is not immediately bioavailable but can become so under changing environmental conditions or through biological interactions, representing its potential bioavailability.

Bioaccumulation – a phenomenon whereby a chemical present in a living organism's environment accumulates in tissues of the organism, i.e., the rate of chemical intake exceeds that of chemical elimination.

Bioavailability/bioavailable – the portion of a pollutant that is available to biota for uptake and accumulation.

Bioconcentration – used to describe the buildup of a pollutant in or on an organism from water as the source of the pollutant.

Biomagnification – is a term that signifies an increase in concentration of a pollutant in organisms at successively higher levels in a food chain.

Biomonitoring – involves assessing the content of pollutants absorbed by test organisms by determining the increase in concentration of pollutants in test organisms before and after exposure or determining the change of pollutant concentration in an environmental media to which the test organisms are exposed before and after exposure.

Bioreporter – living microbial cells that have been genetically engineered to produce a measurable signal in response to a specific chemical or physical agent in their environment.

Biotic ligand – a biotic ligand is a specific receptor within an organism for which complexation by a heavy metal can lead to a toxic response.

Biotic Ligand Model – a speciation-based model that includes expressions to describe heavy metal complexation with biotic ligands. The model is run in speciation and toxicity modes. The speciation mode predicts the distribution of metal ions among various chemical forms (e.g., free ions, complexes, and precipitates) in solution as well as biotic ligand species, i.e., the Biotic Ligand Model run in speciation mode differs from other speciation models because of the addition of biotic ligand reactions. Toxicity mode quantifies heavy metal binding to biotic ligands in the context of a toxicity threshold or endpoint, i.e., this mode may be used to predict the amount of total heavy metal in any given aqueous system that will produce a toxic effect.

Blank – there are different types of blank sample (see Chapter 4), all of which share the common characteristic that no contamination should be found in a blank, i.e., the acceptance criterion for a blank is that all blank results should be below the limits of detection.

Caging – a containment system that allows exposure of test organisms by various routes/multiple sources, while also providing organisms adequate habitat during the exposure period; for bivalve mollusks this can include traditional cage-type structures or alternate containment such as bait/"exposure" bags.

Calibration blank – analyzed along with prepared standards in the calibration of analytical instruments. A calibration blank should be prepared exactly the same as calibration standards, except that no analyte is added or present. Such a blank represents "zero" in calibrating an instrument for instrumental analysis.

Calibration checks – refers to an analysis that is performed periodically after calibration to assess whether the initial calibration has drifted (changed) and for memory effects over time. Memory effects or carryover may be present in some forms of instrumental analysis, and this term refers to a phenomenon whereby an instrument "remembers" analyte from prior samples, in which case the instrument response for standards increases over time, requiring instrument maintenance and recalibration. Calibration check analyses involve reanalysis of a blank, or low, mid-range, or high standard, to check if the result is suitably consistent with the initial calibration. If a calibration check fails, Corrective Action is recalibration, often with reanalysis of samples analyzed under failed calibration conditions.

Calibration curve – a plot of instrument response to an analyte versus known concentrations of the blank and analyte calibration standards. Calibration standards are prepared by successively diluting a standard solution to produce working standards that cover an instrumental working range. Standards and the blank used in calibration should be prepared at the frequency specified in the appropriate instrumental method. Ideally, the matrix of calibration standards should match that of samples analyzed to the greatest degree possible; however, this is often not practical. Methods of instrumental analysis typically provide guidance on preparation of calibration standards.

Carrier – with reference to the Permeation Liquid Membrane (PLM) technique is a critical component that selectively binds and transports target pollutants (e.g., metal ions, organic molecules) from the source phase to the strip phase through the liquid membrane. With reference to spiking organic pollutants, carrier refers to the solvent or solvent mixture used for maintaining the target pollutant in the aqueous phase.

Chain-of-Custody – a chronological documentation of the handling, control, transfer, and storage of samples from the point of collection to analysis and reporting. This process maintains a clear record of every individual or entity that has handled the sample, along with the time, date, location, and purpose of each transfer, for the ultimate purpose of ensuring the integrity and reliability of results. Usually accomplished with Chain-of-Custody forms.

Chelation – is bonding between an organic polydentate ligand and a central metal atom in a manner that results in multiple coordination and/or ionic bonds between organic molecules and metals. Chelation is the process of bonding, and a chelate is the product of a chelation process. The term chelate is derived from the word chela, from the Greek, for claw, to describe how ligands with multiple associating functionalities bind to a central metal atom.

Chelating resin – is a kind of functional material with a cross-linked polymer as the skeleton to which are bound pendant functional groups that chelate to metal ions.

Chemical fate models – also sometimes referred to as environmental multimedia models, are a class of mathematical models used to simulate and predict the behavior, transformation, and transport of chemical substances in and across different environmental compartments, such as air, water, soil, and biota.

Chemoavailability/chemoavailable – availability of target pollutants to chemical proxies. Chemoavailability to chemical proxies is what bioavailability is to biota.

Co-located samples – a type of field duplicate/replicate (usually duplicate) where independent samples are collected as close as possible to the same point in space and time. The co-located samples are collected independently, stored in separate containers, and analyzed independently by the same method, under the same conditions, and at the same time. This type of field duplicate/replicate sample, while dependent upon the precision of the field sampling process, is also useful in documenting field heterogeneities.

Comparability – refers to the equivalency of different sets of data, i.e., the degree to which different methods or data agree or can be represented as similar. Often, different data sets subject to different methods of chemical analysis or other varying conditions are not comparable. Comparable data sets must contain the same variables of interest and must possess values that can be converted to a common unit of measurement. Comparability is primarily qualitative; for example, if two different data sets for studies on the same pollutant were obtained using different analytical methods, the two methods may not be comparable. Data sets should never be combined if doubts concerning comparability exist.

Competitive ligand exchange – refers to a chemical process in which two or more ligands compete for binding to a central metal ion or receptor site. This occurs when a metal complex is exposed to a solution containing a "competing ligand" that can potentially displace the original ligand(s) due to differences in binding affinity, concentration, or reaction conditions.

Completeness – a measure of the amount of valid data obtained from a program of measurement compared with the amount that was expected to be obtained, i.e., the amount of "usable" data. Percent completeness is calculated as

$$\%\text{Completeness} = \frac{n_{\text{data,obtained}}}{n_{\text{data,planned}}} \times 100\%,$$

where $n_{\text{data,obtained}}$ is the amount of valid data obtained, and $n_{\text{data,planned}}$ is the amount of data planned or that was expected to be obtained. A value of 90% is a common acceptance criterion for completeness, i.e., loss of more than 10% of the data deemed needed is taken as compromising the ability of the results to be deemed fit-for-purpose.

Complexation – a term used in chemistry to describe molecules or ensembles formed by coordination bonding through one or more atom, ion, or molecular ligand donating electrons to a metal center.

Component – with reference to chemical speciation modeling, components are the fundamental building blocks used to define a chemical system to be modeled. All possible chemical species in the system can be formed from components, whereas no component can be formed by a combination of other components. Components often include major ions, elements, or molecules that serve as the basis for equilibrium calculations. Components are entered into a chemical speciation model in units of total component concentration. The total concentration of species formed by a component at equilibrium sums to the total concentration of the component.

Concentration gradient – refers to the gradual change in the concentration of a solute (e.g., ions, molecules, or pollutants) between two regions, such as across a membrane, in a solution, or within environmental matrices (e.g., water–soil interfaces).

Contaminant (contamination) – for the purposes of this book, refers to pollutants adventitiously and erroneously introduced during the sampling, handling, processing, or analysis process usually due to, operator error or procedural lapses.

Corrective Action – measure(s) taken to rectify conditions adverse to achieving quality outcomes and, where possible, to preclude their recurrence. Corrective Action is most often prompted by the discovery of a QA/QC Nonconformity.

Criterion Maximum Concentration (CMC) – the highest concentration of a pollutant that an aquatic ecosystem can tolerate for brief durations without significant acute effects. The CMC is derived from Genus Mean Acute Values (definition below) of the most sensitive species in a manner to allow for a safety margin and is used in the development of regulatory criteria for pollutants.

Data Quality Indicators (DQIs) – DQIs refer to the quantitative statistics and qualitative descriptors that are used to interpret the degree of acceptability or utility of data. The principal indicators of data quality are PARCCS data (see below). Multiple metrics may apply to each of these six types (precision, accuracy, representativeness, completeness, comparability, and sensitivity) of DQI.

Data Quality Objectives (DQOs) – DQOs are qualitative and quantitative statements derived from the DQO planning process that clarify the purpose of a study, define the most appropriate type of information to collect, determine the most appropriate conditions from which to collect that information, and specify tolerable levels of potential decision errors. The DQO process is documented via, *inter alia*, a Sampling and Analysis Plan (SAP). See US EPA (2006) and (2014), respectively, for extensive information on the DQO process and SAPs.

Data Validation – this terminology is highly variable globally, nationally, and even regionally. For the purposes of this book, Data Validation refers to evaluation of data subsequent to Data Verification and includes all aspects of data analysis that involve analysis of the structure of the data (e.g., distribution), the appropriate quantitative methods to use in analyzing the data to address information needs, evaluation of trends or biases in the data and associated QC results, a decision as to whether the data is fit-for-purpose for study goals, and conclusion-making.

Data Verification – as with Data Validation, the use of this terminology is also highly variable. For purposes here, Data Verification involves evaluating all of the data against PARCCS Acceptance Criteria that are specified in Standard Operating Procedures (SOPs) and/or as defined for the project during the QA/QC Planning phase (in other words, against all Acceptance Criteria for field and laboratory data QA/QC, as well as any other relevant criteria specified in advance).

Depuration – a process whereby test organisms are placed into a clean environment for a period of time to allow purging of biological or chemical contaminants and/or physical impurities (such as sand and silt). A term that is not limited to, but is frequently used in reference to, bivalve mollusks.

Desalting – a phenomenon whereby a decrease in the solubility of organic compounds results from an increase in the ionic strength of the solution due to dissolved salts.

Diffusive boundary layer – denotes the transition between advective transport, i.e., in aqueous settings with flow, current, or stirring, and diffusion-limited transport.

Diffusive gel – is a hydrogel matrix designed to mimic the diffusion properties of water, allowing solutes (e.g., ions, metals, or organic molecules) to passively diffuse through its pores while excluding larger particles or interfering substances.

Diffusive Gradients in Thin-Films (DGT) – a method of measuring chemoavailability that has become popular in recent years as having a higher performance than many other chemovailability methods with regard to obtaining results that correlate with methods for measuring true bioavailability.

Distribution – as applies to K_d, the soil/sediment water distribution coefficient, refers to the apportionment of pollutant between a complex aqueous phase and a complex solid phase. Similar in usage to partitioning, however, implying control by complex environmental factors such as ionic strength and heterogeneity of surface reaction sites.

Duplicate/replicate (sample/s) – duplicate (two) or replicate (*n*, where *n* is greater than two) refers to a sampling strategy whereby two or more samples representative of the same population and treated identically are compared. Results from analysis of different forms of duplicates/replicates are used to measure precision. There are a variety of types of field and laboratory duplicate samples. See definitions for, e.g., co-located samples, laboratory duplicate/replicate, field duplicate/replicate, field split samples, matrix spike duplicate/replicate, and recovery spike duplicate/replicate.

Earthworms – any one of more than 1,800 species of terrestrial worms of the class Oligochaeta (phylum Annelida) – in particular, members of the genus *Lumbricus*. Earthworms occur in virtually all soils of the world in which the moisture and organic content are sufficient to sustain them.

EC_{50} – the concentration of a pollutant at which 50% of test organisms exhibit an adverse effect. EC_{50} is a key metric in toxicity assessments.

Effects location – exact spatial location and depth where test organisms will be exposed to measure bioavailable pollutants associated with effects, i.e., for use in evaluation of effects.

Elimination phase – with respect to worm studies, is a phase of a study corresponding to the period of time following the transfer of the test organisms from a polluted medium to pollutant-free medium, during which the worms empty their gut contents (egestion) of soil/sediment, i.e., so that the tissue analyzed contains only pollutants that were bioavailable to the worm, and not pollutants sorbed to soil/sediment in the worm digestive tract.

Emerging pollutant – pollutants that are not often monitored in the environment, including many synthetic or natural substances, such as pharmaceuticals and personal care products, hormones, and flame retardants. A pollutant may be "emerging" because its toxicity has recently been discovered or because new pathways to environmental damage from the pollutant are being discovered.

Enchytraeid worms – are an important family of terrestrial Oligochaeta. This group of small, unpigmented worms, also known as "potworms", is classified within the "microdrile" oligochaetes and consists of some 600 species in 28 genera. Species from 19 of these genera are found in soil, the remainder occurring primarily in marine and freshwater habitats.

Environmental effects envelope – a term sometimes used in ***Environmental Impact Assessment*** to signify, including, but not limited to, the temporal and spatial extent of environmental effects to be monitored. It is important to note that the environmental effects envelope should never be limited to a temporal or spatial extent of known or reasonably envisaged effects. There should be an additional safety margin, temporally and spatially, to allow for effects that might occur, but are not definitively foreseen.

Environmental Risk Assessment – a formal process whereby environmental risk associated with pollutant exposure is quantified and assessed.

Equipment blank/equipment rinsate – an equipment blank, sometimes referred to as equipment rinsate, is a sample of analyte-free media that has been passed over or used to rinse field sampling equipment. The medium for the equipment blank is typically water or solvent, dependent on what the final aqueous media are for instrumental analysis, though other media may be applicable for soils or sediment work (e.g., contaminant-free bentonite). An equipment blank is collected after completion of decontamination – this may refer to laboratory decontamination that occurred prior to taking the equipment into the field if the equipment is being used in the field for the first time, or it may refer to decontamination of equipment that must occur in the field between each use of the equipment for sampling. An equipment blank is taken after field decontamination and immediately prior to sampling. Equipment blanks are used to demonstrate whether or not field samples may have been contaminated by field equipment. Contamination of field equipment that transfers to samples may affect both precision and accuracy of results.

Exposure – contact of a biological organism with a chemical or physical agent.

Exposure location – exact spatial location and depth where test organisms will be exposed. Many types of study, for instance biomonitoring with bivalve mollusks, must include a minimum of two exposure locations, one reference location and one effects location.

Fick's first law – explains how molecules travel from areas of higher density to areas of lower density and is crucial for understanding processes such as respiratory gas transfer and target pollutant diffusion.

Field blank – field blanks are samples that are collected in the field and should be analyte-free, assuming that appropriate precautions to avoid contamination, introduced during transport or occurring from incomplete field decontamination, are in place and successful. There are various types of field blanks, depending on the type of field collection technique used and the type of pollutant that is scheduled for analysis. The most common types of field blank are field procedural blanks, equipment blanks, and trip blanks. Field blanks provide information on various types of contamination that may be introduced into samples as a result of various aspects of field operations. Field sample contamination may affect both precision and accuracy of results.

Field duplicates/replicates – field duplicates/replicates are duplicates/replicates collected in the field used to assess precision of results as impacted by environmental heterogeneities in the field and uniform conduct/reproducibility of field sampling procedures. Results from field duplicates show the precision of the field sampling process; however, field duplicates/replicates also reflect laboratory precision. It is not uncommon to collect field duplicates/replicates for all samples; however, two common types of field duplicate additionally collected for QA/QC are co-located samples and field split samples. Because field split samples are homogenized, this substantively reduces field heterogeneity, hence comparison of co-located samples with field split samples enables differentiation of how precision of results is affected by field sampling procedures versus field heterogeneity.

Field split samples – a type of field duplicate/replicate wherein a sample is collected, homogenized, and then divided into two or more portions, each of which is one field replicate. Homogenization is usually performed under clean conditions after transport to the laboratory; however, it is sometimes performed in the field if convenient and feasible. After homogenization, duplicates/replicates (field split samples) are stored in separate containers and analyzed independently by the same method, under the same conditions, and at the same time. This type of field duplicate/replicate sample is useful in documenting the precision of the sampling process. Homogenization may have an adverse impact on sample integrity for some sample types, notably volatile organic compounds, and in such cases, only co-located samples may be used as field duplicates/replicates.

Filter membrane – is a thin, semipermeable barrier designed to selectively separate particles, molecules, or solutes from a fluid (liquid or gas) based on size, charge, or chemical affinity. It acts as a physical or chemical sieve, allowing certain components to pass while retaining others.

Genus Mean Acute Value (GMAV) – the geometric mean of EC_{50} values for a genus (based on an acute toxicity measurement), used to determine toxicity thresholds for use in regulatory frameworks.

Host cell – a cell that hosts foreign molecules or viruses or may have introduced DNA or RNA – a bacterium, for instance. Whole-cell bioreporters are typically based on a host cell being transformed with an expression vector that carries a transcriptional fusion of a gene promoter to a reporter gene encoding one of several possible reporter molecules.

Hydrophobic organic pollutants (HOCs) – non-polar compounds with low water solubility and a relatively high octanol–water partition coefficient.

Impregnation – refers to the process of saturating or permeating a material with a substance, often to enhance its properties.

Independent validation – refers to model validation that is performed using one observational data set to parameterize/calibrate a model and a different, independent, observational data set to test for accuracy of the model's predictions; independent validation represents best practice in validation.

Induction – with respect to bioreporters, is a process by which a pollutant interacts with a regulatory system to affect a change in the status of a promoter, which initiates or increases the activation of transcription factors after exposure to the pollutant(s) to which a reporter responds.

Initial Demonstration of Proficiency (IDP) – is the single most important element of a QA/QC program. Prior to the analysis of samples, an initial demonstration of (analytical) proficiency is accomplished through the successful analysis of reference samples (usually Certified or Standard Reference Materials, see below) to obtain results that are in accord with DQIs stipulated as typical and acceptable, as provided by the suppliers of reference materials. The reference material used should have a matrix as close as possible to the sample matrices. The analysis of reference materials is often stipulated in methods for instrumental analysis, and therefore should always be undertaken as part of instrumental analysis. Users should refer to methods of instrumental analysis for guidance on the topic of analysis of reference materials. Methods that have substantive sample preparation steps prior to instrumental analysis may have additional IDP requirements, and these additional requirements are outlined in individual chapters in this book. The IDP process demonstrates whether or not a laboratory is capable of obtaining a correct result that is fit-for-purpose with respect to DQOs/study objectives and DQI criteria.

Internal standard – internal standards are employed for a variety of QA/QC purposes, are used in many methods, and are typically added as spikes. Such standards usually comprise substances similar to the analyte (perhaps even isotopes of the analyte), that are not found in the environment and are added to each sample, QA/QC sample, and calibration solution. Since internal standards are added in a fixed amount, the signal from internal standards should always be the same; any deviations from expectations indicates some form of compromised analysis. If the internal standard in all samples is effectively constant, as expected, taking the ratio of analyte signal to internal standard signal and plotting it against the analyte concentrations in the calibration solutions produces a calibration curve that can then be used to calculate the analyte concentration. An internal standard is like a surrogate (see below) as it is a pure substance, added at a known amount, and unlikely to be found in a sample. An internal standard, however, differs from a surrogate in its function. Internal standards are primarily used to measure the relative responses of analytes, and sometimes surrogates as well, to confirm that analytical instrumentation is working properly and/or to correct for drift that might occur. An internal standard is normally added at the end of all sample handling and preparation procedures, right before analysis. Occasionally, the interplay of how internal standards and surrogates function in an analysis is more complex (particularly as regards matrix effects), and a few methods call for internal standards to be spiked into samples before intensive preparation. In this case, as for surrogates, the recovery of internal standards may be used to demonstrate the lack of severe matrix effects in the instrumental analysis by setting criteria for the internal standard response in comparison to a response in a sample with a known lack of matrix effect. The typical pattern of use of surrogates is to monitor for analyte loss and or/matrix effects, with internal standards being used to correct for variability in analytical conditions and calibrate for analyte response (internal calibration). The schedule and purpose of using internal standards and surrogates are given in standardized methods.

Ion activity product (IAP) – is the product of the activities of ions that are in solution when a given solid dissolves, raised to their respective stoichiometric coefficients. The IAP has the same mathematical form as the solubility product constant, K_{sp}, which is an equilibrium constant that specifically determines the activities of ions in solution, *at equilibrium*, when a given solid dissolves. IAP reflects the current state of a solution with respect to activities of ions in the potential presence of a solid, such that if IAP $<K_{sp}$, the solution is undersaturated, and more solid can dissolve, if IAP $= K_{sp}$, the solution is at equilibrium, and no more solid will dissolve, and if IAP $>K_{sp}$, the solution is supersaturated, and precipitation should occur.

Ion pairing – a term in physics and chemistry used to describe a paired duplex of charged particles, often charged atoms or molecules, one positive and one negative.

Ionic strength – is a measure of the total concentration of ions in a solution, reflecting the intensity of electrostatic interactions between charged species. It accounts for the influence of all dissolved ions, considering both their concentrations and charges. Higher ionic strength leads to greater inter-ionic interactions, which can affect chemical equilibria, solubility, and activity coefficients. In dilute solutions, ionic strength is low, and the system behaves more ideally, with activity coefficients close to 1. As ionic strength increases, deviations from ideality become significant, altering the behavior of dissolved species and their reactivity in solution.

Labile metals – heavy metals that are in a chemically active or easily exchangeable form and thus more likely to be bioavailable.

Laboratory duplicate/replicate – this term refers to the analysis of two or more subsamples of the same sample. The subsamples are taken from the same container, processed/prepared the same, i.e., under the same conditions and at the same time, and then analyzed in the same batch. Analysis of laboratory duplicates/replicates reflects variability associated with subsampling and the matrix, but not the precision of field sampling, preservation, or storage. As a result, comparison of laboratory and field duplicates/replicates enables differentiation of to what extent laboratory operations versus field operations affect the precision of results. Laboratory duplicates/replicates are more commonly used to assess precision for inorganic and radiological constituents, while precision for organic analyses is usually assessed by determining the precision for matrix spike duplicates analysis.

Ligand – an ion or molecule having one or more chemical functional groups that bind, typically as a Lewis base, through electron-pair donation to a central metal or metalloid atom to form a coordination complex.

Lights-off bioreporters – bioreporters that emit a continuous signal under non-toxic conditions but exhibit a suppressed signal when exposed to bioavailable toxic concentrations of a pollutant. Signal diminishes in proportion to the degree of toxic effect.

Lights-on bioreporters – bioreporters that produce a signal that increases in response to increasing levels of bioavailable pollutants. The mechanism of response could be Class I (response to a specific pollutant or class of pollutants, e.g., Pb or bivalent heavy metal cations) or Class II (response to a specific stressor, e.g., heat shock).

Limit of detection (LOD) – the lowest concentration of analyte that may be statistically affirmed as being detected. The LOD is calculated as

$$\text{LOD} = \overline{y_{\text{blk}}} + 3.3 \times s_{y_{\text{blk}}},$$

where $\overline{y_{\text{blk}}}$ is the mean signal of the blank and $s_{y_{\text{blk}}}$ is the standard deviation around $\overline{y_{\text{blk}}}$. The factor of 3.3 is the t-test probability at $p < 0.05$, ensuring that the LOD is above $\overline{y_{\text{blk}}}$. An alternate, more conservative, approach to calculating the LOD is to use the intercept from linear calibration plus 3.3 times the standard error from linear regression of the calibration curve, respectively, instead of $\overline{y_{\text{blk}}}$ and $s_{y_{\text{blk}}}$. The LOD determined in the above manner will have units of response signal, which can be converted into units of concentration using the calibration curve.

Limit of quantification (LOQ) – the lowest concentration of an analyte that might be encountered in a sample and quantified with sufficient statistical reliability so as to be distinguishable from concentrations in other samples. The LOQ is calculated as

$$\text{LOQ} = \overline{y_{\text{blk}}} + 10 \times s_{y_{\text{blk}}},$$

where the factor of 10 is obtained from statistical considerations, much as the factor of 3.3 is for LOD. The LOQ determined in the above manner will have units of response signal, which can be converted into units of concentration using the calibration curve.

Liquid–liquid extraction – taking advantage of the difference in solubility or partition coefficients of pollutants in two immiscible (or slightly soluble) solvents in order to selectively transfer pollutants from one solvent to the other, often to achieve some purification or concentration during the process.

Macrophytes – are plant taxa that are visible to the naked eye and physiologically dependent on the aquatic environment, with at least a portion of their reproductive cycle occurring in water or on the water surface.

Mass action equation – is a mathematical expression derived from the law of mass action that describes how a chemical reaction at equilibrium is characterized by an equilibrium constant, K, that defines the stability of the species formed under specific conditions of temperature and pressure. Mass action equations take the form of a quotient: the numerator of the quotient is a product of the activities of the chemical reaction products, each raised to the power of their stoichiometric coefficients, and the denominator is a product of the activities of the chemical reactants, also each raised to the power of their stoichiometric coefficients.

Matrix – refers to the chemical components of which a sample is composed minus the target analyte(s), i.e., matrix is the material in which the target analyte is suspended in samples.

Matrix effects – effects exerted by matrix on analyses. Normally, in instrumental analysis, matrix effects are defined as effects that may produce bias in analytical results if not controlled for in QA/QC Planning. In terms of bioavailability studies, matrix effects represent environmental influences that exert control over bioavailability.

Matrix spike – matrix spikes are spikes that are added to a sample before sample preparation, cleanup, and procedures are implemented. Often the matrix spike will be mixed into the sample ahead of sample preparation and processing for analysis and allowed to integrate with the sample (24 hours or more). In some cases, target analyte spiked into sample matrix will be in a more labile form, i.e., will not have time to integrate into refractory matrix components in a sample, leading to a better spike recovery than will be obtained for native analyte in a sample.

Matrix spike duplicates/replicates – matrix spike duplicates/replicates are additional replicates of matrix spike samples. Analysis of matrix spike duplicates/replicates allows for measurement of matrix spike precision, which, in combination with the average matrix spike recovery, provides better information on whether spike recovery is statistically differentiable from target analyte in unspiked samples (an accuracy problem) or not.

Medium (plural, media) – material in which target analyte is suspended. Can include matrix in the case of field samples (environmental media such as water, soils, and sediments), or can also be comprised of media into which target analyte is introduced prior to analysis (e.g., aqueous- or solvent-based extractant).

Model validation – is the process of determining whether a model accurately represents the behavior of the system being modeled.

Nonconformance/Nonconformity – a deviation from requirements of Acceptance Criteria, i.e., when a sample, analytical batch or entire data set is compromised, as indicated by QA/QC results.

Octanol–water partition coefficient (K_{ow}) – is the ratio of the concentration of an organic pollutant in *n*-octanol to its concentration in water at equilibrium, i.e., it is a measure of how an organic pollutant distributes at equilibrium between these two immiscible solvents. Pollutants with a high K_{ow} value are lipophilic and tend to bioaccumulate, which can lead to higher concentrations in organisms higher up the food chain.

Oxidation-reduction potential (ORP) – is a measurement, usually expressed in millivolts, that indicates the inclination of a chemical substance to either oxidize or reduce another chemical substance.

Parameterization – refers to the process of defining a system, model, or phenomenon by a set of measurable or adjustable variables (parameters) that characterize its behavior, structure, or dynamics.

PARCCS (precision, accuracy, representativeness, completeness, comparability, and sensitivity) is a set of key DQIs used to evaluate the reliability and usability of analytical results in environmental testing, laboratory analysis, and regulatory compliance. Precision is a measure of how consistent repeated measurements are under the same conditions. Accuracy is a measure of how close a measurement is to the true or accepted value. Representativeness reflects whether the sample reflects the actual conditions or population. Completeness indicates whether enough valid data was collected to support conclusions. Comparability is whether data can be reliably compared across different methods of data analysis, acquisition, or different acquisition time frames, and sensitivity characterizes the level of an analyte that can be reliably measured, which determines whether a result is fit-for-purpose for intended use. PARCCS help ensure that sampling and analysis is conducted in a manner that is scientifically sound such that resultant data is trustworthy and decisions based on the data (e.g. remediation, compliance) are defensible.

Partitioning – refers to pollutant distribution across two distinct phases, for instance, organic versus aqueous phase, gas versus liquid, or solid versus water. Notable examples of partitioning mentioned in this book include K_{ow} (*n*-octanol–water partition coefficient, describing partitioning of pollutants between *n*-octanol and water) and K_{oc} (soil/sediment organic carbon–water partition coefficient, describing the partitioning of pollutants between soil/sediment and water, as influenced by organic carbon in the solid phase).

Passive sampling – relies on the movement of target pollutant from the sampled medium to a collecting medium over time, providing time-averaged concentrations of chemoavailable pollutants collected by passive sampling over times of hours to weeks.

Phytoextraction – is the adsorption and bioaccumulation of pollutants from the environment (from multiple media) into plant tissues. Unlike phytosorption, this term implies a process by which pollutants are actively taken up by plants.

Phytosorption – refers to the process by which plants sorb and accumulate pollutants from soil, sediments, or water. Akin to the term sorption, which may include, for instance, both adsorption and absorption; often for plant uptake of pollutants, phytosorption is used to denote the initial association of pollutants with the plant root surfaces or tissues.

Phytotransformation – involves biochemical reactions of pollutants in plants that result in changes in pollutant structure or toxicity, in macrophytes typically transforming bioavailable pollutants into less bioavailable or biounavailable forms, thereby increasing or decreasing plant tissue levels of pollutants in a complex manner.

Phytotransport – movement of pollutants within a plant to different tissues or organs such as leaves, stems, and roots.

Polychaete worms –are a diverse class of marine annelid worms (class Polychaeta) characterized by their segmented bodies, paired paddle-like appendages (parapodia) bearing bristles (chaetae), and a well-defined head with sensory organs.

Precision – a measure that quantifies the agreement among a set of duplicate or replicate measurements without assumption of knowledge of accuracy (true value), i.e., precision tells how variable results are, which is necessary to compare two or more results to each other statistically (e.g., testing for differences), however, is not a measure of accuracy (whether a result has the correct value or not). Examples of metrics for precision are as follows:

$$\%\text{RPD} = \frac{2 \times |n_1 - n_2|}{n_1 + n_2} \times 100\%,$$

and

$$\%\text{RSD} = \text{CV} = \frac{s_{\bar{x}}}{\bar{x}} \times 100\%,$$

where %RPD is the relative percent difference between replicate samples, n_1, and n_2; the vertical braces indicate absolute value; %RSD is the relative standard deviation or coefficient of variance (CV); $\bar{x}$ is the mean of the total number of replicate samples; and $s_{\bar{x}}$ is the standard deviation around the mean.

Preconcentration – is a term used in chemical analysis to describe the process of augmenting the concentration of a pollutant to enhance the sensitivity of measurement.

Procedural blank – a procedural blank is a sample that does not contain the matrix, which is brought through the entire measurement procedure and analyzed in the same manner as other samples. When preparing procedural blanks, water is often used in place of the matrix, though sometimes solvent is used if the final samples are subject to instrumental analysis and are in solvent. Procedural blanks may be used to assess any contamination that may occur within the entire process of sample collection, transport, field and laboratory conditions, reagents used, containers used, etc. In some cases, procedural blanks are taken to specifically monitor field contamination or laboratory contamination. For field procedural blanks, containers of reagent water, solvent, or a solid blank material (whatever is used as the basis for a procedural blank) should be opened in the field for a predefined period of time that is approximately the same as the time that a sample is exposed in the field during collection and processing, thus capturing approximately the same ambient contamination, if any is present, that a field sample container would. Individual methods often have guidance concerning procedural blanks. For methods that do not specify procedural blanks, users should consider how best to add these to a SAP.

Promoter – is a region of DNA that functions in a manner to initiate and regulate the transcription of one or more genes downstream.

Quality Assurance/Quality Control (QA/QC) – QA is focused on preventing data quality issues and refers to processes and procedures put in place and conducted in order to ensure quality of data for, e.g., the purpose of measuring bioavailability; QC focuses on reviewing data quality after results are produced (for instance, in the context of Acceptance Criteria) and identifying quality issues that might compromise the utility of the data. QC also focuses on Corrective Action, when possible or applicable.

Quality Assurance/Quality Control sample – a sample that is analyzed to obtain results that will be used in Data Validation, namely with respect to DQI Acceptance Criteria. Examples of QA/QC samples are blanks, duplicates/replicates, and spikes.

Quantitative Structure–Activity Relationship/Quantitative Structure–Property Relationship (QSAR and QSRP models) – is a computational model that establishes the connection between biological activity and chemical structures.

Reagent blank – a blank that is analyzed to confirm that the reagents used in sample preparation are not contaminated. Reagent blanks are prepared by adding all reagents in the same volumes or proportions as used in sample preparation but without any handling such as occurs through sample preparation, cleanup, and other handling procedures. These blanks are often prepared in water, or alternately solvents if analytes are extracted into solvents prior to instrumental analysis.

Recovery spike – sometimes referred to as post-digestion spikes for some forms of analysis, recovery spikes are spikes that are added to a sample after all sample preparation, cleanup, and other procedures that occur leading up to analysis are completed, i.e., a recovery spike is added before instrumental analysis and after all other sample treatment and handling processes are complete. Because of matrix effects and the contact time between matrix and spike for matrix spikes, matrix spike recoveries may differ from 100%. In the case of recovery spikes, because contact time is brief (the recovery spike sample being spiked right before instrumental analysis), the recovery spike results should be close to 100%.

Recovery spike duplicates/replicates – are additional replicates of recovery spike samples, analyzed for the purpose of quantifying precision in spike recovery.

Redox potential (Eh) – when an electrode made of a redox couple (such as SO_4^{2-}/H_2S) is measured in a galvanic cell against a normal hydrogen electrode, the resultant potential is known as the redox potential (Eh).

Reference location – exact spatial location and depth where test organisms will be exposed for control measurements, i.e., to measure the levels of bioavailable pollutants not associated with effects arising from/within an environmental effects envelope.

Reference material – a material containing known quantities of target analytes in a homogenous matrix of solution or solid that is similar to sample matrices. Analysis of reference materials (preferably Certified Reference Materials or Standard Reference Materials, i.e., those obtained by organizations that have determined analyte content or the material very accurately) is used to calculate accuracy. The accuracy of the analysis of a reference material is used as a gauge of whether the analysis of samples containing unknown amounts of analyte is likely to be accurate. The amount of analyte in the reference material provided by the company or organization providing the material is taken as the true value, μ, and the mean of results from analysis of the reference material obtained by the user, $\bar{x}$, is used to estimate accuracy.

Reporter gene – a gene that, when expressed, results in the formation of a reporter molecule (enzyme or protein) that produces a measurable signal, such as a bioluminescent, fluorescent, or visible (colored product) signal in response to promoter activation.

Reporter molecules – in a whole-cell bioreporter, reporter molecules are fluorescent proteins (such as green fluorescent protein) or enzymes (such as luciferase, which catalyzes a reaction producing luminescence, or β-galactosidase, which catalyzes a reaction leading to a color change).

Representativeness – is a qualitative evaluation of how well the results from a study represent actual environmental conditions or circumstances, i.e., the degree to which sampling accurately represents a characteristic population, a parameter variation at a sampling point, a process condition, or an environmental condition. Most aspects of QA/QC Planning and QA/QC Implementation can affect whether results are representative or not, for example, the sampling plan (sampling locations, times, etc.), matrix heterogeneity, and protocols and procedures for collecting, preserving, and transporting samples. Representativeness is addressed through QA/QC Planning; however, if unexpected results are encountered (e.g., bioavailable pollution where none was expected or vice versa), resampling designed to test for representativeness is typically in order.

Rhizosphere – is the narrow region of soil around plant roots directly influenced by root exudates (e.g., sugars, organic acids, enzymes) and associated microbial activity. It extends a few millimeters from the root surface and is a hotspot for biogeochemical processes due to intense interactions between plants, soil, and microorganisms.

Salinity – is defined as a measure of the total amount of dissolved salts in a solution and, as such, affects the bioavailability of pollutants in the same manner as ionic strength (dissolved salts originating from compounds composed of ions).

Sample hold time – refers to the maximum allowable time between when a sample is collected and when it is analyzed or processed.

Sampling and Analysis Plan – a document developed prior to any field work that provides complete and detailed information about what samples will be collected, how many, when, and how. SAPs also detail all aspects of sample handling and analysis, as well as prescribing QA/QC measures needed to ensure consistent and valid data.

Saturation index (SI) – is a quantity related to IAP, as $SI = \log(IAP/K_{sp})$, such that if the $SI < 0$, the solution is undersaturated; if $SI = 0$, the solution is saturated and in equilibrium with the solid phase; and if $SI > 0$, the solution is supersaturated, and precipitation of the solute should occur. The SI is convenient as no knowledge of K_{sp} is required to understand the status of solution with respect to being undersaturated, at equilibrium, or oversaturated.

Selectivity – formally defined by the International Union of Pure and Applied Chemists as "expressing the extent to which a particular method can be used to determine analytes under given conditions in the presence of other components of similar behavior" and according to relative magnitude of parameters in the equation:

$$\text{Signal} = m_a \times C_a + \cdots + m_n \times C_n + S_{\text{blk}},$$

where m is analytical sensitivity, C is concentration, each for components a to n, and S_{blk} is the blank signal. When sensitivity (e.g., instrumental response) for a given analyte is very high compared to others or to S_{blk}, selectivity for the analyte is high, and vice versa.

Sensitivity – is important to consider with respect to the extent and degree of bioavailable pollution. In this context, sensitivity refers to the capability of a method or instrument to detect a given target analyte at a given concentration and reliably quantitate the target analyte at that concentration. For some studies, qualitative (present/absent) or semiquantitative data (low or high concentrations) might provide a sufficient level of sensitivity for basic screening. For quantitative work, samples that contain high levels of bioavailable pollutant require less sensitivity and vice versa. Typical metrics for sensitivity are LOD, LOQ, and m, defined above.

Site – a term used with great variability in environmental science. In some usages, site refers to an exact spatial sampling location. In other usages, site refers to a specific field area under study and includes the entire spatial area within the environmental effects envelope and may additionally include other locations/areas, notably a reference location. The meaning intended in the use of this term may often be inferred from context. With respect to bivalve mollusk studies, for this book, the term site is used in the latter context.

Soil/sediment organic carbon–water partition coefficient (K_{oc}) – is the soil/sediment distribution coefficient normalized to soil/sediment organic matter, according to $K_{oc} = K_d \times (100/\%OC)$, where %OC represents the percent of organic carbon in a soil/sediment.

This normalization compensates for differences in pollutant partitioning that arise from the effect of organic matter variation between soil/sediment samples.

Solubility product constant – denoted K_{sp}, is an equilibrium constant that describes the extent to which a solid dissolves in a solution to form ions. It applies specifically to sparingly soluble salts and provides a measure of their solubility. K_{sp} is the product of the concentrations of dissolved ions resulting from dissolution of a solid, each raised to the power of the applicable stoichiometric coefficient for each ion given in the chemical reaction describing dissolution.

Sorption – a phenomenon that occurs universally in nature and a term that describes the association of a chemical species with a solid phase, i.e., in environmental chemistry, the accumulation of pollutants from solution onto or into surfaces of suspended particulates of soil/sediment.

Source – in the PLM technique, refers to the aqueous phase containing the target analyte (e.g., metal ions, organic compounds) that is to be extracted or separated through the liquid membrane.

Speciation (chemical) – the process of analyzing and understanding the distribution of the various chemical forms in which a pollutant exists in the environment.

Species (chemical) – the various chemical forms of components. Species include free ions, complexes, precipitates, and adsorbates. Their concentrations are determined according to chemical equilibria and are subject to mass and charge balance constraints. A component may undergo reactions to form a variety of chemical species; however, the sum of all species formed by a given component is the same as the total concentration of the component, and the sum of the concentration of positively charged species must match that of negatively charged species.

Spike – a spike is a relatively small volume of solution having a very well-known concentration of target analyte (for instance, as occurs in standard solutions) that is added to a sample. Subsequently, a portion of unspiked sample is analyzed and compared to results of analysis of a spiked sample in order to calculate a spike recovery. Usually the process of spiking is performed by taking two portions of a highly accurate volume of the sample, spiking one, leaving the other unspiked, and then analyzing both. The amount of pollutant in the spiked sample should be higher than the unspiked sample by a predictable amount, as measured by spike recovery. Too high or too low spike recovery might reflect bias (e.g., matrix effects), sample loss/contamination, or a variety of other analytical issues, and is an indication of problems with accuracy. The volume of spiking solution used to make a spike must be no more than 5% of the volume of solution spiked (otherwise, the concentration of analyte is changed if the spike volume is larger) and no less than 1% of the volume of solution spiked (as the relative volume of spiking solution decreases, the accuracy by which the amount of analyte added in the spike becomes compromised). Two common types of spikes distinguished here are matrix spike and recovery spike. Internal standards and surrogates are also usually added as spikes.

Spike recovery – is calculated according to

$$\%\text{Recovery} = \frac{C_{\text{spiked sample}} - C_{\text{unspiked sample}}}{C_{\text{spiking solution}} \dfrac{V_{\text{spike}}}{V_{\text{sample spiked}}}} \times 100\%,$$

where $C_{\text{spiked sample}}$ and $C_{\text{unspiked sample}}$, respectively, are results from analysis of the concentrations of the sample spiked and unspiked, and the denominator is the effective concentration of the target analyte in the spike added to the spiked sample. In the denominator, $C_{\text{spiking solution}}$ is the concentration of analyte in the spiking solution, V_{spike} is the volume of spike, and $V_{\text{sample spiked}}$ is the volume of the sample receiving the spike.

Spiking solution – a spiking solution is usually a target analyte standard, the concentration of which is 20%–50% higher than the concentration in the unspiked sample that will be spiked. This level of concentration, in combination with the low volume of the spike, ensures that the amount of analyte spiked is neither too much nor too little.

Standard Operating Procedure/Standard Method (SOP/SM) – is a detailed, step-by-step instruction for performing routine operations to maintain consistency across repetition. SOPs are similar to SMs, however may be developed by a particular organization or lab for internal use, being customized to specific operational conditions as needed.

Station position – surface location overlying subsurface exposure location, determined by GPS, in addition to other methods such as marking with transponder or buoy.

Strip (stripping phase) – in the PLM technique, refers to the receiving solution where the target pollutant(s) (e.g., metal ions, organic compounds) is ultimately transported and concentrated after crossing the liquid membrane.

Surrogate – is an element or compound added to field samples prior to any substantive sample handling or preparation. As for internal standards, a surrogate must be chemically similar to the analyte in behavior and not found in the environment. Surrogates are often deuterated forms of the target analyte. A surrogate is a type of spike, which, in addition to being added to samples, is also added to procedural blanks. Upon final analysis, a laboratory blank spiked with a surrogate is analyzed side-by-side with a surrogate-spiked procedural blank, and a comparison of the two results enables calculation of surrogate recovery. This reveals whether sample handling, processing, and preparation steps have resulted in any loss of surrogate, and, since the surrogate is chemically similar to the analyte in its behavior, surrogate loss can be assumed to reflect analyte loss and compromised results. Once the recovery of surrogate from the procedural blank is known, this can be compared to surrogate recoveries from samples. In the absence of matrix effects, the surrogate recovery from procedural blanks should be the same as for samples; if this is not the case, then matrix can be assumed to be affecting analysis of analytes in samples.

Target analyte (sometimes also referred to as target pollutant, or shortened to either analyte or pollutant for brevity) – for the purposes of this book, refers to any pollutant or class of pollutants that is the topic of a study and subject to analysis.

Test organism – individual organisms of species to be used in bioavailability measurements.

Toxicity-modifying factors (TMFs) – environmental variables (e.g., dissolved organic matter, major cations) that influence the bioavailability and toxicity of pollutants.

Transformed – refers to the process of introducing foreign DNA (such as a promoter:reporter gene fusion) into a host organism or cell, enabling the host cell to express new genetic traits, such as pollutant-responsive bioreporting capabilities.

Trip blank – is a sample of analyte-free media taken from the laboratory to the sampling site and returned to the laboratory unopened. These blanks are not required for every type of target analyte and are usually used when a study involves volatile organic carbons (VOCs) because these analytes can "seep" into or out of containers. When VOCs are included in a study's analytes, a trip blank should be prepared at a frequency of one per day of sampling during which samples are collected for VOCs. Trip blanks must be prepared prior to the site visit and must accompany field samples throughout all sample collection and transport operations. The trip blank must not be opened during VOC sampling activities or it may be contaminated by same.

Uptake phase (alternately, exposure phase) – with respect to worm studies, is a phase of a study corresponding to the period of time during which the test organisms are exposed to pollutants.

REFERENCES

IUPAC (International Union of Pure and Applied Chemists), Vessman, J., Stefan, R., Van Staden, J.F., Danzer, K., Lindner, W., Burns, D.T., Fajgel, A., Müller, H., 2001. Selectivity in analytical chemistry (IUPAC recommendations). *Pure Appl. Chem.* 73, 1381–1386.

US EPA, 2006. Guidance on Systematic Planning Using the Data Quality Objectives Process, EPA Document EPA/240/B-06/001, US EPA, Washington DC.

US EPA, 2014. Sampling and Analysis Plan – Guidance and Template v.4 – General Projects 04/2014, EPA Document R9QA/009.1, US EPA, Washington DC.

3 Sample Integrity and Avoiding Contamination

3.1 INTRODUCTION

The primary ***Quality Assurance*** (QA) consideration in all work described herein, whether it be laboratory or field-based, is the preservation of sample integrity to ensure the ***accuracy*** of target pollutant analyses. This integrity is achieved by (1) prevention of loss of target pollutant analytes during any process and (2) prevention of extraneous contamination being introduced. For the purposes of this chapter and the remainder of this book, the ***target analyte*** is also referred to as the ***target pollutant***, indicating the specific substance or substances under study that were originally present in a sample at the time of collection (or added to the sample through spiking). The term ***contaminant*** or ***contamination*** refers to pollutants adventitiously and erroneously introduced during the sampling, handling, processing, or analysis process usually due to operator error or procedural lapses. If target analyte loss or extraneous contamination occurs, results are invalidated. Target analyte loss most commonly occurs due to ***sorption*** onto the surfaces of containers or sampling equipment, leading to lower concentrations in the sample than what is actually present in the environment. This occurs when improper materials are used. Other causes of loss include degradation and chemical reactions (both may occur for organic pollutants), which primarily occur as a result of improper handling and storage. Examples of how contamination is introduced include (but are not limited to) (1) cross-contamination – e.g., when sampling equipment or containers are not properly cleaned between uses; (2) environmental ***exposure*** – e.g., when samples exposed to the environment during collection or processing pick up pollutants from the air, surfaces, or surrounding areas; this is especially problematic in areas with high levels of pollution; (3) operator error – e.g., inadvertent introduction of contaminants through handling, touching the inside of containers, sampling equipment, or using contaminated gloves; (4) improper storage – storing samples in inappropriate containers or at incorrect temperatures can lead to contamination through leaching of chemicals from containers or storing samples in areas where they may be cross-contaminated, for instance, with pollutants in other samples. It is always advisable to consult specific analytical/instrumental methods to be used (e.g., Standard Methods of analysis, Chapter 5) for advice on compatible sampling materials and handling procedures.

3.2 AVOIDING POLLUTANT LOSS, CONTAMINATION, AND CHANGES IN BIOAVAILABILITY DURING FIELD OPERATIONS

Sources of extraneous contamination in the field include (1) contamination from sampling gear, grease from winches or cables, spilled engine fuel (gasoline or diesel), engine exhaust, ice chests, films on improperly cleaned handling trays, processing tools, plastic bags, foil, ice packs used for cooling, i.e., every single item used in the field; (2) environmental sources: primary of these is aerosols and dust, though improper procedures in the field can cause other environmental sources of contamination to be substantial; and (3) improper procedures by field team personnel – ultimately contamination from the first two sources are almost always a result of improper procedures by field personnel. In some cases, the approach to avoiding such contamination is mitigation. For example, sampling on a calm rather than windy day will reduce the chances of aerosol contamination; alternately, providing a localized windbreak may be mitigatory.

Ideally, contamination is always controlled rather than mitigated. In matters where control is possible, avoidance must be strictly practiced. All items used in the field and that will contact samples

 DOI: 10.1201/9781003355229-4

must be of materials that can be cleaned easily and that are not themselves potential sources of contamination (refer to Section 3.4). Any field gear that has not been properly cleaned and decontaminated must not be used, i.e., all materials used (ice chests, handling trays, processing tools, plastic bags, foil, etc.) that can be treated, must be subjected to appropriate cleaning and decontamination in the laboratory prior to use under clean room or clean-room equivalent conditions. Further, after cleaning, they must be transported and stowed in a manner to keep them decontaminated. Stowing typically involves carrying cleaned gear in clean, sealed, containers. Typical cleaning and decontamination protocols are given in Section 3.5. In the field, when working at multiple locations, the field team must clean all items used in between locations. To control environmental contamination, when sampling, the amount of time that a given sample is exposed to the environment should be minimized; i.e., transfer from the sampling equipment to a sampling container should occur expeditiously. Closure of all sample containers should be complete in order that no leakage can occur. A sampling container represents primary containment. To avoid sample cross-contamination, at minimum, secondary containment is also needed. Secondary containment often consists of placing sample containers inside bags, e.g., resealable zip-type plastic bags. Control of field contamination entails reduction of the risk of contamination that might otherwise prevail. Individual chapters herein suggest field ***Quality Assurance/Quality Control*** (QA/QC) procedures that are designed to detect failures in mitigation and control of field contamination.

Loss of target pollutants/analytes during field operations is prevented by ensuring sample ***representativeness***. This nominally consists of processing samples correctly by following procedures for collection, field preservation (when applicable), and transport conditions. For instance, a water sample requiring preservation should be placed in a bottle with the correct amount of preservative for the sample size, and the bottle should be filled to the proscribed limit, then closed, and agitated well immediately. If the bottle is overfilled or underfilled, the incorrect preservative-to-sample volume will result, which may adversely affect the subsequent laboratory analysis. For ***bioavailability*** studies, preservative is typically not used, however, the general principle still applies; filling a sample bottle to the brim reduces air and oxidation reactions. Ideally, the container is clean and handled such that no contamination on the outside of the container will be in contact with the sample. The issue of sample ***representativeness*** is particularly critical when collecting organisms in the field. Mollusks, for instance, can lose tissues and/or fluids if their shells are cracked on collection, and this can result in loss of target analyte in the organisms. It is generally best practice to ensure that all samples are collected in sealed packaging/sample containers right away on collection. Individual chapters herein have more information, as pertinent to specific methods and procedures.

Perhaps the greatest risk with measurements of bioavailability in field operations is the proclivity of collecting samples to itself impact bioavailability. This can occur with no net change in target analyte load (no net loss in target analyte, no net gain in contamination). ***Sample hold time*** and sample transport and storage temperature are very important considerations in this regard. Sample hold time refers to the maximum allowable time between when a sample is collected and when it is analyzed or processed. If a field sample is collected under hot conditions, for instance, then transported and stored at a considerably lower temperature, the bioavailability of target analytes measured in the lab may be less than the actual bioavailability under *in situ* field conditions. Alternately, sampling disturbs a sample in a manner that can often result in stimulation of microbial ***activity***, which can alter bioavailability, and the longer the sample hold time, the greater the risk that alterations in bioavailability will occur under hot conditions where microbial activity is not inhibited. Normally, samples are transported and stored under cooling to inhibit microbial activity and prevent changes in samples over the often not inconsiderable time needed for transport back to the laboratory and the time spent in the laboratory awaiting processing. If, however, transport times are envisaged to be rapid, and in-laboratory processing can be undertaken immediately, it may be preferable to transport samples at or near *in situ* warm conditions. Freezing is often used to prolong sample hold times, but this in particular may alter bioavailability and is not best practice. In some cases, the logistics of sample transport and processing make it unavoidable to freeze samples.

Careful consideration of factors that will affect bioavailability post-sampling, and how to mitigate or control for them, must be considered as part of any Sampling and Analysis Plan that involves field operations. Individual chapters herein have additional commentary on this point, as relevant.

3.3 MAINTAINING SAMPLE INTEGRITY IN THE LABORATORY

In the laboratory, potential sources of contamination include, as examples, (1) environmental contamination – e.g., through aerosol or dust deposition; (2) improperly cleaning/decontamination – e.g., of instruments, utensils, work surfaces, containers that may contact the samples, and any other items used; and (3) improper procedures by laboratory personnel. All sample processing for analysis must be done under conditions that approach cleanroom conditions. This entails that work areas and implements should be free of trace metals and trace organic contaminants. Ideally, these areas should be under positive pressure with filtered air (HEPA filter class 100). All instruments, work surfaces, and containers used to process samples must be of materials that can be cleaned easily and that are not themselves potential sources of contamination (refer to Section 3.4). To avoid cross-contamination, all equipment used in sample processing should be cleaned thoroughly between each sample. Cleaning and decontamination protocols are given in Section 3.5. As with field work, the amount of time that a sample container is open or that a sample in the laboratory is being manipulated should be minimized. In some cases, this is difficult, for instance, in handling ***macrophytes*** for analysis (measuring condition, dissecting, Chapter 8). In this case, manipulation should certainly occur in a clean room or under a hood with positive pressure of HEPA-filtered air. Care should continue to be taken to avoid cross-contamination by ensuring sample containers are well-sealed, and, if there are any doubts about cross-contamination in storage, secondary containment should continue to be used. Verification of the efficacy of cleaning procedures should be documented through the analysis of processing ***blanks*** or rinsates (per specific method QA/QC procedures).

For frozen samples, particular care must be taken during thawing to ensure that all liquids from/originally associated with the thawed samples are retained and the sample homogenized exhaustively prior to analysis. Homogenization, in general, is relevant to some extent to all samples for representativeness.

Loss of target pollutant in the laboratory is less of a risk than in the field, primarily because the sample containers and storage conditions (main risk factors) were already determined in field operations. The element of field control that prevents loss of pollutants is continued in the laboratory by continued use of appropriate materials and adherence to method-prescribed sample hold times.

3.4 SUITABLE MATERIALS TO AVOID TARGET ANALYTE LOSS AND CONTAMINATION

In practice, the range of appropriate materials to avoid contamination is narrow. The guidelines for materials listed here also serve the purpose of avoiding loss of target analyte. It cannot be overemphasized that materials themselves must be properly cleaned/decontaminated in order to be contaminant-free (Section 3.5).

3.4.1 Materials Compatible with Organic Target Analytes

Materials to use with samples containing organic target analytes are as follows:

- Stainless steel, anodized aluminum, titanium, borosilicate glass, quartz, or polytetrafluoroethylene (PTFE).
- Quartz utensils are ideal but expensive. For bench liners and bottles, borosilicate glass is often inexpensive and convenient.

- Polypropylene and high-density polyethylene (plastic) surfaces, implements, gloves, and containers are a potential source of contamination by organics, particularly for pollution related to plastics, and should not be used in the lab. These materials may be suitable for limited application in the field, (1) if no suitable alternative is available, (2) if rigorously cleaned, and (3) if it can be demonstrated by analysis that they are not a source of contamination. Occasionally, there may also be limited use of these materials in the laboratory under certain conditions and if contact time is minimized (for instance, filtering an extract that contains organic target pollutants in an organic solvent using a polypropylene syringe filter). If a laboratory chooses to use these materials, there should be clear documentation that they are not a source of contamination. This is often achieved via the use of ***procedural blanks*** (Chapter 4).

3.4.2 Materials Compatible with Heavy Metal(oid) Target Analytes

Materials to use with samples containing heavy metal(oid) target analytes are as follows:

- Polypropylene or polyethylene (high density), PTFE, or titanium.
- The predominant metal contaminants from stainless steel are chromium and nickel. If these metals are not of concern, the use of high-quality, corrosion-resistant stainless steel for sample processing equipment is acceptable.
- Knives with titanium blades and PTFE handles are recommended for performing tissue resections.
- High-purity quartz may be usable in some cases. Sorption tests should be conducted with solutions of known amounts of target analyte prior to bioavailability tests to ensure that any high-purity quartz items used do not lead to loss of target analyte.

3.5 CLEANING/DECONTAMINATION PROCEDURES

Cleaning/decontamination needs vary according to target analyte. General good practice is given below. Individual methods provide additional guidance.

3.5.1 Cleaning/Decontamination of Materials in Contact with Samples having Organic Target Analytes

Borosilicate glass and quartz may be cleaned/decontaminated by baking in a muffle furnace for 30 minutes at 500°C. It is common to cover openings of containers with or wrap entire implements with heavy-duty aluminum foil, and mark either the glass or foil with indelible, Sharpie-type, marker. When decontamination is complete, the marker ink is vaporized.

All other implements are washed with detergent solution, rinsed with ultra-high-purity water, and dried. Next, clean with high-purity (pesticide-grade) isopropanol or acetone, followed by high-purity hexane, and cover openings of containers with or wrap entire implements with clean heavy-duty aluminum foil for storage until use.

Work surfaces should be cleaned with detergent solution, rinsed with ultra-high-purity water, dried, then subsequently cleaned with high-purity isopropanol or acetone and allowed to dry completely.

3.5.2 Cleaning/Decontamination of Materials in Contact with Samples having Heavy Metal(oid) Target Analytes

Plastic or PTFE containers must be cleaned thoroughly with a detergent solution, then rinsed with ultra-high-purity water, and then soaked in 20% HNO_3 for a minimum of 16 hours at room temperature. Acids used for cleaning should be high purity grade. For materials used in the field that might be highly contaminated, 50% HNO_3 may be used. Stainless steel or titanium may be cleaned in the same manner, however, prolonged contact with acid should be avoided, and instead a short acid dip is used. Generally, it is desirable to avoid the use of metals, even stainless steel or titanium, for samples scheduled for heavy metal/metalloid analysis.

3.5.3 Cleaning/Decontamination of Materials in Contact with Samples having both Organic and Heavy Metal(oid) Target Analytes

Cleaning/decontamination is quite different for materials that will come into contact with samples scheduled for analysis of organic pollutants, as compared to samples scheduled for heavy metal/metalloid analysis. In consequence, taking separate subsamples for each (samples for heavy metals versus organic pollutant analysis) is best practice. If a single sample is prepared for the analyses of both organics and heavy metal(oid)s, precautions must be taken to use materials and cleaning procedures that are non-contaminating for both organics and metals.

4 Guidance Notes on Quality Assurance and Quality Control

4.1 INTRODUCTION

There is a great need for increased scrutiny of environmental data for quality because it is impossible to form any conclusions on the meaning of data without properly understanding the uncertainties/limits associated with the data. This chapter provides information to assist with understanding data quality requirements needed for work using the family of methods presented in this book. The basic tenets of characterizing data quality are comprised first in following a structured system of ***Quality Assurance*** (QA) and ***Quality Control*** (QC). The degree of complexity of the QA/QC process may vary considerably depending on whether applied to a research project, on the one hand, or to a project involving pollutants and liability, on the other hand. The latter requires considerable documentation, which is beyond the scope of remarks here, and also very well known by environmental practitioners having the proper national or international credentials to work on such projects. The information in this chapter is thus more geared to researchers, some of whom may wish to have some further background on the precepts of QA/QC for the purpose of assessing what level of QA/QC is needed for specific projects in order to derive defensible and useful conclusions from their data. Many agencies have guidance documentation on QA/QC procedures for Environmental Monitoring; these are typically more oriented to experienced practitioners. Two excellent and practical resources for professionals who are less familiar with QA/QC practices are Ganapathy (2008, laboratory focus) and Quevauviller (1995, field sampling and sample preparation focus; also see US EPA, 2005).

At its simplest, the QA/QC process may be summarized in three phases. To understand the terminology associated with these three phases, refer to the Terms and Definitions given here and in the next section[1]:

1. **QA/QC Planning** – this may be as involved as planning and development of ***Data Quality Objectives*** (DQOs, US EPA, 2006) and documentation of all stages of a project through a formal ***Sampling and Analysis Plan*** (SAP, US EPA, 2014) or simply compiling lists of all information needed, associated ***QA/QC samples***, and development of checklists to support acquisition of needed QA/QC data. Part of planning QA/QC sampling and analysis must, at minimum, involve specification of ***Data Quality Indicators*** (DQIs, US EPA, 2006), ***Acceptance Criteria*** or performance criteria (US EPA, 2006) for evaluation of conformance/***Nonconformance*** (e.g., ISO, 2015) of QA/QC results according to DQIs, ***Corrective Actions*** (ISO, 2015) that might be possible for problematic data, and how QA/QC results will affect reporting of data (*Data Verification* and *Data Validation*, US EPA, 2006). QA/QC Planning also involves establishing in advance all procedures to be conducted and how. Ideally, all procedures used are ***Standard Operating Procedures*** (SOPs, US EPA) or ***Standard Methods*** (SMs, e.g., APHA, 2023; NB, for simplicity here, one term, SOPs is used in this chapter). Examples of classes of SOPs that have been developed include the following:
   - ***Chain-of-Custody*** (e.g., US EPA, 2002) – to ensure the integrity of samples by specifying processes for sample identification, preservation, holding times, handling, security, and tracking

[1] Also given in Chapter 2 for ease of future reference.

DOI: 10.1201/9781003355229-5

- Sample collection (Ganapathy, 2008) – to ensure that samples are collected properly, in a manner that optimizes the integrity of results
- Instrumental analysis (Quevauviller, 1995) – governs all process needed to obtain good results in the laboratory – these methods usually, and should, contain guidelines for reagents used, preparation of standards, equipment calibration and maintenance, and data reporting.

    It should be noted that some SOPs may themselves contain significant guidance on QA/QC procedures specifically associated with procedures within the SOP. NB: Also see SOPs listed in Chapter 5 and comments in individual chapters about additional QA/QC appropriate to the methods in this book.

2. **Implementation** (QA/QC) – the QA/QC Implementation phase consists of executing all processes and procedures developed in the QA/QC Planning phase, e.g., sample collection, documentation of field and laboratory activities and ***Chain-of-Custody*** procedures, analysis of all samples and needed QA/QC samples, and conducting any other planned activities associated with project goals.
3. **Assessment** (QA/QC) – after data are collected in the QA/QC Implementation phase, QA/QC Assessment first consists of Data Verification/evaluating data according to the metrics chosen for DQIs (at minimum, one metric each for ***precision***, bias/***accuracy***, ***representativeness***, ***completeness***, and ***sensitivity***; ***comparability*** may be applicable if more than one data set is to be used). For the different DQI metrics, Acceptance Criteria will have been established during the QA/QC Planning process. If results are within prespecified conformance with Acceptance Criteria, they may be deemed fit-for-purpose on a QA/QC basis. Otherwise, results must be rejected and a determination made if Corrective Action is possible. Corrective Action may be as simple as, for instance, reanalysis of a sample for which results were rejected. A process of Data Validation to further analyze data and use data to form conclusions follows Data Verification.

In the case of studies on ***bioavailability***, there is a paucity of SOPs, ergo the need for this book, the aim of which is to move toward standardization.

4.2 TERMS AND DEFINITIONS

Some specific terms and definitions that are universal to QA/QC as used in this book are contained in this section, grouped according to different points in the phases of QA/QC in which they are most relevant.

4.2.1 General Terms

Acceptance Criteria – preestablished QC criteria that must be met in order for data to be accepted as meeting standards for ***precision, accuracy (bias), representativeness, completeness, comparability, and sensitivity (PARCCS)***, and therefore being fit-for-purpose for use in data analysis and subsequent conclusions about the data. PARCCS standards are often given in Standard Methods.

Analyte (target) – for the purposes of this book, it refers to any pollutant or class of pollutants that is the topic of a study and subject to analysis.

Blank – there are different types of blank samples, all of which share the common characteristic that no contamination should be found in a blank, i.e., the acceptance criterion for a blank is that all blank results should be below limits of detection (LODs).

Chain-of-Custody – a chronological documentation of the handling, control, transfer, and storage of samples from the point of collection to analysis and reporting. This process maintains a clear record of every individual or entity that has handled the sample, along with the time, date, location, and purpose of each transfer, for the ultimate purpose of ensuring the integrity and reliability of results. Usually accomplished with Chain-of-Custody forms.

Corrective Action – measure(s) taken to rectify conditions adverse to achieving quality outcomes and, where possible, to preclude their recurrence. Corrective Action is most often prompted by discovery of a QA/QC Nonconformity.

Data Quality Indicators (DQIs) – DQIs refer to the quantitative statistics and qualitative descriptors that are used to interpret the degree of acceptability or utility of data. The principal indicators of data quality are PARCCS data (see below). Multiple metrics (indicators) may apply to each of these six types (precision, accuracy, representativeness, completeness, comparability, and sensitivity) of DQI.

Data Quality Objectives (DQOs) – DQOs are qualitative and quantitative statements derived from the DQO planning process that clarify the purpose of a study, define the most appropriate type of information to collect, determine the most appropriate conditions from which to collect that information, and specify tolerable levels of potential decision errors. The DQO process is documented via, *inter alia*, an SAP. See US EPA (2006) and (2014), respectively, for extensive information on the DQO process and SAPs.

Data Validation – this terminology is highly variable globally, nationally, and even regionally. For the purposes of this book, Data Validation refers to evaluation of data subsequent to Data Verification and includes all aspects of data analysis that involve analysis of the structure of the data (e.g., ***distribution***), the appropriate quantitative methods to use in analyzing the data to address information needs, evaluation of trends or biases in the data and associated QC results, a decision as to whether the data is fit-for-purpose for study goals, and conclusion-making.

Data Verification – as with Data Validation, this use of this terminology is also highly variable. For purposes here, Data Verification involves evaluating all of the data against PARCCS Acceptance Criteria that are specified in SOPs and or as defined for the project during the QA/QC Planning phase (in other words, against all Acceptance Criteria for field and laboratory data QA/QC, as well as any other relevant criteria specified in advance).

Duplicate/replicate (sample/s) – duplicate (two) or replicate (n, where n is greater than two) refers to a sampling strategy whereby two or more samples representative of the same population and treated identically are compared. Results from analysis of different forms of duplicates/replicates are used to measure precision. There are a variety of types of field and ***laboratory duplicate*** samples. See definitions for, e.g., ***co-located samples***, laboratory duplicate/replicate, field duplicate/replicate, ***field split samples***, ***matrix spike duplicate/replicate***, and ***recovery spike*** duplicate/replicate below.

Initial Demonstration of Proficiency (IDP) – is the single most important element of a QA/QC program. Prior to the analysis of samples, an initial demonstration of (analytical) proficiency is accomplished through the successful analysis of reference samples (usually Certified or Standard Reference Materials, see below) to obtain results that are in accord with DQIs stipulated as typical and acceptable, as provided by the suppliers of reference materials. The reference material used should have a matrix as close as possible to the sample matrices. The analysis of reference materials is often stipulated in methods for instrumental analysis, and therefore should always be undertaken as part of instrumental analysis. Users should refer to methods of instrumental analysis for guidance on the topic of analysis of reference materials. Methods that have substantive sample preparation steps prior to instrumental analysis may have additional IDP requirements, and these additional requirements are outlined in individual chapters in this book. The IDP process demonstrates whether or not a laboratory is capable of obtaining a correct result that is fit-for-purpose with respect to DQOs/study objectives and DQI criteria.

Matrix – refers to the chemical ***components*** of which a sample is composed minus the ***target analyte(s)***, i.e., matrix is the material in which target analyte is suspended in samples.

Matrix effects – effects exerted by matrix on analyses. Normally, in instrumental analysis, matrix effects are defined as effects that may produce bias in analytical results if not controlled for in QA/QC Planning. In terms of bioavailability studies, matrix effects represent environmental influences that exert control over bioavailability.

Medium (plural, media) – material in which target analyte is suspended. Can include matrix in the case of field samples (environmental media such as water, soils, and sediments), or can also be comprised of media into which target analyte is introduced prior to analysis (e.g., aqueous- or solvent-based extractant).

Nonconformance/Nonconformity – a deviation from requirements of Acceptance Criteria, i.e., when a sample, ***analytical batch*** or entire data set is compromised, as indicated by QA/QC results.

QA/QC – QA is focused on preventing data quality issues and refers to processes and procedures put in place and conducted in order to ensure quality of data for, e.g., the purpose of measuring bioavailability; QC focuses on reviewing data quality after results are produced (for instance, in the context of Acceptance Criteria) and identifying quality issues that might compromise the utility of the data. QC also focuses on Corrective Action, when possible or applicable.

QA/QC sample – a sample that is analyzed to obtain results that will be used in Data Validation, namely with respect to DQI Acceptance Criteria. Examples of QA/QC samples are blanks, duplicates/replicates, and ***spikes***.

Sampling and Analysis Plan – a comprehensive document that outlines the procedural and analytical requirements for collecting, handling, and analyzing environmental samples. It typically includes details on the sampling strategy, sample collection methods, QA and QC measures, analytical methods, and data reporting procedures. The SAP ensures that the data collected are reliable, accurate, and suitable for making informed analysis of results and consequent environmental decisions.

Standard Operating Procedure/Standard Method – a detailed, written set of instructions to perform specific tasks consistently and correctly. The purpose is to ensure that processes are carried out in a standardized manner for maintaining quality. Some SOPs/SMs represent established protocols that have been developed and are accepted by, e.g., professional organizations or regulatory bodies overseeing environmental testing and analysis.

4.2.2 PARCCS Terms for Different Categories of DQI

Note: PARCCS is a convenient abbreviation by which to remember different categories of DQI – ***precision***, accuracy (bias), representativeness, comparability, ***completeness***, and sensitivity.

Accuracy – the degree of agreement between an observed value and an accepted reference or "true" value. Inaccuracy results from a combination of a random error and systematic error (bias – an error in accuracy that is systematic and causes a persistent positive or negative deviation from the true value). Accuracy is determined as absolute or relative, respectively, according to

$$\text{Accuracy} = \bar{x} - \mu$$

and

$$\%\text{Accuracy} = \frac{\bar{x} - \mu}{\mu} \times 100\%,$$

where $\bar{x}$ is the mean value of a measurement on a sample and μ is the true value. The standard approach to the evaluation of accuracy is via the use of Certified or Standard Reference Materials.

Comparability – refers to the equivalency of different sets of data, i.e., the degree to which different methods or data agree or can be represented as similar. Often, different data sets subject to different methods of chemical analysis or other varying conditions are not comparable. Comparable data sets must contain the same variables of interest and must possess values that can be converted to a common unit of measurement. Comparability is primarily qualitative; for example, if two different data sets for studies on the same pollutant were obtained using different analytical methods, the two methods may not be comparable. Data sets should never be combined if doubts concerning comparability exist.

Completeness – a measure of the amount of valid data obtained from a program of measurement compared with the amount that was expected to be obtained, i.e., the amount of "usable" data. Percent completeness is calculated as

$$\%\text{Completeness} = \frac{n_{\text{data,obtained}}}{n_{\text{data,planned}}} \times 100\%,$$

where $n_{\text{data, obtained}}$ is the amount of valid data obtained, and $n_{\text{data, planned}}$ is the amount of data planned or that was expected to be obtained. A value of 90% is a common acceptance criterion for completeness, i.e., loss of more than 10% of the data deemed needed is taken as compromising the ability of the results to be deemed fit-for-purpose.

Precision – a measure that quantifies the agreement among a set of duplicate or replicate measurements without assumption of knowledge of accuracy (true value), i.e., precision tells how variable results are, which is necessary to compare two or more results to each other statistically (e.g., testing for differences), however is not a measure of accuracy (whether a result has the correct value or not). Examples of metrics for precision are as follows:

$$\%\text{RPD} = \frac{2 \times |n_1 - n_2|}{n_1 + n_2} \times 100\%,$$

and

$$\%\text{RSD} = \text{CV} = \frac{s_{\bar{x}}}{\bar{x}} \times 100\%,$$

where %RPD is the relative percent difference between replicate samples, n_1, and n_2; the vertical braces indicate absolute value; %RSD is the relative standard deviation or coefficient of variance (CV); $\bar{x}$ is the mean of the total number of replicate samples; and $s_{\bar{x}}$ is the standard deviation around the mean.

Representativeness – is a qualitative evaluation of how well the results from a study represent actual environmental conditions or circumstances, i.e., the degree to which sampling accurately represents a characteristic population, a parameter variation at a sampling point, a process condition, or an environmental condition. Most aspects of QA/QC Planning and QA/QC Implementation can affect whether results are representative or not, for example, the sampling plan (sampling locations, times, etc.), matrix heterogeneity, and protocols and procedures for collecting, preserving, and transporting samples. Representativeness is addressed through QA/QC Planning; however, if unexpected results are encountered (e.g., ***bioavailable*** pollution where none was expected or vice versa), resampling designed to test for representativeness is typically in order.

Sensitivity – is important to consider with respect to the extent and degree of bioavailable pollution. In this context, sensitivity refers to the capability of a method or instrument to detect a given target analyte at a given concentration and reliably quantitate the target analyte at that concentration. For some studies, qualitative (present/absent) or semiquantitative data (low or high concentrations) might provide a sufficient level of sensitivity for basic screening. For quantitative work, samples that contain high levels of bioavailable pollutants require less sensitivity and vice versa. Typical metrics for sensitivity are LOD and ***limit of quantification*** (LOQ), defined below.

4.2.3 Field-applicable QA/QC Sampling

Co-located samples – a type of field duplicate/replicate (usually duplicate) where independent samples are collected as close as possible to the same point in space and time. The co-located samples are collected independently, stored in separate containers, and analyzed independently by the same method, under the same conditions, and at the same time. This type of field duplicate/

replicate sample, while dependent upon the precision of the field sampling process, is also useful in documenting field heterogeneities.

Equipment blank/equipment rinsate – an equipment blank, sometimes referred to as equipment rinsate, is a sample of analyte-free media that has been passed over or used to rinse field sampling equipment. The medium for the equipment blank is typically water or solvent, dependent on what the final aqueous media are for instrumental analysis, though other media may be applicable for soils or sediment work (e.g., ***contaminant***-free bentonite). An equipment blank is collected after completion of decontamination – this may refer to laboratory decontamination that occurred prior to taking the equipment into the field if the equipment is being used in the field for the first time, or it may refer to decontamination of equipment that must occur in the field between each use of the equipment for sampling. An equipment blank is taken after field decontamination and immediately prior to sampling. Equipment blanks are used to demonstrate whether or not field samples may have been contaminated by field equipment. Contamination of field equipment that transfers to samples may affect both precision and accuracy of results.

Field blank – field blanks are samples that are collected in the field and should be analyte-free, assuming that appropriate precautions to avoid contamination, introduced during transport or occurring from incomplete field decontamination, are in place and successful. There are various types of field blanks, depending on the type of field collection technique used and the type of pollutant that is scheduled for analysis. The most common types of field blank are field ***procedural blanks***, equipment blanks, and ***trip blanks***. Field blanks provide information on various types of contamination that may be introduced into samples in result of various aspects of field operations. Field sample contamination may affect both precision and accuracy of results.

Field duplicates/replicates – are duplicates/replicates collected in the field used to assess the precision of results as impacted by environmental heterogeneities in the field and uniform conduct/reproducibility of field sampling procedures. Results from field duplicates show the precision of the field sampling process, however, field duplicates/replicates also reflect laboratory precision. It is not uncommon to collect field duplicates/replicates for all samples, however, two common types of field duplicate additionally collected for QA/QC are co-located samples and field split samples. Because field split samples are homogenized, this substantively reduces field heterogeneity; hence, comparison of co-located samples with field split samples enables differentiation of how precision of results is affected by field sampling procedures versus field heterogeneity.

Field split samples – a type of field duplicate/replicate wherein a sample is collected, homogenized, and then divided into two or more portions, each of which is one field replicate. Homogenization is usually performed under clean conditions after transport to the laboratory; however, it is sometimes performed in the field if convenient and feasible. After homogenization, duplicates/replicates (field split samples) are stored in separate containers and analyzed independently by the same method, under the same conditions, and at the same time. This type of field duplicate/replicate sample is useful in documenting the precision of the sampling process. Homogenization may have an adverse impact on sample integrity for some sample types, notably volatile organic compounds, and in such cases only co-located samples may be used as field duplicates/replicates. Because field split samples are homogenized, this substantively reduces field heterogeneity.

Procedural blank – a procedural blank is a sample that does not contain the matrix, which is brought through the entire measurement procedure and analyzed in the same manner as other samples. When preparing procedural blanks, water is often used in place of the matrix, though sometimes solvent is used if the final samples subject to instrumental analysis are in solvent. Procedural blanks may be used to assess any contamination that may occur within the entire process of sample collection, transport, field and laboratory conditions, reagents used, containers used, etc. In some cases, procedural blanks are taken to specifically monitor field contamination or laboratory contamination. For field procedural blanks, containers of reagent water, solvent, or a solid blank material (whatever is used as the basis for a procedural blank) should be opened in the field for a predefined period of time that is approximately the same as the time that a sample is exposed in the field during collection and processing, thus capturing approximately the same ambient contamination, if any is present, that a field sample container would. Individual methods often have guidance concerning procedural blanks. For methods that do not specify procedural blanks, users should consider how best to add these to a SAP.

Trip blank – is a sample of analyte-free media taken from the laboratory to the sampling ***site*** and returned to the laboratory unopened. Trip blanks are not required for every type of target analyte and are usually used when a study involves volatile organic carbons (VOCs) because these analytes can "seep" into or out of containers. When VOCs are included in a study's analytes, a trip blank should be prepared at a frequency of one per day of sampling during which samples are collected for VOCs. Trip blanks must be prepared prior to the site visit and must accompany field samples throughout all sample collection and transport operations. The trip blank must not be opened during VOC sampling activities, or it may be contaminated by the same.

4.2.4 Laboratory-applicable QA/QC Sampling and Analysis

(Analytical) batch – a group of samples, including QA/QC samples, which are processed together using the same method, the same lots of reagents, and at the same time in continuous, sequential time periods. Batches always contain one or more laboratory QA/QC samples. Samples in each batch should be of similar composition and all be subject to the same treatment. Samples collected from the same area would normally be grouped together for batching purposes, and samples prepared in the same batch would normally be analyzed together on a single instrument. In some cases, a batch may contain similar samples, though from different sites/locales. Field QA/QC samples are site-specific, and a batch having samples from more than one site/locale would require more QA/QC samples to accommodate needs.

Calibration blank – analyzed along with prepared standards in calibration of analytical instruments. A calibration blank should be prepared exactly the same as calibration standards, except that no analyte is added or present. Such a blank represents "zero" in calibrating an instrument for instrumental analysis.

Calibration checks – refers to an analysis that is performed periodically after calibration to assess whether the initial calibration has drifted (changed) and for memory effects over time. Memory effects or carryover may be present in some forms of instrumental analysis, and this term refers to a phenomenon whereby an instrument "remembers" analyte from prior samples, in which case the instrument response for standards increases over time, requiring instrument maintenance and recalibration. Calibration check analyses involve reanalysis of a blank, or low, mid-range, or high standard, to check if the result is suitably consistent with the initial calibration. If a calibration check fails, Corrective Action is recalibration, often with reanalysis of samples analyzed under failed calibration conditions.

Calibration curve – a plot of instrument response to an analyte versus known concentrations of the blank and analyte calibration standards. Calibration standards are prepared by successively diluting a standard solution to produce working standards that cover an instrumental working range. Standards and the blank used in calibration should be prepared at the frequency specified in the appropriate instrumental method. Ideally, the matrix of calibration standards should match that of samples analyzed to the greatest degree possible, however, this is often not practical. Methods of instrumental analysis typically provide guidance on preparation of calibration standards.

Internal standard – internal standards are employed for a variety of QA/QC purposes, are used in many methods, and are typically added as spikes. Such standards usually comprise/substances similar to the analyte (perhaps even isotopes of the analyte) that are not found in the environment and are added to each sample, QA/QC, and calibration solution. Since internal standards are added in a fixed amount, the signal from internal standards should always be the same; any deviations from expectation indicate some form of compromised analysis. If the internal standard in all samples is effectively constant, as expected, taking the ratio of analyte signal to internal standard signal and plotting it against the analyte concentrations in the calibration solutions produces a calibration curve that can then be used to calculate the analyte concentration. An internal standard is like a ***surrogate*** (see below) as it is a pure substance, added at a known amount, and unlikely to be found in a sample. An internal standard, however, differs from a surrogate in its function. Internal standards are primarily used to measure the relative responses of analytes, and sometimes surrogates as well, to confirm that analytical instrumentation is working properly and/or to correct for drift that might occur. An internal standard is normally added at the end of all sample handling and preparation procedures, right before analysis. Occasionally the interplay of how internal standards and surrogates function in an analysis is more complex (particularly as regards matrix effects), and a few methods call for internal standards to be spiked into sample before intensive preparation. In this case, as for surrogates, the recovery of internal standards may be used to demonstrate the lack of severe matrix effects in the instrumental analysis by setting criteria for the internal standard response in comparison to a response in a sample with a known lack of matrix effect. The typical pattern of use of surrogates is to monitor for analyte loss and or/matrix effects, with internal standards being used to correct for variability in analytical conditions and calibrate for analyte response (internal calibration). The schedule and purpose of using internal standards and surrogates are given in standardized methods.

Laboratory duplicate/replicate – this term refers to the analysis of two or more subsamples of the same sample. The subsamples are taken from the same container, processed/prepared the same, under the same conditions, at the same time, and then analyzed in the same batch. The results from field duplicates/replicates reflect variability from all ***sources*** that affects precision. Analysis of laboratory duplicates/replicates reflect variability associated with subsampling and the matrix, but not the precision of field sampling, preservation, or storage. In result, comparison of laboratory and field duplicates/replicates enables differentiation of to what extent laboratory operations versus field operations affect the precision of results. Laboratory duplicates/replicates are more commonly used to assess precision for inorganic and radiological constituents, while precision for organic analyses is usually assessed by determining the precision for matrix spike duplicates analysis.

Limit of detection (LOD) – the lowest concentration of analyte that may be statistically affirmed as being detected. The LOD is calculated as

$$\mathrm{LOD} = \overline{y_{\mathrm{blk}}} + 3.3 \times s_{y_{\mathrm{blk}}},$$

where $\overline{y_{\mathrm{blk}}}$ is the mean signal of the blank and $s_{y_{\mathrm{blk}}}$ is the standard deviation around $\overline{y_{\mathrm{blk}}}$. The factor of 3.3 is the t-test probability at $p < 0.05$, ensuring that the LOD is above $\overline{y_{\mathrm{blk}}}$. An alternate, more conservative, approach to calculating the LOD is to use the intercept from linear calibration plus 3.3 times the standard error from linear regression of the calibration curve, respectively, instead of $\overline{y_{\mathrm{blk}}}$ and $s_{y_{\mathrm{blk}}}$. The LOD determined in the above manner will have units of response signal, which can be converted into units of concentration using the calibration curve.

Limit of quantification (LOQ) – the lowest concentration of an analyte that might be encountered in a sample and quantified with sufficient statistical reliability so as to be distinguishable from concentrations in other samples. The LOQ is calculated as

$$\mathrm{LOQ} = \overline{y_{\mathrm{blk}}} + 10 \times s_{y_{\mathrm{blk}}},$$

where the factor of 10 is obtained from statistical considerations, much as the factor of 3.3 is for LOD. The LOQ determined in the above manner will have units of response signal, which can be converted into units of concentration using the calibration curve.

Matrix spike – matrix spikes are spikes that are added to a sample before sample preparation, cleanup, and procedures are implemented. Often the matrix spike will be mixed into the sample ahead of sample preparation and processing for analysis and allowed to integrate with the sample (24 hours or more). In some cases, target analyte spiked into sample matrix will be in a more labile form, i.e., will not have time to integrate into refractory matrix components in a sample, leading to a better ***spike recovery*** than will be obtained for native analyte in a sample.

Matrix spike duplicates/replicates – are additional replicates of matrix spike samples. Analysis of matrix spike duplicates/replicates allows for measurement of matrix spike precision, which, in combination with the average matrix spike recovery, provides better information on whether spike recovery is statistically differentiable from target analyte in unspiked samples (an accuracy problem) or not.

Reagent blank – a blank that is analyzed to confirm that the reagents used in sample preparation are not contaminated. Reagent blanks are prepared by adding all reagents in the same volumes or proportions as used in sample preparation but without any handling such as occurs through sample preparation, cleanup, and other handling procedures. These blanks are often prepared in water, or alternately solvent if analytes are extracted into solvents prior to instrumental analysis.

Recovery spike – sometimes referred to as post-digestion spikes for some forms of analysis, recovery spikes are spikes that are added to a sample after all sample preparation, cleanup, and other procedures that occur leading up to analysis are completed, i.e., a recovery spike is added before instrumental analysis and after all other sample treatment and handling processes are complete. Because of matrix effects and the contact time between matrix and spike for matrix spikes, matrix spike recoveries may differ from 100%. In the case of recovery spikes, because contact time is brief (the recovery spike sample being spiked right before instrumental analysis), the recovery spike results should be close to 100%.

Recovery spike duplicates/replicates – are additional replicates of recovery spike samples, analyzed for the purpose of quantifying precision in spike recovery.

Reference material – a material containing known quantities of target analytes in a homogenous matrix of solution or solid that is similar to sample matrices. Analysis of reference materials (preferably Certified Reference Materials or Standard Reference Materials, i.e., those obtained by organizations that have determined analyte content or the material very accurately) is used to calculate accuracy. The accuracy of the analysis of a reference material is used as a gauge as to whether the accuracy of analysis of samples containing unknown amounts of analyte is likely to be accurate. The amount of analyte in the reference material provided by the company or organization providing the material is taken as the true value, μ, and the mean of results from analysis of the reference material obtained by the user, $\bar{x}$, is used to estimate accuracy.

Spike – a spike is a relatively small volume of ***spiking solution***, having a very well-known concentration of target analyte (for instance, as occurs in standard solutions), that is added to a sample. Subsequently, a portion of unspiked sample is analyzed and compared to the results of analysis of a spiked sample in order to calculate a spike recovery. Usually, the process of spiking is performed by taking two portions of highly accurate volume of the sample, spiking one, leaving the other unspiked, and then analyzing both. The spiked sample should be higher than the unspiked sample by a predictable amount, as measured by spike recovery. Too high or too low spike recovery might reflect bias (e.g., matrix effects), sample loss/contamination, or a variety of other analytical issues, and is an indication of problems with accuracy. The volume of spiking solution used to make a spike must be no more than 5% of the volume of solution spiked (otherwise, the concentration of analyte is changed if the spike volume is larger), and no less than 1% of the volume of solution spiked – as the relative volume of spiking solution decreases, the accuracy by which the amount of analyte added in the spike becomes compromised. Two common types of spike distinguished here are matrix spike and recovery spike. Internal standards and surrogates are also usually added as spikes.

Spike recovery – is calculated according to

$$\%\text{Recovery} = \frac{C_{\text{spiked sample}} - C_{\text{unspiked sample}}}{C_{\text{spiking solution}} \dfrac{V_{\text{spike}}}{V_{\text{sample spiked}}}} \times 100\%,$$

where $C_{\text{spiked sample}}$ and $C_{\text{unspiked sample}}$, respectively, are results from analysis of the concentrations of the sample spiked and unspiked, and the denominator is the effective concentration of the target analyte in the spike added to the spiked sample. In the denominator, $C_{\text{spiking solution}}$ is the concentration of analyte in the spiking solution, V_{spike} is the volume of spike, and $V_{\text{sample spiked}}$ is the volume of the sample receiving the spike.

Spiking solution – a spiking solution is usually a target analyte standard, the concentration of which is 20%–50% higher than the concentration in the unspiked sample that will be spiked. This level of concentration, in combination with the low volume of the spike, ensures that the amount of analyte spiked is neither too much nor too little.

Surrogate – is an element or compound added to field samples prior to any substantive sample handling or preparation. As for internal standards, a surrogate must be chemically similar to the analyte in behavior and not found in the environment. Surrogates are often deuterated forms of the target analyte. A surrogate is a type of spike, which, in addition to being added to samples, is also added to procedural blanks. Upon final analysis, a laboratory blank spiked with a surrogate is analyzed side-by-side with a surrogate-spiked procedural blank, and a comparison of the two results enables calculation of surrogate recovery. This reveals whether sample handling, processing, and preparation steps have resulted in any loss of surrogate, and, since the surrogate is chemically similar to the analyte in its behavior, surrogate loss can be assumed to reflect analyte loss and compromised results. Once the recovery of surrogate from the procedural blank is known, this can be compared to surrogate recoveries from samples. In the absence of matrix effects, the surrogate recovery from procedural blanks should be the same as for samples; if this is not the case, then matrix effects can be assumed to be affecting analysis of analytes in samples.

4.3 SPIKING METHODOLOGY

When spiking a sample[2] with a target analyte below the LOD to achieve a specific final concentration, the process involves careful calculation and precise addition of the spike solution. First, determine the desired final concentration of the analyte in the sample to be spiked, C_f. Next, calculate the volume of the spike, V_{spike}. Generally, spike volume should be 1%–5% of the total sample volume, V_{Tot}, e.g., 3% is suggested. For example, if the sample volume is 10 mL, the value of V_{spike} would be 0.3 mL or 300 μL. Then, calculate what concentration of target analyte standard solution, C_{std}, is needed for the ***spiking solution***, according to

$$C_{\text{std}} = \left(C_f \times V_{\text{Tot}}\right)/V_{\text{spike}}.$$

Dilute a certified standard stock solution appropriately (never exceeding a dilution factor of 10) serially until a standard have concentration C_{std} is obtained, then add V_{spike} to the sample and mix thoroughly to ensure homogeneity. The final concentration of the target analyte in a sample having no target analyte (below LOD) spiked in this way will be C_f.

For recovery testing, where the sample already contains the target analyte, the goal is to spike the sample to increase the analyte concentration by 20%–50%, for the purposes of this example, 35%. First, quantify the initial concentration of the analyte in the unspiked sample, C_{unspiked}, using a validated analytical method. Based on C_{unspiked}, calculate C_f as $C_f = 1.35 \times C_{\text{unspiked}}$. Since V_{Tot} is known and $V_{\text{spike}} = 0.03 \times V_{\text{Tot}}$, C_{std} is calculated as above, a suitable standard is prepared, and the sample is spiked. The final concentration of the target analyte in the sample will be C_f, in this case C_f being the amount of target analyte originally in the sample, plus the 35% (in this example) added from the spike.

[2] In regard to spiking, the term sample usually refers to some portion of sample or subsample that is taken aside for spiking, leaving the bulk of the original sample (which may itself be a preconcentrate) unspiked.

4.4 PLANNING FOR QA/QC

Project planning preferably occurs via the use of a SAP. Guidance and templates are available to standardize the SAP process (US EPA, 2014) of documenting the procedural and analytical requirements for projects involving the collection of environmental media. Bioavailability is readily incorporated into such planning via the use of methodologies as described herein. A SAP should also combine elements of a field sampling plan and QA/QC (laboratory and field). Table 4.1 gives an

TABLE 4.1
Example of a Field Data Quality Checklist

Checks	Yes	No	NA
Prior to Field Event:			
Do field staff understand what procedures are required and have enough experience implementing the procedures?	☐	☐	☐
Do field staff know the correct names and descriptions for field sites, locations, sample names, and descriptions to be recorded on paperwork and field logs?	☐	☐	☐
Do field staff have a schedule of the types, frequency, and locations of samples to be collected, including QA/QC samples?	☐	☐	☐
Do field staff have correct forms/paperwork to collect all the information needed?	☐	☐	☐
Is any relevant historical data prepared (printed or downloaded) for field reference?	☐	☐	☐
Is there a schedule of all field equipment to be used, and is that field equipment properly decontaminated and prepared for field transport in a decontaminated state?	☐	☐	☐
Has all field equipment been calibrated and are there complete records of calibration?	☐	☐	☐
Is there a schedule of all consumables and reagents needed for all aspects of field operations?	☐	☐	☐
During the field event:			
Has all field equipment used in sampling been properly and thoroughly decontaminated prior to each sample collection?	☐	☐	☐
For field documentation			
• Has it been double-checked to be complete?	☐	☐	☐
• Has it been double-checked to be accurate?	☐	☐	☐
• Is the information consistent and legible across all documents?	☐	☐	☐
• Is the spelling consistently correct?	☐	☐	☐
• Are dates and times in same and correct format?	☐	☐	☐
For all data collected			
• Has it been double-checked to be in acceptable/expected range?	☐	☐	☐
• Is it consistent with historical data?	☐	☐	☐
• If data are not consistent with historical, has this been noted/explained?	☐	☐	☐
If field calculations were performed, have all calculations been double-checked to be correct?	☐	☐	☐
Are all scheduled activities double-checked and verified as being completed?	☐	☐	☐
If documents require signatures, have all been signed?	☐	☐	☐
After the field event:			
Has field documentation been reviewed, transcribed if necessary or converted into digital form, recorded and backed up digitally, and stored in a locatable and retrievable manner?	☐	☐	☐
Have QA/QC issues been identified by the review, and, if so, what Corrective Action, if any, are taken?	☐	☐	☐
Are QA/QC Corrective Action needed for future field events, and, if so, have these been recorded?	☐	☐	☐

Such a checklist would be accompanied by a schedule of environmental samples and QA/QC samples to be collected, labels, and any Chain-of-Custody forms.

example of a checklist of the type that is usually developed to support field activities, with schedules of all environmental samples and associated QA/QC samples to be collected being developed as part of the SAP. The specific QA/QC samples to be collected and associated metrics that must be planned for in advance are given in Section 4.6.

4.5 IMPLEMENTATION

QA/QC Implementation ideally occurs according to project-specific guidelines formulated during the QA/QC Planning stage, laid out in the SAP.

4.6 ASSESSMENT

QA/QC Assessment evaluates whether the data are of the type, quantity, and quality needed to meet project requirements. The terms Data Verification and/or Data Validation are commonly used and represent the first step in the QA/QC Assessment process. The manner in which these terms are used internationally and even regionally by different organizations and agencies can vary considerably, and the terms are used here as defined in Terms and Definitions section. Data Verification involves evaluating all of the data against PARCC Acceptance Criteria that are specified in SOPs and/or as defined for the project during the QA/QC Planning phase (in other words, against all criteria that are based on field and laboratory data QA/QC, as well as other criteria specified in advance). Validation refers to the evaluation of data subsequent to Data Verification and may include, for instance, statistical or other analysis. Data Verification and/or Data Validation may be performed at various phases throughout a project. For instance, if a project involves monitoring quarterly over the period of a year, Data Verification and Validation should be performed after each quarterly monitoring exercise. This should be incorporated into and specified in the SAP. Data Verification/Validation needs to be documented in order to satisfy the precepts of the requirements of QA/QC.

In effectively all analytical schemes, samples are run in batches, and each sample batch includes QA/QC samples. The purpose of running samples in ***analytical batches*** is to allow for verification of instrument performance at intervals (e.g., blank checks, ***calibration checks***, ensuring the instrument is "clean" or without crossover between samples), i.e., in between batches. Generally, consecutive analysis of large numbers of samples is known to introduce increasingly unacceptable variation in results (e.g., Zelena et al., 2009).

When validating data, individual data and or whole data sets may be compromised if Acceptance Criteria are failed. For instance, a single sample might fail Acceptance Criteria for a surrogate spike, while other samples in the same analytic or field batch might not. Other criteria, such as % completeness, apply to the data set as a whole. If the data set as a whole does not pass the minimum acceptance criterion for % **completeness**, the entire data set is compromised.

Any failure of Acceptance Criteria represents a QC Nonconformance, which calls for Corrective Action, when possible. Corrective Action is not always possible, particularly for events that occur in the field. For this reason, it is particularly important that field operations are well planned and executed. The nature of Corrective Action will be determined by the circumstances. For instance, if ***procedural blanks*** suggest that samples have been heavily contaminated in the field, there may be no possibility to recover the data, and the only recourse may be to discard data associated with contaminated sample batches. In contrast, if a few processed samples fail surrogate recovery, there may be a possibility to reprocess stored field samples and obtain data that meet Acceptance Criteria on reanalysis.

After Data Verification, the results should be documented with a clear delineation of which data passed QA/QC Acceptance Criteria (listing all samples and analytes and how each sample/analyte pair performed with respect to QC criteria). Any exceedances of QC Acceptance Criteria should be documented as part of the Data Verification/Validation process, as well as the extent of the exceedance and whether Corrective Action was taken. If Corrective Action was taken, the results should be documented as to whether actions enabled recovery of any data and how these affected QC performance with respect to Acceptance Criteria. Finally, it is necessary to document a conclusion as to which data passed Data Verification as suitable for use in Data Validation. Any observations that might impact the integrity of data should be documented prior to Data Validation. Table 4.2 gives examples of different QA/QC elements that are typically included in QA/QC Planning, representative Acceptance Criteria, notes on what constitutes Nonconformance and consequences thereof, possible causes of Nonconformance, and the PARCCS categories of DQI criteria affected by Nonconformance.

TABLE 4.2
Examples Illustrating QA/QC Data Usability Guidelines

QA/QC Element	Acceptance Criteria	Nonconformance	Possible Causes	PARCCS Affected
Methods	Appropriate methods used for collection, preparation, and analysis of analytes	All data compromised or unreliable from use of inappropriate methods	Lack of familiarity with best practice	All
Sample labeling	Present, readable, attached to sample container, accurate	Inability to unambiguously identify sample	Failure to label accurately, to protect label, to use appropriate marker or labels	Completeness Representativeness
Sample containers	Use of proper containers made of proper material, appropriately decontaminated, for each target analyte	High chance of contamination or loss of analyte	Failure to read and follow guidelines/SOPs	Accuracy Completeness Representativeness
Sample preservation	Select the appropriate approach based on the pollutant and sample type[a]	No preservative added, wrong amount, wrong preservative, leads to loss of analyte or contamination	Varies	Accuracy Comparability Completeness Representativeness
Handling during transport	Correct temperature and sample packaging used; transport to storage facility within method-specified times (usually less than 24 hours)	Degradation of analyte; loss of analyte; cross contamination; change in bioavailability	Too high or low temperature during transport; samples not packaged properly	Accuracy Completeness Representativeness
Equipment blank	Sufficient frequency of collection (usually one after each equipment decontamination); no evidence of contamination	Lack of data to rule out equipment contamination; sample contamination	Contaminated or insufficiently decontaminated field equipment	Accuracy Representativeness
Procedural blank	Sufficient frequency of collection (method specific, e.g., 1 per 20 samples); no evidence of contamination	Lack of data to rule out procedural contamination; sample contamination	Contaminated containers, reagents, preservatives, ambient contamination, poor field handling technique	Accuracy Completeness Representativeness
Trip blank	Sufficient frequency of collection (one trip or site); no evidence of contamination	Lack of data to rule out volatile cross contamination; sample contamination	Cross contamination of volatiles during shipment or storage	Accuracy Completeness Representativeness
Field co-located samples	Sufficient frequency of collection (method specific, e.g., 1 per 20 samples); acceptable precision (no greater than 50% RPD for soil/sediment samples; no greater than 30% RPD for other sample types)	Lack of data to evaluate precision; unacceptable precision	Sample heterogeneity (soils/sediments); inadequacy of field collection technique	Precision Representativeness

(*Continued*)

TABLE 4.2 (*Continued*)
Examples Illustrating QA/QC Data Usability Guidelines

QA/QC Element	Acceptance Criteria	Nonconformance	Possible Causes	PARCCS Affected
Field split samples	Sufficient frequency of collection (e.g., 1 per site); acceptable precision (no greater than 35% RPD for soil/sediment samples; no greater than 20% RPD for other sample types)	Lack of data to evaluate precision; unacceptable precision	Inadequate homogenization of sample (soils/ sediments)	Precision Representativeness
Field duplicates/ replicates	Highly variable; for bioavailability studies, common for each sample to be collected in duplicate/ replicate. RPD or RSD criteria are highly variable due to site and method variability	RPD or RSD for field duplicate/replicate is higher than set acceptance criterion	Improper field techniques, contamination	Mainly precision
Hold times[a]	Analysis completed within holding time for target analytes	Holding time expires before analysis complete – loss or degradation of target analyte; changes in bioavailability	Inadequate experience of laboratory facility	Accuracy Completeness Representativeness
Matrix spikes	Sufficient frequency of analysis (e.g., 1 per 20 samples); acceptable recovery (variable, 75%–125% is common)	Lack of data to evaluate matrix effects; high or low recovery, matrix compromise of accuracy; matrix effects are often operant in bioavailability studies	Matrix effects, inadequate sample preparation/cleanup, improper spiking	Accuracy
Matrix spike duplicates	Sufficient frequency of analysis (e.g., 1 per 20 samples); acceptable precision (variable, 20% RPD if analyte concentration is 5 × LOD is common)	Lack of data to evaluate matrix effects precisely; high or low recovery, matrix compromise of accuracy	Matrix effects, inadequate sample preparation/cleanup, improper spiking	Accuracy
Calibration curve (including blank)	Often linear, with R^2 of >0.99 evaluated in terms of p-value or Analysis of Variance (ANOVA); highly variable since some methods are nonlinear. Subject to method guidance	Deviation of goodness of calibration from method criteria, i.e., outside method criteria for calibration	Contamination, improper preparation of standards	All
Calibration checks	Often once per batch; highly variable and subject to method criteria for how well the check should agree with original calibration	Lack of evidence of stable calibration; check is outside the method criteria for agreement with original calibration	Instrumental drift; problems with instrument	All

(*Continued*)

TABLE 4.2 (*Continued*)
Examples Illustrating QA/QC Data Usability Guidelines

QA/QC Element	Acceptance Criteria	Nonconformance	Possible Causes	PARCCS Affected
Reagent blank	Once per lot of reagents used, all analytes in reagent blank should be below LOD	One or more analytes above LOD	Contamination	All
Laboratory duplicate/ replicate	e.g., once per batch; RPD or RSD below method criteria; 5%–10% is common, highly variable and subject to method criteria for RPD/RSD of laboratory duplicate	RPD or RSD for laboratory duplicate/ replicate is higher than method criteria	Improper analytical techniques, instrument drift, contamination	Mainly precision
Recovery spike	Sufficient frequency of analysis (e.g., 1 per 20 samples); acceptable recovery (variable, 85%–115% is common)	Lack of data to evaluate matrix effects; high or low recovery, matrix compromise of accuracy	Matrix effects, inadequate sample preparation/cleanup, improper spiking	Accuracy
Recovery spike duplicate	Sufficient frequency of analysis (e.g., 1 per 20 samples); acceptable precision (variable, 15% RPD if analyte concentration is 5 × LOD is common)	Lack of data to evaluate matrix effects precisely; high or low recovery, matrix compromise of accuracy	Matrix effects, inadequate sample preparation/cleanup, improper spiking	Accuracy
Reference material	Analyze one reference sample per continuous analytical run or per day, subject to instrumental method guidelines	Result for analysis of reference material is inaccurate/outside standards of accuracy for material and method	Insufficient overall skill of analytical facility	Accuracy
Limit of detection (and LOQ)	Determined from calibration curve; valid as long as calibration is valid (subject to calibration checks); Acceptance Criteria for LOD is method-specific and given in SOPs/SMs	LOD is too high, above method criteria	Contamination, insufficient measures to combat interferences (i.e., cleanup, background correction); high dilution factor; wrong or inappropriate techniques	Comparativity Completeness Sensitivity
Dilution factors	Unless specified by method, never greater than a factor of 10 for dilution of standards and samples	Greater than a factor of 10 dilution is a QA/QC Nonconformity	Lack of care by analyst	Accuracy Comparativity Completeness
Internal standards	Internal standards, when used, must be added to every single sample analyzed	Internal standard not added to every single sample analyzed or variability of some samples is greater than/ outside method criterion	Matrix effects, contamination, loss of analyte during preparation and handling	All

(*Continued*)

TABLE 4.2 (*Continued*)
Examples Illustrating QA/QC Data Usability Guidelines

QA/QC Element	Acceptance Criteria	Nonconformance	Possible Causes	PARCCS Affected
Surrogates	Surrogates, when used, must be spiked into every single sample analyzed and must be within a recovery range specified in the instrumental method specified variation (e.g., 70%–130%)	Surrogates not spiked into every single sample analyzed or variability of recovery is greater than/less than the method criterion	Matrix effects, inadequate sample preparation/cleanup, loss of sample in preparation and handling, contamination improper spiking	Accuracy Completeness

Notes:

- Acceptance Criteria associated with PARCCSs DQIs are included in any SAP; examples given for Acceptance Criteria are indicative. Actual Acceptance Criteria are dependent upon methods used, any method guidance given, and the needs of individual projects. Acceptance Criteria may be relaxed or tightened based on project needs. All Acceptance Criteria should be transparently reported.
- The % completeness goal for critical samples is generally 100%. Overall project completeness goals may be set below 100%, however, should typically not be set at lower than 90% in the worst-case scenarios (lower than 100% should be incorporated into project planning in order to provide for enough samples collected to compensate for losses due to unintended issues with sample collection (e.g., spillage, loss) or to account for quality issues that might render some samples unusable (e.g., QA/QC Nonconformances).
- All PARCCSs are at issue for Data Verification, according to content of this table; in practice, comparability (as applicable) and representativeness must be reviewed with particular scrutiny during Data Validation.
- What constitutes contamination in blanks is important to carefully predefine. Field blank contamination typically renders results in the field batch invalid. For studies that might involve screening, and in cases where field contamination may be difficult to prevent, some rules of thumb might be adopted in the SAP: if the concentration of an analyte in a sample is within 3× the concentration of analyte in equipment, trip, or procedural blanks, then it is unlikely that the analyte is authentically in the sample. If the concentration of an analyte in a sample is 3–10× the concentration of analyte in equipment, trip or procedural blanks, the presence of the analyte in the sample is considered real. If the concentration of an analyte is authentically in the sample is >10× the concentration of analyte in equipment, trip, or procedural blanks, the impact of the blank on the sample results is likely to be negligible. In all cases, blank contamination and Corrective Actions associated with same must be carefully documented.
- If the laboratory duplicates/replicates are within Acceptance Criteria, but field duplicates are not, a likely source of field Nonconformance is heterogeneity of the matrix or field sampling technique. If the laboratory duplicates/replicates are not acceptable, laboratory method performance is the likely source for the lack of reproducibility.

[a] Preservative is rarely used for samples that will be subject to bioavailability measurements because it may alter bioavailability. For this reason, hold times for samples to be tested should be as short as possible.

The complexity of Data Validation might vary considerably depending upon the goals of a project. Generally speaking, only verified data should be used, and Data Validation should involve, at minimum, decisions about the structure of the data (e.g., statistical distribution), the appropriate quantitative methods to use in analyzing the data to address information needs, evaluation of trends or biases in the data and associated QC results, and a decision as to whether the data is fit-for-purpose. As with Data Verification, Corrective Action may be needed in Data Validation as well. For instance, if analysis with Gaussian statistics was planned in the SAP, and later the data are found to be non-normally distributed, Corrective Action might involve identification of suitable non-parametric methods. Finally, after data have been reviewed, methods of statistical/numerical analysis identified (and tested for appropriateness), data analysis is conducted, and results from data analysis are used to draw conclusions from the data.

REFERENCES

APHA/AWWA/WEF (American Public Health Association, American Water Works Association, Water Environment Federation), 2023. In: Lipps, W.C., Braun-Howland, E.B., Baxter, T.E. (eds.) *Standard Methods for the Examination of Water and Wastewater*, 24th Edition, APHA Press, Washington, DC.

Ganapathy, S. (ed), 2008. *Quality Assurance in Environmental Monitoring: Instrumental Methods*. John Wiley & Sons, New York.

ISO (International Organization for Standardization), 2015. Environmental management systems – Requirements with guidance for use. ISO/DIS Standard No. 14001. ISO, Geneva.

Quevauviller, P. (ed), 1995. *Quality Assurance in Environmental Monitoring: Sampling and Sample Pretreatment*. John Wiley & Sons, New York.

US EPA (United States Environmental Protection Agency), 2002. Standard operating procedures for chain-of-custody of samples revision. US EPA, Washington, DC.

US EPA, 2005. QA field activities procedure. Directive No: CIO 2105-P-02.1, reviewed 2020. US EPA, Washington, DC.

US EPA, 2006. Guidance on systematic planning using the data quality objectives process. EPA Document EPA/240/B-06/001. US EPA, Washington, DC.

US EPA, 2014. Sampling and analysis plan – Guidance and template v.4 – General Projects 04/2014. EPA Document R9QA/009.1. US EPA, Washington, DC.

Zelena, E., Dunn, W.B., Broadhurst, D., Francis-McIntyre, S., Carroll, K.M., Begley, P., O'Hagan, S., Knowles, J.D., Halsall, A., Wilson, I.D., Kell, D.B., 2009. Development of a robust and repeatable UPLC-MS method for the long-term metabolomic study of human serum. *Anal. Chem.* 81, 1357–1364. https://doi.org/10.1021/ac8019366.

5 Standard Methods for the Instrumental Analysis of Heavy Metals and Organic Compounds and Ancillary Procedures

5.1 NOTES ON METHODS AND TYPES OF POLLUTANTS

Many of the chapters in this book comprise a pretreatment or ***preconcentration*** of environmental samples in order to isolate the ***bioavailable*** or ***chemoavailable*** species, subsequent to which instrumental analysis is still required. The methods listed in this chapter cover the extraction, digestion, or detection of pollutants in various matrices including food (similar to some environmental samples), tissue, environmental aqueous samples, soils and sediments, and some related materials (e.g., sludges). For some of the instrumental analytical methods given, sample digestion and/or extraction is included in the method, which may not be needed if samples have already been subject to preconcentration. Users may refer to the information in this chapter to select the appropriate method for their application based on the type of sample ***matrix*** and the type of target pollutant(s) in the matrix. Standard Methods for the detection of dry-to-wet weight and lipids are also listed. The lists provided herein are by no means complete, and for some applications, users may need to seek a more specialized method and/or choose methods from the literature. Website URLs are given for the convenient download of methods, where applicable; however, URLs are often not stable. For this reason, each method is also listed in the references section.

5.1.1 List of Heavy Metals/Trace Element Pollutants

The heavy metal or trace element pollutants of greatest concern in environmental matrices include antimony (Sb), arsenic (As), beryllium (Be), cadmium (Cd), chromium (Cr), copper (Cu), lead (Pb), mercury (Hg), nickel (Ni), selenium (Se), silver (Ag), titanium (Ti), and zinc (Zn).

5.1.2 List of Types of Organic Pollutants

There is a vast and increasing range of organic pollutants of concern. The list here includes most, but not all. These pollutants fall into several different classes, some of which include Persistent Organic Pollutants (POPs, as defined by the Stockholm Convention, Feng et al., 2023) and/or organic Contaminants of Emerging Concern (NOAA, 2024).

1. Polychlorinated biphenyls (PCBs) – used in electrical equipment, paints, and other industrial applications.
2. Dioxins and furans – by-products of industrial processes and combustion.
3. Hexachlorobenzene (HCB) – used as a pesticide and in the production of synthetic rubber.
4. Chlorinated pesticides – such as aldrin, dichloro-diphenyl-trichloroethane (DDT), dieldrin, endrin, and chlordane.

DOI: 10.1201/9781003355229-6

5. Polybrominated diphenyl ethers (PBDEs) – used as flame retardants in various consumer products.
6. Per- and polyfluoroalkyl substances (PFAS) – used in non-stick cookware, water-repellent clothing, and firefighting foams.
7. Polycyclic aromatic hydrocarbons (PAHs) – include compounds like benzo[a]pyrene, naphthalene, and anthracene, found in fuels, coal tar, tobacco smoke, and grilled foods.
8. Pharmaceuticals and personal care products (PPCPs) – include antibiotics, anti-inflammatory medications, hormones, and cosmetics.
9. Endocrine disrupting chemicals – such as bisphenol A (BPA) and phthalates.
10. Organochlorine pesticides – such as lindane and heptachlor.
11. Polychlorinated naphthalenes (PCNs) – used in wood preservatives, paints, and electrical insulators.
12. Hexabromocyclododecane – used as a flame retardant in polystyrene foam.
13. Organophosphorus pesticides – such as malathion and parathion, used in agriculture.
14. Carbamate pesticides – such as carbaryl and aldicarb, used in agriculture.

5.2 DIGESTION AND EXTRACTION METHODS

5.2.1 Freshwater and Saltwater

TABLE 5.1
Standard Methods for the Digestion and Extraction of Pollutants in Freshwater and Saltwater

Method	Sample Type/Pollutant	Source	Website URL
ISO 15587-1:2002: Digestion of water samples using aqua regia (a mixture of nitric acid and hydrochloric acid) for the determination of heavy metals.	Drinking water, surface water, groundwater, soil, sediment, and solid waste/Ag, Al, As, B, Ba, Be, Ca, Cd, Co, Cr, Cu, Fe, Hg, K, Mg, Mn, Mo, Na, Ni, P, Pb, Sb, Se, Sn, Sr, Tl, V, and Zn	ISO (2002a)	https://www.iso.org/standard/31354.html
ISO 15587-2:2002: Digestion of water samples using nitric acid (HNO_3) for the determination of heavy metals.	Surface water, groundwater, drinking water, wastewater, and seawater/Ag, Al, As, B, Ba, Be, Ca, Cd, Co, Cr, Cu, Fe, Hg, K, Mg, Mn, Mo, Na, Ni, P, Pb, Sb, Se, Sn, Sr, Tl, V, and Zn	ISO (2002b)	https://cdn.standards.iteh.ai/samples/31355/8b940412659e4f468ddbfc539c76b3ee/ISO-15587-2-2002.pdf
US EPA Method 3510C: Extraction of organic pollutants from water samples using separatory funnel ***liquid–liquid extraction*** method.	Surface water, groundwater, drinking water, wastewater, and seawater/Organochlorine pesticides, phthalate esters, and phenols	US EPA (1996b)	https://www.epa.gov/sites/default/files/2015-12/documents/3510c.pdf
US EPA Method 3520C: Extraction of organic pollutants from water samples using continuous liquid–liquid extraction method.	Surface water, groundwater, drinking water, wastewater, and seawater/Organochlorine pesticides, phthalate esters, and phenols	US EPA (1996c)	https://www.epa.gov/sites/default/files/2015-06/documents/epa-3520c.pdf
JIS K 0102-4: Determination of organic pollutants in water using various extraction methods.	Surface water, groundwater, drinking water, wastewater, and seawater/Organochlorine pesticides, phthalate esters, phenols, and polychlorinated biphenyls	JIS (2024)	https://webdesk.jsa.or.jp/preview/pre_jis_k_00102_004_000_2024_j_ed10_ch.pdf

Abbreviations: ISO, International Organization for Standardization; JIS, Japanese Industrial Standards; US EPA, United States Environmental Protection Agency.

5.2.2 Soils, Sediments, and Other Solid Environmental Samples

TABLE 5.2
Standard Methods for the Extraction and Detection of Pollutants in Soils and Sediments

Method	Sample Type/Pollutant	Source	Website URL
BSI 7755-3.9:1995: Digestion of water samples using nitric acid (HNO_3) and hydrochloric acid (HCl) for the determination of heavy metals.	Soil/Trace elements	BSI (1995)	https://knowledge.bsigroup.com/products/soil-quality-chemical-methods-extraction-of-trace-elements-soluble-in-aqua-regia
ASTM D3974-09: Extraction of trace elements from soil using nitric acid (HNO_3) and hydrochloric acid (HCl).	Soils, bottom sediments, suspended sediments, and waterborne materials/Al, B, Ba, Cd, Ca, Cr, Co, Cu, Fe, Pb, Mg, Mn, Mo, Ni, K, Na, Sr, V, and Zn	ASTM (2023)	https://www.astm.org/d3974-09r23.html
US EPA Method 3050B: Acid digestion of sediments, sludges, and soils using nitric acid (HNO_3) and hydrogen peroxide (H_2O_2) for the determination of heavy metals.	Sediments, sludges, and soils/Al, Sb, Ba, Be, Cd, Ca, Cr, Co, Cu, Fe, Pb, V, Mg, Mn, Mo, Ni, K, Ag, Na, TI, V, Zn, As, and Se	US EPA (1996a)	https://www.epa.gov/sites/default/files/2015-12/documents/3050b.pdf
US EPA Method 3051A: Microwave-assisted acid digestion of sediments, sludges, soils, and oils using nitric acid (HNO_3) and hydrochloric acid (HCl) for the determination of heavy metals.	Sediments, sludges, and soils/Al, Sb, As, Ba, Be, B, Cd, Ca, Cr, Co, Cu, Fe, Pb, Mg, Mn, Hg, Mo, Ni, K, Se, Ag, Na, Sr, TI, V, and Zn	US EPA (2007a)	https://www.epa.gov/sites/default/files/2015-12/documents/3051a.pdf
NF X31-147: Digestion of soil and sediment samples using nitric acid (HNO_3) and hydrochloric acid (HCl) for the determination of heavy metals.	Soils and sediments/Pb, Cd, Cr, Cu, and Zn	AFNOR (1996)	https://www.boutique.afnor.org/en-gb/standard/nf-x31147/soil-quality-soils-sediments-total-solubilizing-by-acid-attack/fa038958/14416
US EPA Method 3545A: Extraction of water-insoluble or slightly water-soluble organic pollutants from soils, clays, sediments, sludges, and waste solids by pressurized fluid extraction (PFE) method.	Soils, clays, sediments, sludges, and waste solids/Semivolatile organic compounds, Organophosphorus pesticides, Organochlorine pesticides, chlorinated herbicides, PCBs, polychlorinated dibenzo-*p*-dioxins and polychlorinated dibenzofurans, and diesel range organics	US EPA (2007b)	https://www.epa.gov/sites/default/files/2015-12/documents/3545a.pdf
US EPA Method 3550C: Extraction of nonvolatile and semivolatile organic compounds from soils, sludges, and wastes by ultrasonic extraction method.	Soils, sludges, and wastes/Nonvolatile and semivolatile organic compounds	US EPA (2007c)	https://www.epa.gov/sites/default/files/2015-12/documents/3550c.pdf

Abbreviations: ASTM, American Society for Testing and Materials; AFNOR, Association Française de Normalisation; BSI, British Standards Institution; ISO, International Organization for Standardization; US EPA, United States Environmental Protection Agency.

5.3 METHODS OF INSTRUMENTAL ANALYSIS

5.3.1 Analysis of Heavy Metal/Trace Element Pollutants

TABLE 5.3
Standard Methods for the Instrumental Analysis of Heavy Metal/Trace Element Pollutants in Environmental Matrices

Method	Sample Type/Pollutant	Source	Website URL
ISO 8288:1986: Water quality – Determination of cobalt, nickel, copper, zinc, cadmium and lead – Flame atomic absorption spectrometric methods.	Freshwater and saltwater/Co, Ni, Cu, Zn, Cd, and Pb	ISO (1986)	https://www.iso.org/standard/15408.html
ISO 12846:2012: Water quality – Determination of mercury – Method using atomic absorption spectrometry (AAS) with and without enrichment.	Freshwater and saltwater/Hg	ISO (2012)	https://www.iso.org/standard/51964.html
ISO 17294-2:2023: Water quality – Application of Inductively coupled plasma – mass spectrometry (ICP–MS) – Part 2: Determination of selected elements including uranium isotopes.	Freshwater and saltwater/Al, Sb, As, Ba, Be, Bi, B, Cd, Cs, Ca, Ce, Cr, Co, Cu, Dy, Er, Gd, Ga, Ge, Au, Hf, Ho, In, Ir, Fe, La, Pb, Li, Lu, Mg, Mn, Hg, Mo, Nd, Ni, Pd, P, Pt, K, Pr, Rb, Re, Rh, Ru, Sm, Sc, Se, Ag, Na, Sr, Tb, Te, Th, Tl, Tm, Sn, Ti, W, U, V, Y, Yb, Zn, and Zr	ISO (2023a)	https://www.iso.org/standard/82245.html
ISO 17378-2:2014: Water quality – Determination of arsenic and antimony – Part 2: Method using hydride generation atomic absorption spectrometry (HG–AAS).	Drinking water, surface water, groundwater, and rainwater/As and Sb	ISO (2014a)	https://www.en-standard.eu/bs-iso-17378-2-2014-water-quality-determination-of-arsenic-and-antimony-method-using-hydride-generation-atomic-absorption-spectrometry-hg-aas/
ISO 5961:1994: Water quality – Determination of cadmium by atomic absorption spectrometry.	Water and wastewater/Cd	ISO (1994a)	https://www.iso.org/standard/12157.html
ISO/TS 19620:2018: Water quality – Determination of arsenic (III) and arsenic (V) species – Method using high-performance liquid chromatography (HPLC) with detection by Inductively coupled plasma – mass spectrometry (ICP–MS) or hydride generation atomic fluorescence spectrometry (HG–AFS).	Freshwater, saltwater, and wastewater/As(III) and As(V)	ISO (2018)	https://www.iso.org/standard/65512.html
ISO 11083:1994: Water quality – Determination of chromium (VI) – Spectrometric method using 1,5-diphenylcarbazide.[a]	Raw water, drinking water, surface water, aqueous eluates, cooling water, and treated wastewater/Cr (VI)	ISO (1994b)	https://www.iso.org/standard/19070.html

(Continued)

TABLE 5.3 (*Continued*)
Standard Methods for the Instrumental Analysis of Heavy Metal/Trace Element Pollutants in Environmental Matrices

Method	Sample Type/Pollutant	Source	Website URL
ISO 20280:2007: Soil quality – Determination of arsenic, antimony and selenium in aqua regia soil extracts with electrothermal or hydride-generation atomic absorption spectrometry.	Soil/As, Sb, and Se	ISO (2007)	https://www.iso.org/standard/44538.html
ISO 22036:2024: Environmental solid matrices – Determination of elements using inductively coupled plasma optical emission spectrometry (ICP–OES).	Solid matrices/Al, Sb, As, Ba, Be, Bi, B, Cd, Ca, Ce, Cr, Co, Cu, Dy, Er, Eu, Ga, Gd, Ge, Au, Hf, Ho, In, Ir, Fe, La, Pb, Li, Lu, Mg, Mn, Hg, Mo, Nd, Ni, Pd, P, Pt, K, Pr, Rh, Ru, Sm, Sc, Se, Si, Ag, Na, Sr, S, Ta, Te, Tb, Tl, Tm, Th, Sn, Ti, W, V, Y, Yb, Zn, and Zr	ISO (2024a)	https://www.iso.org/standard/82632.html
ASTM D3557-17: Standard Test Methods for Cadmium in Water.	Freshwater, saltwater, and wastewater/Cd	ASTM (2017)	https://www.astm.org/d3557-17.html
ASTM D5673-16: Standard Test Method for Elements in Water by Inductively coupled plasma – Mass Spectrometry.	Fresh and saltwater/Al, Sb, As, Ba, Be, Bi, Cd, Ca, Ce, Cr, Co, Cu, Dy, Er, Ga, Ge, Au, Hf, Ho, In, Ir, Fe, La, Pb, Li, Lu, Mg, Mn, Hg, Mo, Nd, Ni, Pd, Pt, K, Pr, Rb, Re, Rh, Ru, Sm, Sc, Ag, Na, Sr, Tb, Th, Tl, Tm, Sn, Ti, W, U, V, Y, Yb, Zn, Zr, B, P, Se, Si, S, and Te	ASTM (2025)	https://www.astm.org/d5673-16.html
US EPA Method 200.8: Determination of trace elements in waters and wastes by Inductively coupled plasma–mass spectrometry (ICP–MS).	Drinking water, surface water, groundwater, soil, sediment, and solid waste/Ag, Al, As, Ba, Be, Cd, Co, Cr, Cu, Hg, Mn, Mo, Ni, Pb, Sb, Se, Th, Tl, U, V, and Zn	US EPA (1994b)	https://www.epa.gov/esam/epa-method-2008-determination-trace-elements-waters-and-wastes-inductively-coupled-plasma-mass
US EPA Method 200.9: Determination of trace elements by stabilized temperature graphite furnace atomic absorption (STGFAA).	Drinking water, surface water, groundwater, soil, sediment, and solid waste/Ag, Al, As, Be, Cd, Co, Cr, Cu, Mn, Ni, Pb, Sb, Se, Tl, Fe, and Sn	US EPA (1994c)	https://www.epa.gov/sites/default/files/2015-08/documents/method_200-9_rev_2-2_1994.pdf
US EPA Method 200.7: Determination of metals and trace elements in water and wastes by inductively coupled plasma–atomic emission spectrometry (ICP–AES).	Water, wastewater, and solid waste/Al, Sb, As, Ba, Be, B, Cd, Ca, Ce, Cr, Co, Cu, Fe, Pb, Li, Mg, Mn, Hg, Mo, Ni, P, K, Se, and Ag	US EPA (1994d)	https://www.epa.gov/sites/default/files/2015-06/documents/epa-200.7.pdf
US EPA Method 245.1: Determination of mercury in water by cold vapor atomic absorption spectrometry.	Drinking water, surface water, groundwater, sea water, brackish water, industrial and domestic wastewater/Hg	US EPA (1994e)	https://www.epa.gov/esam/epa-method-2451-determination-mercury-water-cold-vapor-atomic-absorption-spectrometry
US EPA Method 218.6: Determination of dissolved hexavalent chromium in drinking water, groundwater, and industrial wastewater effluents by ion chromatography.	Drinking water, groundwater, and industrial wastewater effluents/Cr (VI)	US EPA (1994f)	https://www.epa.gov/sites/default/files/2015-08/documents/method_218-6_rev_3-3_1994.pdf

(*Continued*)

TABLE 5.3 (*Continued*)
Standard Methods for the Instrumental Analysis of Heavy Metal/Trace Element Pollutants in Environmental Matrices

Method	Sample Type/Pollutant	Source	Website URL
Chinese Standard GB 17378.5-2007: This standard provides methods for analyzing a variety of contaminants, including organic contaminants and heavy metals in sediments.	Seabed sediments, coastal sediments/ As, Cd, Cr, Cu, Pb, and Se	SAC (2007)	https://www.chinesestandard.net/PDF/English.aspx/GB17378.5-2007
Chinese Standard GB 5750.1-2023: This standard provides methods for analyzing a variety of contaminants, including organic contaminants and heavy metals in drinking water.	Domestic drinking water, source water, treated, stored, and distributed drinking water/Al, Fe, Mn, Cu, Zn, As, Se, Hg, Cd, Cr(VI), Pb, Ag, Mo, Co, Ni, Ba, Ti, V, Sb, Be, Tl, Na, Sn, and Pb	MOH (2023)	https://www.codeofchina.com/standard/GBT5750.1-2023.html
GB/T 20380.1-2006: Starch and derived products – Heavy metals content – Part 1: Determination of arsenic content by atomic absorption spectrometry.	Food/As	SAC (2006a)	https://openstd.samr.gov.cn/bzgk/gb/newGbInfo?hcno=315DA7FF9CA42DF7DAC9FC7649B25616
GB/T 20380.2-2006: Starch and derived products – Heavy metals content – Part 2: Determination of mercury content by atomic absorption spectrometry.	Food/Hg	SAC (2006b)	https://openstd.samr.gov.cn/bzgk/gb/newGbInfo?hcno=5DC0F5373995C390E1D246A3921BB542

[a] At the time of publication, a new method, "ISO/DIS 18724: Water quality – Determination of dissolved chromium(VI) in water – Photometric method" is under development to replace ISO 11083:1994.

Abbreviations: ASTM, American Society for Testing and Materials; ISO, International Organization for Standardization; MOH, Ministry of Health, China; SAC, Standardization Administration of China; US EPA, United States Environmental Protection Agency.

5.3.2 Analysis of Organic Pollutants

TABLE 5.4
Standard Methods for the Instrumental Analysis of Organic Pollutants in Environmental Matrices

Method	Sample Type/Pollutant	Source	Website URL
ISO 6468-1996: Water quality – Determination of certain Organochlorine insecticides, polychlorinated biphenyls and chlorobenzenes – Gas chromatographic method after liquid–liquid extraction.	Drinking water, groundwater, surface water, and wastewater/ Chlorobenzenes (except the mono- and dichlorobenzenes), PCBs	ISO (1996)	https://www.iso.org/standard/12826.html
ISO 9377-2:2000: Water quality – Determination of hydrocarbon oil index – Part 2: Method using solvent extraction and gas chromatography.	Freshwater and saltwater/ Petroleum hydrocarbons and mineral oils	ISO (2000)	https://www.iso.org/standard/27604.html

(Continued)

TABLE 5.4 (*Continued*)
Standard Methods for the Instrumental Analysis of Organic Pollutants in Environmental Matrices

Method	Sample Type/Pollutant	Source	Website URL
ISO 18475:2023: Determination of polychlorinated biphenyls (PCB) by gas chromatography – mass selective detection (GC–MS) or electron-capture detection (GC–ECD).	Soil, sludge, sediment, treated biowaste, and waste/Seven selected polychlorinated biphenyls	ISO (2023b)	https://www.iso.org/obp/ui/en/#iso:std:iso:18475:ed-1:v1:en
ISO 21135:2024: Chemicals for the leather tanning industry – Determination of the total content of certain bisphenols.	Leather tanning industry/Bisphenol A, bisphenol AF, and bisphenol F	ISO (2024b)	https://cdn.standards.iteh.ai/samples/86660/22b86233a1ed458f9bb7dfeaad23ff3e/ISO-21135-2024.pdf
ISO 25101:2009: Determination of perfluorooctanesulfonate (PFOS) and perfluorooctanoate (PFOA) in water by solid-phase extraction and liquid chromatography/mass spectrometry.	Drinking water, groundwater, and surface water (freshwater and seawater)/PFOS and PFOA	ISO (2009)	https://www.iso.org/standard/42742.html
ISO/TS 16780:2015: Determination of polychlorinated naphthalenes (PCN) – Method using gas chromatography (GC) and mass spectrometry (MS).	Water and wastewater/PCNs	ISO (2015)	https://ndls.org.cn/standard/detail/94d4c0a079ec8e7edafd2a1f9f17c095
ISO 13859:2014: Soil quality – Determination of polycyclic aromatic hydrocarbons (PAH) by gas chromatography (GC) and high-performance liquid chromatography (HPLC).	Sludge, soil, and treated biowaste/16 PAHs	ISO (2014b)	https://www.iso.org/standard/54337.html
ISO 15680:2003: Water quality – Gas-chromatographic determination of a number of monocyclic aromatic hydrocarbons, naphthalene and several chlorinated compounds using purge-and-trap and thermal desorption.	Freshwater, saltwater, and wastewater/Monocyclic aromatic hydrocarbons, naphthalene, and chlorobenzene	ISO (2003)	https://www.iso.org/standard/27983.html
ISO 18856-1:2004: Water quality – Determination of selected phthalates using gas chromatography/mass spectrometry.	Groundwater, surface water, wastewater, and drinking water/Dimethyl phthalate, dipropyl phthalate, and di-*n*-octyl phthalate	ISO (2004a)	https://www.iso.org/standard/33680.html
ISO 16703:2004: Soil quality – Determination of content of hydrocarbon in the range C10–C40 by gas chromatography.[a]	Oil/*N*-alkanes from $C_{10}H_{22}$ to $C_{40}H_{82}$, of isoalkanes, cycloalkanes, alkylbenzenes, alkylnaphthalenes, and polycyclic aromatic compounds	ISO (2004b)	https://www.iso.org/standard/39937.html
ISO 7981–2:2005: Water quality – Determination of polycyclic aromatic hydrocarbons (PAH) – Part 2: Determination of six PAH by high-performance liquid chromatography with fluorescence detection after liquid–liquid extraction.	Drinking, mineral and table water, and ground and surface water/Fluoranthene, benzo[*b*]fluoranthene, benzo[*a*]pyrene, benzo[*k*]fluoranthene, indeno[1,2,3-*cd*]pyrene, and benzo[*ghi*]perylene	ISO (2005)	https://www.iso.org/standard/33884.html

(*Continued*)

TABLE 5.4 (*Continued*)
Standard Methods for the Instrumental Analysis of Organic Pollutants in Environmental Matrices

Method	Sample Type/Pollutant	Source	Website URL
ISO 17993:2002: Determination of PAHs in water by high-performance liquid chromatography (HPLC) with fluorescence detection.	Drinking and groundwater, wastewater/PAH	ISO (2002c)	https://www.iso.org/standard/31666.html
ISO 18635:2016: Water quality – Determination of short-chain polychlorinated alkanes (SCCPs) in sediment, sewage sludge and suspended (particulate) matter – Method using gas chromatography–mass spectrometry (GC–MS) and electron capture negative ionization (ECNI).	Sediment and suspended matter, sewage sludge, and soil/SCCPs	ISO (2016)	https://www.iso.org/obp/ui/en/#iso:std:iso:18635:ed-1:v1:en
ISO 22032:2006: Water quality – Determination of selected polybrominated diphenyl ethers in sediment and sewage sludge – Method using extraction and gas chromatography/mass spectrometry.	Sediment and sludge/PBDE	ISO (2006)	https://www.iso.org/obp/ui/en/#iso:std:iso:22032:ed-1:v1:en
ASTM D7065-17: Standard Test Method for Determination of Nonylphenol, Bisphenol A, *p*-tert-Octylphenol, Nonylphenol Monoethoxylate and Nonylphenol Diethoxylate in Environmental Waters by Gas Chromatography–Mass Spectrometry.	Fresh and saltwater/Nonylphenol ethoxylate, octylphenol, and BPA	ASTM (2024)	https://www.astm.org/d7065-17r24.html
ASTM D5317–20: Standard Test Method for Determination of Chlorinated Organic Acid Compounds in Water by Gas Chromatography with an Electron Capture Detector.	Freshwater/Bentazon, dicamba, and dichlorprop	ASTM (2020)	https://www.astm.org/d5317-20.html
US EPA 549.2-1997: Determination of diquat and paraquat in drinking water by liquid–solid extraction and high-performance liquid chromatography with ultraviolet detection.	Drinking water/Diquat and paraquat	US EPA (1997)	https://www.epa.gov/esam/epa-method-5492-determination-diquat-and-paraquat-drinking-water-liquid-solid-extraction-and
US EPA 8260D: Volatile organic compounds by gas chromatography/mass spectrometry (GC/MS).	Solid waste/Acrylonitrile, allyl alcohol, carbon disulfide, 2-chloroethanol, 1,2-dicholoroethane, ethylene oxide, 2-fluoroethanol, methyl acrylonitrile, propylene oxide, and 1,4-thioxane; nondrinking water samples for allyl alcohol, carbon disulfide, 2-chloroethanol,	US EPA (2006)	https://www.epa.gov/esam/epa-method-8260d-sw-846-volatile-organic-compounds-gas-chromatography-mass-spectrometry-gcms

(*Continued*)

TABLE 5.4 (*Continued*)
Standard Methods for the Instrumental Analysis of Organic Pollutants in Environmental Matrices

Method	Sample Type/Pollutant	Source	Website URL
	1,2-dicholoroethane, ethylene oxide, 2-fluoroethanol, propylene oxide, and 1,4-thioxane; drinking water samples for allyl alcohol, 2-chloroethanol, ethylene oxide, 2-fluoroethanol, and propylene oxide; and wipe samples for acrylonitrile, methyl acrylonitrile, and 1,4-thioxane		
US EPA Method 539: Determination of hormones in drinking water by SPE and LC/MS/MS.	Drinking water, finished drinking water and surface water/16α-Hydroxyestradiol (estriol),17β-estradiol, 17α-ethynylestradiol, testosterone, estrone, 4-androstene-3, 17-dione, equilin	US EPA (2010a)	https://f.hubspotusercontent00.net/hubfs/6549100/EPA%20Method%20539.pdf
US EPA Method 1694: Determination of PPCPs in water, soil, sediment, and biosolids by liquid chromatography/tandem mass spectrometry (LC/MS/MS).	Surface water, groundwater, drinking water, wastewater, and sediment/PPCPs	US EPA (2007d)	https://www.epa.gov/sites/default/files/2015-10/documents/method_1694_2007.pdf
US EPA Method 608.3: Organochlorine pesticides and PCBs by GC/HSD.	Industrial discharges/PCBs	US EPA (2014b)	https://www.epa.gov/sites/default/files/2017-08/documents/method_608-3_2016.pdf
US EPA Method 531.2: Measurement of *N*-methylcarbamoyloximes and *N*-methylcarbamates in water by HPLC with Post-Column Derivatization.	Water/*N*-Methylcarbamoyloximes and*N*-methylcarbamates	US EPA (2001)	https://www.epa.gov/sites/default/files/2015-06/documents/epa-531.2.pdf
US EPA Method 8318A (SW-846): *N*-Methylcarbamates by high-performance liquid chromatography (HPLC), Revision 1.	Soil, water, and waste matrices/Aldicarb, aldicarb sulfone, bendiocarb, carbaryl, carbofuran, *m*-cumenyl methylcarbamate, dioxacarb, formetanate hydrochloride, 3-hydroxycarbofuran, methiocarb, methomyl, metolcarb, mexacarbate, oxamyl, promecarb, propoxur, and thiodicarb	US EPA (2000)	https://www.epa.gov/esam/epa-method-8318a-sw-846-n-methylcarbamates-high-performance-liquid-chromatography-hplc
US EPA Method 537.1: Determination of selected PFAS in drinking water by solid-phase extraction and liquid chromatography/tandem mass spectrometry (LC/MS/MS).	Drinking water/PFAS	US EPA (2018)	https://f.hubspotusercontent00.net/hubfs/6549100/METHOD%20537_1_FINAL.pdf

(*Continued*)

TABLE 5.4 (*Continued*)
Standard Methods for the Instrumental Analysis of Organic Pollutants in Environmental Matrices

Method	Sample Type/Pollutant	Source	Website URL
US EPA Method 8310: Determination of PAHs in groundwater and wastes by high-performance liquid chromatography (HPLC) with ultraviolet (UV) and fluorescence detectors.	Groundwater and wastes/ Acenaphthene, acenaphthylene, anthracene, benzo(a)anthracene, benzo(a)pyrene, benzo(b) fluoranthene, benzo(ghi)perylene, and benzo(k)fluoranthene	US EPA (1986)	https://www.epa.gov/sites/default/files/2015-12/documents/8310.pdf
US EPA Method 1613B: Determination of dioxins and furans by high-resolution gas chromatography/ high-resolution mass spectrometry (HRGC/HRMS).	Water, soil, sediment, sludge, tissue, and other sample matrices/ Octa-chlorinated dioxins and furans	US EPA (1994a)	https://www.epa.gov/sites/default/files/2015-08/documents/method_1613b_1994.pdf
US EPA Method 8141B: Organophosphorus compounds by gas chromatography.	Surface water, groundwater, drinking water, and wastewater/ Foods of plant origin, such as fruits (including dried fruits), vegetables (including dried vegetables), cereals, and many processed products	US EPA (2007e)	https://www.epa.gov/sites/default/files/2015-12/documents/8141b.pdf
US EPA Method 8270D: Determination of semi-volatile organic compounds, including HCB in solid waste matrices, soils, air sampling media, and water samples by gas chromatography/mass spectrometry (GC/MS).	Solid waste matrices, soils, air sampling media, and water samples/PAHs, PCBs, Organochlorine pesticides, and nitroaromatic compounds	US EPA (2014a)	https://archive.epa.gov/epa/sites/production/files/2015-12/documents/8270d.pdf
US EPA Method 8081B: Determination of Organochlorine pesticides in extracts from solid and liquid matrices by gas chromatography with electron capture detection (GC/ECD).	Pesticides in extracts from solid and liquid matrices/ Organochlorine pesticides	US EPA (2007f)	https://www.epa.gov/sites/default/files/2015-12/documents/8081b.pdf
US EPA Method 1614A: Determination of PBDEs in water, soil, sediment, and tissue by high-resolution gas chromatography/high-resolution mass spectrometry (HRGC/HRMS).	Water, soil, sediment, biosolids, tissue, and other sample matrices/ PBDEs	US EPA (2010b)	https://www.epa.gov/sites/default/files/2015-08/documents/method_1614a_2010.pdf
US EPA Method 8082A: Polychlorinated biphenyls (PCBs) by gas chromatography.	Solid, tissue, and aqueous matrices/ PCBs	US EPA (2007g)	https://www.epa.gov/sites/default/files/2015-12/documents/8082a.pdf
GB/T 25225-2010: Animal and vegetable fats and oils – Detection and identification of a volatile organic contaminant by GC/MS.	Tissue samples/Esters, nitrogen-containing compounds, and acrylates	SAC (2010)	https://openstd.samr.gov.cn/bzgk/gb/newGbInfo?hcno=35DD6619BB1E6B9803C251FD483EF122

[a] At the time of publication, expected to be replaced by ISO/DIS 16703.

Abbreviations: ASTM, American Society for Testing and Materials; ISO, International Organization for Standardization; MOH, Ministry of Health, China; SAC, Standardization Administration of China; US EPA, United States Environmental Protection Agency.

5.4 DRY-TO-WET RATIOS AND TISSUE LIPIDS

TABLE 5.5
Standard Methods for Dry-to-Wet Ratios and Tissue Lipids

Method	Sample Type/Pollutant	Source	Website URL
ISO 14156:2001 – IDF 172:2001: Milk and milk products – Extraction methods for lipids and liposoluble compounds.	Food/Lipids and liposoluble compounds	ISO (2001)	https://www.iso.org/standard/23746.html
ASTM D7263-21: Standard Test Methods for Laboratory Determination of Density and Unit Weight of Soil Specimens.	Soil/Density and unit weight	ASTM (2021)	https://www.astm.org/d7263-21.html
GB/T 5535.1-2008: Animal and vegetable fats and oils – Determination of unsaponifiable matter – Part 1: Method using diethyl ether extraction.	Animal and vegetable fats and oils/Unsaponifiable substances	SAC (2008)	https://std.samr.gov.cn/gb/search/gbDetailed?id=71F772D75AA3D3A7E05397BE0A0AB82A
A rapid method for total lipid extraction and purification.	Frozen fish/Lipid	Bligh and Dyer (1959)	https://cdnsciencepub.com/doi/pdf/10.1139/o59-099
Comparison of extraction methods or the isolation of lipids and PCBs from mussel homogenate.	Mussel/Lipid	Latimer and Quinn (1990)	https://cfpub.epa.gov/si/si_public_record_report.cfm?Lab=NHEERL&dirEntryId=48391

Abbreviations: ASTM, American Society for Testing and Materials; ISO, International Organization for Standardization; SAC, Standardization Administration of China.

REFERENCES

American Society for Testing and Materials (ASTM), 2025. Standard Test Method for Elements in Water by Inductively Coupled Plasma-Mass Spectrometry. ASTM D5673-16. ASTM International, West Conshohocken, PA.

ASTM, 2017. Standard Test Methods for Cadmium in Water. ASTM D3557-17. ASTM International, West Conshohocken, PA.

ASTM, 2020. Standard Test Method for Determination of Chlorinated Organic Acid Compounds in Water by Gas Chromatography with an Electron Capture Detector. ASTM D5317-20. ASTM International, West Conshohocken, PA.

ASTM, 2021. Standard Test Methods for Laboratory Determination of Density and Unit Weight of Soil Specimens. ASTM D7263-21. ASTM International, West Conshohocken, PA.

ASTM, 2023. Standard Practices for Extraction of Trace Elements from Sediments. ASTM D3974-09. ASTM International, West Conshohocken, PA.

ASTM, 2024. Standard Test Method for Determination of Nonylphenol, Bisphenol A, p-tert-Octylphenol, Nonylphenol Monoethoxylate and Nonylphenol Diethoxylate in Environmental Waters by Gas Chromatography Mass Spectrometry. ASTM D7065-17. ASTM International, West Conshohocken, PA.

Association Française de Normalisation (AFNOR), 1996. Soil quality – Extraction of trace elements soluble in aqua regia. NF X31-147. AFNOR, Paris.

Bligh, E.G., Dyer, W.J., 1959. A rapid method for total lipid extraction and purification. *Can. J. Biochem. Physiol.* 37, 911–917.

BSI, 1995. Soil quality – Extraction of trace elements soluble in aqua regia. BS 7755-3.9:1995. BSI, Brussels.

Feng, W., Deng, Y., Yang, F., Miao, Q., Ngien, S.K., 2023. Systematic review of contaminants of emerging concern (CECs): Distribution, risks, and implications for water quality and health. *Water* 15(22), 3922.

International Organization for Standardization (ISO), 1986. Water quality – Determination of cobalt, nickel, copper, zinc, cadmium and lead – Flame atomic absorption spectrometric methods. ISO 8288:1986. ISO, Geneva.

ISO, 1994a. Water quality – Determination of cadmium by AAS. ISO 5961:1994. ISO, Geneva.
ISO, 1994b. Water quality – Determination of chromium (VI) – Spectrometric method using 1,5–diphenylcarbazide. ISO 11083:1994. ISO, Geneva.
ISO, 1996. Water quality – Determination of certain organochlorine insecticides, polychlorinated biphenyls and chlorobenzenes – Gas chromatographic method after liquid-liquid extraction. ISO 6468:1996. ISO, Geneva.
ISO, 2000. Water quality – Determination of hydrocarbon oil index. Part 2: Method using solvent extraction and gas chromatography. ISO 9377-2:2000. ISO, Geneva.
ISO, 2001. Milk and milk products – Extraction methods for lipids and liposoluble compounds. ISO 14156:2001. ISO, Geneva.
ISO, 2002a. Water quality – Digestion for the determination of selected elements in water – Part 1: Aqua regia digestion. ISO 15587-1:2002. ISO, Geneva.
ISO, 2002b. Water quality – Digestion for the determination of selected elements in water – Part 2: Nitric acid digestion. ISO 15587-2:2002. ISO, Geneva.
ISO, 2002c. Water quality – Determination of 15 polycyclic aromatic hydrocarbons (PAH) in water by HPLC with fluorescence detection after liquid-liquid extraction. ISO 17993:2002. ISO, Geneva.
ISO, 2003. Water quality – Gas-chromatographic determination of a number of monocyclic aromatic hydrocarbons, naphthalene and several chlorinated compounds using purge-and-trap and thermal desorption. ISO 15680:2003. ISO, Geneva.
ISO, 2004a. Water quality – Determination of selected phthalates using gas chromatography/mass spectrometry. ISO 18856:2004. ISO, Geneva.
ISO, 2004b. Soil quality – Determination of content of hydrocarbon oil index in soil by gas chromatography. ISO 16703:2004. ISO, Geneva.
ISO, 2005. Water quality – Determination of polycyclic aromatic hydrocarbons (PAH) – Part 2: Determination of six PAH by high-performance liquid chromatography with fluorescence detection after liquid-liquid extraction. ISO 7981–2:2005. ISO, Geneva.
ISO, 2006. Water quality – Determination of selected polybrominated diphenyl ethers in sediment and sewage sludge – Method using extraction and gas chromatography/mass spectrometry. ISO 22032:2006. ISO, Geneva.
ISO, 2007. Soil quality – Determination of arsenic, antimony and selenium in aqua regia soil extracts with electrothermal or hydride-generation atomic absorption spectrometry. ISO 20280:2007. ISO, Geneva.
ISO, 2009. Water quality – Determination of perfluorooctanesulfonate (PFOS) and perfluorooctanoate (PFOA) – Method for unfiltered samples using solid phase extraction and liquid chromatography/mass spectrometry. ISO 25101:2009. ISO, Geneva.
ISO, 2012. Water quality – Determination of mercury – Method using atomic absorption spectrometry (AAS) with and without enrichment. ISO 12846:2012. ISO, Geneva.
ISO, 2014a. Water quality – Determination of arsenic – AAS method (hydride technique). ISO 17378-2:2014. ISO, Geneva.
ISO, 2014b. Soil quality – Determination of polycyclic aromatic hydrocarbons (PAH) by gas chromatography (GC) and high performance liquid chromatography (HPLC). ISO 13859:2014. ISO, Geneva.
ISO, 2015. Water quality – Determination of polychlorinated naphthalenes (PCN) by GC/MS. ISO/TS 16780:2015. ISO, Geneva.
ISO, 2016. Water quality – Determination of short-chain polychlorinated alkanes (SCCPs) in sediment, sewage sludge and suspended (particulate) matter – Method using gas chromatography-mass spectrometry (GC-MS) and electron capture negative ionization (ECNI). ISO 18635:2016. ISO, Geneva.
ISO, 2018. Water quality – Determination of arsenic(III) and arsenic(V) species – Method using high performance liquid chromatography (HPLC) with detection by inductively coupled plasma mass spectrometry (ICP-MS) or hydride generation atomic fluorescence spectrometry (HG-AFS). ISO/TS 19620:2018. ISO, Geneva.
ISO, 2023a. Water quality – Application of inductively coupled plasma mass spectrometry (ICP-MS). Part 2: Determination of selected elements including uranium isotopes. ISO 17294-2:2023. ISO, Geneva.
ISO, 2023b. Environmental solid matrices – Determination of polychlorinated biphenyls (PCB) by gas chromatography – mass selective detection (GC-MS) or electron-capture detection (GC-ECD). ISO 18475:2023. ISO, Geneva.
ISO, 2024a. Environmental solid matrices – Determination of elements using inductively coupled plasma optical emission spectrometry (ICP-OES). ISO 22036:2024. ISO, Geneva.

ISO, 2024b. Chemicals for the leather tanning industry – Determination of the total content of certain bisphenols. ISO 21135:2024. ISO, Geneva.
Japanese Industrial Standards (JIS), 2024. Testing methods for industrial wastewater. JIS K 0102-4. JIS, Tokyo.
Latimer, J.S., Quinn, J., 1990. Comparison of extraction methods or the isolation of lipids and PCBs from mussel homogenate. EPA/600/3-90/092 (NTIS PB91127787). U.S. Environmental Protection Agency, Washington, DC.
Ministry of Health, China (MOH), 2023. Standards for drinking water quality. GB 5750-2023. MOH, Beijing.
NOAA Sea Grant, 2024. Contaminants of emerging concern – CECs. NOAA Sea Grant.
Standardization Administration of China (SAC), 2010. Animal and vegetable fats and oils – Detection and identification of a volatile organic contaminant by GC/MS. GB 25225-2010. SAC, Beijing.
SAC, 2006a. Starch and derived products – Heavy metals content – Part 1: Determination of arsenic content by atomic absorption spectrometry. GB 20380.1–2006. SAC, Beijing.
SAC, 2006b. Starch and derived products – Heavy metals content – Part 2: Determination of mercury content by atomic absorption spectrometry. GB 20380.2–2006. SAC, Beijing.
SAC, 2007. The specification for marine monitoring – Part 5: Sediment analysis. GB 17378.5–2007. SAC, Beijing.
SAC, 2008. Animal and vegetable fats and oils – Determination of unsaponifiable matter – Part 1: Method using diethyl ether extraction. GB/T 5535.1–2008. SAC, Beijing.
US EPA, 1986. Method 8310: Polynuclear aromatic hydrocarbons. EPA/625/R-96/010b. US EPA, Washington, DC.
US EPA, 1994a. Method 1613B: Tetra-through octa-chlorinated dioxins and furans by isotope dilution HRGC/HRMS. Revision B. EPA/821/B-94/005. US EPA, Washington, DC.
US EPA, 1994b. Method 200.8: Determination of trace elements in waters and wastes by ICP-MS. EPA/600/R-94/111. US EPA, Washington, DC.
US EPA, 1994c. Method 200.9: Determination of trace elements by STGFAA. EPA/600/R-94/111. US EPA, Washington, DC.
US EPA, 1994d. Method 200.7: Determination of metals and trace elements in water and wastes by ICP-AES. EPA/600/R-94/111. US EPA, Washington, DC.
US EPA, 1994e. Method 245.1: Determination of mercury in water by cold vapor atomic absorption spectrometry. EPA/600/R-94/111. US EPA, Washington, DC.
US EPA, 1994f. Method 218.6: Determination of dissolved hexavalent chromium in drinking water, groundwater, and industrial wastewater effluents by ion chromatography. EPA/600/R-94/111. US EPA, Washington, DC.
US EPA, 1996a. Method 3050B: Acid digestion of sediments, sludges, and soils. Revision 2. EPA/600/R-94/111. US EPA, Washington, DC.
US EPA, 1996b. Method 3510C: Separatory funnel liquid-liquid extraction. Revision 3. EPA/600/R-94/111. US EPA, Washington, DC.
US EPA, 1996c. Method 3520C: Continuous liquid-liquid extraction. Revision 3. EPA/600/R-94/111. US EPA, Washington, DC.
US EPA, 1997. Method 549.2: Determination of diquat and paraquat in drinking water by liquid-solid extraction and high performance liquid chromatography with ultraviolet detection. Revision 1.0. Cincinnati, OH. US EPA, Washington, DC.
US EPA, 2000. Method 8318A (SW-846): *N*-Methylcarbamates by high performance liquid chromatography (HPLC). Revision 1. US EPA, Washington, DC.
US EPA, 2001. Method 531.2: Measurement of *N*-methylcarbamoyloximes and *N*-methylcarbamates in water by HPLC with post-column derivatization. EPA/600/R-01/056. US EPA, Washington, DC.
US EPA, 2006. Method 8260D (SW-846): Volatile organic compounds by gas chromatography/mass spectrometry (GC/MS). Revision 3. US EPA, Washington, DC.
US EPA, 2007b. Method 3545A (SW-846): Pressurized fluid extraction (PFE). Revision 1. US EPA, Washington, DC.
US EPA, 2007c. Method 3550C: Ultrasonic extraction. Revision 3. EPA/625/R-96/010b. US EPA, Washington, DC.
US EPA, 2007d. Method 1694: Pharmaceuticals and personal care products in water, soil, sediment, and biosolids by HPLC/MS/MS. EPA/821/R-08/002, US EPA, Washington, DC.
US EPA, 2007e. Method 8141B: Organophosphorus compounds by gas chromatography. Revision 2. EPA/625/R-96/010b. US EPA, Washington, DC.

US EPA, 2007f. Method 8081B: Organochlorine pesticides by gas chromatography. Revision 2. EPA/625/R-96/010b. US EPA, Washington, DC.

US EPA, 2007g. Method 8082A: Polychlorinated biphenyls (PCBs) by gas chromatography. Revision 1. EPA/625/R-96/010b. US EPA, Washington, DC.

US EPA, 2010a. Method 539: Determination of hormones in drinking water by SPE and LC/MS/MS. EPA/600/R-10/146. US EPA, Washington, DC.

US EPA, 2010b. Method 1614A: Brominated diphenyl ethers in water, soil, sediment, and tissue by HRGC/HRMS. Revision A. EPA/821/R-10/005, US EPA, Washington, DC.

US EPA, 2014a. Method 8270D: Semivolatile Organic Compounds by Gas Chromatography/Mass Spectrometry. Revision 4. EPA/625/R-96/010b, US EPA, Washington, DC.

US EPA, 2014b. Method 608.3: Organochlorine pesticides and PCBs by GC/HSD. US EPA, Washington, DC.

Section 2

Biotic Methods

Measurement of True Bioavailability

6 Environmental Monitoring of Bioavailability Using Caged Bivalve Mollusks

6.1 INTRODUCTION AND SCOPE

One pervading issue that complicates the use of ***bioavailability*** methods, as opposed to ***chemoavailability*** methods, is that there is an ever-present possibility that high concentrations of pollutants will exert a toxic effect. Toxic effects may adversely compromise an organisms' ability to be useful for monitoring bioavailability. Bivalve mollusks are filter feeders evolved to exist in environments with higher chemical background than a pristine water column. They are distributed from the North to South Pole and are also sessile. These qualities of being robust to chemical effects and staying in one spot constituted part of the original rationale for the development of methods to capitalize on using bivalve mollusks for ***biomonitoring***, most famously the Mussel Watch program of the US National Oceanic and Atmospheric Administration (Farrington and Tripp, 1993; Farrington et al., 2016). The method here amalgamates aspects of best practice developed over almost 40 years for using bivalve mollusks for biomonitoring of pollutants.

While some methodological approaches to measuring bioavailability give an instantaneous measurement, bivalve mollusks ***bioaccumulate***[1] ***bioavailable*** pollutants over time and have demonstrated excellent use in situations wherein the absolute concentration of pollutants in the water column might at any instantaneous moment be low, albeit the flux of pollutants, e.g., from sediments, could be high. In such cases, the pollutant load in bivalve mollusks reflects the cumulative flux of bioavailable pollutants, which has been shown to be linked to biological effects (Hornberger et al., 2000; Kim et al., 2008).

The scope of this method includes guidance and procedures for performing biomonitoring studies utilizing caged bivalve mollusks. This includes obtaining ***test organisms*** (individual organisms of the species of bivalve mollusk to be used in this method); preparing them for deployment, ***exposure***, and retrieval; and preparation of test organism tissue homogenates for chemical analysis of bioavailable pollutants concentrated by test organisms. The method described here applies to both organic and heavy metal pollutants and is specifically for surface and water column studies. Modification of this procedure might be justified by special needs or circumstances; however, modification should be conducted with a specific rationale that is cognizant of why modifications are needed and how they might affect the validity of results.

Caution must be exercised in the use of this method for some applications and for some pollutants. For instance, pollutants with low ***octanol–water partition coefficients***,[2] including some charge-neutral organics, are not significantly bioaccumulated by bivalve mollusks (Ernst, 1977; Geyer et al., 1982; Gosset et al., 1982; van Gestel et al., 1985). Likewise, this method has been most frequently used for known, high-priority pollutants whose behavior with respect to bivalve mollusks is known. The method cannot be assumed *a priori* to be applicable to the measurement of bioavailability of some

[1] A phenomenon whereby a chemical present in a living organism's environment accumulates in tissues of the organism.

[2] The octanol–water partition coefficient, K_{ow}, is the ratio of the concentration of an organic pollutant in *n*-octanol to its concentration in water at equilibrium, i.e., it is a measure of how an organic pollutant distributes at equilibrium between these two immiscible solvents. Pollutants with a high K_{ow} value are lipophilic and tend to bioaccumulate, which can lead to higher concentrations in organisms higher up the food chain.

DOI: 10.1201/9781003355229-8

emerging pollutants. If organisms are not properly prepared, notably depurated, prior to use, some pollutants in samples for analysis may represent sediment-associated pollutants in the gut (Phillips, 1980). Life/reproductive stage and seasonal variations may also cause large variations in results; thus, these factors must be controlled for and taken into account when designing field studies and interpreting results (e.g., Shindo and Otsuki, 1999). Chapters containing Terms and Definitions, guidance on avoiding contamination of samples in field and lab work, guidance on Quality Assurance (QA)/Quality Control (QC), and ***Standard Methods*** of instrumental analysis are also within the scope of this chapter.

6.2 PRINCIPLE

In this method, test organisms are sourced, prepared, and caged, and then cages are deployed at one or more locations for a month. ***Caging*** refers to a containment system that allows exposure of test organisms by various routes/multiple ***sources***, while also enabling organisms adequate habitat during the exposure period. A minimum of two ***exposure locations***, i.e., exact spatial location and depth where test organisms will be exposed, are required. At least one exposure location must be at a ***reference location***, i.e., an exact spatial location and depth where test organisms will be exposed for control measurements. There must also be at least one ***effects location*** representing an exact spatial location and depth where test organisms will be exposed to measure bioavailable pollutants associated with effects. The number of test organisms required for the purposes of chemical analysis is placed in one or more cages and deployed into water at each exposure location. During exposure, test organisms bioaccumulate bioavailable pollutants. After exposure, the test organisms are retrieved/collected, processed, and subsequently analyzed at a laboratory to determine the bioavailable pollutants bioaccumulated over the exposure period.

6.3 EQUIPMENT, REAGENTS, EXPOSURE LOCATIONS, AND LOGISTICS

6.3.1 Equipment

This method is particularly field-intensive; hence, it is important to plan both field and laboratory operations carefully to manage risks of both contamination and loss of ***target analytes***. All field gear and laboratory materials must be properly cleaned and decontaminated prior to use. For field gear, this may be particularly challenging, i.e., all materials used (ice chests, handling trays, processing tools, plastic bags, foil, etc.) that can be treated, must be subjected to appropriate cleaning and decontamination in the laboratory prior to use under clean room or clean-room equivalent conditions (see Chapter 3). Further, after cleaning, field gear must be transported and stowed in a manner to keep it decontaminated. Stowing typically involves carrying cleaned gear in clean, sealed containers. In the field, when working at multiple locations, the field team must clean each item used between locations. Laboratory items subject to reuse must also be carefully cleaned in between each use. The information provided in this section is intended as guidance rather than a comprehensive list of all items that might be useful to have. Users should modify this list to meet specific needs. Additionally, it is assumed that vessels needed (boat or ship) and associated equipment are available, along with dive gear, if needed.

Test organism collection/field-processing related equipment:

Test organisms – for exposure testing, basis of bioavailability method.

Multiparameter water quality meter or sonde – to measure temperature, pH, conductivity/***salinity***, and other water quality parameters at exposure locations.

Gloves – to prevent manual contamination during test organism collection.

Stainless-steel blade or pry bar – to break mollusks off substrate to which they might attach or be attached.

Buckets – cleaned/***contaminant***-free and of appropriate material for target analytes; used for collecting ambient water and temporary placement of test organisms.
Thermo-insulated box – cleaned/contaminant-free and of appropriate material for target analytes; must have insulating characteristics sufficient to maintain test organisms at ambient temperature during handling and transport.
Field clothing items – e.g., waders; all items in contact with water and/or test organism must be cleaned/contaminant-free and of appropriate material for target analytes.
Scrubber – for removal of epibonts from test organisms; must be cleaned/contaminant-free and of appropriate material for target analytes.
Ruler or caliper – for measuring bivalve dimensions.

Field data records-related equipment:

Indelible pens – to record field observations without loss of fidelity in a wet environment and for labeling samples.
Field logbook – to document information on sample collection activities, hydrologic conditions, weather conditions, boat or equipment operations, or any other unusual activities observed or problems encountered, that would be useful in evaluating the quality of data collected using this method.
Field record ledger – a type of sample tracking form that enables an at-a-glance summary of details of all samples deployed and retrieved during a biomonitoring program. Each line of the form usually includes (1) project number and/or name, and/or short description; (2) unique test organism batch identifier; (3) deployment/retrieval date and time (yyyy-mm-dd and 24-hour clock format); (4) deployment/retrieval location latitude/longitude (***station position***, i.e., surface location overlying subsurface exposure location, determined by GPS), and potentially including exposure location name, geographic description, water body name/description; (5) deployment/retrieval depth(s) (with units); (6) information on deployment/retrieval/collection method; (7) if ***Chain-of-Custody*** is applicable, collectors' names and signatures; and (8) contact for responsible person or entity.
Specimen identification labels – a type of sample tracking form that is completed in indelible ink for each individual batch of test organism collected, deployed, or, for post-deployment sample collection for each group of organisms constituting a sample. This label is for unique identification and should contain (1) a unique identifier (per field record ledger), (2) species scientific name and type (e.g., *mussel*, *oyster*, *clam*), (3) date collected/deployed/retrieved, (4) location of collection/deployment/retrieval, and (5) sample description (purpose, parameters/pollutants to be analyzed).
Several sizes of resealable watertight (press-seal/ziplock type, polyethylene) plastic bags – for storage of field records and forms.

Test organism preparation-related equipment:

Depuration water and system – contaminant-free water with water quality characteristics similar to reference locations; used in the ***depuration*** process. Requires a system to ensure sufficient food, aeration, and circulation are provided for health of test organisms. Depuration involves placing test organisms into a clean water environment for a period of time to allow purging of biological or chemical contaminants and/or physical impurities. A separate dosing system may be needed if feeding is required.
Water sample containers and sample forms – to collect samples of depuration water for analysis prior to depuration of test organisms. Appropriate sample containers with preservatives and transport containers with instructions are provided by accredited testing labs;

labels for containers should comply with lab Chain-of-Custody requirements, along with completion of Chain-of-Custody forms.
Multiparameter water quality meter or sonde – as above, for use with depuration water.
Gloves – as above.
Buckets – as above; used for temporary placement during transfer of test organisms.

Deployment/caging-related equipment:

Anchors or mooring materials – for lines to secure cages with test organisms in position at exposure locations.
Lines – to attach/secure cages to anchors/mooring and/or subsurface stabilization buoys. Additional line-type material may be needed for other purposes in securing cages to deployment equipment.
Buoys – surface buoys and/or subsurface buoys to aid in the location of cages and/or stabilization of cages at exposure locations.
Exposure cages – which, depending upon target pollutants scheduled for analysis, may be metal or hard/inert plastic (polypropylene preferred), or appropriate netting cages.
Flags, signals, and/or beacons – visual or electronic markers to assist with locating the position of cages at exposure locations and/or alert others to the location of cages in order to prevent damage, collisions, etc.
Electronic or physical devices – to ensure the exact siting of exposure locations (e.g., GPS) and exposure depths (e.g., depth gauge).
Thermo-insulated boxes – as above.
Depuration water – to cover test organisms in thermo-insulated boxes while transported to deployment locations(s).
Buckets – as above.
Gloves – as above.
Sample collection materials – to collect test organisms at deployment for analysis; this initial collection of organisms is used to establish the initial biological and chemical status of test organisms immediately prior to deployment (see test organism retrieval sample packaging and transport-related equipment below).
Sack – clean/contaminant-free, of appropriately inert plastic and large enough to cover deployment cage; for placement over exposure cages to avoid adherence of surface contaminants during deployment and retrieval.

Test organism retrieval sample packaging and transport-related equipment:

Holding trays – to place test organism samples in between retrieval and packaging. Assists in reducing the chances of contamination.
Aluminum foil (extra heavy duty) – hexane-rinsed, for primary packaging of test organisms sampled after deployment for analysis of organic pollutants.
Several sizes of resealable watertight (press-seal/ziplock type, polyethylene) plastic bags – contaminant-free, for primary packaging of test organisms sampled after retrieval for analysis of heavy metal pollutants, and for secondary packaging of test organisms sampled after deployment for analysis of organic pollutants that have been wrapped in foil.
String/twine – for securing packaging and/or labels to packaging.
Ice/dry ice – ice packets are preferred if it is possible to keep them well-frozen until used; alternatives include dry ice (subject to certain shipping requirements) or wet ice (which increases chances of sample cross-contamination when melting). When using dry ice, thermo-insulated boxes must be drilled in places to allow pressure regulation as the dry ice evaporates.
Filament-reinforced freezer tape – to seal thermo-insulated boxes for test organism transport.

Gloves – as above.
Thermo-insulated boxes – as above; for transport of test organisms/samples.
Sack – as above.

Test organism preparation, processing, and sample measurement-related equipment:

Muffle furnace – for high-temperature decontamination of glassware (organic pollutants).
Scrubber – clean/contaminant-free and of appropriate material for pollutants to be analyzed, e.g., stainless steel or polytetrafluoroethylene (PTFE), a commonly known brand of which is Teflon.
Lab gloves – clean/contaminant-free.
Vernier caliper – for measuring shell length.
Specimen trays – clean/contaminant-free, of appropriate material for pollutants to be analyzed, and large enough to hold an entire sample.
Thermometer – for monitoring water temperatures.
Shucking knife/blade – used for dissection.
Scraping implement/lab spatula – clean/contaminant-free, of appropriate material (e.g., stainless steel, PTFE) for pollutants to be analyzed; to remove test organism soft tissue from shell.
Small lab forceps – clean/contaminant-free, of appropriate material (e.g., stainless steel, PTFE) for pollutants to be analyzed; to remove test organism soft tissue from shell.
Sample containers – clean/contaminant-free, of appropriate material (e.g., stainless steel, PTFE) for pollutants to be analyzed; different sizes, for both shucked samples and tissue homogenate samples.
High-energy ball mill/shaker – with PTFE vials and balls, for homogenization of samples for heavy metals analysis.
Lab tissue homogenizer – with titanium generator, for homogenization of samples for analysis or organic pollutants.
Several sizes of resealable watertight (press-seal/ziplock type, polyethylene) plastic bags – as above; for secondary packaging to prevent sample cross-contamination.
Balance – to measure tissue masses (metric units, accurate to 1 mg).
Freezer – −20°C, for storing test organism samples and/or processed sample homogenates.
Freeze-dryer or lab oven – used in the preparation of dry mass samples. For the digestion of samples scheduled for heavy metals analysis, freeze-drying is recommended.
Mortar and pestle or knife mill – to homogenize freeze-dried tissue samples prior to digestion for instrumental analysis.

6.3.2 Reagents

Ultrapure water – also known as Type I water, has a resistance of 18 MΩ·cm.
Ultrapure hexane – pesticide grade, equivalent, or better.
Ultrapure acetone or isopropyl alcohol – acetone is preferred; grade should be pesticide grade, equivalent, or better.
Depuration water and reagents as needed to adjust hardness, pH, or salinity of depuration water; high purity/contaminant-free.

6.3.3 Exposure Locations and Logistics

6.3.3.1 Exposure Locations

A minimum of two exposure locations must be used: (1) one at a reference or control location and (2) one at an effects location (US EPA, 1983). Additional exposure locations may be necessary or desirable if the spatial extent of the likely ***environmental effects envelope*** is extensive. The environmental effects envelope signifies an area including, but larger than, the temporal and spatial extent of environmental effects to be monitored. The environmental effects envelope should be larger than the known or surmised spatial extent of effects, recognizing that actual effects may be more widespread than initially supposed. Multiple exposure locations may also be desirable because multiple pollutant sources can affect the overall ***bioaccumulation*** levels, and having multiple exposure locations enables this to be determined and taken into account in interpreting results.

The reference exposure location(s) must have hydrographic and water quality characteristics similar, i.e., matched, to those at the effects exposure location(s) (US EPA, 1983). Bivalve mollusks can be sensitive to variations in salinity and other ***components*** of water quality; hence, variations in salinity or water quality have the potential to affect the degree of bioaccumulation. More than one reference exposure location might be indicated, for instance, in studies involving estuarine environments where salinity gradients are present.

Exposure locations should be chosen, inasmuch as possible, to target a single event or potential impact within an environmental effects envelope. For instance, if the purpose of a study is to monitor for pollutant bioavailability resulting from a chemical spill, it is undesirable to have an exposure location near a shipping lane or dredging ***site***, which might have their own, separate effects.

Exposure locations should be chosen with maintenance considerations in mind. For instance, the exposure system used must be able to hold the cages in a fixed and secure position for the duration of the deployment and to ensure that the test organisms have a sufficiently accommodating habitat to remain viable over the deployment duration.

Station positions that mark exposure locations must be established in a manner to secure easy location of cages, via, e.g., surface buoys, shore transects or tethers, fathometers, acoustic transducers, and GPS devices.

One additional consideration associated with site delineation and exposure locations is whether threatened or endangered species are present. Studies should not be conducted in such areas. In most countries, whether and how to proceed with effects/impact studies in such cases is governed by law.

6.3.3.2 Exposure Depth

Exposure must be conducted at a minimum of one depth; however, in some instances, it may be desirable to conduct exposures at multiple depths at a single exposure location. The maximum permissible exposure depth is ideally 1 m above the sediment surface to avoid undue influences from substances released or resuspended from sediments. For each reference exposure location, corresponding effects location depths should be matched; otherwise, differential depth effects may bias the measurement outcome.

In cases where pollutant exposure might vary as a function of depth, it is desirable to deploy cages at different depths to ensure that the depth of maximum exposure is not missed. In the case of using multiple exposure depths, for each depth, there should be both reference and effects cages at each. In some cases, the likely spatial distribution of effects may be provisionally assessed, for example, by using modeling or easily tracked pollutants that might be associated with an environmental effects envelope (e.g., nutrients such as ammonia or phosphate).

6.3.3.3 Exposure Systems

The description of exposure systems here is not proscriptive; any system that fulfills the method requirements for security of placement and preservation of test organism viability would be suitable. Systems described here have been used successfully. In some environments, duplicate placements should be considered. Even the most secure placements are subject to loss from storms/weather, accidents, and vandalism, and the costs of replication are sometimes justified by the ***sensitivity*** and need for data collection and Environmental Impact Assessment.

Anchors are needed to ensure that cages stay put after placement. Examples of inexpensive anchors that can be used include the following (Note: Minimum anchor mass should be 25 kg, and for some environments 20 times this would be in order):

- Train wheels (clean, axles removed)
- Cast concrete blocks (secure attachment for placement needed, can be imbedded or blocks can be cast with hole for looping line)
- Fence anchors, Auger type, minimum 1 m length (soft sediment bottoms).

Lines are needed to connect cages to anchors and buoys. A minimum diameter of 16 mm polypropylene or 8 mm polypropylene encased steel cable is recommended.

Surface buoys are typical choices to mark exposure locations. Subsurface buoys are often used to support test organism cages or to add additional stability, for instance, attaching one or more lines connecting anchors to subsurface buoys. Most countries have regulations concerning buoy type, placement, and use. Commercially sourced buoys compliant with such standards and with the ability to withstand physical stresses needed are recommended.

For the purposes of this method, the term cage refers to any appropriate enclosure used for bivalve mollusk biomonitoring. It is important to consider which pollutants will be monitored and to choose a material that is compatible with same. If heavy metal(oid)s are to be monitored, plastic containers are preferable. Since most plastic materials for underwater use are not of appropriate composition, use of highly inert and rigorously decontaminated materials is the goal. Polypropylene or high-density polyethylene materials may be usable if demonstrated to be contaminant-free. In contrast, if organic pollutants are to be monitored, metal containers are preferable. Since metal materials for underwater use are potentially problematic, these must also be demonstrated not to cause contamination in the monitoring process. In all cases, the same material must be used at all locations. In practice, it is often the case that it is easier to find plastic materials that can be used without causing contamination than metal. Since aluminum is not typically of interest in pollutant studies, it may be a suitable cage material.

The most common types of exposure cages are either aluminum metal cages, which in many cases can be fabricated to suit, or bait bags (exposure bags) made of high-quality polypropylene. Cages must be supported at depth, whereas bait bags are often attached to one or more lines connected to the anchor and/or a submerged buoy. Figure 6.1 shows examples of cages, and Figure 6.2 shows examples of representative deployment configurations. Increasingly, a large number of suppliers to the aquaculture industry offer ropes, lines, buoys, cages, and other accessories for farming

FIGURE 6.1 Examples of cages suitable for use with test organisms. Left: metal cage; Center: high-strength polypropylene bait bags; Right: aluminum-frame cage with free-flow plastic inserts.

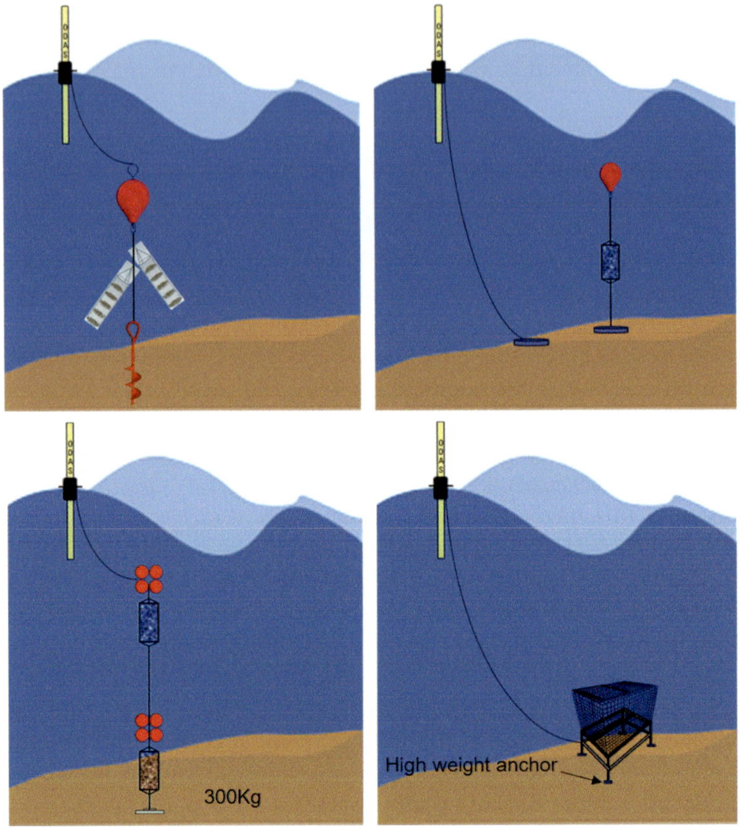

FIGURE 6.2 Examples of deployment configurations suitable for use for this method. Clockwise from upper left: test organisms in cages or bags tethered to line held fast by Auger-type anchor and tensioned with subsurface buoy; single small cage anchored with weight to bottom, line tensioned by subsurface buoy; multi-depth configuration featuring multiple subsurface buoys and high-weight anchorage to maintain stable positioning on the water body bed; large cage integrated with stilts and multiple high weight anchors.

bivalves that are made of food-safe materials. Such materials are often contaminant-free and thus suitable for use in biomonitoring applications. These suppliers are conveniently identified via internet search.

6.4 TEST ORGANISMS: SPECIES, SIZE, AND NUMBER

6.4.1 Species Suitable for Use as a Test Organism

A list of species relative to locations where these species are typically used for biomonitoring is given in Table 6.1 for marine applications (Bendell et al., 2020; Beyer et al., 2017; Evariste et al., 2018; Krishnakumar et al., 2018 [and references therein]; Lehtonen et al., 2019; Ramu et al., 2007). The list contains mostly mussel species and is by no means comprehensive. Many other species, including some oysters and clams and other mussels, may be suitable. Concerning freshwater environments, no listing is made due to some crucial aspects of species choice. On one hand, invasive species such as Zebra Mussel have been used (Richman and Somers, 2005) when already well-established and common; however, it would be unethical and an environmental disaster to use these species outside of their native environment (southern Russia and Ukraine) where potential for invasion exists, or in locations where eradication efforts are underway. On the other hand, some

TABLE 6.1
Bivalve Mollusk Species Used in Biomonitoring, with Applicable Geographic Areas for Each

Species	Geographic Areas of Typical Use in Mussel Biomonitoring
Mytilus edulis (Blue Mussel)	North America to North Carolina; northern Europe to Bay of Biscay; Argentina, Falklands, southern Chile
Mytilus trossulus (Bay Mussel)	High latitude Atlantic-Pacific-Baltic; California to Alaska; Asia – Hokkaido north
Mytilus galloprovincialis (Mediterranean Mussel)	Black Sea, Mediterranean, Atlantic Coast of Europe; Chile, Australia, New Zealand; West Coast of US; South Africa; China, Japan, Korea
Perna viridis (Green Mussel)	Southeast Asia, India, Persian Gulf
Crassostrea virginica (American Cupped Oyster)	Found in many places on the East Coasts of North and South America
Crassostrea gigas (American Cupped Oyster)	East coast of China, Korea, Russia, West coasts of North and South America, West coast of Africa, much of Europe, New Zealand, Australia

species that are known to be suitable for freshwater use are endangered. Raising or purchasing test organisms is preferable; only species that are abundant, noninvasive, and hold the conservation status of Least Concern should be considered.

6.4.2 Size and Number of Test Organisms Needed

Bioaccumulation of pollutants in bivalve mollusks is well known to vary with size and life stage (e.g., Richir and Gobert, 2014). It is desirable for measurement purposes and to comply with best practice in fisheries to use mature organisms within a tight size range (WDNR, 1988). The criterion of maturity determines the minimum size. In terms of size range, one early method, for instance, recommended using *Mytilus edulis* in a size range of 50–70 mm shell length, i.e., 60 mm ± 17% or a ratio of small to large of 71% (US EPA, 1983). This is from early work, whereas later the US EPA, responsible for the origination of the bivalve mollusk monitoring approach, recommended that the total length of the smallest individual in any composite sample should be no less than 75% of the total length of the largest individual in the composite sample (US EPA, 1990).

Size of test organism is usually determined based on shell length, as illustrated in Figure 6.3, which is measured with a vernier caliper. The difficulty in recommending a size range is multifold. First, different species grow to different sizes. Second, for any given species, size distributions of mature individuals vary according to conditions in the growth environment. Third, organisms are measured according to shell length; however, the number of individuals needed depends on mass, and the mass-to-shell length ratio is variable for different species and growth environments.

The matter of ensuring that test organisms of an appropriate size are used is another reason why purchasing test organisms from a culture facility is preferable to collecting test organisms. Figure 6.4 shows representative data needed when collecting organisms for the mussel species *Choromytilus chorus* (Ruiz-Velásquez et al., 2017). According to this figure, mussels with a shell length of over 51 mm would be suitable for use (the lowest shell length associated with full maturity of the population). Allowing a safety margin of 3 mm entails the smallest acceptance length of 53 mm, and by the 75% rule above, the largest acceptance length is 53 mm/0.75 = 71 mm, with an average shell length of 62 mm, resulting in a size range of 53–71 mm, which is quite similar to that from the 1983 guidance for *M. edulis*. From the graph, this size range would ensure that 100% of test organisms collected would be mature.

A difficulty with this approach is that the maturity-versus-size relationship (as in Figure 6.4) must be known; however, this relationship itself is variable, being a function of water quality parameters

FIGURE 6.3 Illustration depicting how to measure shell length for different types of bivalve mollusk. Top left and center: two different types of clams. Bottom left: oyster. Top right: mussel. Bottom right: scallop.

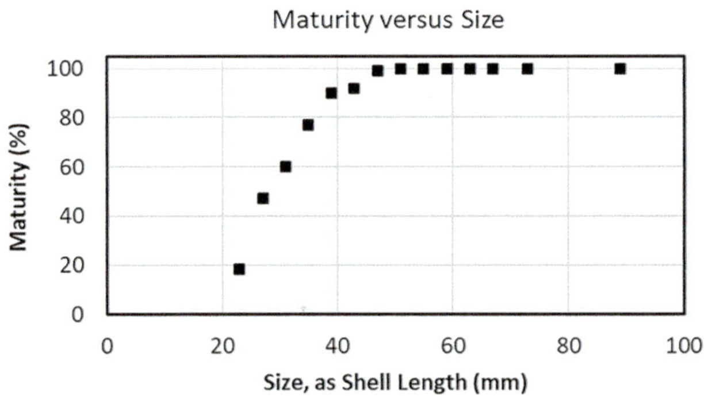

FIGURE 6.4 Example of maturity-versus-size (as shell length) relationship for one study on species *C. chorus* (data from Ruiz-Velásquez et al., 2017). Maturity is reported as the percent of individuals at maturity for each given size range surveyed. The smallest size at which all individuals are mature is a function of growing conditions and can vary geographically.

and temperature. For some test organism species, this information on the variability of maturity versus size relationships is well known for given regions, in which case the available data will guide a choice of size range. In other cases, the manner in which the maturity-versus-size relationship for a test organism changes with water quality and temperature is not known. In this case, the only alternative is to construct a curve such as in Figure 6.4. This requires making studies of test organisms' gonads to determine what percentage of organisms in any given size range are mature (e.g., per methods used by Ruiz-Velásquez et al., 2017).

The total number of test organisms needed for a given size depends on the analytes scheduled in the ***Sampling and Analysis Plan*** and the method of analysis. Some analytical methods require larger sample sizes than others. For instance, US EPA method 3052 for digestion of tissue samples for heavy metals analysis calls for ~0.5 g of dry sample, which is ~5 g wet tissue (Mo and Neilson, 1992; US EPA, 2004). It is first necessary to know what the average mass of a test organism is for the size used (typically the shell length is proportionate to mass of flesh). If this average mass of a test organism is not known, a representative sample of test organisms needs to be shucked and individually weighed in order to obtain the information. The total number of test organisms, $N_{\text{T,sample}}$, needed for analysis of "one sample" is

$$N_{\text{T,sample}} = \sum_j \left(\text{ceiling}\left(\frac{m_{\text{sample},1}}{m_{\text{org},1}} \right) + \text{ceiling}\left(\frac{m_{\text{sample},2}}{m_{\text{org},2}} \right) + \cdots + \text{ceiling}\left(\frac{m_{\text{sample},j}}{m_{\text{org},j}} \right) \right),$$

where the summation over numbers 1, 2, …, j represents a total of j different types of analyses to be performed on the sample (e.g., heavy metals, polycyclic aromatic hydrocarbons, etc.), m_{sample} is the wet mass of sample required for each different type of analysis, m_{org} is the average wet mass of a single test organism, and the function "ceiling" rounds up to produce an integer number of test organisms needed for each different type of analysis scheduled. Once $N_{\text{T,sample}}$ is determined, the total number of test organisms needed for each exposure location, $N_{\text{T,location}}$, is

$$N_{\text{T,location}} = \left(\left(N_{\text{T,sample}} \times N_{\text{replicates}} \times N_{\text{depths}} \right) + \left(N_{\text{i/f}} \times N_{\text{replicates}} \right) \right) \times F_{\text{margin}},$$

where $N_{\text{replicates}}$ denotes the number of field replicates to be analyzed in order to quantify ***precision*** from sample to sample in field conditions (typically a minimum of three), N_{depths} is the number of depths at which test organisms will be deployed at each exposure location, and F_{margin} is a factor that increases $N_{\text{T,location}}$ by some safety margin, for instance, allowing for damage to organisms and mortality that might occur during deployment/exposure/retrieval. This margin should be a minimum of 10% (ie., $F_{\text{margin}} = 1.1$) and under some field conditions could be much higher. The quantity $N_{\text{i/f}}$ is calculated the same as $N_{\text{T,sample}}$,

$$N_{\text{i/f}} = \sum_k \left(\text{ceiling}\left(\frac{m_{\text{sample},1}}{m_{\text{org},1}} \right) + \text{ceiling}\left(\frac{m_{\text{sample},2}}{m_{\text{org},2}} \right) + \cdots + \text{ceiling}\left(\frac{m_{\text{sample},k}}{m_{\text{org},k}} \right) \right) + N_{\text{QC}},$$

except that k is larger than j (discussed below), representing additional analyses needed to ***determine*** test organism health/condition when deployed ("initial", i), and retrieved ("final", f). Some of the measurements needed for initial and final will be collected from test organisms used in chemical analytical studies (j sets). The first j sets of $N_{\text{i/f}}$ samples are for testing the baseline/initial quantities of pollutants to be included in the study that are present in test organisms initially. The quantity N_{QC} reflects the number of samples needed for QC purposes, which is variable by analysis type, as described below and stipulated in Standard Methods for chemical analysis.

6.5 PROCEDURE

6.5.1 Collection or Acquisition of Test Organisms

Test organisms may be collected or purchased from an aquaculture facility. Purchasing test organisms is vastly preferable when possible, for numerous reasons. When purchased, there is no need to identify species; the point-of-purchase purveyor clearly knows which species they are cultivating and selling. Purchasing ensures that all organisms are the same species, and commercial organisms are typically in a narrow size range, both of which are highly beneficial to measurement outcomes (e.g., measurement precision). Purchasing test organisms saves labor and expense because the need to collect, process, and prepare organisms is eliminated or greatly minimized. Because aquaculture facilities raise mollusks for food, purchased test organisms are typically much lower in pollutant load, because they must be safe for human consumption, and are thus subject to testing to verify pollutant loads. Purchased organisms will almost always be depurated by the aquaculture facility, as depurated organisms are the standard for food-quality bivalve mollusks.

In many places in the world, many species of mollusk are endangered and/or overharvested (Zieritz et al., 2022). The problem of extinction of mussel species is particularly acute in freshwater environments and exacerbated by the rise of invasive species such as *Zebra* and *Quagga* mussels and the Asiatic clam *Corcibula* (Matthews et al., 2014). Purchasing test organisms is more ecologically responsible and lowers liability risks. Both marine and freshwater mollusks are farmed in many countries of the world. Another possibility is to raise the test organisms (Gosling, 2003). Purchased and raised organisms do not need to be depurated; however, in some cases might benefit from ***acclimation***, depending on how different the exposure location conditions are from the water in which purchased or raised test organisms have been living prior to deployment.

If it is not possible to purchase or raise test organisms, they must be collected from an area that is relatively free of contaminants and has water of similar chemical composition (salinity, major constituents) to the exposure locations. Only species that are abundant and hold the conservation status of Least Concern should be considered. If the concentration of pollutants scheduled for chemical analysis in the organisms at the collection location is not known, this should be assessed by prior collection and testing. Note: In many countries, restrictions on the collection of such organisms apply for various purposes of ecological protection. It is necessary to first determine applicable laws and regulations pertaining to the collection of test organisms prior to doing so. Next, it is important to verify the species of the organisms where collection is to take place, and to ensure that a single species (one not endangered or subject to legislative protection) can be identified and collected. Identification of many bivalve mollusk species is straightforward from morphological characteristics, and many handbooks exist for assistance in identification (e.g., Ellis, 1962; Gosling, 2003, 2015). Because such organisms vary greatly worldwide, specific guidance for the applicable area, particularly for freshwater bivalve mollusks, should be sought by anyone who wishes to initiate studies using this method.

Care is required to ensure that test organisms are not contaminated during collection – this entails that all materials used should be clean and contaminant-free. Protocols for cleaning laboratory ware in pollutant analysis should be followed for cleaning field gear to the greatest extent possible.

Organisms should be collected from natural substrates, preferably rocky or hard substrates, using a clean stainless-steel blade or implement to break or pry the organism's shell off of the substrate, by either trained scientific scuba divers or from wading in near-surface water (appropriately clean waders and boots are needed to ensure absence of contamination). Care needs to be taken not to damage organisms during collection. Organisms should be collected from a single relatively well-defined source area, at the same time, and at the same or similar depth. Organisms should in no circumstance be collected from steel, wood, or other man-made structures.

Immediately after collection, test organisms for caging experiments should be placed in ambient water (i.e., water taken from same location as where test organisms were collected) and transported in a contaminant-free thermo-insulated box, for instance a large cooler or other container that can hold the temperature of the collection water, and transported to a laboratory or field installation for depuration. Prior to depuration, epibonts should be removed from shell surfaces (using a stainless-steel blade or other contaminant-free scrubber), and the test organisms need to be rinsed with ambient water or depuration water. Subsequent to this step, test organisms must be prepared for exposure and assessed for initial condition.

6.5.2 Preparation of Test Organisms for Exposure and Initial Condition Assessment

Prior to deployment for exposure studies, organisms must be properly prepared and subjected to an initial assessment of their condition. Preparation involves depuration and acclimation. Depuration is, by definition, a process whereby mollusks are placed in a clean water environment for a period in order for them to purge biological, chemical, and physical contaminants in their gut tract. It is needed in order to ensure that pollutant body load at the end of the measurement originates from the environmental effects envelope only. Problematically, if not performed correctly, this process can introduce contamination. The water used in depuration needs to be similar in composition to that of the exposure locales and, contaminant-wise, very clean. Depuration conditions should be similar to that of test organism exposure locations and not so different from the conditions at the test organism source as to compromise the test organisms' ability to acclimate. Depuration performed in this manner, with water chemistry and temperature representative of exposure locations, also effects acclimation.

A number of guidelines, summarized here, exist for depuration of bivalve mollusks in a manner that avoids introduction of contamination (ASTM, 2022; Barrento et al., 2013). Depuration is conducted in flow-through tanks. An elevated headbox enables depuration water to be gravity-fed into the depuration tank where test organisms are depurated. Strainers and air traps should be included in the depuration water supply system. Headboxes and depuration tanks should be equipped for temperature control and aeration. The facility where depuration is conducted should be well-ventilated and free of fumes, and air used for aeration must be free of fumes, oil, and water. This is achieved with filters to remove oil and water. To further reduce the possibility of the contamination of test organisms, depuration should not be conducted in a room where laboratory work is underway, equipment is stored or cleaned, or any other ***activity*** that might transmit contamination.

All equipment that might contact depuration water must be nonporous, nontoxic, contaminant-free, and not contain substances that can be leached or dissolved in a manner that would contaminate test organism during depuration. Glass, Type 316 stainless steel, nylon, and fluorocarbon plastics are recommended materials to minimize leaching, with the proviso that stainless steel should not be used if test organisms will be used in studies of heavy metals bioavailability in saltwater. Cast-iron pipe, copper, brass, lead, galvanized metal, neoprene rubber, and natural rubber should specifically never be used in water supply systems or be in contact with depuration water. Concrete and rigid (unplasticized) plastics may be used for tanks and in the water-supply system, but they should be appropriately decontaminated and left in flowing water, with a chemical composition equal or similar to depuration water to be used, for at least a week before use. Tanks used for depuration must be internally smooth, capable of being completely emptied (i.e., no residual water left behind), easily cleaned, and of a suitable volume to hold the needed number of test organisms at or below a specified loading limit. All parts of the water supply system, headbox, depuration tank, etc., and equipment used must be thoroughly cleaned/contaminant-free before use and between each cycle of test organism depuration.

Depuration water must be in adequate supply, compatible with needs for supporting test organisms, and of uniform quality. The working criterion for compatibility and quality of depuration water is that it should be able to support test organisms well enough that they could survive, grow, and reproduce acceptably in it. This working criterion is usually met by matching depuration water quality with what is known about test organisms needs. Natural water is preferable for depuration. If a natural water is used, it should be obtained from an uncontaminated, uniform-quality source. For freshwater, well water or spring is sometimes preferable to a surface water. If surface water is used for fresh or salt depuration water, the water intake should be positioned to minimize fluctuations in water quality and the possibility of contamination. If reconstituted water is used, chlorinated water should not be used because total residual chlorine and chlorine-produced oxidants are quite toxic to aquatic organisms and because dechlorination is almost always incomplete. Municipal drinking water also often contains unacceptably high concentrations of copper, lead, zinc, and fluoride, and water quality is variable. Depuration water must be characterized to demonstrate that it meets Water Quality Criteria. This involves analysis of depuration water for the following analytes:

- All waters – ***alkalinity***, pH, conductivity, particulate matter, total organic carbon (TOC), total Organophosphorus pesticides, organic chlorine (or total Organochlorine pesticides plus polychlorinated biphenyls [PCBs]), chlorinated phenoxy herbicides, ammonia, cyanide, sulfide, calcium, magnesium, potassium, bromide, fluoride, iodide, nitrate, phosphate, sulfate, aluminum, arsenic, beryllium, boron, cadmium, chromium, cobalt, copper, iron, lead, manganese, mercury, molybdenum, nickel, selenium, silver, zinc, and all pollutants that are scheduled for biomonitoring studies.
- Additional analysis for freshwater depuration – hardness, chloride, and sodium.
- Additional analysis for saltwater depuration – salinity.

Some of the analytes above are toxic to bivalve mollusks at low concentrations (e.g., chlorine and chorine containing compounds), and some are quantities that relate to test organism adaptation ranges for optimum viability. It has been shown that hardness, salinity, and pH may be adjusted by the addition of appropriate reagent-grade chemicals (sea salt, acid, base, and deionized or distilled water, well mixed before in contact with test organisms) without adversely affecting test organisms.

In some cases, it might also be needed to ensure an appropriate food source in depuration water. Natural sources of water, e.g., seawater or river water, often have enough algae/phytoplankton and detritus in them to meet food source requirements. If the depuration water is inadequate as a food source, it will be necessary to provide food. Examples of food include fresh algae, Shellfish Diet 1800 (Reed Mariculture, a blend of five marine microalgae species), and the algae-based Nannochloropsis 3600. Fresh algae are preferred, and algal mixes are generally superior to single algae as a food source. A feeding rate maintained at 50,000–70,000 cells/mL is common in a continuous feed disbursed by a peristaltic pump or dosing pump. For very actively feeding bivalves, the cell count can drop rapidly – cell count should be monitored once per hour until the correct dosing rate is known. Due to the increase in bivalve aquaculture, in many places, food may be purchased commercially, complete with supplier recommendations for feeding. Indications that bivalve mollusks have adequate food supply include evidence of active feeding (siphons extended into water) and production of waste. If depuration water is amended with food, it is important to include whole water samples in testing analysis to ensure that the food source does not introduce contamination.

Depuration water should be aerated intensively by using air stones, surface aerators, or column aerators. Adequate aeration will bring the pH and concentrations of dissolved oxygen and other gases into equilibrium with air. The concentration of dissolved oxygen in dilution water should be between 90% and 100% saturation. Supersaturation by dissolved gases, which can occur when depuration water is heated, should be avoided. Water filtration should be avoided as bivalve mollusks are adapted to environments with particulate matter. Other factors to be especially controlled to be compatible with test organisms' preferences include temperature (in particular, temperature must not be so low as to suppress test organism metabolism), and particulate matter concentration, composition, and size.

It is important that all regions of the depuration tank experience flow. Tanks with an aspect ratio or flow intake/outtake designs that result in low flow zones are not to be used for depuration. The system for placement of test organisms in the depuration tank must also allow separation of test organisms from fecal matter settling to the bottom of the tank. Commercially available small-scale depuration systems exist, such as that shown in Figure 6.5, and may be cost-effective. Depuration tanks associated with these often have stackable baskets, with open sides and bottoms that have superior water flow characteristics, and also allow separation of test organisms from fecal matter.

The flow rate through the depuration tank is determined by test organism loading in the tank and volume of the tank. In no circumstance should test organisms be loaded into a depuration tank at greater than 50 kg/m^2; however, much lower loadings are typically needed. The depth of loading of

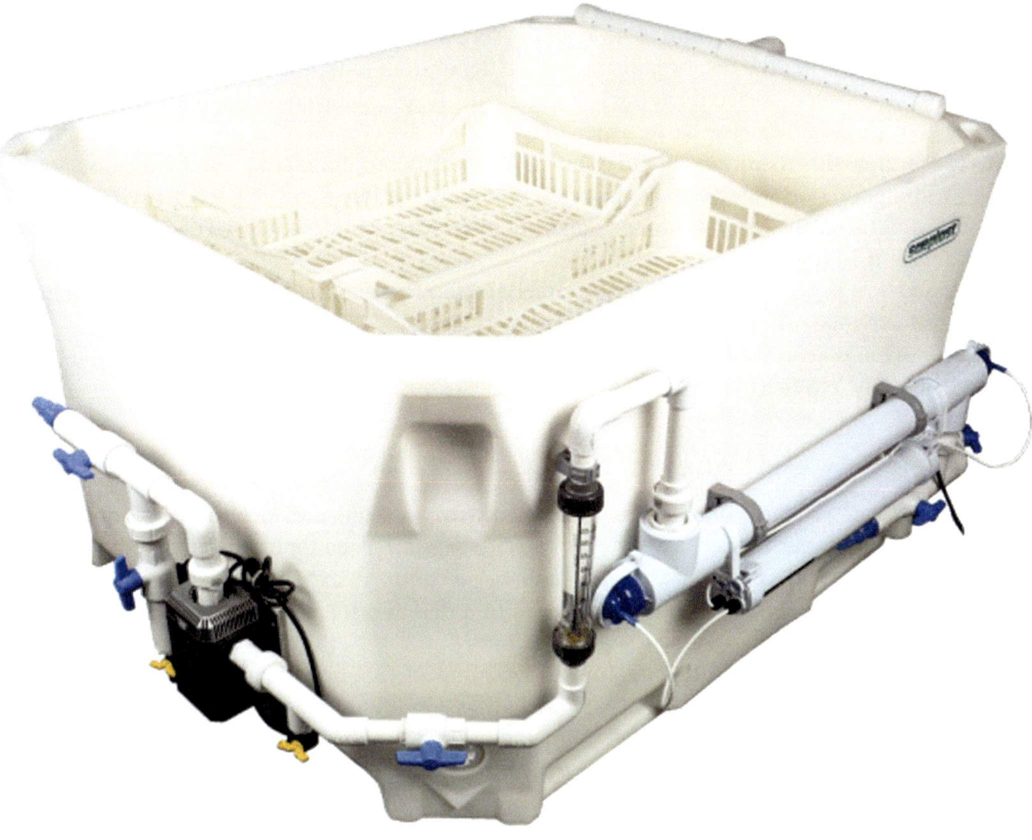

FIGURE 6.5 Example of a commercially available small-scale depuration system. Some such systems are flow-through, and some operate on a recirculating basis.

test organisms in the depuration tank must provide sufficient space to enable the shell to open fully. With these test organisms loading limits in mind, and given the volume of the depuration tank, the flow rate through the depuration tank must at minimum be 110 L/min/m^3 test organism (liters per minute per cubic meter of test organism, i.e., for a layer having 0.05 m^3 of test organisms, the minimum flow rate should be (0.05 m^3) * (110 L/min/m^3 test organism) = 5.5 L/min). In practice, flow rates may be five or ten times higher than the minimum in order to provide sufficient oxygen to test organisms. The minimum necessary flow rate might also depend on the amount of food available in the depuration water. The flow rate must not be so high that fecal or other matter settling at the bottom of the tank is resuspended. The minimum height of water above test organisms in a depuration tank is 8 cm.

Bivalve mollusks require natural light cycles such that depuration should occur under natural light, or a 16-hours light and 8-hours dark photoperiod is provided. A 15–30-minutes transition period is recommended when using artificial light, to reduce test organism stress from instantaneous changes in illumination. The depuration tank should not be placed in a location exposed to strong/direct sunshine.

During depuration, handling of test organisms should be minimized as possible in order to prevent unnecessary stress. Before loading into depuration tanks, test organisms must be cleaned if they are not already clean. This involves washing organisms in depuration water so that they are free of dirt, silt, and debris. All test organisms placed in the depuration tank must be alive. Dead organisms and empty shells can adversely affect the batch of test organisms. Only one species of test organism is to be in any batch, all test organisms in each batch must originate from the same source at the same time, and only one batch of test organisms is to be loaded in a depuration tank at any one time. A depuration time of 48 hours is standard for use of bivalve mollusks in biomonitoring. Before removing the test organisms from a depuration tank, the depuration water should be drained from the system to avoid having test organisms ingest material on the bottom of the depuration tank that might otherwise be resuspended. Small-scale commercial depuration systems have sumps to facilitate this requirement. On removal from the depuration tank, test organisms should be washed with depuration water, and reinspected, with any that are dead or have broken shells being removed from the batch. After 48-hour's depuration, test organisms should be deployed at exposure locations within 24 hours.

Purchased test organisms are much less likely to require depuration since such organisms are often raised for food. A similar process as what is described above for depuration may still be needed, however, if acclimation is required, i.e., so that the test organisms begin adaptation to water that is chemically and thermally similar to that of the exposure locations.

6.5.3 Test Organism Deployment, Exposure, and Retrieval

After transport of properly prepared test organisms to an exposure location, the organisms are placed in exposure cages for deployment. If exposure bags are used for caging, these should be constricted every 15–20 cm to prevent bunching and ensure uniform exposure of organisms to ambient water. Test organisms must be arranged in cages in a manner that does not impede the ability for shells to open.

At the time of deployment, a randomized sampling of test organisms should be collected to make measurements of "initial" conditions, including condition assessment, for control purposes. These samples represent the "i" in $N_{i/f}$ and include samples needed for measurements of (1) shell length, (2) mass, (3) condition index (CI), (4) stage in gametogenic cycle (gonadal index, or GI here, also referred to as gonado-somatic index, GSI), and (5) body burden of pollutants of interest (i.e., baseline of all pollutants scheduled for biomonitoring, j analyses), i.e., to cover measurement needs initially, $j + 4$ sets of test organisms are collected. At the time of deployment, water quality parameters are measured in the field.

Test organisms need to be protected from surface water contamination on deployment (most water bodies have a concentrated film on the surface, which may be contaminant-rich). To accomplish this, a clean sack of appropriately inert plastic may be placed over the cage, then removed once the cage is below the surface. At each exposure depth, the depth profile of the cages should not exceed 1 m of target, i.e., all organisms deployed should be within ±0.5 m of the targeted exposure depth.

Organisms are normally deployed for 1-month exposure periods. The choice of deployment date will vary geographically and according to study needs. Most importantly, using organisms that are in the process of spawning or have spawned, as reflected by a low or decreasing GI, can result in very poor measurement precision (GI and CI often correlate; e.g., Babarro and Reiriz, 2010; Brown et al., 1982). The gametogenic cycle of bivalve mollusks can be described as (1) onset of gonial mitosis, (2) active growth of gonad cells in preparation for spawning, and (3) spawning. Spawning is seasonal and hence, associated with changes in water temperature (Barrento et al., 2013; Gosling, 2003, 2015; Yang et al., 2021). Ideally, deployments occur during periods when organisms are not in the process of active gonad development or spawning, notably in autumn (Barrento et al., 2013) when mature test organisms are not spawning or readying to spawn and when water temperature is such that the test organisms are metabolically active. For some water bodies, maximum pollutant exposure can occur at higher temperatures (Baines et al., 2006; Orazio et al., 2022), however, lower and more variable pollutant load has been associated with spawning (Shindo and Otsuki, 1999). In some cases, it may be advisable to also deploy mussels a second time in colder seasons. In situations where pollutant flux is low, it may be desirable to deploy for more than 1 month if it is possible to do so during periods of the low CI associated with spawning.

Retrieval is the reverse of deployment; ideally the mussel cage is placed in a sack subsurface prior to lifting from the water in order to avoid adhesion of surface contaminants to the test organisms.

Upon retrieval, field notes on fouling should be made according to the estimated percentage of mesh opening (exposure cage or exposure bag) occluded by fouling during deployment. Fouling may interfere with mussel filter feeding, and therefore this measurement may be used to later determine if fouling had an effect on subsequent measurements.

Mortality rate must also be assessed at earliest convenience in the field and recorded in field notes. If mortality at the effects exposure location is greater than at the reference exposure location, this invalidates the measurement for the purposes of studying bioavailability via bioaccumulation due to the confounding factor of significant toxicity (albeit the study might then become about the effect of pollutant loads on observed toxicity). A simple assessment of mussel mortality is to measure all organisms with open or even slightly open shells that are unresponsive. When alive, bivalve mollusks may hold their shells open, slightly closed, or tightly closed. If the shell is open or slightly open and the test organism is alive, the shell will close after a gentle tap to the shell. If the shell does not close after a gentle tap, this is counted as a dead test organism. Mortality for organisms with tightly closed shells is harder to assess. For these organisms, a live specimen should have a fully intact shell and will sink in water. Organisms with damaged shells or that float are likely dead. Dead organisms are discarded and not part of a valid sample. After recording mortality in the field, it is subsequently calculated as % mortality according to

$$\%\text{Mortality} = \frac{N_{\text{dead}}}{N_{\text{D+A}}} \times 100\%,$$

where N_{dead} and $N_{\text{D+A}}$ are the number of dead and dead + alive (D+A) test organisms, respectively, per exposure location and depth.

After retrieval and field observations, test organisms for each exposure location and exposure depth should be placed in separate sample packages (i.e., $N_{\text{T,sample}}$ – groups of test organisms in a sample, usually either double-bagged or foil-wrapped and bagged) for interim transport back to a lab facility. It is important that organisms from different exposure locations and/or depths are not mixed to avoid any potential cross-contamination. Likewise, for different classes of analyte, j portions of $N_{\text{T,sample}}$ might be segregated, depending on analyte type (usually foil-wrapped packages in bags for those that will be analyzed for organic pollutants, and double-bagged for those that will be analyzed for heavy metal pollutants). Data for many of the condition measurements needed to reflect "final" conditions, namely, shell length, mass, and CI are obtained during laboratory processing of test organisms for pollutant analysis. The one measurement for which this is not true is GI; hence, on retrieval, for each sample, additional test organisms must also be sampled for GI measurements, hence making the k portion of $N_{\text{i/f}}$ $k = j + 4$ (initial) + 1 (final) $= j + 5$ (Section 6.4.2). The mean shell length, mass, CI, GI, and the body burden of pollutants of interest for different samples at the time of retrieval will be later compared to initial condition baselines.

Test organisms scheduled for metal analysis should be placed in clean press-seal/ziplock-style plastic bags (i.e., to ensure no transfer of moisture into or out of the bag), preferably double-bagged, and immediately placed on ice or dry ice in a thermo-insulated box. Samples scheduled for analysis of organic pollutant analytes should be double-wrapped in precleaned, hexane-rinsed aluminum foil. Aluminum foil-wrapped samples may then be placed in clean press-seal/ziplock-style bags, which are immediately placed on ice or dry ice in a thermo-insulated box. There must not be any excess water in the transport container. If the test organism is a species with a flat side and rounded side, they should be packed and transported with the rounded shell down and the flat shell up. Samples wrapped in aluminum foil are ideally transported with contiguous foil wrapping facing down.

6.5.4 Sample Preparation for Analysis

On arrival at the lab, samples are frozen at –20°C until processing for analysis. When processing begins, test organisms from a single sample are removed from the freezer, then removed from storage wrappings and, one at a time, scrubbed well with a clean implement (stainless-steel scrubber or similar) in 18 MΩ purity water to remove debris or adhered films. This cleaning should be performed while samples are still frozen to ensure no loss of analytes from test organisms during rinsing. Clean polyethylene gloves (or similar, contamination-free) should be worn. After scrubbing, any excess water is shaken off and each test organism is placed in a clean polyethylene, borosilicate glass, or stainless-steel tray (composition as appropriate for analysis type). Each group of test organisms comprising a sample field replicate (and by type of analysis, heavy metals, or organics) needs its own tray. A vernier caliper is used to measure shell length of each individual test organism for later calculation of mean shell length per sample. Allow cleaned, frozen test organisms to thaw to ~5°C and then shuck them using a clean stainless-steel shucking knife or blade; for heavy metals analysis, it is very important to touch the flesh as little as possible with the shucking implement while shucking.

For shucking oysters, place the oyster on a surface with the flat side facing up and the hinge facing the hand holding the shucking knife. Wiggle the tip of a shucking knife into the gap between the two shells, at or near the hinge – this should be a matter of finesse, not force. When the tip of the shucking knife is securely penetrating the hinge, twist the knife until the hinge pops. Turn the knife perpendicular to create space between the top and bottom shells, then return the knife parallel to the oyster and insert it and drag it across the roof of the top shell, after which it should be possible to remove the top shell. Then cut the adductor muscle that tethers the flesh to bottom shell. The flesh can then be removed and the shells scraped with a suitable implement.

To shuck mussels, clams, and scallops, a blunt tipped shucking blade is used. For mussels, take the mussel in one hand and use thumbs to press the shells sideways until a small gap appears between the shells. Insert the shucking blade into the gap and scrape along the inside of one shell to cut the adductor muscle. The mussel can now be prised open and a suitable scraping implement used to scrape the flesh out of the shells. Note: The flesh or soft tissue of bivalve mollusks includes viscera, meat, and body fluids, but not the byssal threads of mussels, i.e., for mussels, the byssal threads should be removed/excluded from the sample (Smith, 1985).

For clams and scallops, the flat side is up as for oysters, only, unlike oysters, the shucking blade should be inserted in between the two shells on the side opposite the hinge. Wiggle the shucking blade into the gap until it penetrates slightly, then begin to twist the blade to separate the two shells. Grip with fingers to stop the shell from closing. Scrape along the inside of the flat shell to cut away the adductor muscle, open and twist off the flat shell, and then use a suitable implement (spatula, small lab forceps) to remove the flesh from both shells. Figure 6.6 shows some illustrations relevant to shucking.

For heavy metals analysis, soft tissue should be collected with clean plastic/PTFE forceps and/or scraper, whereas for organic analysis, clean stainless-steel forceps/scrapers are used. Collect soft tissue for all test organisms comprising a sample field replicate and sample analysis type in a pre-weighed, clean container. All fluids accumulated during removal of soft tissue should be retained as part of the sample. The sample container should be weighed on each addition to calculate the mass of each bivalve mollusk's soft tissue, and the corresponding shell mass should be taken at the same time if CI will be calculated based on wet mass and shell mass. After collecting and pooling the soft tissue of all test organisms in a sample, pooled samples may be frozen for storage at this time (at –20°C, in a sample container of appropriate material and cleanliness for the type of analysis intended), or homogenized. The entire pollutant load in exposed bivalves is operationally bioavailable pollutant, hence freezing will not alter the measurement at this point.

After shucking and weighing, each group of test organisms comprising a sample is then composited via homogenization. For samples to be analyzed for heavy metals, the preferred approach is to place each entire pooled sample in an appropriately cleaned PTFE jar with three

FIGURE 6.6 Photos illustrating some facets of shucking test organisms. Top left: insertion of the tip of the oyster knife into hinge; a twist of the blade at this point will pop the hinge. Top middle: after the hinge is popped, the top/flat shell can be raised slightly to reach in and cut the adductor muscle to separate the top shell completely. Bottom left: in this photo, a mussel is gripped in order to slide the bottom half of the shell away from the top half; in the photo, this creates a small overhang of the top shell, into which the blunt shucking knife is inserted. Byssal threads, characteristic of mussels, are visible in the middle of the photo to the left of the thumb, just in between the grip of the index and middle fingers. All byssal threads need to be removed and discarded; these should not be including with sample soft tissue. Bottom middle: scallops usually have a slight gap opposite the hinge into which the shucking blade is inserted. Right: shucking a clam; after the shucking blade is finessed into the gap between the two shells, fingers can be used to force the shell open enough to slide the blade into position to cut the adductor muscle. Note: These example photos are relevant to shucking for food preparation. For biomonitoring purposes, all materials must be clean/contaminant-free and of an appropriate material for analysis type (i.e., gloves, holding trays, etc.).

large PTFE balls. After securely tightening a PTFE lid, each jar is placed in a trace-metal clean press-seal/ziplock bags and shaken in an industrial paint shaker or laboratory high-energy ball mill/shaker for 20 minutes. This should thoroughly homogenize the bulk sample. For organics analysis, homogenization is performed using a lab tissue homogenizer, wand type, until fully and homogenously blended. Typically, the homogenizer should come equipped with a titanium generator, suitably cleaned before use, to prevent contamination. Composite tissue homogenates prepared in either of these ways should be homogenized to a paste-like consistency. No chunks should remain in the sample because these may not be extracted or digested efficiently. At this point, the composite homogenates of samples may be processed for analysis or frozen and stored at −20°C. Representative homogenate masses needed are a minimum of 5 g for heavy metals analysis and 20–50 g for organics (refer to analytical method protocols and sample mass needed for each pollutant or pollutant class). If samples are frozen prior to further analysis, when subsequently thawed samples must be re-homogenized by stirring vigorously with a contaminant-free implement of appropriate material (e.g., stainless-steel or PTFE spatula, for organics and heavy metals, respectively).

For organics analysis, homogenates are digested directly. For heavy metals analysis, a portion of homogenate is first freeze-dried or oven-dried; this process ensures that there is no water to dilute and/or compromise the effects of acids needed to digest samples for heavy metals analysis. Freeze-drying is more commonly used in modern times; freeze-dried samples may sometimes be used for other analytes, such as neurotoxins (e.g., McCarron et al., 2007), whereas oven-drying may destroy some potential analytes. To freeze-dry (Reksten et al., 2020), the needed quantity of wet tissue homogenate sample (typically 5 g or more) is placed in a contaminant-free container of appropriate material. The depth of the sample in the container needs to be a small fraction of the height of the container. Prior to freeze-drying, a lid is placed on the sample container and samples prepared this way are frozen at –20°C for a minimum of 12 hours to ensure they are properly frozen to start freeze-drying. Subsequently, samples are placed in a freeze-dryer with lids loosened, and primary drying/ice sublimation is conducted for a minimum of 24 hours with a shelf temperature of –20°C and collector temperature of –50°C under freeze-dryer vacuum (0.2–0.01 mbar). Immediately following this, the shelf temperature is reset to +25°C, and secondary or desorption drying is conducted for a minimum of 48 hours. After removal from the freeze-dryer, the samples are immediately placed in a desiccator to avoid drawing humidity from the air. After thorough freeze-drying or oven-drying, the dry samples must be homogenized prior to analysis to ensure ***representativeness*** of freeze-dried samples. Homogenization of dry tissue homogenate is accomplished by grinding using a metals-clean mortar and pestle or other means, such as a knife mill or high-energy ball mill. Whatever method is used to achieve homogenization of dry tissue homogenates, care must be taken to ensure that all materials contacting the sample have been cleaned in between use in order to avoid contamination. When handling samples, in all cases care must be taken to use contaminant-free gloves of an appropriate material. Table 6.2 summarizes some elements of best practice for sample container materials, preservation, and hold time for storage of tissue homogenates, according to type of target pollutant, prior to laboratory analysis. Hold time indicates how long samples may be stored without invalidating results due to potential loss of analyte over long storage periods under the conditions specified.

Standard practice for reporting results for heavy metals is to report concentrations in terms of dry mass, and usually results are reported for organic pollutants as normalized to lipids. Studies have shown that gravimetry is preferred for dry mass and that freeze-drying is not the best method to determine percent dry mass (Mo and Neilson, 1992). Methods for determination of dry mass/dry-to-wet mass ratio and total lipids are given in Chapter 5.

6.5.5 Measurements of Bivalve Mollusk CI and GI

It is necessary to have measurements reflecting changes in condition of bivalve mollusks across the exposure period, because ideally, for bioavailability measurements, exposure would not result in substantive diminution of test organism health. Length of individual organisms can be measured at any time, in the lab (more conveniently) or the field, for later calculation of mean shell length, $\bar{l}$, according to

TABLE 6.2
Sample Container Materials, Preservation, and Hold Time for Storage of Tissue Homogenates Prior to Laboratory Analysis, According to Type of Target Pollutant

Sample Analysis Type	Recommended Container Material	Sample Preservation	Sample Hold Time
Mercury	PTFE (preferred), plastic, quartz	Freeze at ≤–20°C	28 days
Other heavy metals	PTFE (preferred), plastic, quartz	Freeze at ≤–20°C	6 months
Organic pollutants	Borosilicate glass or quartz (preferred), aluminum foil	Freeze at ≤–20°C	1 year
Heavy metals and organics	Borosilicate glass, quartz, PTFE	Freeze at ≤–20°C	As above for each different type of heavy metal or organic
Lipids	Plastic, borosilicate glass, quartz, PTFE	Freeze at ≤–20°C	1 year

Generally, hold times for volatiles are the same as for mercury, and for semivolatiles three times the hold time for volatiles (after US EPA, 2000).

$$\bar{l} = \frac{\sum_n l_1 + l_2 + \cdots + l_n}{n},$$

where *l*s are the lengths of each individual test organism in a sample, and *n* is the total number of test organisms in the sample. The mean wet mass, $\bar{m}_{\text{wet}}$, of test organisms is

$$\bar{m}_{\text{wet}} = \frac{\sum_n m_{1,\text{wet}} + m_{2,\text{wet}} + \cdots + m_{n,\text{wet}}}{n},$$

where *m*s are the masses of each individual test organism in a sample and *n* is the total number of test organisms in the sample, can be determined based on the wet mass of each individual test organism (determined during shucking as described above). There are many methods for determining the CI ("fatness") and GI of bivalve mollusks (for CI, see Zeng and Yang, 2021). A commonly used method for CI is to take the ratio of the wet or soft tissue mass (flesh) from each test organism, m_{wet}, to total mass, $m_{\text{wet+shell}}$, i.e., soft tissue mass + shell mass (US EPA, 1983; Zeng and Yang, 2021):

$$\text{CI} = \frac{m_{\text{wet}}}{m_{\text{wet+shell}}} \times 100\%.$$

The mean CI for a sample is then calculated based on individual CIs as for mean length and mean wet mass above. CI is an ecophysiological indicator used to assess the response of test organisms under different environmental pressures.

Bivalve mollusk GI has been determined, for instance, visually, gravimetrically, seriologically, and histological (Gosling, 2003; US EPA, 1983). Histological approaches involve gonad sampling, fixing, slide-mounting, and microscopic examination to determine stage of gonad development. The simplest approach to determination of GI is to remove and weigh the gonads, then determine GI based on the ratio of gonad mass, m_{gonad}, to total soft tissue mass, m_{wet} (as above), of each individual (US EPA, 1983):

$$\text{GI} = \frac{m_{\text{gonad}}}{m_{\text{wet}}} \times 100\%.$$

Mean sample GI is then calculated from individual GIs. Bivalve mollusk gonads may be difficult to identify, and various technical manuals with diagrams are of assistance (Howard et al., 2004). While this method is straight forward in theory, for some species, the gonad develops within the mantle, is diffusely distributed and often interwoven with other organs, and cannot be easily separated, in which case an alternative method is to geometrically estimate the gonad fraction within the gonad/mantle complex. The most widely accepted method for determination of GI in bivalve mollusk biomonitoring studies is histological (Hillman, 1993; TDI/NOAA, 2007). Many more methods, including for species other than oysters and mussels, are available in reports of various environmental agencies and in the literature (Jaramillo et al., 2022; Jiang et al., 2023; Kim et al., 2006; Urban and Riascos, 2001). The GI increases during gametogenesis and decreases during spawning; hence, if the GI has decreased substantially over the exposure period, this may reduce precision of results.

6.5.6 Chemical Analysis

Once samples for analysis are collected, processed, homogenized, dried and re-homogenized, if applicable, and stored appropriately, it remains to perform chemical analysis for analytes of interest, as determined in a Sampling and Analysis Plan. Digestion techniques for dry mass (heavy metal analysis) and extraction techniques for wet homogenates (analysis of organic pollutants) vary, however, a number of Standard Methods exist to provide guidance on best practice (some are given in Chapter 5).

Note: Due to the need to minimize the interference of water in digestion of samples for heavy metals analysis, these samples are typically digested after first freeze-drying. An exception is for the analysis of mercury. Due to the high volatility of mercury, better practice is to perform a wet digestion, followed by instrumental analysis using the cold vapor technique.

Subsequent to digestion or extraction, there are Standard Methods for instrumental analysis of bioavailable/bioaccumulated pollutant, including methods for heavy metals (As, Cd, Cr, Hg, Ni, Pb, Se, Zn), Organochlorine pesticides, organophosphate pesticides, chlorophenoxy herbicides, polycyclic aromatic hydrocarbons, polychlorinated biphenyls, dioxins/furans, and tributyl tin. Chapter 5 lists one or more Standard Methods for digestion, extraction, and analysis of these pollutants.

6.6 VALIDITY OF RESULTS – QUALITY ASSURANCE

A QA/QC plan is needed for both laboratory and field activities to ensure the quality of data collected, hence the usefulness and validity of data in environmental management and decision support (US EPA, 2000, and references therein). Natural variations in chemical and biological data collected from test organisms can be large, and rigorous QA processes (e.g., using mature organisms of a similar size, CI, and GI) reduce such variation. Factors in the field that require QA Planning include, but are not necessarily limited to, the following:

- Information on field conditions including steps to ensure that field contamination of test organisms was avoided (indicators such as descriptive statistics on baseline pollutant loads) that deployment depth and location were measured accurately and that water quality parameters at the test-organism collection site (if applicable), exposure locations, and exposure depths were measured and recorded.
- Uniformity of size of test organisms deployed (to within a minimum shell length not less than 0.75 * longest shell length for a given study).
- Condition index – CI can be highly variable; care must be taken in deployment to randomize organisms at different exposure locations and depth so that all test organisms have statistically undifferentiable CI across locations (and depths if applicable) at the time of deployment. Because CI affects test organism's ability to bioaccumulate, post-deployment differences in CI are scrutinized as part of the QC process.
- Gonad index – comments above regarding CI are also applicable to GI.
- Fouling – it can interfere with the flow of water to test organisms and adversely affect test organism condition (feeding) and/or adversely affect pollutant exposure; post-deployment differences in fouling are scrutinized as part of the QC process.
- Mortality – the purpose of quantifying mortality is to assess bioavailability via bioaccumulation; differences in mortality rates from location to location, or even sample to sample may reflect toxic response, a different type of measurement.
- The collection of specific samples for QA evaluation of field QA via laboratory analysis – the need for these samples is varied and analyte/method-specific, and therefore dictated by stipulations of Standard Methods used in analysis. Some examples of field QA samples include ***field blanks*** (materials handling, bottles, field reagents) and field replicates.

For laboratory QA, minimum requirements consist of

1. ***Initial Demonstration of Proficiency***/laboratory capability, and
2. routine analyses of appropriate QA/QC samples to demonstrate continued acceptable performance and to document data quality.

Initial demonstration of laboratory capability occurs prior to analysis of any biomonitoring program samples and includes demonstration of the ability to analyze samples for pollutants of interest within accepted ranges for ***accuracy*** and precision. This in turn involves successful instrument calibration, documentation of detection and quantitation limits, documentation of accuracy and precision, and analysis of an accuracy-based performance evaluation sample (Certified Reference Material or Standard Reference Material). Ongoing demonstration of acceptable laboratory performance and documentation of data quality includes routine calibration and ***calibration checks***, routine assessment of accuracy and precision, routine monitoring of interferences and contamination, and regular QA/QC Assessment of performance through successful analysis of accuracy-based performance evaluation samples.

Examples of additional chemical or data analyses performed in support of laboratory QA include the following (US EPA, 2000):

- Calibration assessment
  - External calibration samples (calibration standards)
  - Internal calibration (***internal standards*** in samples)
  - Calibration verification (calibration check standards)
- Method detection limit and limit of quantitation calculations
- Accuracy and precision assessment
  - ***Matrix spikes*** to assess for interferences (affects accuracy)
  - Matrix spike replicates (precision of ***spike*** measurement for interferences)
  - Laboratory replicates of homogenate from a single field sample replicate (method precision)
  - Analytical replicates (replicate instrumental analysis for the same laboratory replicate sample, analytical/instrument precision)
  - Accuracy-based performance evaluation sample (Certified or Standard Reference Material, measure for accuracy)
- Contamination assessment (blanks – e.g., method, processing, bottle, reagent)
- Loss of analyte assessment (***surrogate*** spikes, applicable for some organic pollutants)
- External QA Assessment
  - Split sample analysis, one half of split analyzed by external certified laboratory (tests for accuracy and precision).

Development and adherence to a Sampling and Analysis Plan, such as the templates provided by the US EPA (US EPA, 2014), is highly recommended in order to ensure Quality Assurance.

6.7 QUALITY CONTROL – DATA ANALYSIS, INTERPRETATION, AND REPORTING

The overall objective of most caged mollusk biomonitoring programs is to understand if pollutants, often associated with an environmental effects envelope, are significantly bioavailable, a precursor to understanding biological/ecological effects. Once all field and laboratory activities are complete, the first step toward this is data analysis, which consists at minimum of analysis for QC and statistical analysis of differences and/or trends in results. Statistical analysis typically first involves tests for normality of data and choice of appropriate parametric or non-parametric tests (e.g., *t*-test, Mann–Whitney test, ANOVA, Kruskal–Wallis test, etc.). QC requires analysis of all information that might suggest defects in the data and/or confounding factors that might affect validity and interpretation of results.

6.7.1 Quality Control, Reporting, and Interpretation of Biological Data

Fouling – reported as percent for each exposure location and depth. If differences in fouling are observed, further tests to assess if fouling has adversely affected CI or GI are needed.

Mortality – reported as percent of test organisms surviving for each exposure location and depth. As rules of thumb, if less than 90% of test organisms survive at the reference location (any depth), toxic conditions exist and results from monitoring for bioavailability/bioaccumulation are invalid. If the percent of surviving test organisms in all exposure locations/depths is different at $p < 0.05$, results from monitoring for bioavailability/bioaccumulation are invalid (US EPA, 1983).

CI – reported as percent for each sample. CI can vary considerably from location to location and as a function of depth. Test organisms with a higher CI are likely to have a higher capacity to bioaccumulate bioavailable pollutants (Pantea et al., 2020); however, under high loads of bioavailable pollutants, CI may decrease (Aguilar et al., 2012). CI is evaluated for each sample, and differences at $p < 0.05$ are flagged for comparison with fouling, GI, and chemical data to assess potential causes for differences in CI (e.g., fouling, high bioavailable pollutant load) and to assess for consequences of CI (e.g., higher bioaccumulation of bioavailable pollutant as a function of high CI).

GI – it, like CI, can vary considerably from location to location and as a function of depth. GI is another reflection of the overall physiological condition of test organisms and as such may also be affected by bioavailable toxic pollutants. GI is evaluated for each sample, and differences at $p < 0.05$ are flagged for comparison with fouling, CI, and chemical data, similar to comments above for CI.

6.7.2 Quality Control, Reporting, and Interpretation of Chemical Data

Pollutants – typically reported as mass pollutant per mass dry mass of sample (heavy metal pollutants) or mass dry mass of pollutant per mass total lipid of sample (organic pollutants). QC requirements for the analyses of pollutants is based on specific performance criteria (i.e., warning or control limits) for ***Data Quality Indicators*** such as accuracy and precision. Warning limits are numerical criteria that serve to alert data reviewers and data users that data may be of limited quality. Control limits are ***Acceptance Criteria*** – numerical data criteria that, when exceeded, render the results invalid and require ***Corrective Action***, when possible (e.g., if the original sample is sufficient and is not contaminated, reanalysis may be possible). As pertains to aspects of pollutant extraction from test organism homogenates or freeze-dried homogenates, full guidance on QA/QC (including interferences), data analysis, interpretation of results, and reporting of results are included in these methods. Analytical methods must be selected on the basis of technical merit (based on current, validated analytical techniques that are widely accepted by the scientific community) and data quality (the method detection limit, limit of quantitation, method accuracy, and method precision should all be fit-for-purpose to allow reliable quantitation of pollutant to meet the objectives of the biomonitoring program).

Water quality data – reported in standard units (e.g., °C for temperature, pH units for pH, µS/cm or mS/cm for conductivity, etc.). Large differences in water quality between test organisms' collection/raising conditions, as compared to depuration conditions (as applicable) and conditions on deployment can adversely affect test organisms. These factors should be controlled for during biomonitoring activities and verified as controlled during the QC process. Any control failure might relate to effects such as unacceptably high mortality or a consistently significant decrease in CI between deployment and retrieval at all exposure locations/depths. Assuming no control failure occurred, tests for differences in water quality data at different exposure locations/depths are needed to evaluate relationships with water quality data and all other data; any such relationship represents a possible confounding effect that must be considered in data interpretation.

Significant differences in bioavailable/bioaccumulated pollutant between the reference exposure location and one or more effects locations (subject to the ability to rule out or account for confounding factors that might exert differences on results) are a basis for assigning environmental effects.

Increase in bioavailable pollutant does not itself represent an impact. In some cases, regulatory guidance on pollutant loads in test organisms may serve as a basis for identifying adverse impacts on the basis of caged bivalve mollusk biomonitoring studies. Relationships between results from chemical analysis and biological analysis, as described here, could indicate adverse impacts. Additionally, results from bivalve mollusk biomonitoring are increasingly used in conjunction with other biological tests to develop a Weight of Evidence understanding of environmental impacts (Lehtonen et al., 2019).

6.8 CONCLUSIONS AND OUTLOOK

The technique of using caged bivalve mollusks for bioavailability monitoring has been in codified environmental practice for at least five decades. Using this approach is attractive for several reasons, importantly because this approach is capable of detecting chemicals in water that are well below the detection limits from direct instrumental analysis of water. Additional positive features are that bivalve mollusks often sequester toxins that might kill other organisms or be degraded by other organisms, and that these test organisms are also convenient for collection of a large array of other biological data, expanding the scope of understanding results in the context of impacts. The method is simple and requires only basic laboratory capabilities (samples can be sent to contract laboratories for chemical analysis), however, it is quite time- and labor-intensive in some other respects. As compared to newer and emerging techniques, for instance ***Diffusive Gradients in Thin-Films*** and ***bioreporter*** technology, the technique is very cumbersome. Because the approach is now so ingrained, with many countries having very well-established longitudinal monitoring programs based on bivalve mollusks, it is sure to remain relevant in the future.

REFERENCES

Aguilar, C.A., Montalvo, C., Rodríguez, L., Cerón, J.G., Cerón, R.M., 2012. American oyster (*Crassostrea virginica*) and sediments as a coastal zone pollution monitor by heavy metals. *Int. J. Environ. Sci. Technol.* 9, 579–586. https://doi.org/10.1007/s13762-012-0078-y

ASTM (ASTM International/American Society for Testing and Materials), 2022. *Standard Guide for Conducting Bioconcentration Tests with Fishes and Saltwater Bivalve Mollusks*. ASTM E 1022-22, ASTM, West Conshohocken, PA.

Babarro, J.M., Reiriz, M.J., 2010. Secretion of byssal threads in *Mytilus galloprovincialis*: Quantitative and qualitative values after spawning stress. *J. Comp. Physiol.* 180, 95–104. https://doi.org/10.1007/s00360-009-0392-y

Baines, S., Fisher, N.S., Kinney, E.L., 2006. Effects of temperature on uptake of aqueous metals by blue mussels *Mytilus edulis* from Arctic and temperate waters. *Mar. Ecol. Prog. Ser.* 308, 117–128. https://doi.org/10.3354/meps308117

Barrento, S., Lupatsch, I., Keay, I., Shields, S., 2013. Protocol on best practice guide on holding and conditioning mussels. Technical Report, European Commission Seventh Framework Programme (2007–2013), Project No. 243452, FP7-SME/MusselsAlive. University of Swansea, Swansea, UK.

Bendell, L.I., LeCadre, E., Zhou, W., 2020. Use of sediment dwelling bivalves to biomonitor plastic particle pollution in intertidal regions: A review and study. *PLoS One* 15, e0232879. https://doi.org/10.1371/journal.pone.0232879

Beyer, J., Green, N.W., Brooks, S., Allan, I.J., Ruus, A., Gomes, T., Lise, I., Brate, I.L.N., Schøyen, M., 2017. Blue mussels (*Mytilus edulis* spp.) as sentinel organisms in coastal pollution monitoring: A review. *Mar. Environ. Res.* 130, 338–365. https://doi.org/10.1016/j.marenvres.2017.07.024

Brown, D.A., Perkins, E.M., Jenkins, J.D., Oshida, P.S., Bay, S.M., Alfafara, J.F., Raco, V.E., 1982. Seasonal changes in mussels. In: Bascom, W. (ed.) *Southern California Coastal Water Research Project Biennial Report, 1981–1982*. Southern California Coastal Water Research Project, Long Beach, CA.

Ellis, A.E., 1962. *British Freshwater Bivalve Molluscs. With Keys and Notes for the Identification of the Species*. Linnean Society, London.

Ernst, W., 1977. Determination of the bioconcentration potential of marine organisms: A steady state approach. I. Bioconcentration data for seven chlorinated pesticides in mussels (*Mytilus edulis*) and their relation to solubility data. *Chemosphere* 11, 731 740. https://doi.org/10.1016/0045-6535(77)90144-8

Evariste, L., David, E., Cloutier, P.L., Brousseau, P., Auffret, M., Desrosiers, M., Groleau, P.E., Fournier, M., Betoulle, S., 2018. Field biomonitoring using the zebra mussel *Dreissena polymorpha* and the quagga mussel *Dreissena bugensis* following immunotoxic responses. Is there a need to separate the two species? *Environ. Pollut.* 238, 706–716. https://doi.org/10.1016/j.envpol.2018.03.098

Farrington, J.W., Tripp, B.W., 1993. International mussel watch. *Oceanus* 36, 2.

Farrington, J.W., Tripp, B.W., Tanabe, S., Subramanian, A., Sericano, J.L., Wade, T.L., Knap, A.H., 2016. Edward D. Goldberg's proposal of "the mussel watch": Reflections after 40 years. *Mar. Pollut. Bull.* 110, 501–510. https://doi.org/10.1016/j.marpolbul.2016.05.074

Geyer, H., Sheehan, P., Kotzias, D., Freitag, D., Korte, F., 1982. Prediction of ecotoxicological behaviour of chemicals: Relationship between physico-chemical properties and bioaccumulation of organic chemicals in the mussel *Mytilus edulis*. *Chemosphere* 11, 1121–1134. https://doi.org/10.1016/0045-6535(82)90122-9

Gosling, E., 2003. *Bivalve Molluscs: Biology, Ecology and Culture*. Fishing New Books, Blackwell Publishing, Oxford, UK.

Gosling, E., 2015. *Marine Bivalve Molluscs*, 2nd Edition, Wiley-Blackwell, Hoboken, NJ.

Gosset, R.W., Brown, D.A., Young, D.R., 1982. Predicting the bioaccumulation and toxicity of organic compounds. In: Bascom, W. (ed.) *Southern California Coastal Water Research Project Biennial Report, 1981–1982*. Southern California Coastal Water Research Project, Long Beach, CA.

Hillman, R. E. (1993) Gonadal index and histopathology for the East and West coasts used in the National Status and Trends Mussel Watch Project. In: Sampling and Analytical Methods of the National Status and Trends Program National Benthic Surveillance and Mussel Watch Projects 1984-1992. Vol. II Comprehensive descriptions of complementary measurements. NOAA Tech. Mem. NOS ORCA 71. NOAA/NOS/ORCA, Silver Spring, MD. 45–53.

Hornberger, M.I., Luoma, S.N., Cain, D.J., Parchaso, F., Brown, C.L., Bouse, R.M., Wellise, C., Thompson, J.K., 2000. Linkage of bioaccumulation and biological effects to changes in pollutant loads in south San Francisco Bay. *Environ. Sci. Technol.* 34, 2401–2409. https://doi.org/10.1021/es991185g

Howard, D.W., Lewis, E.J., Keller, B.J., Smith, C.S., 2004. *Histological Techniques for Marine Bivalve Mollusks and Crustaceans*. NOAA Technical Memorandum NOS NCCOS 5, NOAA Tech. Memo. NOS NCCOS 5, NOAA, Oxford, MD.

Jaramillo, R., Prida, V., Rubilar, P.S., Cardenas, L., Prieto, V., Astorga, M.P., 2022. Gonadal cycle, reproductive indices and detection of parasitism in the clam *Ameghinomya antiqua* in natural beds of importance for fisheries. *Ploe One* 17(4), e0266538. https://doi.org/10.1371/journal.pone.0266538

Jiang, W., Shi, W.J., Li, N.N., Zhang, W.K., Chen, L.Z., Liu, Q., Ye, C.Y., Fan, R.L., Quan, W.M. 2023. Gonad development and recruitment pattern of the oysters *Crassostrea ariakensis* and *Crassostrea sikamea* in a subtropical bay in China. *Regional Studies in Marine Science* 58, 102781. https://doi.org/10.1016/j.rsma.2022.102781

Kim, Y., Ashton-Alcox, K.A., Powell, E.N., 2006. *Histological Techniques for Marine Bivalve Molluscs: Update*. NOAA Technical Memorandum NOS NCCOS 27, NOAA, Silver Spring, MD.

Kim, Y., Powell, E.N., Wade, T.L., Presley, B.J., 2008. Relationship of parasites and pathologies to contaminant body burden in sentinel bivalves: NOAA Status and Trends "Mussel Watch" Program. *Mar. Environ. Res.* 65, 101–127. https://doi.org/10.1016/j.marenvres.2007.09.003

Krishnakumar, P.K., Ali Qurban, M.A., Sasikumar, G., 2018. Biomonitoring of trace metals in the coastal waters using bivalve mollusks. In: Saled, H. (ed.) *Trace Elements–Human Health and Environment*, pp. 153–176. InTech Open, London. https://doi.org/10.5772/intechopen.76938

Lehtonen, K.K., d'Errico, G., Korpinen, S., Regoli, F., Ahkola, H., Kinnunen, T., Lastumäki, A., 2019. Mussel caging and the weight of evidence approach in the assessment of chemical contamination in coastal waters of Finland (Baltic Sea). *Front. Mar. Sci.* 6, 688. https://doi.org/10.3389/fmars.2019.00688

Matthews, J., Van der Velde, G., Bij de Vaate, A., Collas, F.P.L., Koopman, K.R., Leuven, R.S.E.W., 2014. Rapid range expansion of the invasive quagga mussel in relation to zebra mussel presence in The Netherlands and Western Europe. *Biol. Invasions* 16, 23–42. https://doi.org/10.1007/s10530-013-0498-8

McCarron, P., Emteborg, H., Hess, P., 2007. Freeze-drying for the stabilisation of shellfish toxins in mussel tissue (*Mytilus edulis*) reference materials. *Anal. Bioanal. Chem.* 387, 2475–2486. https://doi.org/10.1007/s00216-006-1104-z

Mo, C., Neilson, B., 1992. A comparative study of dry weight measurements of oyster soft tissue. Data Report No. 40. Virginia Institute of Marine Science, College of William and Mary, Gloucester Point, VA.

Orazio, V.G., Baldacci, S., Desiante, F., Basile, F.C., Franco, E., Fragnelli, G.R., Diletti, G., Conversano, M., 2022. PCDD/Fs and PCBs in *Mytilus galloprovincialis* from a contaminated area in Italy: The role of mussel size, temperature and meteorological factors. *Food Addit. Contam. A* 39, 1123–1135. https://doi.org/10.1080/19440049.2022.2059108

Pantea, E.D., Oros, A., RoŞioru, D.N., RoŞoiu, N., 2020. Condition Index of mussel *Mytilus galloprovincialis* (Lamarck, 1819) as a physiological indicator of heavy metals contamination. *Ann. Acad. Rom. Sci. – Biol. Sci.* 9, 2036. https://doi.org/10.56082/annalsarscibio.2020.1.20

Phillips, D.J.H., 1980. *Quantitative Aquatic Biological Indicators – Their Use to Biomonitor Trace Metal and Organochlorine Pollution*. Applied Science Publishers, London.

Ramu, K., Kajiwara, N., Sudaryanto, A., Isobe, T., Takahashi, S., Subramanian, A., Ueno, D., Zheng, G.J., Lam, P.K.S., Takada, H., Zakaria, M.P., Viet, P.H., Prudente, M., Tana, T.S., Tanabe, S., 2007. Asian mussel watch program: Contamination status of polybrominated diphenyl ethers and organochlorines in coastal waters of Asian countries. *Environ. Sci. Technol.* 41, 4580–4586. https://doi.org/10.1021/es070380p

Reksten, A.M., Bøkevoll, A., Frantzen, S., Lundebye, A.K., Kögel, T., Kolas, K., Aakre, I., Kjellevold, M., 2020. Sampling protocol for the determination of nutrients and contaminants in fish and other seafood – The EAF-Nansen Programme. *MethodsX* 7, 101063. https://doi.org/10.1016/j.mex.2020.101063

Richir, J., Gobert, S., 2014. The effect of size, weight, body compartment, sex and reproductive status on the bioaccumulation of 19 trace elements in rope-grown *Mytilus galloprovincialis*. *Ecol. Indic.* 36, 33–47. https://doi.org/10.1016/j.ecolind.2013.06.021

Richman, L., Somers, K., 2005. Can we use zebra and quagga mussels for biomonitoring contaminants in the Niagara River? *Water Air Soil Poll.* 167, 155–178. https://doi.org/10.1007/s11270-005-0083-6

Ruiz-Velásquez, M., Zapata, M., Gonzalez, M.T., Ordenes, D., Escalona, M., 2017. Sexual differentiation and size at first maturity of the mussel *Choromytilus chorus* (Molina, 1782) (Mollusca, Bivalvia) in Northern Chile. *Am. Malacol. Bull.* 35, 31–41. https://doi.org/10.4003/006.035.0104

Shindo, K., Otsuki, A., 1999. Establishment of a sampling strategy for the use of blue mussels as an indicator of organotin contamination in the coastal environment. *J. Environ. Monit.* 1, 243–250. https://doi.org/10.1039/a809669d

Smith, R.L., 1985. Guidance on sampling aquatic organisms for tissue analyses during FY 1986. Environmental Services Division, Region 3. United States Environmental Protection Agency, Philadelphia, PA.

TDI/NOAA (TDI-Brooks International Inc. and National Oceanic and Atmospheric Administration. (2007). *Standard Operating Procedure for Gonadal Index Analysis in Bivalves*. www.tdi-bi.com/analytical_services/environmental/NOAA_methods/s&tmethod-gonad.pdf. TDI-Brooks International Inc., College Station, TX.

Urban, H.J., Riascos, J.M., 2001. Estimating gonado-somatic indices in bivalves with fused gonads. *J. Shellfish Res.* 21, 249–253.

US EPA (United States Environmental Protection Agency), 1983. *Methods for Use of Caged Molluscs for In Situ Biomonitoring of Marine Sewage Discharges*, Weber, C.I. (ed.). Environmental Monitoring and Support Laboratory, US EPA, Cincinnati, OH.

US EPA, 1990. *Work Plan for FY 91 Regional Ambient Fish Tissue Monitoring Program Activity No. ELR 80*. US EPA Environmental Monitoring and Compliance Branch, Region 7, Kansas City, KS.

US EPA, 2000. *Guidance for Assessing Chemical Contaminant Data for Use in Fish Advisories: Volume 1, Fish Sampling and Analysis*, 3rd Edition. EPA 823-B-00-007, US EPA, Washington, DC.

US EPA, 2004. Method 3052: Microwave assisted acid digestion of siliceous and organically based matrices. EPA SW-846, Revision 3. US EPA, Washington, DC.

US EPA, 2014. *Sampling and Analysis Plan – Guidance and Template V.4 – General Projects*. R9QA/009.1. US EPA Office of Environmental Information, Washington, DC.

Van Gestel, C.A.M., Otermann, K., Canton, J.H., 1985. Relation between water solubility, octanol/water partition coefficients, and bioconcentration of organic chemicals in fish: A review. *Regul. Toxicol. Pharmacol.* 5, 422–431. https://doi.org/10.1016/0273-2300(85)90007-8

Wisconson Department of Natural Resources (WDNP), 1988. Fish contaminant monitoring program – Field and laboratory guidelines. Report No. 1005.1., WDNR, Madison, WI.

Yang, H.S., Yoon, S.J., Kang, D.H., 2021. The annual gametogenesis of the mussel, *Mytilus unguiculatus* (*coruscus*) (Valenciennes, 1858), occurring in the shallow subtidal of Ulleungdo Island in Korea. *Ocean Sci. J.* 56, 287–296. https://doi.org/10.1007/s12601-021-00021-3

Zeng, Y., Yang, H., 2021. Review of molluscan bivalve condition index calculations and application in Northern Quahogs *Mercenaria mercenaria*. *Aquac. Res.* 52, 23–36. https://doi.org/10.1111/are.14866

Zieritz, A., Sousa, R., Aldridge, D.C., Douda, K., Esteves, E., Ferreira-Rodríguez, N., Mageroy, J.H., Nizzoli, D., Osterling, M., Reis, J., Riccardi, N., Daill, D., Gumpinger, C., Vaz, A.S., 2022. A global synthesis of ecosystem services provided and disrupted by freshwater bivalve molluscs. *Biol. Rev.* 97, 1967–1998. https://doi.org/10.1111/brv.12878

7 Using Worms for the Measurement of Pollutant Bioavailability

7.1 INTRODUCTION AND SCOPE

Rapidly expanding industrial manufacturing activities resulted in the release of large quantities of heavy metals and organic pollutants into soils and sediments with increasing levels of pollution, posing significant environmental and human health risks (Khan et al., 2021; Kumar et al., 2022). Globally, about 30,000 tons of chromium (Cr) and 800,000 tons of lead (Pb) have been released and mostly accumulated in soils over the last 50 years (Zhao et al., 2022). In addition, varying degrees of heavy metal pollution with cadmium (Cd), Cr, and Pb have been detected in the sediments of major rivers of the world, with high levels of ecological risk (Debnath et al., 2021). Soil and sediment are also long-term reservoirs of persistent pollutants, with the presence of 96 compounds identified in 13 soil samples and three sediment samples from Switzerland alone (Chiaia-Hernández et al., 2020). For example, atrazine, a widely used herbicide, was present in soil/sediment in different parts of the world at levels up to 2,400 and 980 ng/g, respectively (Chang et al., 2022). Worms, because they are in many places ubiquitous in soil and sediment, are strong natural candidates for detecting the ***bioavailability*** of heavy metals and organic pollutants (Degtyarev et al., 2024; Fründ et al., 2011; Mwevura et al., 2020; Xing et al., 2023; Zhang and Van Gestel, 2017). Figure 7.1 shows examples of three different types of worms that are at the heart of the bioavailability method in this chapter.

Worms are widely distributed in various ecological environments. ***Earthworms*** are found in terrestrial soil, polychaetes are found predominantly in marine sediments, and enchytraeids are found in both terrestrial soils and brackish sediments (Degtyarev et al., 2024; Gajda et al., 2017; Hutchings, 1998), which makes it easy to use worms in many environmental pollution monitoring areas. Moreover, the special physiological structure and lifestyle of worms enable them to respond quickly to environmental changes. For example, oligochaetes have a thin and permeable epidermis, which facilitates their uptake of petroleum hydrocarbon pollutants (Veena et al., 2023). In contrast, ***polychaete worms*** are rapidly responsive to pollutants on their body surface and gills, among others (Mwevura et al., 2020). As shown in Table 7.1, this responsiveness to, and at the same time ability to survive, pollutant ***exposure*** coupled with a wide ***distribution*** means that worms can be used as a generalized indicator organism capable of detecting the bioavailability of pollutants in different geographic areas.

As examples of the use of worms in bioavailability studies, Šmídová and Hofman (2014) reported that the uptake of PAHs in different soils by earthworms (*Eisenia fetida*) can be used to evaluate the bioavailability of PAHs. Another study showed that by analyzing the types and levels of heavy metals in the body of ***enchytraeid worms*** after exposure to polluted soil, the degree of pollution and effect of aging could be assessed (Zhang and Van Gestel, 2017). Polychaete worms are one of the most common and species-rich invertebrates in the world (Sizmur et al., 2013). They have been shown to be important ***components*** involved in ***biomagnification*** of chemicals in a variety of food webs in the oceans, and are therefore well suited for use in pollutant bioavailability studies. It has been reported that polychaete worms have been used as benthic ***test organisms*** for ***biomonitoring bioavailable*** pollutants, including metals and Persistent Organic Pollutants (POPs), in sediments (Méndez et al., 1998; Thathola et al., 2025).

DOI: 10.1201/9781003355229-9

FIGURE 7.1 Photographs showing three different types of worms discussed in this chapter: (a) polychaetes, (b) enchytraeids, and (c) earthworms.

TABLE 7.1
Representative Studies Using Worms to Detect the Bioavailability of Pollutants

Worm Type	Species	Pollutants Detected	Detection Matrix	Reference
Earthworm	*Eisenia fetida*	As	Soil	Xing et al. (2023)
		Lindane		Miao et al. (2018)
Enchytraeid	*Enchytraeus crypticus*	Cd, Cu, Pb, Zn	Soil	González-Alcaraz and van Gestel, (2015)
	Enchytraeus crypticus	Pb		Zhang and Van Gestel, (2019)

(Continued)

TABLE 7.1 (*Continued*)
Representative Studies Using Worms to Detect the Bioavailability of Pollutants

Worm Type	Species	Pollutants Detected	Detection Matrix	Reference
	Cognettia sphagnetorum, Cognettia glandulosa	PAHs		Uffindell et al. (2005)
Polychaete	*Armandia agilis, Branchiomma* sp., *Eunice* sp., *Goniada echinulate, Laeonereis acuta, Lumbrineris* sp., *Scolelepis chilensis, Scolelepis* sp., *Terebellidae gen.* sp.	Hg	Sediment	Sola et al. (2022)
	Capitellidae spp., Corophium volutator, *Glyceridae* spp., *Maldanidae* spp., *Nereidae* spp., *Orbiniidae* spp., *Paraonidae* spp., *Phyllodocidae* spp., *Spionidae* spp.	Hg		Sizmur et al. (2013)
	Eurythoe complanata	Cd, Co, Cr, Cu, Mn, Ni, Pb, Zn		Méndez et al. (1998)
	Capitella capitata	Organochlorine pesticides, PAHs		Mwevura et al. (2020)

This chapter will describe how to obtain and use worms for measuring the bioavailability of heavy metals and organic pollutants in soil/sediment. The methods and steps needed for biomonitoring studies are described in detail, including the acquisition, cultivation, and ***acclimation*** processes to be used for different worms, the collection of field samples, uptake and ***elimination phase*** work, and tissue sample collection and handling. Chapters containing Terms and Definitions, guidance on avoiding contamination of samples in field and laboratory work, methods of environmental sample extraction and instrumental analysis, and guidance on Quality Assurance (QA)/Quality Control (QC) are also within the scope of this chapter.

7.2 PRINCIPLE

In this method, either earthworms, enchytraeid worms, or polychaete worms, used as ***test organisms***, are procured, cultivated, or collected and prepared, and the soils/sediments needed for controls and as test samples are collected in the natural environment. The test organisms are placed in containers with collected soil/sediment and undergo ***exposure***, i.e., the process wherein the test organisms are in contact with bioavailable pollutants in soil/sediment, and bioavailable pollutants are ***bioaccumulated*** into the test organisms for a period. After this uptake phase, subsequently, test organisms are removed and placed into separate containers to eliminate their gut contents (***elimination phase***), so that only bioavailable pollutants in test organism tissues are collected. After the elimination phase, test organism tissue is collected, then processed via extraction and instrumental analysis to determine the bioavailable pollutants that were bioaccumulated during exposure.

7.3 EQUIPMENT, CONSUMABLES, AND REAGENTS

The list of equipment, consumables, and reagents given in this section is intended to be fairly comprehensive; users should, however, read the entire method before finalizing needs. For example, if a user intends to study soils, a grab sampler to collect sediments is not needed. Similarly, there may be some items of small labware that might be useful and are not specifically mentioned. Users will wish to modify this indicative list accordingly.

7.3.1 Equipment and Consumables

For this method, it is important to properly organize both field and laboratory and operations in order to minimize the danger of sample contamination or loss of target pollutants in samples. Before being used, all field and laboratory equipment and consumables need to be thoroughly cleaned and decontaminated. Field gear also needs to be carried and stored in a way that keeps it decontaminated after cleaning. Carrying decontaminated equipment in pristine, sealed containers is the standard stowing procedure. Refer to Chapter 3 for additional information on decontamination.

Field-related:

Auger and/or shovel – for soil sampling.
Buckets – for surface water grabs, and if a sonde is not being used, to take surface water samples and assess the water quality at field ***sites*** by taking measurements of temperature, pH, conductivity/salinity, and other determinands.
Corer or grab sampler – used for sediment collection for work involving sediments; also used to collect sediment from worm collection site if worms are being collected in the field.
GPS/electronic or physical devices – to ensure the exact siting of sample locations.
Field logbook – to record information about pH, temperature, and events that may affect the results of the experiment at the site of sample collection or in the lab to assist in subsequent analysis of the results.
Glass or rigid plastic containers – for collection of field soil/sediment samples.
Indelible pens – to record relevant data without losing fidelity in wet field environments.
Muffle furnace – for high-temperature decontamination of glassware (organic pollutants).
Multiparameter water quality meter or sonde – used to assess the water quality at sites by taking measurements of temperature, pH, conductivity/salinity, and other determinands.
Plastic bags, several sizes, resealable/watertight (press-seal/ziplock type, polyethylene) – for storing forms and field records and for secondary containment of smaller sample containers.
Sample containers – suitable material (such as stainless steel or polytetrafluoroethylene [PTFE]) that is clean and free of ***contaminants***; different sizes.
Solvents and/or cleaners – as needed for decontamination of field equipment.
Tank(s) – made of appropriate, contaminant-free material, for collecting seawater samples required for polychaete worm exposure experiments.
Thermo-insulated box with ice packs, ice, or dry ice – clean, free of contaminants, and made of an appropriate material for the target pollutant(s); must be able to keep samples and/or test organisms at ambient temperature while being handled and transported.

Test organism processing and sample measurement-related:

Aquarium tanks and aerators – used in transport, storage, and in exposure experiments and during maintenance to provide sufficient oxygen for the survival of polychaete worm test organisms.
Autoclave – used to sterilize the food of test organisms.
Balance – to measure the masses of reagents, soil/sediments, wet and dry worms/soft tissue.
Beaker and watch glasses or similar containers – clean and sterile, made of suitable material for target pollutants; used to hold experimental soft tissues, reagents, and/or as containers for worm uptake and elimination phase work.
Freeze-dryer or lab oven – used to dry soil/sediments and/or wet tissues for dry-to-wet weight analysis and other uses.

Glass or Teflon rod(s) – clean, devoid of contaminants, used for mixing tissue and/or soil/sediments.

Incubator or environmental chamber – used to control temperature in worm exposure and elimination phase work; optional for laboratory environments that have a reasonable level of temperature control.

Lab tissue homogenizer – for homogenizing tissue samples in preparation for instrumental analysis.

Medicine spoon – made of clean and contaminant-free materials (PTFE, stainless steel, etc.); used for handling soft tissue.

pH meter – for testing pH.

Pipettes and/or syringes – For accurately measuring and transferring small volumes of liquids, including reagents and calibration standards, for diluting samples, or adding precise amounts of water to soil or sediment samples to maintain appropriate moisture levels.

Plastic bags, several sizes, resealable/watertight (press-seal/ziplock type, polyethylene) – for secondary packaging to avoid cross-contamination of samples.

Sample containers – suitable material (such as stainless steel or PTFE) that is clean and free of contaminants; different sizes, suitable for the preparation of test organism samples/tissue homogenate samples and for lab processing of soil/sediment and other biological samples (lipids, dry weight).

Scalpel, stainless steel or Teflon-coated – used for dissection.

Soil/sediment mixing tool – to homogenize soil or sediment samples; e.g., rolling mill, feed mixer, or gardener's soil mixer (e.g., DaRos MX1000 or similar).

Small lab forceps – clean and free of contaminants, made with the proper material (PTFE, stainless steel, etc.) for the target pollutants under study.

Specimen trays – clean and devoid of contaminants, made of the correct materials for the analysis of target pollutants and big enough to accommodate worms during various processing steps.

Thermometers – for monitoring temperatures.

Personnel protection-related:

Laboratory gloves – contaminant-free, with good acid, alkali, and corrosion resistance, and effective protection against the hazards of organic solvents.

Field clothing items – items such as waders that come into contact with water or test organisms need to be appropriately cleaned, to be devoid of contaminants, and made of materials suitable for the pollutant(s) being targeted.

7.3.2 Reagents

Note: All reagents must be of guaranteed reagent (GR) quality, or better to ensure that the reagents used are of the highest purity to avoid contamination.

Certified calibration standards – suitable for target pollutant.

Cultivation media, earthworms – a 50:50 mixture of horse or cattle manure:peat, a commercial gardening soil free of additives, an artificial soil, or a 50:50 mixture of commercial gardening soil: artificial soil.

Ethyl alcohol, 75% – used to disinfect labware when autoclaving is not required.

Fish food, chopped clam, chopped shrimp, or other food suitable to feed polychaete worms.

Liquid nitrogen or freezer – used to kill worms for subsequent analysis.

Manure – organic/contaminant-free; used to feed earthworms.

Rolled oats – milled and autoclaved to prevent mite infestation. Used to feed enchytraeid worms.

Certified Reference Material or Standard Reference Material – suitable for environmental ***medium*** (soil/sediment or tissue) and target pollutant.
Ultrapure water – commonly known as Type I water, with a resistance of 18 MΩ·cm.
Vitamin protein powder – used as a nutrient ***source*** for culturing enchytraeid worms.

Additional reagents, and potentially equipment, will be needed to extract pollutants from worms in preparation for instrumental analysis; considerations associated with this process are discussed in Chapter 6, and some ***Standard Methods*** of extraction are given in Chapter 5. Some instrumental methods have specific guidelines for extraction. For organics, Soxhlet extraction is a well-known approach, and for heavy metals, acid digestion is used.

7.4 PROCEDURE

7.4.1 Acquisition and Preparation of Test Organisms

The availability, ***sensitivity*** to pollutants of interest, tolerance to test conditions (temperature, ***salinity***, soil/sediment grain size), and ease of handling in the laboratory should all be taken into consideration when choosing a test species of worm to be used in studies on bioavailable pollutants. The species used should be known with assurance/taxonomic authority for the purposes of reporting.

Commonly used terrestrial oligochaetes include earthworms (more commonly used for soils) and enchytraeids (able to tolerate some salinity, depending on species). Species commonly used are *E. fetida*, *Eisenia andrei*, *Enchytraeus albidus*, *Enchytraeus crypticus*, and *Enchytraeus luxuriosus*. Another, less commonly used, type of worm used for bioavailability testing is polychaete worms. Polychaetes are a robust and widespread type of worm in marine environments, across all ranges of temperature and depths, making them particularly well suited to studies on marine sediments. This type of worm is uncommon in freshwater environments. Many different species are used, top examples of which include *Nereis succinea*, *Capitella capitata*, and *Hediste diversicolor* (Annabi et al., 2025; Checon et al., 2021; Driscoll and McElroy, 1996).

Purchasing test organisms is appealing for a number of reasons. Aside from saving many person-hours in cultivation, vendors often know what species they are selling, and, for vendors who raise the worms they are selling, commercial organisms may have a tight size distribution and be effectively the same age. These advantages will not necessarily be captured with every vendor, and it is worthwhile to make detailed enquiries. Worms obtained commercially cannot be assumed to be contaminant-free. For users wishing to purchase worms for bioavailability experiments, potential sources include the following:

- Biological supply companies – There are companies that specialize in providing marine organisms for research and educational purposes. These suppliers often offer a variety of worms and marine invertebrates, including polychaetes, for purchase. In the USA, prominent biological supply companies include Carolina Biological Supply Company, Ward's Science, and Aquatic Biosystems.
- Research institutions and universities – Some research institutions or university biology/marine biology departments may be able to provide worms for experimental use.
- Local fishing shops – Local suppliers who sell worms for fishing bait and/or fish food.
- Online suppliers – There are online platforms and websites that specialize in providing live organisms for scientific research and educational purposes. Because of the difficulty of importing organisms in most countries, an internet search for such companies should target the country in which studies will be performed.
- A number of specialist providers to the aquaculture industry have begun to cultivate polychaete worms, providing a good source if available.

All individual test organisms in a given study should be adult organisms, from the same source, and procured or collected at the same time in order to have similar sensitivities to ***bioaccumulation*** of target bioavailable pollutants. For a given species, the age or size class should be selected so that factors such as seasonality, reproduction, maturity state, etc. do not affect the sensitivity to target pollutants. The age and size of test organisms should be as similar as possible. Worms used in bioavailability testing must be free from observable diseases, abnormalities, and parasites. In the case of cultivated worms, this is normally achieved through control of cultivation conditions. In the case of purchase or collection, this must be assessed at the time of receipt or during processing after collection.

Perhaps the most common species in use for bioavailability work are *E. fetida* and *E. andrei*. These species are now distributed worldwide and are often commercially available as fish bait. After hatching from cocoons, these worms reach maturity within ~2–3 months. They do well on moist substrates with a pH ~6–7, high content organic material substrate, and a temperature range of 20–25°C (OECD, 1984; Venter and Reinecke, 1988). Recommended cultivation media include (1) a 50:50 mixture of horse or cattle manure:peat, (2) a commercial gardening soil free of additives, (3) artificial soil, or (4) a 50:50 mixture of commercial gardening soil:artificial soil (ISO, 2023; OECD, 1984). Caution is advised in obtaining contaminant-free manure, e.g., from an organic source. Cultivation media should have a low ionic conductivity (less than 6 mS/cm or less than 0.5% salt concentration), should not be contaminated with pollutants, and should not have high amounts of ammonia or urine. Worms can be kept in boxes of 10–50 L volume. As it is important to have individuals of similar age and mass, if worms meeting these criteria cannot be purchased, it may be worthwhile for some studies to cultivate worms.

To cultivate, adult worms are added to a box containing fresh substrate to produce cocoons. Approximately 100 adult worms per kg substrate (wet weight – cultivation medium must be kept moist) have been shown to lead to good reproduction rates. After 28 days, the adult worms are removed, and the earthworms hatched from cocoons are used for testing when mature, after at least 2 months but less than 12 months. *E. fetida* and *E. andrei* worms are considered healthy if they move through the substrate, do not try to leave the substrate, and reproduce continuously. Very slow motioning or a yellow posterior end is an indication of substrate exhaustion, which calls for either the addition of fresh cultivation substrate and/or a reduction of the number of worms per box. Feeding fresh manure should be performed periodically (weekly), and substrate should be amended or refreshed as indicated by the appearance of worm health.

Like earthworms, enchytraeid worms are distributed worldwide. *E. albidus* is found in marine, limnic, and terrestrial habitats; which diversity of habitat is highly desirable for bioavailability studies. Like earthworms, *E. albidus* is commercially available as fish food. The same provisos concerning the use of commercially available organisms, as above, apply. Other species of the *Enchytraeus* genus may be used, examples of which include *E. luxuriosus* (Bruns et al., 2001) and *E. crypticus* (Zhang and Van Gestel, 2017).

These species are of different size (with implications as to how many worms are needed for tests) and mature at different rates. For cultivation of *E. albidus* (analogous approaches are used for other species), large adult specimens are selected to start a new culture. *E. albidus* reaches maturity in 33 days at 18°C; thus, for cultured worms only cultures which have been kept in the laboratory for at least 5 weeks (one generation) without problems should be used for a test. *E. albidus* (as well as other *Enchytraeus* species) can be bred in large plastic boxes (1–2 L volume) filled with a mixture of artificial soil (OECD, 2010) and commercially available, uncontaminated garden soil free of additives. Compost material should be avoided due to the frequent presence of contaminants in compost. Before use, fauna that might be present in cultivation media should be removed by deep-freezing and thawing the cultivation media three times. The cultivation medium should be kept moist, but not wet, and at a temperature of 15°C–20°C without light (OECD, 2010). These worms typically prefer pHs of 5–6, however, if test exposure conditions will involve samples with very different pHs than this, the effect of tolerance of different pHs should be tested on adult worms first.

Anoxic conditions must be avoided (e.g., if a lid is used, the number of lid holes should be high enough to provide sufficient exchange of air), and the cultivation media should be aerated by carefully mixing it once per week. The worms should be fed at least once per week with rolled oats that have been autoclaved, ground, and then stored in sealed vessels to avoid infections by mites. The oats are placed into a cavity on the soil/sediment surface and covered with soil/sediment, to be routinely checked for rate of disappearance (sufficiency), and topped up when at or near exhaustion. Other food sources include baker's yeast or the fish food TetraMin. If food from the last feeding date remains after one week, the amount of food given should be adjusted accordingly, and any old food not exhausted should be replaced by new. The rolled oats may be supplemented with commercially available, vitamin-amended protein powder every 2 weeks. For longer term maintenance, worms are transferred to a freshly prepared cultivation medium after three months. Enchytraeid worms are considered healthy if they move through the substrate, do not try to leave the substrate, reproduce continuously, exhibit a shiny outer surface without soil/sediment particles clinging to it, and are more or less whitish colored.

In recent years, some companies have begun to cultivate certain species of polychaete worms as an aquaculture food source; however, most bioavailability studies use worms gathered from the marine environment. Purchase or cultivation of worms is always preferable, both as a matter of best practice in QA and also to avoid ecological damage, however, while it is possible to cultivate polychaete worms (Kannappan et al., 2021), the process is typically such that collection or purchase is the primary avenue for such experiments.

Collecting worms from the natural environment is to be avoided for most users. It is important to know definitively what the species is and what the local, regional, and global conservation status of the species is (ensuring only to collect species that are very common), and to adhere to all applicable laws and policies safeguarding ecological standards; this level of specialist knowledge and expertise will be lacking for many users. If polychaetes cannot be procured and therefore need to be collected from the field, expert advice should be sought to ensure the above provisos are fulfilled, and collection should occur as soon as possible prior to use. Worms must be collected from relatively contaminant-free waters/sediments. If the contaminant concentration of a chemical analyte/pollutant in organisms at the collection site is unknown, this should be assessed through prior collection and testing. Next, it is important to verify the species of the organisms to be collected and ensure that a single species can be identified and collected. Care is needed to ensure that the test organisms are not contaminated during collection, which means that all materials used should be clean and free of contamination. Cleaning protocols for laboratory equipment used in contaminant analysis should be followed whenever possible to maximize the cleanliness of field equipment. Organisms should be collected from a single, relatively well-defined source area at the same time and at the same or similar depth. Under no circumstances should organisms be collected from near-steel, wood, or other man-made structures. A shovel can be used to collect polychaetes intertidally, and they can be collected by grab or scuba diving subtidally. Wading and/or diving should be done by trained personnel or scientific scuba divers (clean wading shoes and boots required as appropriate to ensure no contamination; the same is true for scuba gear and grab equipment). The least amount of and most gentle handling should be done with polychaetes. When handling is required, it should be done swiftly, carefully, and gently to avoid undue stress on the organisms.

When being transported to the laboratory, polychaetes should be kept in the sediment in which they were collected at a temperature that is the same or similar to that of the collection site. Aeration may be required during transportation. Sediments taken from the collection site on the day of collection must be analyzed to confirm the absence of significant levels of pollution, and a large amount of seawater must be collected. The ambient seawater from the collection site will be used in holding of worms and also for separation. Sediment containing worms may be stored in aquarium-style tanks topped with ambient seawater and equipped with aeration. In most cases, when holding under these conditions, feeding polychaetes over a 10-day period is not required. Over an extended amount of time feeding may be required; many species of polychaete worm are famous for having a broad diet. Many types of fish food, chopped clam, or chopped shrimp are acceptable for food. Ideally, test conditions should mimic conditions under which polychaetes were collected (ambient seawater and temperature). Healthy polychaetes may either stay on the sediment's surface or burrow into it, however, will be active in either case.

7.4.2 Number of Test Organisms Needed

For many pollutants, the amount of tissue sample needed for instrumental analysis is on the order of 2 g, wet weight. In the case of *Eisenia*, a single adult worm might weigh this under some growth conditions. For many species of *Enchytraeus*, the individual weight is very low, e.g., possibly on the order of 10 mg wet weight per individual for *E. albidus*. This entails that 200 worms or more may be needed for a single sample. Users should (1) consult the Standard Method to be used for instrumental analysis of the pollutants of interest to verify how many grams tissue is needed per sample (see Chapter 5) and then (2) ascertain the average grams wet weight per worm for the species to be used. The number of worms used will depend on these two factors, and additionally the number of replicates and QA/QC samples needed (Section 7.5) and how many worms are needed for dry-to-wet weight and lipids analysis used in reporting. Lipids are variable from worm to worm and can have a strong effect on uptake of target pollutants; therefore, the percent lipids is mandatory for the purposes of reporting. Further, a reasonable rule of thumb is to plan for approximately one-third more worms than is expected to be needed based on samples, replicates, and QA/QC samples. Once these factors are known, the number of worms needs is calculated as

$$n_{\mathrm{Total}} = 1.33 \times \left(\mathrm{ceiling}\left(w_{\mathrm{sample}} / w_{\mathrm{worm}}\right) \times \left(n_{\mathrm{samples}} \times n_{\mathrm{reps}} + n_{\mathrm{QA/QC}} + n_{\mathrm{bio}} + n_{\mathrm{bkg}}\right)\right),$$

where *ceiling* is a mathematical function designating that a quotient is rounded up to the nearest integer value, w_{sample} is the wet weight tissue needed for each sample to meet the needs of instrumental analysis, w_{worm} is the average weight of an adult worm of the species under the growth conditions to be used in the test, n_{samples} is the number of test samples, n_{reps} is the number of replicates for each sample, $n_{\mathrm{QA/QC}}$ is the total number of QA/QC samples that will be needed, n_{bio} is the number of worms needed for replicate biological measurements (dry-to-wet weight measurement, lipids analysis), and n_{bkg} is for replicate analysis of background concentrations of pollutants in worms, i.e., samples of worms taken for background analysis at the beginning of uptake phase work.

7.4.3 Collection and Handling of Soil/Sediments

A minimum of two types of test soil/sediment are needed, one of which must be a control (clean/contaminant-free). The test soil/sediment should have physical characteristics that enable test organisms to survive for the duration of the test. Test organisms need to burrow into/not avoid the test soil/sediment. Chemically, the sample(s) on which bioaccumulation tests will be done must not be toxic to test organisms; otherwise, the test is a toxicity, rather than bioavailability, test. Natural soil/sediment from unpolluted sites may serve simultaneously as both a control and culture or holding soil/sediment.

Because of the manner in which soil/sediments are spatially and temporally variable (e.g., Stemmer et al., 1990), maintaining the *in situ* integrity of soil/sediment during collection, transport, and laboratory testing is extremely difficult. The sampling process itself often causes disruption of, for instance, redox gradients, rates or amounts of microbial degradation, chemical ***sorption***, and ***partitioning***. Replicate samples should be collected in all instances to determine variance in soil/sediment characteristics. While samples should be collected with as little disruption as possible, nonetheless, subsampling, compositing, or homogenization of samples will be necessary for many experimental designs. Coring or digging is used for soils. For sediments, coring or a benthic grab, see Figure 7.2, is preferable to sampling dredged material. Whichever type of sampler is used, decontamination and collection of rinsate ***blanks*** must be performed between each use.

Coring sediment samplers consist of a core tube or liner and a weight used to push the core tube into the sediment. The sediment sampler is dropped into the water and then relies on its own weight to enter the sediment to a target depth and then sealing the sample in the sampler by means of a valve at the bottom of the sampler. The sampler is then withdrawn back from the bottom of the water column, ensuring the integrity of the sample (Tuit and Wait, 2020).

Grab samplers can also be used, but only if the water being sampled is still or slow-moving. Grab samplers are not recommended for use in strong currents or deep waters. The sampler must be open on the deck of the vessel when the sampler is dropped and the jaws of the sampler should close, collecting sample, on impact with sediments. Once the sample is loaded, the sampling device should be raised slowly from the bottom to avoid having collected sediments wash out as the sampler rises through the water column (Tuit and Wait, 2020).

Exposure to direct sunlight during collection should be minimized, especially if the sediment contains photolytic compounds. Sediment samples should be cooled to 4°C in the field before transport, with ice, ice packs, or dry ice; however, samples should never be frozen (ASTM, 1999; US EPA, 1999). Soil/sediment samples should be stored at 4°C in the dark and used in bioavailability testing as soon as possible after collection. Samples should not be stored unused for more than two weeks (ASTM, 1999). Even short storage periods may change the concentrations of bioavailable pollutants.

Use of control soil/sediment is best practice and may fulfill a variety of purposes including to have a comparator for the health and quality of the test organisms, to ensure that test conditions (e.g., organism handling and environmental parameters) are suitable for the test, and to provide a sort of baseline for interpreting results. When collecting soil/sediment for the control, an appropriate reference site must be selected. Soil/sediment parameters should be as closely matched as possible (see notes on characterization below) to expectations of soil/sediments from test field sites. Otherwise, differences between these parameters between the control and test soil/sediments may influence the

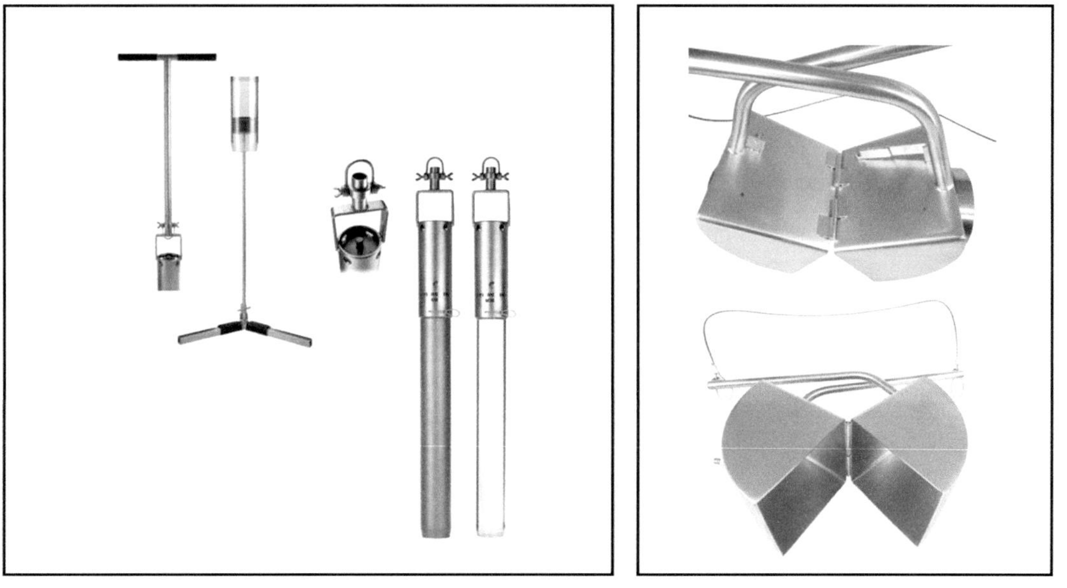

FIGURE 7.2 Sediment samplers: (a) sediment corer and (b) grab sampler.

outcome of the test. It is also important to select an appropriate reference site for obtaining a control that is "clean". In practice, fulfilling all test requirements is impossible, and so it is a matter of making the best choice possible. Presampling/analysis of soil/sediments from sites where test samples and the proposed reference site for the control soil/sediment are to be collected is fundamental to determining whether, or to what extent, an adequate reference site exists for the collection of control soil/sediments.

Soil/sediment samples should be transported to the laboratory as expeditiously as possible, in insulated boxes cooled to 4°C, and need to be processed in the laboratory before use. Soil/sediment, particularly sediment, tends to settle during shipment, sometimes separating pore waters from solid material. Separated water should not be discarded but should be mixed back into soil/sediment. Soil/sediment should be mixed prior to use by stirring or using a rolling mill, feed mixer, or other suitable apparatus. Care should be taken to mix the soil/sediment without altering its physicochemical characteristics substantively. Samples should not be sieved unless indigenous organisms that might influence the response of the test organism are present, in which case such organisms should be removed, preferably by a manual method such as forceps, if possible. If samples must be sieved, this may affect soil/sediment chemistry, which must be discussed as a study limitation in reporting.

All soil/sediment used should be characterized by, at minimum, location, depth, and other characteristics of location site (environmental attributes); qualitative sample information (color, texture, odor, etc.); pH; macronutrient levels (nitrogen, phosphorous); organic carbon content; particle size distribution (percent sand, silt, and clay); and bulk density. Further, the control and soil/sediment in which organisms are cultivated (if different from the control) must be analyzed for the background concentrations of pollutants of interest, i.e., to ensure that the baseline level of pollutants in this material is low. Additional attributes that might be desirable to measure include maximum water holding capacity (for soils), percent water content, cation exchange capacity, exchangeable cation concentrations (potassium, calcium, sodium, magnesium), salinity (as assessed by electrical conductivity), biological oxygen demand, chemical oxygen demand, Eh, total inorganic carbon, total volatile solids, acid volatile sulfides, metals, synthetic organic compounds, oil and grease, petroleum hydrocarbons, and interstitial water analyses for selected physicochemical parameters of interest. In the case of sediment collection, the bottom temperature, salinity, and sediment temperature should be recorded in addition to post-collection characterization of the sediment samples themselves. All equipment must be fully decontaminated before use and between each use, and ***equipment rinsate*** blanks must be taken after each decontamination. Measurement of soil/sediment properties and instrumental analysis of sediments (Chapter 5) are outside the scope of this method; standalone Standard Methods for these measurements should be used as a matter of best practice.

7.4.4 Preparation for Uptake Phase Experiments

Often, test conditions differ from worm cultivation, holding, or collection conditions. In this or other circumstances where there is to be an alteration between cultivation or holding conditions and test conditions, worms must be conditioned or acclimated prior to using in testing. For collected worms, i.e., polychaetes, the worms are gently rinsed from collected sediment with a continuous stream of collected seawater and gently removed as encountered. It may be possible to manually pick other worms from cultivation or holding media. The acclimation media should be a clean soil/sediment substrate that resembles the substrates to be used in testing. Once the acclimation containers are prepared, worms are gently removed from their cultivation or holding containers, rinsed gently in water ambient to the conditions in which they have been living, and are then transferred to the clean acclimation soil/sediment. Food should be provided. Worms are left in acclimation conditions for 24–72 hours.

7.4.5 Prescreening Experiments

Users undertaking complex field experiments or experiments that are not repeatable in case of substantive problems in implementation should consider optional prescreening experiments. The purpose of these is to assess the reaction of test organisms to exposure, and, in the case of organic pollutants, to account for metabolic capabilities of the test species toward the target pollutant, in other words, whether the test organism will experience adverse effects from planned or representative levels of exposure to target pollutants, and whether the test organism metabolizes organic pollutants rapidly or not. Obviously, rapid metabolism is incompatible with a long uptake phase of experiments, in which case the uptake kinetics might be studied during the prescreening uptake phase in order to determine a more suitable time period for uptake. Typically, prescreening experiments are conducted with more limited samples, e.g., representative high and low exposure scenarios, and may involve more replicates so that samples can be taken at different times in the uptake phase.

7.4.6 Process for Uptake Phase Experiments

For uptake phase experiments, individual containers for controls and samples must be prepared. A minimum of three replicates is required for each control or sample, though four or five would be preferable; replicate refers to sample replicates – i.e., enough worms may be collected from one container and aliquoted separately for three replicates (alternately separate identical treatment containers per replicate is possible as well). For each test container, a high soil/sediment-to-worm ratio should be used in order to minimize the decrease of the test substance concentration in soil/sediment during the uptake phase. For most experiments, a wet weight of soil/sediment ~200 times the wet weight of worms to be used is adequate. The soil/sediment vessels should contain soil/sediment at minimum thick enough for worms to burrow (~2–5 cm depending on the size of the worm). Soil/sediment must be moist, but not waterlogged, i.e., worms need air, except if polychaetes are the test organism species, in which case the solid material in test containers needs ample overlaying water (on the order of three times the depth of water as the depth of solids at the bottom of the containers). In this case, the overlying water must be kept oxygenated, and the usual way is by introduction of a small air stone bubbler/aerator in each container. For all worm experiments, a cover is needed for each container; soil moisture will evaporate and bubblers will increase evaporation – for earthworms and enchytraeids, a cover with holes to ensure good air exchange is appropriate, whereas a watch glass or similar is typically sufficient as a cover for polychaetes when aerators are used. Ideally, uptake phase work should be conducted in an incubator or a constant-temperature laboratory environment to control the temperature to a constant value. Figure 7.3 shows a schematic illustration of control and sample sets for different worm uptake experiments.

Once testing containers are prepared, worms are removed from holding or acclimation containers, rinsed carefully as for transfers discussed above, and weighed before they are added to the test soil/sediment containers. Prior to weighing, excess water should be removed from the worms by gently touching them against the edge of a dish/beaker or by blotting them cautiously dry by using a slightly moistened paper towel. It is very important that worms are not blotted with a dry towel or dry matter. This may damage a worm. After weighing, worms are added to test containers at the rate of nreps × *ceiling* (w_{sample}/w_{worm}) per container (if one test container is to be used for all replicates) or *ceiling* (w_{sample}/w_{worm}) per container if using separate containers for each replicate as shown in Figure 7.3; the latter is generally preferable. Ideally, each sample has the same number of worms and as close as possible to the same total weight of worms. Initial measurements and sampling that must be done at this time includes, at minimum, measurement of soil/sediment temperature and pH, sampling of soil/sediment samples for determination of initial pollutant concentration, and sampling for worm wet weight/dry weight, lipid content, and for background pollutant content. Additionally, all test vessels with either earthworms or enchytraeids need to be weighed to track/maintain moisture content during experiments. For the biological observations (dry-to-wet weight

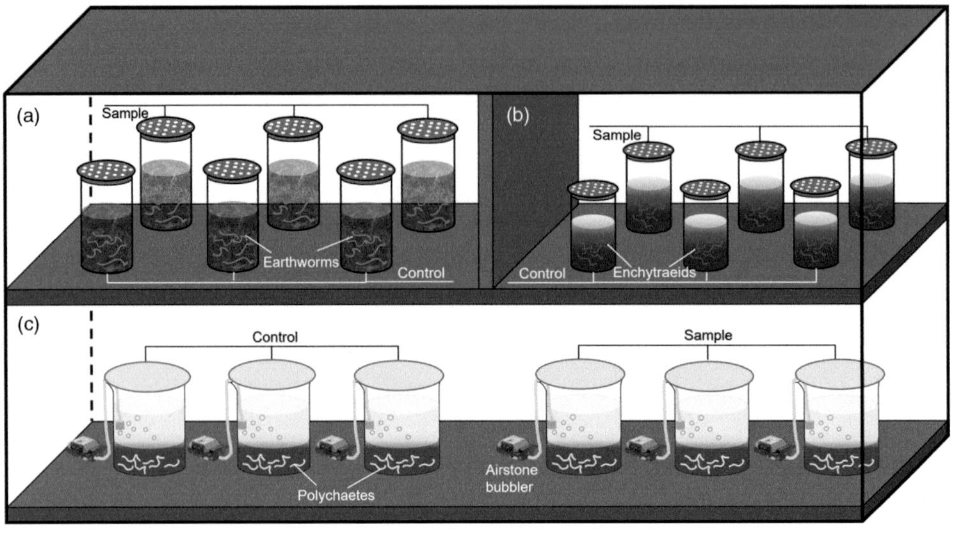

FIGURE 7.3 Schematic illustration of control and test sample replicates for different worm pollutant uptake experiments: (a) earthworms, (b) enchytraeid worms, and (c) polychaete worms. Ideally, these experiments are performed in an incubator, or at least a fairly constant-temperature laboratory environment.

ratio, lipid content) and for the analysis of background concentrations of target pollutants, which is needed for both worms and soil/sediment, at least four replicate measurements are required for each, and these measurements must be made both immediately before the uptake experiments and immediately after the elimination phase. Methods for the determination of dry-to-wet weight ratio and lipids are given in Chapter 5.

For tests involving soil/sediment with low organic content, and depending on the length of exposure time, feeding may be necessary. Any test vessels to receive food rations should have an initial ration at this time. Feeding, when done, should consist of the same type and amounts as in cultivation/holding (Section 7.4.1). Ambient light is appropriate (16/8 hours light/dark), and temperature should be regulated to a constant level that is consistent with the needs of the test species used.

After worms are added to test vessels, burrowing behavior of the test organisms should be observed and recorded. Earthworms normally burrow within a period of a few hours; this should be checked. For bioavailability studies using earthworms, if more than ~10% fail to burrow for as much as half of the uptake phase, this indicates a problem with the test organisms or conditions and invalidates results. Enchytraeids and polychaetes have somewhat different behavior, and failure of these test organisms to burrow in itself does invalidate the test. These organisms, however, also need to be monitored daily for signs of ***activity***. If 10% or more of test organisms appear inactive or with compromised health, or if there is a substantive difference between the control and test containers, this invalidates the test.

The uptake phase of experiments typically lasts on the order of 21 days for earthworms and 14 days for enchytraeids. For polychaetes the uptake phase time is more variable, with 6–18 days being reasonably representative (Driscoll and McElroy, 1996). During experiments, temperature and worm behavior should be recorded daily. On a weekly basis, food should be replenished and moisture/waterchecked. For earthworms and enchytraeids, this involves weighing test vessels and adding back enough Type I water to restore the total vessel weight to the weight at the initiation of uptake experiments. For polycheates, collected seawater can be added to beakers according to the level present at the beginning of uptake experiments. Under particularly hot and/or dry conditions, moisture levels may need to be monitored and adjusted more frequently. At the end of the uptake period, temperature, pH, and soil/sediment moisture content (as applicable) must be measured again, along with taking soil/sediment samples from each container for measurement of target pollutant.

7.4.7 Elimination and Post-elimination Processing

After uptake, worms need to purge their gut content before sampling of tissues. Worm mortality should be recorded at the beginning and the end of the elimination phase for later use (*NB*: Beginning of elimination phase ~ end of uptake phase). Implementing the elimination phase involves removing dead worms, removing and cleaning live worms, as in prior worm manipulation steps, and placing the cleaned worms either in containers with moistened filter papers (well-moistened, for earthworms and enchytraeids) or in beakers with ambient seawater and no sediment (polychaetes). In all cases, media in elimination containers (i.e., filters or seawater) should be changed periodically (at least twice during elimination) in order to prevent reingestion of excreted material. Earthworm gut content should be nominally purged in 24 hours and certainly by 48 hours; Enchytraeids will take less time (12–24 hours), and 12 hours has been used for polychaetes (Arnold and Hodson, 2007; Díaz de Alba et al., 2016; Giessing et al., 2003; Santos et al., 2024). After elimination, the wet weight of the worms for each control or sample is determined and then the worms are killed immediately, e.g., using liquid nitrogen or freezing at temperatures below −18°C. Samples of worm tissue should be processed for instrumental analysis within one week. At the end of the elimination phase, samples should also be taken for lipid content, dry-to-wet weight ratio, and for QA/QC analyses. For specific extraction and sample analysis methods for pollutants in worms, please refer to Chapter 5.

7.5 VALIDITY OF RESULTS, DATA ANALYSIS, INTERPRETATION, AND REPORTING

7.5.1 Notes on QA/QC Planning Requirements

Chapter 4 provides detailed guidance on QA/QC that is applicable to this chapter, and Chapter 5 gives information useful in deciding upon methods of tissue and soil/sediment extraction and methods of instrumental analysis needed to quantify bioavailable target pollutants; these Standard Methods will have their own QA/QC requirements. Specific QA/QC information pertinent to this chapter and not already discussed above or found in these other Standard Methods is as follows:

- All analysts performing this method must undergo an ***Initial Demonstration of Proficiency*** (IDP). For this method, IDP consists of being able to demonstrate proficiency in extraction and subsequent instrumental analysis to obtain suitable results for Certified or Standard Reference Materials, i.e., obtaining results that fall within accepted ranges for ***accuracy*** and ***precision*** for analysis of soil/sediments and biological tissues, with respect to target pollutants.
- Routine instrumental calibration and ***calibration blank*** checks are required; guidelines for ***Acceptance Criteria*** for these are stipulated in the appropriate Standard Methods used for instrumental analysis.
- Samples that are analyzed as having concentrations at or near the upper limit of instrument calibration must be reprocessed and reanalyzed. For this method, this should only involve dilution – dilution should always be conducted at a factor of 10 or less.
- Batch sizes for instrumental analysis shall not exceed 20 samples; samples batched include sediment and tissue digests/extracts and replicate samples; QA/QC samples are additional and should not exceed 5 per batch. Instrument check samples assess calibration stability and are not part of batching.
- Sample order in batches is to be randomized over all samples and sample types.
- One ***field split sample*** should be analyzed for every 25 field soil/sediment samples, and one laboratory split sample should be analyzed for every 25 laboratory tissue samples.

- For every batch, one randomly chosen equipment rinsate blank, OR one equipment rinsate blank per replicate sample, is required, whichever is smaller. These samples enable evaluation of whether field contamination occurred.
- For every group of soil/sediment and tissue samples extracted at the same time, one ***procedural blank*** is required. The contents of the procedural blank will be determined by the methods of extraction and instrumental analysis used.
- One ***recovery spike*** is required per batch. This enables evaluation of whether artifactual contamination (recoveries significantly above 100%) or analyte loss (recoveries significantly below 100%) is occurring.
- One ***matrix spike*** is required per batch unless otherwise indicated by the instrumental method used. Results from this indicate the short-term role of matrix in affecting bioavailability.
- One matrix ***spike*** duplicate and one recovery spike duplicate per instrument run (i.e., assumes multiple batches run sequentially) OR one matrix spike duplicate and one recovery spike duplicate every 12 hours is required, whichever is more frequent, or unless otherwise indicated by the instrumental method used.
- One ***reagent blank*** for analysis type (soil/sediment, tissue, and pollutant) is required for every group of sample extracts prepared on the same day, with the same technique, and with the same reagent batches.
- Reference material analyses are used to validate the accuracy of laboratory analytical methods. It is recommended that analytical runs be performed every day, however, guidelines for instrumental methods may supersede guidelines here. If the analytical results do not meet the accuracy criteria for the reference material and method, there is an operator QA/QC issue, and ***Corrective Action*** is needed.

7.5.2 Data Analysis, Interpretation of QA/QC Results (Acceptance Criteria), Corrective Action, and Reporting

Data need to be analyzed, interpreted in the context of QA/QC requirements, and if needed subjected to Corrective Action prior to reporting. These include, at minimum, the following:

- Mortality or problems with test organisms – mortality of control and sample worms and any observed abnormal behavior must be reported. A significantly lower mortality of worms in the control invalidates the test (i.e., implies toxicity). If the overall mortality during uptake and elimination phase exceeds 10% for earthworms or 20% for enchytraeids or polychaete worms, this invalidates the test.
- The dry-to-wet weight ratio of soil/sediment and the test organisms must be reported.
- Wet weights of worms at start of the test immediately prior to uptake and immediately following elimination should be reported. While the latter may be less than the former, a significant drop is indicative of a problem with the test. A mean mass loss of more than 20% over the uptake phase invalidates the test.
- The lipid content of the test organisms must be reported, and, for organic pollutants, the final tissue concentrations of target pollutants should be reported in absolute units and normalized to lipid content.
- Dilution factors (DFs) – unless specified by an instrumental method, DFs are never greater than a factor of 10 for dilution of standards and samples. The basic dilution formula is $C_1V_1 = C_2V_2$. The dilution factor is given by DF = V_2/V_1.
- ***Calibration checks*** and metrics – perform calibration checks according to directions specified in the method used for instrumental analysis. This will include guidance on using certified standards and ensuring regular check intervals. Metrics such as linearity, sensitivity, and precision are assessed for stability according to Acceptance Criteria given in methods

of instrumental analysis. If instability occurs, Corrective Action may be taken. Only data collected under validated instrumental calibration conditions (stability of calibration) should be reported. Should the metrics associated with the instrumental ***calibration curve*** not meet requirements/Acceptance Criteria, or if calibration checks fail, as stipulated in the Standard Method chosen for instrumental analysis, it is mandatory to take Corrective Action to establish or re-establish proper calibration. Any data collected under failed calibration is not reportable.

- Precision for replicates – report precision of replicates as relative standard deviation (RSD), where RSD = stdev × 100%/average. Acceptable RSDs for instrument replicate analysis should always be below 10%, with Acceptance Criteria for pollutants analyzed in sediments or tissues being larger and more variable. Poor precision for replicates using this method (>20% RSD) should be investigated. If the cause of the poor precision is not discovered, Corrective Action is to reanalyze samples if enough sample material is available.
- Equipment, procedural, and reagent blanks – results for blanks must be calculated and reported in units of absolute concentration (not bioavailable) for all blank samples. The acceptance criterion for blanks is a value below the ***limit of detection*** (LOD), and it should be readily possible to achieve this metric with this method. Blanks above the LOD are indicative of field contamination (***equipment blanks***) or laboratory contamination (procedural blanks). In either case, no Corrective Action is possible, and all samples must be regarded as potentially contaminated. In the case of procedural and reagent blanks, if enough sample is available, samples integrity may allow reanalysis. If routine analysis of a small sampling of equipment blanks reveals contamination, then all equipment blanks must be analyzed and descriptive statistics performed to assess degree of contamination. If the contamination is effectively constant around an average value that is low, compared to field samples, then this should be reported with results and considered in interpretation of results. For highly varying contamination, in the case of field sample contamination, reporting is not possible and field work must be repeated, taking stronger measures for cleaning of field equipment and/or replacement of equipment parts responsible for contamination.
- LOD and LOQ (***limit of quantification***) – these critical parameters define the lowest concentration of heavy metal that can be reliably detected (LOD) and quantified (LOQ) using a given method of instrumental analysis. LOD is often defined as three times the standard deviation of the calibration blank divided by the slope of the calibration curve. LOQ is often defined as three times the LOD. Standard Methods for instrumental analysis commonly contain guidelines concerning what LOD and LOQ are typical. Substantial deviation from such reported guidelines may typically reflect issues with user competency in performing analysis. LOD and LOQ must be reported, as they reflect upon competence of analysis and interpretation of results (e.g., lower precision typically occurs between the LOD and LOQ).
- Split samples (lab and field) – precision is reported as relative percent difference (RPD) and RPD = 200% × abs $(R_1 - R_2)/(R_1 + R_2)$, where subscripts 1 and 2 represent the two duplicate samples. Precision acceptance is no greater than 20% RPD for sediment samples and no greater than 15% RPD for tissues. Corrective Actions include reviewing the sample handling and preparation processes to ensure consistency.
- Recovery spikes and matrix spikes – these samples and their method of preparation are discussed in more detail in Chapter 4. Acceptable recovery spikes for this method should be stated in the Standard Methods for extraction and instrumental analysis used. If recovery spike values fall outside of this, Corrective Action involves reviewing and adjusting the sample preparation and analytical procedures to identify potential sources of error.
- Recovery and matrix spike duplicates – these samples should be evaluated by RPD for the duplicate samples, with results reported and explained. The recovery spike duplicate acceptance criterion is 15% RPD or less. Matrix spike duplicates may be higher if ***matrix effects*** are severe.

7.6 CONCLUSIONS AND OUTLOOK

The use of worms of one species or another to study bioavailability, via bioaccumulation, has been in practice as early as the beginning of the 20th century (Smith, 1915). The long-recognized role of earthworms, in particular, in soil health, and their often high thresholds of toxicity with respect to many pollutants have made them, and more lately enchytraeids and polychaetes, valuable cost-effective and ecologically relevant indicators of soil and sediment pollution. As with any method, despite the commanding position of the advantages of using worms in bioavailability studies, there are a number of disadvantages. Uptake of pollutants may be variable, with some pollutants less suitable for worm studies, depending on the chemical behavior of the pollutant. The measurement is, by nature, a time-integrated measurement. This is advantageous for some applications, but not others. Notably, capturing field conditions accurately in a laboratory is unrealistic. In contrast, however, field studies of *in situ* worms present difficulties in interpretation due to issues with ***representativeness*** (widespread in worm age distribution, lipid content, health status, etc., from sampling site to sampling site). Overall, the advantages of using worms as a relatively simple way to measure true bioavailability in the environment far outweighs disadvantages, ensuring that this method will continue to be highly relevant in the future.

REFERENCES

Annabi, A., Ben Ameur, W., Akermi, N., Marini, M., 2025. Monitoring and assessment of the trace element accumulation in the *Polychaete Hediste* diversicolor from Tunisian Coastal Localities (Southwest of Mediterranean Sea). *J. Mar. Sci. Eng.*, 13(7), 1353. https://doi.org/10.3390/ jmse13071353

Arnold, R.E., Hodson, M.E., 2007. Effect of time and mode of depuration on tissue copper concentrations of the earthworms *Eisenia andrei*, *Lumbricus rubellus* and *Lumbricus terrestris*. *Environ. Pollut.* 148(1), 21–30. https://doi.org/10.1016/j.envpol.2006.11.003

ASTM (American Society for Testing and Materials), 1999. Standard guide for collection, storage, characterization, and manipulation of sediments for toxicological testing. E1391-94. In: *Annual Book of ASTM Standards, Vol. 11.05*. ASTM, Philadelphia, PA.

Bruns, E., Egeler, P., Roembke, J., Scheffczyk, A., Spoerlein, P., 2001. Bioaccumulation of lindane and hexachlorobenzene by the oligochaetes Enchytraeus luxuriosus and Enchytraeus albidus. *Hydrobiologia* 463(1–3), 185–196. https://doi.org/10.1023/A:1013159810067

Chang, J., Fang, W., Chen, L., Zhang, P., Zhang, G., Zhang, H., Liang, J., Wang, Q., Ma, W., 2022. Toxicological effects, environmental behaviors and remediation technologies of herbicide atrazine in soil and sediment: A comprehensive review. *Chemosphere* 307, 136006. https://doi.org/10.1016/j.chemosphere.2022.136006

Checon, H.H., Corte, G.N., Silva, C.F., Bícego, M.C., Amaral, A.C.Z., 2021. Using the Capitella complex to investigate the effects of sympatric cryptic species distinction on ecological and monitoring studies in coastal areas. *Mar. Biodivers.* 51(3), 1–15. https://doi.org/10.1007/s12526-021-01185-w

Chiaia-Hernández, A.C., Scheringer, M., Müller, A., Stieger, G., Wächter, D., Keller, A., Pintado-Herrera, M.G., Lara-Martin, P.A., Bucheli, T.D., Hollender, J., 2020. Target and suspect screening analysis reveals persistent emerging organic contaminants in soils and sediments. *Sci. Total Environ.* 740, 140181. https://doi.org/10.1016/j.scitotenv.2020.140181

Debnath, A., Singh, P.K., Chandra Sharma, Y., 2021. Metallic contamination of global river sediments and latest developments for their remediation. *J. Environ. Manage.* 298, 113378. https://doi.org/10.1016/j.jenvman.2021.113378

Degtyarev, M.I., Saifutdinov, R.A., Korobushkin, D.I., Bastrakov, A.I., Danilova, M.A., Davydov, I.D., Gorbunova, A.Y., Guseva, P.A., Karlik, E.I., Koshchanova, R.E., Kuznetsova, K.G., Lebedev, I.M., Medvedev, D.A., Obolenskiy, R.R., Popova, A.V., Pronina, N.A., Rybalov, L.B., Surov, A.V., Tadzhimov, A.B., Tarasov, A.I., Vasiliev, V.A., Zaitsev, A.S., Zvychaynaya, E.Y., Gongalsky, K.B., 2024. Taxonomic diversity and abundance of enchytraeids (*Annelida*, *Clitellata*, *Enchytraeida*) in the Northern Palaearctic. 1. Asian part. *Biodivers Data J.* 12, e114682. https://doi.org/10.3897/BDJ.12.e114682

Díaz-de-Alba, M., Huerta-Diaz, M.A., Delgadillo-Hinojosa, F., Hare, L., Galindo-Riaño, M.D., Siqueiros-Valencia, A., 2016. Trace metals partitioning among different sedimentary mineral phases and the deposit-feeding polychaete *Armandia brevis*. *Sci. Total Environ.* 543, 248–266. https://doi.org/10.1016/j.scitotenv.2015.11.033

Driscoll, S.K., McElroy, A.E., 1996. Bioaccumulation and metabolism of benzo[A]pyrene in three species of polychaete worms. *Environ. Toxicol. Chem.* 15(8), 1401–1410. https://doi.org/10.1002/etc.5620150820

Fründ, H.C., Graefe, U., Tischer, S., 2011. Earthworms as bioindicators of soil quality. In: Karaca, A. (ed.) *Biology of Earthworms. Soil Biology*, vol. 24. Springer, Berlin, Heidelberg.

Gajda, Ł., Gorgoń, S., Urbisz, A.Z., 2017. Food preferences of enchytraeids. *Pedobiologia* 63, 19–36. https://doi.org/10.1016/j.pedobi.2017.06.002

Giessing, A.M.B., Mayer, L.M, Forbes, T.L., 2003. 1-hydroxypyrene glucuronide as the major aqueous pyrene metabolite in tissue and gut fluid from the marine deposit-feeding polychaete *Nereis diversicolor*. *Environ. Toxicol. Chem.* 22(5), 1107–1114. https://doi.org/10.1002/etc.5620220519

González-Alcaraz, M.N., van Gestel, C.A.M., 2015. Climate change effects on enchytraeid performance in metal-polluted soils explained from changes in metal bioavailability and bioaccumulation. *Environ. Res.* 142, 177–184. https://doi.org/10.1016/j.envres.2015.06.027

Hutchings, P. 1998. Biodiversity and functioning of polychaetes in benthic sediments. *Biodivers. Conserv.* 7, 1133–1145. https://doi.org/10.1023/A:1008871430178

ISO (International Standard Organization), 2023. *Soil Quality – Effects of Pollutants on Earthworms*, 3rd Edition, No. 11268. ISO, Geneve.

Kannappan, S., Sivagnanam, S., Poochirian, J.K., Praveena, P.E., Balasubramanian, C.P., Kizhakedath, V.K., 2021. Development of indoor grow-out practices for polychaete, *Marphysa gravelyi* with a note on biochemical composition. *Aquac. Res.* 52(2), 4278–4287. https://doi.org/10.1111/are.15266

Khan, S., Naushad, M., Lima, E.C., Zhang, S., Shaheen, S.M., Rinklebe, J., 2021. Global soil pollution by toxic elements: Current status and future perspectives on the risk assessment and remediation strategies – A review. *J. Hazard. Mater.* 417, 126039. https://doi.org/10.1016/j.jhazmat.2021.126039

Kumar, V., Pandita, S., Setia, R., 2022. A meta-analysis of potential ecological risk evaluation of heavy metals in sediments and soils. *Gondwana Res.* 103(7), 487–501. https://doi.org/10.1016/j.gr.2021.10.028

Méndez, N., Páez-Osuna, F., 1998. Trace metals in two populations of the fireworm *Eurythoe complanata* from Mazatlán Bay: Effect of body size on concentrations. *Environ. Pollut.* 2–3(102), 279–285. https://doi.org/10.1016/S0269-7491(98)00061-X

Miao, J., Chen, X., Xu, T., Yin, D., Hu, X., Sheng, G.D., 2018. Bioaccumulation, distribution and elimination of lindane in *Eisenia foetida*: The aging effect. *Chemosphere* 190, 350–357. https://doi.org/10.1016/j.chemosphere.2017.09.138

Mwevura, H., Bouwman, H., Kylin, H., Vogt, T., Issa, M.A., 2020. Organochlorine pesticides and polycyclic aromatic hydrocarbons in marine sediments and polychaete worms from the west coast of Unguja island, Tanzania. *Reg. Stud. Mar. Sci.* 36, 101287. https://doi.org/10.1016/j.rsma.2020.101287

OECD (Organization for Economic Cooperation and Development), 1984. *Test No. 207: Earthworm, Acute Toxicity Tests*. OECD Guidelines for the Testing of Chemicals, Section 2, OECD Publishing, Paris.

OECD, 2010. *Test No. 317: Bioaccumulation in Terrestrial Oligochaetes*. OECD Guidelines for the Testing of Chemicals, Annex 4: Artificial Soil – Preparation and Storage Recommendations, OECD Publishing, Paris.

Santos, F.C.F., Verweij, R.A., Soares, A.M.V.M., Scott-Fordsmand, J.J., van Gestel, C.A.M., Amorim, M.J.B., 2024. Multigenerational exposure of Ag materials (nano and salt) in soil – environmental hazards in Enchytraeus crypticus (Oligochaeta). *Nanoscale Adv.* 6(3), 826-831. https://doi.org/10.1039/D3NA00487B

Sizmur, T., Canário, J., Gerwing, T.G., Mallory, M.L., O'Driscoll, N.J., 2013. Mercury and methylmercury bioaccumulation by polychaete worms is governed by both feeding ecology and mercury bioavailability in coastal mudflats. *Environ. Pollut.* 176, 18–25. https://doi.org/10.1016/j.envpol.2013.01.008

Šmídová, K., Hofman, J., 2014. Uptake kinetics of five hydrophobic organic pollutants in the earthworm *Eisenia fetida* in six different soils. *J. Hazard. Mater.* 267, 175–182. https://doi.org/10.1016/j.jhazmat.2013.12.063

Smith, H.L., 1915. The accumulation of heavy metals in earthworms. *J. Agric. Res.* 29, 123–135.

Sola, M.C.R., de Jesus, R.M., Nascimento, M.M., da Rocha, G.O., de Andrade, J.B., 2022. Occurrence of mercury in *Polychaete* species (Annelida) and their associated sediments from an important Southern Atlantic Ocean Bay. *Sci. Total Environ.* 851, 157965. https://doi.org/10.1016/j.scitotenv.2022.157965

Stemmer, B.L., Burton, G.A., Sasson, B.G., 1990. Effect of sediment spatial variance and collection method on cladoceran toxicity and indigenous microbial activity determinations. *Environ. Toxicol. Chem.* 9(8), 1035–1044. https://doi.org/10.1002/etc.5620090810

Thathola, P., Bhowmik, M., Haldar, S., 2025. Tracking toxic threats: Bioaccumulation of persistent organic pollutants in benthic macroinvertebrates of the Gulf of Khambhat, India. *Mar. Pollut. Bull.* 218, 118207. https://doi.org/10.1016/j.marpolbul.2025.118207

Tuit, C.B., Wait, A.D. 2020. A review of marine sediment sampling methods. *Environ. Forensics* 21, 291–309. https://doi.org/10.1080/15275922.2020.1771630

Uffindell, L.A., Ineson, P., Meharg, A.A., 2005. Enchytraeid worms retard polycyclic aromatic hydrocarbon degradation in a coniferous forest soil. *Soil Biol. Biochem.* 37(1), 27–34. https://doi.org/10.1016/j.soilbio.2004.07.023

US EPA (United States Environmental Protection Agency), 1999. Methods for collection, storage, and manipulation of sediments for chemical and toxicological analyses. US EPA, Washington, DC.

Veena, V., Kathireswari, P., Preethee, S., Saminathan, K., Deepthi, M.P., 2023. Detoxification of petroleum hydrocarbon contaminated soil through vermicomposting and its impact on gut profile of earthworm *Eudrilus eugeniae*. *Water. Air. Soil Pollut.* 234(3), 199. https://doi.org/10.1007/s11270-023-06212-z

Venter, J.M., Reinecke, A.J., 1988. The life-cycle of the compost-worm *Eisenia fetida* (Oligochaeta). *S. Afr. J. Zool.* 23, 161–165. https://doi.org/10.1080/02541858.1988.11448096

Xing, W., Geng, H., Wang, Y., Zhao, L., Yang, Y., Wang, Y., Tian, S., Cao, Y., Zhang, Z., Li, L., 2023. Accumulation and speciation of arsenic in *Eisenia fetida* in sodium arsenite spiked soils – A dynamic interaction between soil and earthworms. *Chemosphere* 319(5), 137905. https://doi.org/10.1016/j.chemosphere.2023.137905

Zhang, L., Van Gestel, C.A.M., 2017. The toxicity of different lead salts to *Enchytraeus crypticus* in relation to bioavailability in soil. *Environ. Toxicol. Chem.* 36(8), 2083–2091. https://doi.org/10.1002/etc.3738

Zhang, L., Van Gestel, C.A.M., 2019. Effect of percolation and chemical form on Pb bioavailability and toxicity to the soil invertebrate *Enchytraeus crypticus* in freshly spiked and aged soils. *Environ. Pollut.* 247, 866–873. https://doi.org/10.1016/j.envpol.2019.01.089

Zhao, H., Wu, Y., Lan, X., Yang, Y., Wu, X., Du, L., 2022. Comprehensive assessment of harmful heavy metals in contaminated soil in order to score pollution level. *Sci. Rep.* 12(1), 3552. https://doi.org/10.1038/s41598-022-07602-9

8 Assessing the Bioavailability of Pollutants Using Macrophytes

8.1 INTRODUCTION AND SCOPE

Since the 1990s, work on monitoring ***bioavailability*** using plants has matured and progressed (Doust et al., 1994; Patel et al., 2024). Some plants are better for bioavailability studies, as opposed to toxicity studies, because they deal with pollutant toxicity by sequestration and ***bioaccumulation*** of pollutants from multiple environmental media (water, soil, or sediment) into different plant tissues (Doust et al., 1994; Patel et al., 2024; Tarish et al., 2024). Plant uptake of organic or heavy metal pollutants in water, soil, and sediment alters pollutant bioavailability and biotoxicity such that the natural growth capacity and biochemical processing of pollutants by plants may partially or fully remove harmful substances from the environment (Raskin et al., 1994). ***Macrophytes*** are plant taxa that are visible to the naked eye and physiologically dependent on the aquatic environment, with at least a portion of their reproductive cycle occurring in or on the water surface (Liu, 1999). Macrophytes are well suited to pollutant bioavailability studies. Macrophytes growing in receiving waters near effluent outfalls have been shown to have a high tolerance to the persistence of pollutants in discharged effluent (Guittonny-Philippe et al., 2015). Some macrophytes bioaccumulate metals in their tissues at levels up to 1,000 times higher than the corresponding concentrations in the environment (Nguyen et al., 2021). Organic pollutants can also be absorbed by macrophytes through the plant cell membrane and, e.g., released from plant leaves through volatilization/transpiration, or metabolically decomposed to a certain extent (Gerhardt et al., 2009).

Macrophyte processes relevant to their interactions with pollutants include ***phytosorption***, ***phytoextraction***, ***phytotransport***, and ***phytotransformation*** (metabolism or change of chemical speciation of pollutants by plants). Phytosorption refers to the physical ***sorption*** of pollutants on living or dead plant tissues. Phytoextraction is the ***absorption*** and bioaccumulation of pollutants from the environment (from multiple media) into plant tissues (Suman et al., 2018). Unlike phytosorption, this term implies a process by which pollutants are actively taken up by plants. Phytotransport moves and distributes pollutants within the plant, transporting them to different tissues or organs such as leaves, stems, and roots. Phytotransformation involves biochemical reactions of pollutants in plants that result in changes in their structure or toxicity, in macrophytes typically transforming ***bioavailable*** pollutants into less bioavailable or biounavailable forms, thereby increasing or decreasing plant tissue levels of pollutants in a complex manner (Dhir et al., 2009).

The form in which macrophytes grow (e.g., free-floating or rooted, at the surface or submerged) plays a key role in whether there is water–sediment contact (water–soil contact for macrophytes that grow, for instance, in submerged soils). In cases wherein the focus of bioavailability study is pollutants in the water column, using macrophytes is of particular interest for ***bioconcentration*** studies (uptake directly from water) if the concentrations of pollutants are too low to be easily detected by direct instrumental analysis of water (Polechońska et al., 2018). Soils and sediments are often the focus of bioavailability study since they are a very important reservoir or sink for organic and heavy metal pollutants. The pollutants bound to soil/sediments can be taken up by rooted macrophytes, and aquatic plants rooted in sediments have been widely used to determine their bioavailability of organic pollutants in sediments in recognition that, while chemical analysis can detect the total load of pollutants in sediments, this total load is often far higher than the actual portion that is biologically active (Diepens et al., 2014). Similarly, when assessing heavy metal

DOI: 10.1201/9781003355229-10

pollution in water bodies and sediments, the bioaccumulation characteristics of aquatic plants can also effectively reflect their bioavailability (Mishra and Tripathi, 2008). Macrophytes are also useful as indicators of the temporal and spatial ***distribution*** of pollution, to characterize water quality in aquatic systems.

Many factors may exert control over bioavailability and macrophyte bioaccumulation, for instance, biological nutrient availability and requirements, the manner in which uptake capacities differ as a function of biological growth rates, and external conditions such as temperature, pH, light, and composition of environmental microorganism consortia. Using macrophytes to monitor bioavailability via bioaccumulation offers a way to biomonitor for varying pollutant bioavailability over time (Wang et al., 2014). ***Biomonitoring*** involves assessing the content of pollutants absorbed by organisms by determining the concentration of pollutants in organisms or determining the change of pollutant concentration in environmental media to which the organisms are exposed before and after ***exposure***. Because of its effectiveness, biomonitoring is well recognized as an approach to assessing environmental quality through the use of indicator organisms (Gecheva et al., 2023). As such, biomonitoring also provides a reliable basis for environmental management and decision-making (Angerer et al., 2006).

This chapter provides a method that may be used to measure the bioavailability of non-volatile organic and heavy metal pollutants in water, submerged soils, and sediments via pollutant uptake to macrophytes. The method is suitable for laboratory and laboratory-adjacent experiments and does not cover field deployments of ***test organisms***. The method covers the acquisition, cultivation, and handling processes to be used for macrophyte test organisms, how to perform bioavailability experiments, and subsequent tissue sample collection and handling. Three types of experiments are supported, including studies on (1) pollutant-spiked waters, (2) pollutant-spiked submerged soils or sediments, and (3) field-polluted submerged soils or sediments. Chapters containing Terms and Definitions, guidance on avoiding contamination of samples in field and laboratory work, methods of environmental sample extraction and instrumental analysis, and guidance on Quality Assessment (QA)/Quality Control (QC) are also within the scope of this chapter.

8.2 PRINCIPLE

In this method, macrophytes as used as ***test organisms*** and are exposed to organic and/or inorganic pollutants in the presence of multiple media (water, submerged soil or sediment, and under aeration) in small-scale ***aquatic microcosms***, i.e., small-scale, controlled environments replicating some aspects of natural aquatic ecosystems. After an exposure period, the bioavailability of pollutants is determined via bioaccumulation of pollutants in test organism tissues. In this method, test organisms are procured, cultivated, or collected and prepared. Sediments (hereafter referring to sediments and submerged soils), when used, are collected in the natural environment or spiked. After the test duration, test organism tissue is collected and then processed via extraction and instrumental analysis to determine the bioavailable pollutants that were bioaccumulated in microcosms. The method strikes a balance between the realism and complexity of full-scale field deployments and a modicum of control that is achieved in non-field settings. The experimental approach inherently incorporates different exposure pathways through which macrophytes come into contact with pollutants (direct water exposure, sediment contact, and natural environmental interactions) in a realistic yet manageable setting.

8.3 EQUIPMENT, CONSUMABLES, AND REAGENTS

The lists of equipment, consumables, and reagents given in this section is intended to be reasonably comprehensive; users should, however, read the entire method before finalizing needs and modify this indicative list accordingly.

8.3.1 Equipment and Consumables

Note: All equipment and consumables must be of materials that are compatible with target pollutants and appropriately clean and pollutant-free (refer to Chapter 3).

Field and field-sample processing-related (sediment and/or macrophyte sampling; associated water sampling if performed):

Aluminum foil – for wrapping samples that will be used in experiments studying bioavailability of organic pollutants.
Bucket with line and/or Niskin-type sampling bottles – for collecting water samples at surface or depth (optional – if water samples are collected).
Camera – for documenting the sampling process and ***site*** conditions.
Decontamination equipment – pollutant-specific and including reagents and/or detergent; for decontaminating field equipment between each use.
Field logbook and sample collection log sheet – to log sample collection and related information regarding sample collection activities, water quality conditions, pH, oxygen conditions, temperature, conductivity, or any anomalies observed or problems encountered that would be useful in evaluating the quality of the data collected using this method.
Foam or other secondary packing material – used to protect macrophyte test organism samples from damage during transportation.
Gloves – ***contaminant***-free; to prevent contamination of equipment, test organisms, and other samples during collection.
GPS/electronic or physical devices – to ensure unambiguous location of sample sites (e.g., GPS, beacons) and depths (e.g., depth gauge).
Holding trays – to place test organism samples in between collection and packaging.
Multiparameter water quality meter or sonde – used to assess the water quality at sites by taking measurements of temperature, pH, conductivity/salinity, and other water quality determinands.
Net, hand-held – used to collect floating macrophytes.
Pens – indelible, to record field observations in a wet environment.
Plastic resealable bags – preferably zip-type; for storing samples or secondary containment of samples.
Sample containers – clean containers with lids or doubled zip-lock bags for storing samples.
Sample labels – for each batch of macrophyte samples collected or for batch of sediments and/or water samples. The label is a unique identifier and should contain (1) a unique sample name (based on the sample collection log sheet), (2) for organisms, the species' scientific name and type, (3) the date of collection or deployment, (4) the type of collection or deployment, and (5) sample description (purpose).
Secchi disk – for measuring water transparency.
Sediment corer or grab sampler – for collecting deeper sediment cores.
Sediment mixing tool – to homogenize soil or sediment samples; e.g., rolling mill, feed mixer, or gardener's soil mixer (e.g., DaRos MX1000 or similar).
Sharp knife, spatula, and/or forceps – for cutting and handling test organism samples.
Spade, shovel, and/or trowel/scoop – for digging/collecting surficial sediment samples, for transferring samples to sample containers, and for digging rooted macrophytes.
Tape measure and/or depth meter – for measuring distances and depths.
Thermo-insulated box with ice packs, ice, or dry ice – for keeping samples at required temperature during transport.
Wide-bore sediment sampler – used to collect macrophytes intact with *in situ* sediments (large plants with roots and associated sediment).
Other field equipment – e.g., waders, boats, scuba gear; as needed depending on field site(s).

Test organisms cultivation-related:

Aquaria or cultivation tanks – made of suitable inert and contaminant-free materials, and with sufficient depth to enable the test organisms being cultivated to grow vertically.
Aquatic microcosms with lids and polytetrafluoroethylene (PTFE) tubing – used to conduct macrophyte bioavailability experiments.
Bubble stone aerator – to introduce air that is suitable for the growth of macrophytes into the cultivation containers and aquatic microcosms.
Diatomite filter – to remove phytoplankton such as algae generated during the cultivation process.
Lamps – to provide the light required for the growth of macrophytes if natural light is not used; must produce photosynthetically active radiation: wavelengths 400–700 nm, providing submerged plants with >10 mol/m^2/d.
Mesh cover – to prevent excessive solar radiation and the rapid proliferation of algae.
pH meter – used to measure the pH value of the cultivation substrate for macrophytes.
Pots – made of suitable materials, which are free of contamination with respect to the target pollutants and non-compostable. They are used to hold sediments and cultivate macrophytes.
Thermometer – used to measure the ambient temperature suitable for the growth of the experimental plants.

Test organism processing, sample measurement, and bioavailability experiment-related:

Aluminum foil and several sizes of containers, resealable watertight plastic bags – for continued sampling of water, sediments, or macrophytes during lab work.
Chlorophyll meter – used to measure the chlorophyll content of test organisms (optional, if measured).
Electronic balance – used to measure the weight of test organisms or weigh reagents.
Freeze dryer – for processing macrophyte tissues as specified in some methods of instrumental analysis.
Freezer – −20°C, for storing test organism samples and/or processed sample homogenates.
Grid paper, planimeter, or camera/digital scanner – used to measure leaf areas.
Holding trays – to place test organism samples while processing. Assists in reducing chances of contamination.
Long-handled spoon, lab scoop, or paddle – used to mix sediment and sample it.
Long-handled tong or forceps – used to handle macrophytes.
Lab spatula – used to handle macrophytes or reagents.
Muffle furnace – for high-temperature decontamination of glassware (organic pollutants).
Refrigerators – for storage of collected test organisms and other samples.
Ruler or vernier caliper – used to measure leaf length and width, stem diameter, and root length/extent.
Several sizes of resealable watertight plastic bags – for primary packaging of test organisms sampled after deployment for analysis of heavy metal pollutants, and for secondary packaging of test organisms sampled after deployment for analysis of organic pollutants that have been wrapped in foil.
Stainless-steel knives – used to slice test organisms after exposure experiment.
Tissue homogenizer – used to homogenize test organisms prior to extraction of pollutants for instrumental analysis. Used for certain organic pollutants.
Wide rubber bands or Teflon tape – used to secure the lids for bioavailability test microcosms.

8.3.2 Reagents

Note: All reagents must be of guaranteed reagent quality, or better, to ensure that the reagents used are of the highest purity to avoid contamination.

Calcium carbonate – used to increase the ***alkalinity*** in some macrophyte growth media.
Calcium chloride dihydrate – used for the preparation of a general-purpose, chemically stable, culture ***medium***.
Heavy metal or organic pollutant standards – for water dosing and sediment spiking experiments; added for macrophyte uptake to determine the bioavailability of contaminants but pollutant standards in acid or basic solution much be buffered (refer to Chapter 9 or 13).
Magnesium sulfate heptahydrate – used for the preparation of a general-purpose, chemically stable, culture medium.
Potassium bicarbonate – used to increase the alkalinity in some macrophyte growth media.
Potassium sulfate – used to increase the alkalinity in some macrophyte growth media.
Sodium bicarbonate – used for the preparation of a general-purpose, chemically stable, culture medium.
Sodium sulfate – used to increase the alkalinity in some macrophyte growth media.
Solvents – in the determination of organic pollutants, a number of solvents may be required to extract, dilute, or dissolve the sample. Alternately used for water dosing and sediment spiking experiments; added for macrophyte uptake to determine the bioavailability of contaminants. Common organic solvents that can be biocompatible include glycerol, dimethyl sulfoxide, methanol, and acetone.
Ultrapure water – also known as Type I water, has a resistance of 18 MΩ·cm.
Water conditioner – used to remove chlorine or other disinfectants from tap water when the water in which macrophytes are cultivated is tap water.

8.4 PROCEDURE

8.4.1 Collection or Acquisition of Test Organisms

Macrophyte test organisms, i.e., individual organisms of one species chosen to be used in a biomonitoring study, can be collected, purchased commercially, or cultured/grown. Culturing is the most difficult and labor-intensive; however, it is feasibly achieved by a careful nonexpert. In whatever manner test organisms are obtained, it is very important that the environment in which the macrophytes are grown prior to experiments is contaminant-free, i.e., free of pollutants (organic pollutants and/or heavy metals, whichever pollutants are the subject of bioavailability studies). Purchasing test organisms saves labor and expense because the collection, handling, and preparation of pre-culture-grown macrophytes is eliminated or greatly minimized. Purchase also ensures that all plants are of the same species and that characteristics such as size, height, number of branches and leaves, and root growth of macrophytes are as similar as possible, all of which are very beneficial to the measurement results (e.g., measurement accuracy) (Knauer et al., 2008). Usual venues for purchase include online suppliers (for instance, in the USA, Carolina Biological Supply; in China, China Garden Network, https://www.yuanlin.com/).

If macrophytes cannot be purchased, options remaining are collection and culturing. Of these, irrespective of the effort involved, culturing is preferable in order to obtain macrophytes that are of the same age, size, and growth stage, which greatly aids in obtaining good experimental ***precision***. Culture vessels should be of an appropriate inert and contaminant-free material and deep enough to allow for the vertical growth of cultured test organisms. Aeration is required. Macrophytes need enough light to achieve photosynthesis (photosynthetically active radiation: wavelengths

400–700 nm, providing submerged plants with >10 mol/m^2/d); however, temperature must be controlled; hence, excessive solar heating is to be avoided. Additionally, high light levels may disproportionately stimulate excessive algal growth, which is to be avoided. Often a temperature range of 20°C–25°C is adequate; however, this varies from species to species, with some requiring lower or higher temperatures. Figure 8.1 shows an example of outdoor cultivation tanks, placed in such a manner as to capture sufficient natural light, but with fine netting covers to prevent excessive solar radiation and algae proliferation.

It is generally accepted that sediments provide sufficient nitrogen and phosphorous to support adequate macrophyte growth. The best sediments to use are clean (i.e., contaminant-free) fine-textured lake sediments with an organic content of less than 20% dry weight. Such sediments are generally able to provide the necessary nitrogen, phosphorous, and an array of micronutrients needed for growth. Commercially available potting soils and related materials are likely to produce poor results. Amending growth media with nitrogen or phosphorous may stimulate growth; however, it is better to avoid such additions, if possible, due to the consequent, excessive growth of phytoplankton/algae that occurs with such enrichment. Natural sediments should be allocated to growth containers several days before planting to allow for settling. During cultivation, water needs to be clear. If problems arise with this, a thin layer of clean sand can be added to the top of the sediment layer.

The number of macrophytes that can be cultivated per square area and volume of culture container depends on the species; a high solution: biomass ratio is favored and overcrowding should be avoided. When possible, using the largest practical culture vessels (tanks) instated with sediment containers at the bottom, instead of a layer of sediment, enables a design with more space around each test organism, and targeting a density of 10% of the bottom surface area covered in sediment pots with macrophytes minimizes crowding and should provide a sufficiently high solution: biomass ratio. Lower-volume systems will likely require either continuous flow or partial solution changes. Shorter duration culture times also relax the need for high solution: biomass ratios, in which case sediment pots can occupy as much as 50% of the bottom surface area of the culture vessel. When using the sediment pot method, contaminant-free non-compostable pots of a suitable material, with respect to target pollutants, may be used. For most experiments, the cultivated macrophytes must be transplanted prior to inception of bioavailability studies, and it is important to be able to remove the

FIGURE 8.1 Representative culture tank – located outdoors to capture maximum light, but covered in shade cloth to prevent excessive algal growth.

test organisms from cultivation containers without damaging them. Figure 8.2 shows an illustration of macrophyte cultivation in aquaria, either with sediment at the bottom of the aquarium or by the sediment pot method.

Water for cultivating macrophytes can be sourced from the sediment collection site (i.e., a clean site) or, if a sufficiently contaminant-free supply is available, tap water may be used. Tap water often has chlorine or other disinfectants in it, in which case, a water conditioner must be added (available at most aquarium supply stores). While it is common to cultivate macrophytes on low nutrient water when using a sediment as a cultivation substrate, several ions and dissolved inorganic carbon can be limiting. As a result, it is sometimes advantageous when using tap water to add ions and dissolved organic carbon as cultivation media. A general-purpose, chemically stable, culture medium is formulated by adding $CaCl_2{\cdot}2H_2O$, $MgSO_4{\cdot}7H_2O$, $NaHCO_3$, and $KHCO_3$ to final concentrations of, respectively, 91.7, 69.0, 58.4, and 15.4 mg/L in suitably contaminant-free and conditioned tap water (the final solution has, approximately, 0.85 meq/L alkalinity, 3.9 mM ***ionic strength***, 280 S/cm conductivity at 25°C, and 7.9 pH after equilibrium with ambient air). For some macrophyte species and/or experimental conditions, higher alkalinities may be required, which may be formulated by adding $CaCl_2{\cdot}2H_2O$, $CaCO_3$, $MgSO_4{\cdot}7H_2O$, Na_2SO_4, and K_2SO_4, to final concentrations of, respectively, 91.6, 87.5, 102.3, 52.0 and 13.6 mg/L in contaminant-free conditioned tap water (final solution is approximately 1.75 meq/L alkalinity, 7.5 mM ionic strength, 490 S/cm conductivity at 25°C, and 8.3 pH after equilibrium with ambient air). For this higher alkalinity culture medium, the salts will not dissolve if simply added to water. To effect the dissolution of carbonate species, the water must first be bubbled thoroughly with CO_2 to form carbonic acid, H_2CO_3, and then the salts are added, such that the H_2CO_3 mimics the natural weathering process/acid dissolution of $CaCO_3$. Following the dissolution of salts, prolonged aeration with ambient air is required to subsequently equilibrate the pCO_2 in the culture medium with atmospheric levels. For marine studies, artificial seawater may be used. It is important to culture under conditions that minimize solution exhaustion/changes in solution composition. Using a lower alkalinity culture solution, as possible, assists with this.

After tanks are prepared with sediment substrates and overlying water (from the sediment field site or prepared culture medium) and left to equilibrate for several days, macrophytes are cultured. These may be grown from seed or rooted from cuttings, or, for some species, tubers. Germination of most species occurs in a few weeks. If planting from seed, ensure to remove any unwanted non-test species that might be present. Cuttings should consist of fresh apical cuttings (apices are intact),

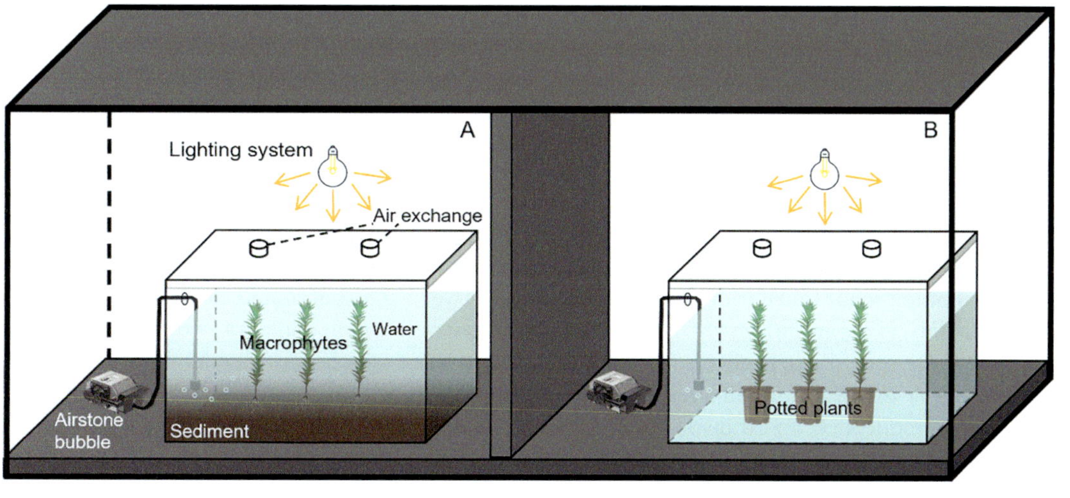

FIGURE 8.2 Schematic diagram showing representative examples of setup for cultivating macrophytes in aquaria. (a) Macrophytes grown in an aquarium with sediments on the bottom. (b) Potted macrophytes grown in an aquarium.

10–15 cm in length, taken from healthy, non-flowering stems, and having numerous leaf nodes/axillary buds on the stem from which roots or additional branches can develop. These cuttings are then inserted to a depth of about 5–10 cm into the sediment and secured with a little pressure. Root formation generally begins within 1 week. For species that produce underground tubers or rhizomes, these can be planted directly into the sediment.

Once macrophyte growth is initiated, algal control is important. If carefully following the protocol here, over time scales of 1–2 months, this should not be a problem. Phytoplankton can be controlled through the use of diatomaceous earth filters such as those used for aquariums. Algicides should be avoided as many are toxic to aquatic plants.

After prolonged periods of growth, cultured macrophytes may lose viability. This can occur from senescence; however, dissolved inorganic carbon in culture media may also be depleted. Increasing aeration assists with maintaining dissolved organic carbon; ultimately, however, the best way to ensure viable cultures over long periods is to subculture periodically (i.e., to use extant macrophyte stock in culture containers prepared with fresh/new sediment and water). For flowering species, subcultures should be initiated prior to flowering. Both fresh sediment and water should be used for subcultures. For further detailed information on culturing macrophytes, publications by Sculthorpe, and Smart are suitable resources (Sculthorpe, 1967; Smart and Barko, 1985).

Collection of macrophyte test organisms from the natural environment is often the least desirable for a number of reasons. While simple in execution, collection requires consideration of diverse factors including plant availability, the need to comply with ecological sourcing principles, the ability of the ***source*** environment to sustain collection, issues with biosecurity, regulatory requirements or permissions, and, most importantly, the challenge in obtaining test organisms of the same size and growth stage and to do so without inadvertent contamination (chemical or biological). Record keeping on the macrophyte collection is very important to demonstrate best practice in ecological sourcing is adhered to. Only non-threatened species may be responsibly collected. For sustainable collection, no more than a maximum of 10% of the plants present in the targeted harvest area should be removed, i.e., to ensure that the macrophyte community that is harvested can recover. Harvesting should not be from a locality that might be critical for population recovery or in the context of the surrounding ecology. Macrophytes must only be collected from areas that are relatively free of pollutants and have water with a similar chemical composition (***salinity***, major constituents) to the test exposure conditions in bioavailability experiments. If the pollutant concentrations of chemical analytes in test organisms at the collection site are not known, they should be assessed by prior collection and testing. Care needs to be taken to ensure that all materials used in the collection of macrophytes are clean and free of pollution. Macrophytes should be collected using clean tools of an appropriate material (refer to Chapter 3). Care also needs to be taken not to damage the test organisms during collection, and test organisms should be collected from the same area at the same time and at the same or similar depth.

When collected in the field, test organisms will be subsequently transported and either recultured in a facility or used directly in experiments. In any case, biosecurity issues exist. Nuisance pests developing under culture conditions or being transferred directly should be considered. Any risk of spreading pest plants or fish should be minimized by identifying plants carefully and minimizing the unintentional transfer of water or sediment between the source and the eventual recipient environments. It is particularly difficult to collect organisms without collecting potentially harmful algae, and for rooted macrophytes, insects, and snails, all of which may proliferate in subsequent reculturing or use of test organisms. Many countries require various permissions and/or approvals for both the collection and transport of submerged macrophytes, for purposes of biosecurity and ecological preservation. It is important to check local, regional, and country-wide requirements before collecting macrophyte test organisms.

To collect floating macrophytes, a net may be used. For rooted macrophytes, plugs of sediment containing rooted macrophytes may be dug either with a digging implement or a wide-bore corer, being careful to avoid damaging root systems; an example of the latter approach is shown in Figure 8.3. Samples collected this way can be transported as is with sediment, macrophytes, and overlying water. The advantage of collecting test organisms intact in sediment is minimizing the stress on the test organism. The disadvantage of the collection with sediments is inconvenience in transport and the propensity to simultaneously collect unwanted organisms. If macrophytes are to be separated from *in situ* sediment before transport, the sediment in which they were collected must be carefully rinsed from test organism roots. It is generally desirable when collecting macrophytes to try and remove hitch-hiking algae. This can be done mechanically, i.e., by gentle rubbing and rinsing. Some protocols call for the use of a disinfection solution; however, in practice, it is difficult to formulate a uniform safe protocol that will not damage test organisms; hence, this is not recommended here.

When handling macrophytes, it is important to be as gentle as possible. Test organisms can be placed into a sample container filled with ambient water, or alternately plastic bags filled with ambient water (i.e., as done for fish sold by aquarium stores). In either case, it is important to ensure that transport containers will not leak or spill during transport. Test organisms collected in this way should be wrapped in secondary material, such as foam cushion or padding in the transport container, for protection from damage during the transport of the samples and in order to avoid severe shocks or bumps to minimize mechanical damage to the samples. Avoid exposing test organisms to bright light during transport, and control the temperature during transport. For macrophytes, the appropriate temperature during transport needs to be determined according to their species. Some macrophytes can be transported relatively safely for short periods of time at temperatures around 20°C–25°C. *Elodea canadensis*, for example, can survive and be transported over a wide range of temperatures (ANHP, 2011; Łoboda et al., 2018). Some species of macrophytes, however, may be more sensitive to temperature. For example, *Myriophyllum verticillatum* is a hardy aquatic plant that generally holds up better at lower temperatures for transport and short-term storage (Yuan et al., 2023). Therefore, lower temperatures may be considered for storage and transport, e.g., refrigerated conditions of ~4°C–8°C may be a more suitable option. Collected test organisms should be brought back to the laboratory and processed for storage (i.e., as for cultivation) or used as soon as possible to try and reduce shock/plant stress. Avoid direct contact with samples as much as possible and ensure the use of uncontaminated tools and equipment. Additionally, since the freshness and shelf life of macrophytes is short,

FIGURE 8.3 Collection of macrophytes using a wide-bore sediment corer.

reduce transport distance and time to get them to the laboratory as possible. In general, large test organisms can be kept fresh for up to 24 hours during transport. All transport containers should be clearly labeled with temperature requirements, sample name, date, and time of transport.

8.4.2 Species Suitable for Use as a Test Organism

A representative list of macrophytes suitable for determining the bioavailability of pollutants, along with information about the growth cycle and advantages and disadvantages for research and biomonitoring, is given in Table 8.1. Macrophytes that have been demonstrated as suitable for use with specific heavy metal or organic pollutants may be assumed to have good potential for other, similar, heavy metal or organic pollutants. As the list in Table 8.1 is representative, there are other species that would also be suitable; however, prior to using any species that have not been used for biomonitoring previously, control experimentation would be needed to verify suitability for bioavailability studies. Note: Many of the species listed in Table 8.1 are invasive; extreme care must be taken to contain such species to the laboratory and prevent spread in the natural environment.

TABLE 8.1
Representative List of Macrophytes Suitable for Use in Bioavailability Studies[a]

Macrophyte Species	Notes on Growth	Pollutants Studied	Advantages and Disadvantages	References
Myriophyllum verticillatum	Whorled Water Milfoil; feathery whorled leaves, growing in dense mats	Fe, Cr, Zn, Ni, Cu, and fluoride	Favors warm, moist, sunny climates, does not tolerate the cold. Nutrients from the substrate can be absorbed through the roots, and certain heavy metals and organic pollutants are detoxified and stored and/or degraded (some organic pollutants) in macrophyte. Considered moderately invasive in some areas, but less invasive than some other water milfoils.	Caillat et al. (2014), Sapci and Ustun (2015), and Zhou et al. (2012)
Elodea canadensis	Canadian Waterweed; robust growth, used in aquariums, tolerant to slow-moving waters	Ni, Zn, Cu, Cd, and Dichlorodiphenyl trichloroethane (DDT)	Aggressive and competitive plants, tubers can lie dormant in waterways and estuaries causing buildup problems, clogging waterways and estuaries, and costing labor to remove. Suitable for contained bioavailability studies; unsuitable for field studies as potentially invasive.	Gao et al. (2000), Maleva et al. (2016), and Nyquist and Greger (2007)
Potamogeton crispus	Curly-leaf Pondweed; wavy, serrated leaves, grows in many freshwaters, including lakes and rivers	Polycyclic aromatic hydrocarbons, Cu, Cr, Pb, As, and Cd	Economic value as green manure and fodder, with a high enrichment capacity for zinc and strong purification capacity for arsenic. Potentially problematic in field settings; reproduces rapidly and jeopardizes the growth of rice in paddy fields.	Meng and Chi (2017), Meng et al. (2015), and Norouznia and Hamidian (2014)

(Continued)

TABLE 8.1 (*Continued*)
Representative List of Macrophytes Suitable for Use in Bioavailability Studies[a]

Macrophyte Species	Notes on Growth	Pollutants Studied	Advantages and Disadvantages	References
Najas marina	Spiny Naiad or Water Nymph; annual, grows in dense clusters, sometimes tolerates brackish waters	Al, Ba, Se, Sr, Ti, and pentachlorophenol	Can tolerate high pollutant bioavailability; its proliferation is often a manifestation of the deterioration of aquatic ecosystems. Problematic in field settings; highly invasive and damaging in many ecosystems.	Ammeri et al. (2022), and Xing et al. (2013)
Vallisneria spp.	Eelgrass; can form dense underwater meadows	As, Cd, Hg, Pb, and polycyclic aromatic hydrocarbons	Requires a high temperature for growth. Considered noninvasive in many environments.	Chi and Liu (2016) and Qian et al. (2019)
Ceratophyllum demersum	Hornwort; whorled leaves and free-floating	Ba, Cd, Cr, Cu, Fe, Mn, Ni, Pb, Rb, Zn, and pyrene	Can grow and be collected year-round in many places. A wide range of growing temperatures is tolerated, however, sensitive to icing and less tolerant of low temperatures. Moderately invasive in some environments.	Polechońska et al. (2018) and Yin et al. (2008)
Potamogeton maackianus	Maack's Pondweed; found in deeper freshwater lakes and rivers	Se, Ba, and dibutyl phthalate	More effective in removing organic pollutants. Generally considered noninvasive in most environments.	Li et al. (2006) and Liu et al. (2015)
Hydrilla verticillata	Hydrilla; whorled leaves, rapid growth, sometimes tolerates brackish waters	Zn, Ni, and perfluoroalkyl acid	Invasive. Reproduces quickly and needs regular cleaning and long periods of sunlight.	Li et al. (2022), Song et al. (2018), and Srivastava et al. (2009)
Leersia oryzoides (L.) Swartz.	Cutgrass; growing in dense clusters along the edges of water bodies; wetland plant	Cu, Fe, Mn, and permethrin	Requires high temperature and light levels for growth. Typically found in wetland areas and not considered an invasion threat.	Coimbra and Borges (2023) and Moore et al. (2009)
Elodea nuttallii	Nuttall's Waterweed; dense underwater growth in nutrient-rich waters	Hg and bisphenol A	Requires low temperatures; can overwinter in some cold environments. Reproduces rapidly; susceptible to rot when water quality is poor or in hot conditions. Considered highly invasive in some environments.	Regier et al. (2013) and Zhang et al. (2007)
Zostera marina	Marine Eelgrass, euryhaline and grows in low saline systems (2‰–5‰)	Co, Zn, Pb, Hg, polynuclear aromatic hydrocarbons, and polychlorinated biphenyls	Well-developed root system; relatively difficult to cultivate under laboratory conditions; noninvasive.	Huesemann et al. (2009) and Lee et al. (2019)
Halophila ovalis	Paddleweed; tolerates a wide range of salinities and grows in relatively low-salinity	Co, Cr, Cu, Mn, Ni, Pb, Zn, Al, As, and Fe	Relatively small size for easy cultivation and experimental manipulation in the laboratory; fragile and needs to be handled with care during operations; noninvasive.	Kilminster (2013) and Nazneen et al. (2022)

[a] All species listed are herbaceous. All are perennial, rooted, and grow in freshwater environments unless otherwise indicated.

8.4.3 Growth Stage, Condition, Size, and Number of Test Organisms Needed

8.4.3.1 Growth Stage

While there is not a universally agreed-upon number of growth stages for macrophytes, the general stages often described (Pujar et al., 2006) include the following:

- Germination/sprouting – when seeds or spores begin to grow.
- Seedling stage – when young macrophytes start to develop roots, stems, and leaves.
- Vegetative growth – a time when macrophytes increase in size, producing more leaves and stems.
- Maturity – when vegetative growth begins to abate and macrophytes reach their full reproductive potential.
- Reproductive stage – when plants produce seeds or spores.
- Senescence – the final stage where plants start to decline and die.

For this method, it is important to select test organisms at the same stage of growth to ensure that any differences observed during experiments are caused by the variables being tested in the experiment and not by differences in the growth stage. For studying bioavailability, vegetative growth is often used, because during this stage, macrophytes are actively developing and uptake is greater. Alternately, macrophytes that have reached maturity are also used as test organisms because, on average, they have more leaves, stems, and well-developed root systems, which are useful for handling and to more confidently identify the test organisms' growth characteristics and response to environmental changes under experimental conditions.

In practice, and particularly for some species, it can be difficult to differentiate maturity, reproductive, and senescence stages. This is another reason why either purchasing commercially available macrophytes or cultivating them is preferable to field collection. When purchasing macrophytes commercially, information on the growth stage is typically available from the supplier. When cultivating macrophytes, it is straightforward to choose test organisms in the vegetative or early mature growth phase, as these are growing well and have not yet begun to produce spores or seeds. For most species, reproductive and post-reproductive mature and senescent stages should be avoided due to variability in test organism responsiveness.

8.4.3.2 Condition and Size

Obtaining macrophytes in the same growth stage does much to ensure uniform condition and size, albeit the size and condition of test organisms naturally vary from individual to individual at the same stage of growth, necessitating due care in documenting the condition and size of test organisms at the beginning of bioavailability experiments. In the case of test organisms that have been collected, careful scrutiny of condition and size is even more important, as this is the primary conduit toward trying to choose macrophytes that are similar/in the same growth stage. Practice for how test organisms are characterized is highly variable and should remain flexible, as what characterization is desirable or needed depends in part on the species of macrophyte used and in part on the goals of the bioavailability study. Characterization may consist of recording observations of disease and or/damage, physical characteristics, biomass, chlorophyll, and other chemical properties, growth parameters (rate and form), and characteristics of the habitat in which test organisms were grown.

Table 8.2 provides a representative example of characteristics to record. Ideally, a table like this is set up as a spreadsheet for ease of recordkeeping and calculation of descriptive statistics. Due to the ease of making and storing digital photos, it is also recommended to document all visual observations with photos. When taking photos, ensure that key growth indicators are captured, including, for instance, an overall view of the test organism, close-ups of localized areas, and any special

TABLE 8.2
Representative Example of a Macrophyte Growth Status Record Sheet; Items Recorded May Vary Significantly from this Depending on Study Objectives

			Date: dd/mm/yyyy
Macrophyte Growth Status Record Sheet[a]			
Type of Observation		**Indication**	**Description**
Visual		Growth condition	Signs of disease or damage, e.g.: chlorosis (yellowing of leaves or stems from nutrient deficiencies or disease), necrosis (dead tissue, often appearing as brown or black spots), wilting, lesions of localized decay or damage on leaves or stems, signs of fungal spores or bacterial ooze, apparent abnormalities in root structures, physical damage such as broken stems or torn leaves, and damage from insects such as holes in leaves.
Structures	Leaves	Color	Hue, shade, and intensity of the leaf color, e.g., dark green, green, yellow, red, variegated.
		Shape	The overall form or outline of the leaf, e.g., elliptical, lanceolate, ovate, linear, distorted.
		Number	Leaf count – estimated total leaf count per stem, branch, or macrophyte test organism, e.g., four leaves per whorl.
		Size	Dimensions of leaves, e.g., length, width, and area, typically in mm, cm or mm^2, cm^2.
	Stems	Color	The color of the stem, e.g., usually ranges from shades of green to brown to red.
		Diameter	The thickness of the stem, e.g., typically measured at the base, middle, and top, in mm or cm.
		Texture	The surface quality of the stem, e.g., smooth, rough, hairy, spiny, hard.
		Curvature	Refers to the bending or deviation of a plant stem from its vertical or horizontal axis. Usually recorded as the angle between the original straight stem and the part of a stem after it curves. Measured by visual estimation (e.g., 30°) using a protractor, or by taking digital photographs followed by image analysis.
	Roots	Color	The color of the roots, e.g., typically ranges from white to brown to black.
		Length/ extension	The total length of the root system, e.g., from the base of the plant to the tip of longest root, including primary and secondary roots.
		Texture	The surface quality of the roots, e.g., smooth, rough, fibrous, spongy, transparent webbing.
		Root hair density	The number and density of fine root hairs on the root surface, e.g., low, medium, high, possibly more quantitative descriptors.
Plant height			
Water quality status		pH	
		DO	
		Temperature	
		Conductivity	

[a] As examples, *Myriophyllum verticillatum* leaves in the mature stage are linear to oval in shape, usually between 1 and 4 cm in length, they are dark green in color and have well-defined veins. The stems are usually dark green to dark brown in color, darkening with age. The stems will form multiple branches, showing a fan-shaped growth pattern. The angle and number of branches may vary depending on the growing environment and conditions. The texture of the stem is tough and hard. It is rough to the touch and may be slightly prickly. The root system is fibrous, white, and usually between 1 and 2 cm in length, which helps to anchor the plant in the substrate and absorb nutrients and water (Zhao, 2005; Zhao and Liu, 2009). *Elodea canadensis* in its mature state can reach a plant height of 60–180 cm. In the mature state, the plant has slender and branched stems, usually green in color. The single leaves are whorled and sessile, with four leaves per whorl, about 1.9 cm long and 6 mm wide, shaped with the apex curved outward, the margins serrulate, and dark green in color. The stems are woody and hard to the touch, and *Elodea canadensis* has a well-developed green-colored root system that stabilizes the plant and absorbs nutrients and water (Zhao, 2005; Zhao and Liu, 2009).

growth or damage features. Visual signs of disease or damage and plant height should always be recorded for each test organism used.

Plant height is measured from the base of the macrophyte to the tip of the longest shoot, as shown in Figure 8.4, and depending on the size of the test organism, can be measured using a tape measure, ruler, or vernier caliper.

In addition to recording overall visual observations of disease or damage and plant height, some suite of measurements to characterize the leaves, stems, and roots of each test organism should also be recorded. A representative set of these, along with a description of each, as well as examples, is given in Table 8.2. Common measurements for leaves include leaf length, width, and area. For stems and roots, common measurements include, respectively, stem diameter and root length/extent. A ruler or vernier caliper can be used to measure leaf length and width, stem diameter, and root length/extent. Figure 8.5 shows examples of length and width measurements on macrophytes.

There are many options for estimating leaf area, ranging from simple manual techniques to advanced digital analysis. A few common approaches include gridding, planimeter, and digital analysis of images. For the grid method, the leaf is placed on a piece of graph paper, where the distances between grid points are known exactly, and the outline of leaves is traced. After the leaves are removed, the number of grid squares covered by each leaf is counted and multiplied by the area of each grid square. This method is straightforward but can be time-consuming. In the planimeter method, the outline of the leaf is traced by a planimeter device, which then gives the area. This method is more accurate than the grid method, but is still quite time-consuming and additionally requires a planimeter. The best approach involves the use of digital images, which can be obtained with a digital camera (smartphone) or by digital scanning of leaves. Subsequently, software is used to analyze each image and give the area. Note that when taking images, the horizontal and vertical scale of each image must be accurately known in order to obtain accurate results. Adding a ruler to the image in the case of photos is one way to deal with this. For digital images, scanners are usually calibrated to cover different standard sizes (e.g., letter, A4). Freely available software packages such as ImageJ are available for image analysis. With some care in planning, the digital analysis approach can be automated. The digital method is accurate, can be time-saving, and is best for documentation; however, for users unfamiliar with digital analysis, there is a barrier to getting started.

FIGURE 8.4 Plant height measurement for macrophytes. Top row, left to right: *Ulva lactuca*, *Fucus vesiculosus*, and *Myriophyllum verticillatum*; bottom row: *Elodea canadensis* and *Vallisneria* spp.

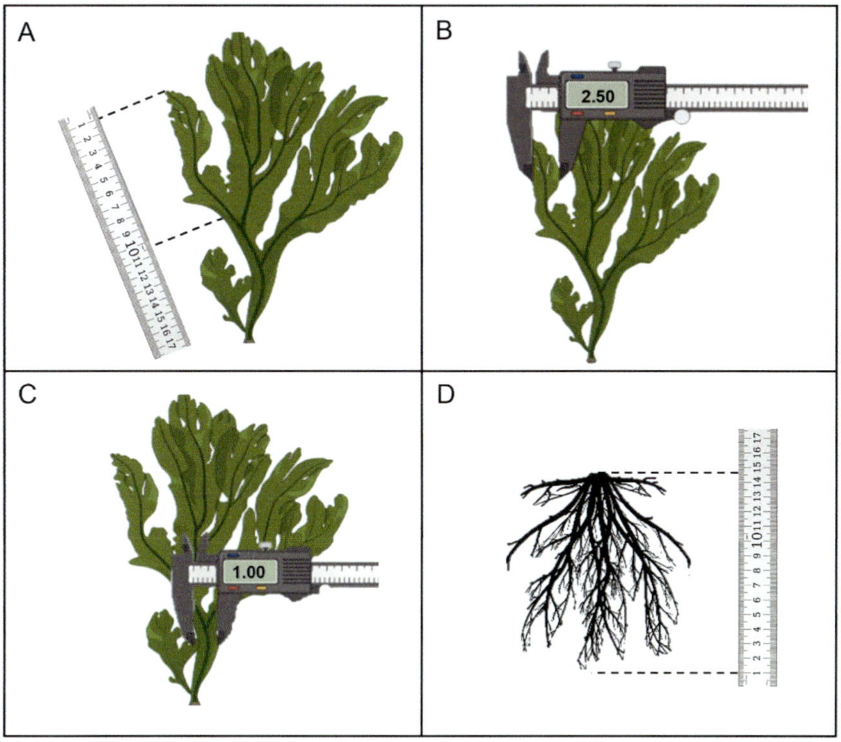

FIGURE 8.5 Examples of length and width measurements to measure macrophyte growth condition. (a) Leaf length; (b) leaf width; (c) stem width; and (d) root length.

Table 8.2 is representative, and it is important to modify it as appropriate to the species of test organism and the needs of the bioavailability study. For instance, with respect to test organisms, not all macrophytes have traditional leaves; some have fronds, blades, thalli, or indistinct leaves and whorls. These structures are not always amenable to the measurement of "leaf width", "leaf length", or "leaf area". As a result, Table 8.2 should be modified to delete these, perhaps adding other measurements as appropriate. The same is true for water quality determinands, whether measured in the field (i.e., for test organisms that are collected), in the laboratory (e.g., culture media), or both. Measuring pH, dissolved oxygen (DO), temperature, and conductivity, as shown in Table 8.2, is standard, however, for some experiments it may be desirable to measure other quantities as well, for instance, cations and anions or chemical oxygen demand as examples. Finally, there are a number of other morphological and physiological quantities that are often measured in macrophyte studies that are not listed in Table 8.2 and should be added for some studies. The most common of these include chlorophyll content, measured by a chlorophyll meter (e.g., DIVING-PAM-II, Walz, Germany or similar, also see, e.g., Van Beek et al., 2024) or via an instrumental method (ISO, 1992; Yu et al., 2024), and rate of photosynthesis, measured via the chlorophyll fluorescence method (Jiang et al., 2018). Standard reporting involves, at minimum, condition and size characteristics for each test organism at the time of initiating and ending a bioavailability experiment. For some experiments, it may be desirable to record the same observations at intervals during experiments as well.

8.4.3.3 Number of Test Organisms Needed

The number of test organisms needed depends on many variables, first and most importantly of which is the weight of tissue needed per sample for the extraction of target pollutants and subsequent instrumental analysis. In some cases, the weight of one test organism may be inadequate for the needs of chemical analysis of pollutants. Another important factor in trying to estimate the weight of sample needed when considering macrophyte experiments is what particular part of the test organism will be analyzed (e.g., composite whole-plant tissue sample, leaves, stems, or roots). The choice of test organism tissues to choose for analysis is discussed briefly in Section 8.4.4 with respect to experimental design. With these points in mind, the number of test organisms needed per sample is calculated as

$$n_{\text{Comp}} = \text{ceiling}\left(\frac{w_{\text{ExIA}}}{w_{\text{TstOrg}}}\right),$$

for studies where tissues will be sampled as a whole-plant composite, and

$$n_{\text{NonComp}} = \text{ceiling}\left(\frac{w_{\text{ExIA}}}{w_{\text{LowWtTissue}}}\right),$$

for studies where multiple tissue types will be sampled. The mathematical function *ceiling* designates that a quotient is rounded up to the nearest integer value. For calculating n_{Comp}, w_{ExIA} is the wet weight tissue needed for each sample to meet the needs of extraction and instrumental analysis, and w_{TstOrg} is the average weight of a test organism. For calculating n_{NonComp}, the denominator $w_{\text{LowWtTissue}}$ is the average weight of a particular type of tissue from a test organism, representing the lowest weight of tissue in the organism of the different tissue types to be sampled. For instance, if a study plan involves measuring roots and stems, and the total weight of stems on a test organism is, on average, lower than the total weight of roots, then $w_{\text{LowWtTissue}}$ represents the average weight of stem for test organisms. If there is enough of the limiting tissue, in this case stems, for collection, this ensures that there will be an adequate weight of roots available from the same test organisms from which stems are sampled.

Generally, even if one test organism is sufficient, better precision results from using three or more, and often experimental designs that require different tissue types will require more than one test organism. With this in mind, even if the total number of test organisms needed per sample is one, users may choose to arbitrarily increase the number of test organisms used per sample to three. Offsetting this consideration, however, is that more test organisms deplete the overlying water faster, which is also undesirable and should also be considered when deciding whether to use more than the minimum number of test organisms per sample needed.

For determining the total number of samples needed for a given experimental design, other considerations include the number of sample replicates, the number of samples needed for Quality Assurance/Quality Control, the number of samples needed for ancillary measurements, and contingency samples. Based on this, the total number of test organisms needed is

$$n_{\text{Total}} = 1.25 \times n_{\text{Comp;NonComp}} \times \left(n_{\text{reps}} + n_{\text{QA/QC}} + n_{\text{anc}}\right),$$

where $n_{\text{Comp;NonComp}}$ is either n_{Comp} or n_{NonComp}, depending on whether an experiment calls for whole-plant tissue composites or not, n_{reps} is the number of replicates for each sample, $n_{\text{QA/QC}}$ is the total number of QA/QC samples that will be needed, n_{anc} is the number of samples needed for ancillary analysis, and the value 1.25 is a contingency factor to allow for loss of test organisms in handling and other unforeseen losses. The value for n_{reps} ranges between three and five, with a minimum of three, four being preferred. The number of QA/QC samples needed depends on the scale of the bioavailability experiment – how many total samples and batches for analysis and should be determined according to the guidance given in Section 8.5.1. There may be a variety of ancillary samples needed for various additional measurements in parallel to the samples needed for pollutant analysis; at minimum replicate samples for dry-to-wet weight analysis and analysis of background pollutants should be taken at the beginning of a bioavailability experiment (see Chapter 5 for methods). Additional ancillary samples should be collected for dry-to-wet weight analysis at the end of experiments.

8.4.4 Experimental Design: Notes on Sediments, Dosing Experiments, and Tissue Sampling

Bioavailability studies may entail exposure to pollutants in the laboratory or the field. This method is oriented toward laboratory experiments. The most common and well-accepted method for bioavailability studies in field settings is described in Chapter 6. Within the purview of laboratory studies, this method is suitable for a variety of approaches to exposure, three of which are described here, and all of which involve the use of sediments. While it is possible to conduct macrophyte experiments, using non-rooted species, without sediments, performing studies without sediments can present certain problems (Smart and Barko, 1985), and it is more realistic to include sediments, even if the macrophyte test organism is a non-rooted species. Sediments used in experiments may be clean, i.e., contaminant-free, or polluted. The three basic protocols for experiments described here, shown in Figure 8.6, involve (1) using clean sediments and dosing the overlying water with pollutants, (2) using clean sediments that are pre-dosed with pollutant prior to experiments, and (3) using polluted sediments. As examples of when each different approach might be used, the first approach may be of greater interest to simulate uptake in scenarios such as a pollutant spill. The second approach could be interesting when the goal is to understand bioavailability prior to aging (the effect whereby pollutants interact with sediment ***components*** in a manner that leads to a decrease in their bioavailability over time), or when the goal is to add pollutants to sediments and perform multiple bioavailability studies over time to study the effects of aging. The third approach is chosen when the goal is to test multiple polluted field sediments of varying levels of pollution for pollutant bioavailability.

The remainder of this method will assume that users have conducted a design of experiment according to one of these three approaches. Chapter 7 contains information on the basic method for obtaining and homogenizing clean/contaminant-free sediments and sampling polluted field sediments. For the latter, it is very important to decontaminate equipment after each sample (Chapter 3). For dosing experiments (Protocols 1 and 2 described above), either water (Protocol 1) or sediments (Protocol 2) are spiked. Spiking means to add a small volume of higher concentration such that the total volume of the medium receiving the ***spike*** is negligibly changed and the final concentration after mixing the spike into the receiving medium is the target final concentration of pollutant desired (see methodology in Section 4.3).

For water, an alternate approach to spiking is to first determine concentration of pollutant, C_f, that is desired after dosing and the total volume of water, V_T, needed for experiments. From this, calculate the mass of pollutant needed for dosing according to $m = C_f \times V_T$, converting moles into mass if C_f, is in molar. Accurately measure the needed volume of test water (e.g., from clean/uncontaminated natural water or laboratory simulated natural water) into a clean container, and add the needed mass of pollutant slowly under constant mixing. Stir continuously for a period of time (e.g., 15–30 minutes) to allow the pollutant to fully dissolve and come to equilibrium. The mass of pollutant used should be at least 50% less than that which reflects the pollutant's total solubility in water of the ionic strength used in experiments. Alternatively, if the mass, m, needed is very low, a concentrated solution should be made and diluted serially tenfold until a convenient volume can be used to dose the water.

Sediment samples are dosed or spiked wet. For heavy metals in an acid ***carrier solution***, i.e., a solution that carries target pollutant(s) and can effect even distribution throughout a sample, the carrier solution must be first buffered in a metal-compatible buffer that can serve as a pH appropriate

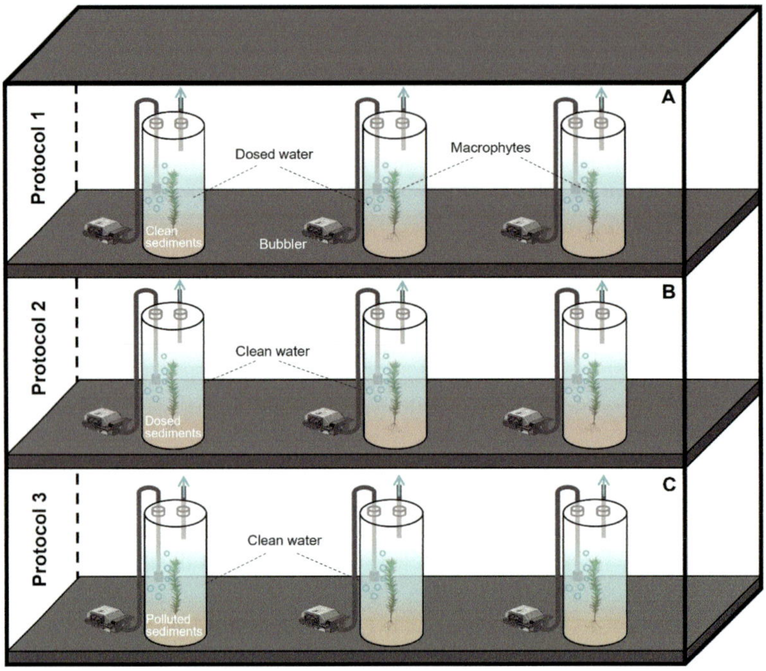

FIGURE 8.6 Schematic diagram showing the basic setup for bioavailability experiments with macrophytes using this method (Windisch et al., 2022). (a) Protocol 1, using clean sediments and dosing the overlying water with pollutant. (b) Protocol 2, using clean sediments that are pre-dosed with pollutant prior to experiments. (c) Protocol 3, using polluted sediment.

carrier (see Chapters 9 and 13). In the case of organics, the carrier solution serves to dissolve the organic contaminant. Because many organic contaminants have low solubility in water, organic solvents are needed to make them dissolve and be able to distribute them uniformly in water or other samples. Common organic solvents such as glycerol, dimethyl sulfoxide, methanol, and acetone can be used as a carrier solution to dissolve organic contaminants in a manner suitable for spiking. Glycerol and dimethyl sulfoxide are typically less toxic/more biocompatible for use with organisms; however, they may not provide enough solubilization. If solvents such as acetone or methanol are used, care must be taken to ensure either that the amount used in the spike has no biological effect or that the spiking solvent has a chance to evaporate prior to use with macrophytes.

As an example of preparation of a spike solution, when the final spiked concentration of sediment is designed to be C_f mg/kg, calculate the total mass of required pollutant, $m_{pollutant} = C_f \times m_{sediment}$. Based on the concentration of the available pollutant standard, C_{std}, calculate the volume of the standard solution to be taken, V_{std}, according to $V_{std} = m_{pollutant}/C_{std}$. For example, if the final concentration $C_f = 5$ mg/kg (5 ppm) and the sediment wet weight, $m_{sediment}$, is 100 g, then $m_{pollutant} = 5$ mg/kg × 0.10 kg = 0.5 mg. If the concentration of the available pollutant standard C_{std} = 1,000 mg/L, then V_{std} = 0.5 mg/1,000 mg/L = 0.0005 L, or 0.5 mL. Ideally, a spike represents between 1% and 5% of the volume of the sediment being spiked. Smaller volumes are too inaccurate. In this example, 0.5 mL, or ~0.5 g, is only 0.5% of the sediment being spiked; hence, this spike volume is unacceptably low. To address this, the standard solution is diluted. For instance, diluting sixfold (1 mL standard added to 5 mL Type I water or solvent if organic) gives a 166.67 mg/L solution. Using this solution, V_{std} = 0.5 mg/166.67 mg/L = 0.003 L, or 3 mL. A spike of 3 mL in 100 g of sediment falls within the required volume for a spike.

Pollutant in carrier solution is added directly to sediment (Loonen et al., 1997), however, it is very important to mix the sediment very well. This can be achieved with a gardener's soil mixer (e.g., DaRos MX1000 or similar) or a small drum mixer (e.g., concrete mixer), but it is very important that the sediment is not contaminated by the materials from which such apparatus is composed, or from inadequate decontamination of apparatus (refer to Chapter 3), and that the sediment is mixed extremely well. Depending on what the carrier is for the spike (e.g., buffer, solvent), it is important to ascertain whether the residue is likely to cause a problem for the test organism. In the case of organics, toxicity data is frequently available. For Protocol 2, spike and mix spike into sediments immediately before using the sediment in experiments for which aging is not studied. Alternately, for aging studies, the spiked sediment is stored at 4°C until aged for a time determined in experimental planning (usually for weeks to months, sometimes years).

For all dosing and spiking experiments, the final concentration of pollutant in dosed water or spiked sediment must be confirmed by an appropriate method of instrumental analysis for the targeted pollutants.

The issue of which test organism tissues to sample must also be considered in the design phase. Composite whole-plant tissues are used in some studies to provide an overall measure of bioavailability. Other studies focus on, e.g., roots (for direct sediment uptake) or leaves (to test for direct foliar uptake), and still others analyze roots, stems, and leaves to try and ascertain something about pollutant transport and fate within test organisms. Which tissues to sample must be decided in the context of bioavailability study aims and objectives.

Macrophytes are much reported in the literature for their use in phytoremediation, and for organic pollutants in particular, various metabolic processes can degrade organic pollutants after uptake. For this reason, experimental design for bioavailability studies on organic pollutants should also consider information on the degradation potential of the specific target pollutants. For some studies, degradation metabolites may also be included on the schedule for analysis, and degradation potential may be of as much interest as bioaccumulation.

8.4.5 Conducting Bioavailability Experiments

The experimental setup, and possible variations, for the macrophyte bioavailability experiment is as illustrated in the schematic in Figure 8.6. The setup is similar to those used in literature reports (Mascaró et al., 2009; Windisch et al., 2022) and consists of individual aquatic microcosms that have sediment at the bottom, overlain by water that is aerated, typically with a bubble stone aerator (or similar, as available at aquarium supply shops). Alternately, particularly for experiments following Protocol 1, instead of sediments at the bottom, it is possible to have small pots with rooting macrophytes if the pots will fit within the diameter of the microcosm. To reduce evaporation, aquatic microcosms are fitted with a removable lid having two holes for gas inflow and outflow. The equipment required is simple, however, requires some thought and planning for setup. First, the materials used must be compatible with the pollutants studied. Next, the material must be able to transmit light usable to macrophytes. The bottom of the microcosms used must be sealed, or is preferably contiguous, and the lid must fit well enough to prevent substantive exchange of gases with ambient air, other than that passing through the air inlet and outlet. The setup shown in Figure 8.6 has been constructed using commercial wide-bore tubing, closed at one end and open at the other. It is possible to purchase such tubing made out of materials suitable for decontamination, such as glass and polycarbonate. Lids may be cut from polycarbonate, and some suppliers of wide-bore tubing also offer plastic caps, both of which can be drilled to accommodate tubing (of a pollutant-compatible material, e.g., silicon or PTFE), either directly or threaded through a cork. The fit of the tubing in the cap should be snug. Lids may be held on with, for example, very wide rubber bands or Teflon tape. Other investigators have used buckets with lids (buckets come in a surprising variety of materials, including glass), food storage containers with lids, and buckets from home improvement stores (of suitable materials, and decontaminated). For most experiments, the setup shown with wide-bore tubing is desirable, both for cost savings, feasibility for multifactorial testing, and given the relatively modest size of many test organism species.

Depending on how macrophytes were obtained, there are two ways in which bioavailability experiments are initiated. If rooted macrophyte test organisms were cultivated in pots, water is added to the aquatic microcosms, then pots containing sediment and macrophyte are added. This approach is only practically possible if the experiments to be performed involve spiking water, in which case it might still undesirable as the amount of sediment used in the experiment is low. Alternately, polluted or sediments or spiked sediments are loaded in aquatic microcosms, and, if rooted macrophytes will be used, $n_{TstOrg/Sample}$ test organisms are planted in sediments for each microcosm. This involves first rinsing the organisms well in Type I or contaminant-free natural water to remove much or all of the sediment from the roots, then instating the roots into the sediment in the aquatic microcosm, ensuring that there is good root-to-sediment contact, but without damaging the test organism. Next, water is carefully decanted into the microcosms to avoid suspending loaded sediment as much as possible. Care must be taken at this point not to cross-contaminate samples in handling spiked water or spiked/polluted sediment samples. If non-rooted/floating test organisms are to be used, these are added after the water.

Once loaded, the aquatic microcosms need to be left in an area with adequate, but not a surfeit of, light for the duration of the bioavailability experiment, per comments in Section 8.4.1. Artificial light may be provided if it fulfills these requirements. If temperature is controlled, this should be recorded. For experiments utilizing natural light, i.e., conducted outdoors, this temperature is typically not controlled, but must be in a range suitable for test organisms. Test organism condition (see comments on Table 8.2) is recorded and, ideally at the time of initiating bioavailability experiments, the wet weight of each test organism is taken if possible (obviously not practical if using potted organisms, but in practice rooting or floating test organisms will more commonly be used). Samples must also be taken for analyses of dry-to-wet weight ratio and background pollutants in test organisms. Samples of water and sediment, corresponding to each treatment, are taken for instrumental analysis. In some cases, it may not be needed to collect water samples as users may already have a general idea that pollutants are present in water below the limits of analytical detection. The duration of the experiment is usually between 4 and 8 weeks. Longer experiments are problematic due to nutrient exhaustion and/or undue change of growth phase in test organisms. Shorter experiments risk inadequate time for uptake of detectable levels of bioavailable pollutant to occur.

At the end of the bioavailability experiment, samples are removed from microcosms in reverse order of placement. It is useful to set up an assembly line on a bench with all needed tools and sample containers ahead of time. If water samples are collected, these must be collected first. Next, macrophytes are sampled. If the macrophyte test organisms used were non-rooted, these can be sampled directly. A long-handled tong or forceps is useful for handling macrophytes. If the test organisms are rooted, it is advisable to pump out the water overlying sediments and then gently remove the macrophytes, taking care not to lose roots to the sediment, next gently and thoroughly rinse macrophytes in Type I water prior to placement in sample containers to await further processing. Both water and macrophytes should be sampled with due care to avoid disturbing sediments. Finally, using a long-handled spoon or lab scoop, mix the remaining sediment and sample it.

To process macrophytes, first take the wet weight mass of each test organism, as well as recording other data that were recorded at the beginning of the experiment (Table 8.2). Test organisms must then be sectioned according to which portions are scheduled for instrumental analysis (roots, stems, leaves), and the sectioned portions weighed and placed in separate containers for homogenization and extraction of pollutants. Samples must also be taken at this time for dry-to-wet weight analysis. For samples that are scheduled for instrumental analysis, test organism samples must first be homogenized with a tissue homogenizer, and subsequently extracted (for analysis of organic pollutants) or freeze-dried and acid-digested (for analysis of heavy metal pollutants). ***Standard Methods*** for the extraction and instrumental analysis of pollutants in environmental samples are given in Chapter 5. After collection, samples should be stored at 4°C until further processing and subjected to extraction/digestion and analysis as soon as possible. Some investigators freeze samples at –20°C to extend hold times; this is a useful practice for long-term storage of samples that may require secondary reanalysis, however, the best practice for bioavailability experiments is immediate processing for primary analysis.

8.5 VALIDITY OF RESULTS, DATA ANALYSIS, INTERPRETATION, AND REPORTING

8.5.1 Notes on QA/QC Planning Requirements

Specific QA/QC information pertinent to this chapter, in addition to information in Chapter 4 and Standard Methods for instrumental analysis in Chapter 5, is as follows:

- Users performing this method must undergo an Initial Demonstration of Proficiency (IDP). For this method, IDP consists of being able to demonstrate proficiency in extraction/digestion and subsequent instrumental analysis to obtain suitable results for standard reference materials (specifically, pollutants of interest in plant tissues), i.e., obtaining results that fall within accepted ranges for ***accuracy*** and precision for analysis of sediments and biological tissues, with respect to target pollutants.
- Routine instrumental calibration and ***calibration blank*** checks are required; guidelines for ***Acceptance Criteria*** for these are stipulated in appropriate Standard Methods used for instrumental analysis.
- Dilutions, when required, should always be conducted at a factor of 10 or less.
- Batch sizes for instrumental analysis shall not exceed 20 samples; samples batched include sediment and tissue digests/extracts and replicate samples; QA/QC samples are additional and should not exceed 5 per batch. Instrument check samples assess calibration stability and are not part of batching.
- Sample order in batches is to be randomized over all samples and sample types.
- For sediments, one split sample should be analyzed for every 25 sediment samples, and one laboratory split sample should be analyzed for every 25 laboratory tissue samples.
- For every batch, one randomly chosen ***equipment rinsate*** blank, OR one equipment rinsate blank per ***duplicate/replicate sample***, is required, whichever is smaller, if collection of polluted (field) sediments was performed. These samples enable evaluation of whether field contamination occurred. Notes below on QA/QC actions for these samples are to be disregarded if polluted sediments were not collected.
- For every group of sediment and tissue samples digested or extracted at the same time, one ***procedural blank*** is required. The contents of the procedural blank will be determined by the methods of digestion/extraction and instrumental analysis used.
- One ***recovery spike*** is required per batch. This enables evaluation of whether artifactual contamination (recoveries significantly above 100%) or analyte loss (recoveries significantly below 100%) is occurring.
- One ***matrix spike*** is required per batch unless otherwise indicated by the instrumental method used. Results from this indicate the short-term role of the matrix in affecting bioavailability.
- One matrix spike duplicate and one recovery spike duplicate per instrument run (i.e., assumes multiple batches run sequentially), OR one matrix spike duplicate and one recovery spike duplicate every 12 hours, are required, whichever is more frequent, or unless otherwise indicated by the instrumental method used.
- Reference material analyses are used to validate the accuracy of laboratory analytical methods. It is recommended that analytical runs be performed every day, however, guidelines for instrumental methods may supersede guidelines here. If the analytical results do

not meet the accuracy criteria for the reference material and method, there is an operator QA/QC issue, and ***Corrective Action*** is needed.

- If test organisms are collected in the field, to ensure the integrity of the samples during transport, transport containers with good thermal insulation should be used to ensure that test organisms are transported within the specified temperature range. Different macrophytes are transported at different temperatures and suitable transport containers should be selected according to the samples collected. Other field samples (water, sediment) should be transported at 4°C.

8.5.2 Data Analysis, Interpretation of QA/QC Results (Acceptance Criteria), Corrective Action, and Reporting

Data are to be analyzed, interpreted in the context of QA/QC requirements, and if needed subjected to Corrective Action prior to reporting. The information needed includes, at minimum, the following:

- Mortality or problems with test organisms – under normal circumstances, there should be no mortality of macrophyte test organisms. Any more than occasional or adventitious mortality is an indication that something is wrong with the experiment, for instance, pollutant concentrations are toxic, disease is present and affecting test organisms, or experimental conditions are not sufficient to sustain test organism life. In case of more than adventitious mortality (2%), results cannot be reported unless the cause of mortality can be explained, along with some understanding of how results and conclusions of the study are affected.
- The dry-to-wet weight ratio of test organisms should be reported; in the case of experiments where specific tissues are separated for analysis, the dry-to-wet weight ratio of different test organ tissues should be reported (leaves, stems, roots). Refer to Chapter 5 for relevant test methods.
- Information in Table 8.2 (as edited to be appropriate for user-specific bioavailability experiments) should be reported, along with wet weights of test organisms at the start and finish of bioavailability experiments, the background concentrations of pollutants in test organisms at the beginning of experiments, and results from analysis of water (if performed) and sediments. In particular, results relating to indicators of test organism health should be discussed to ascertain whether health changes occurred during the experiment and whether any health changes observed might relate to pollutants.
- Dilution factors (DFs) – unless specified by an instrumental method, DFs are never greater than a factor of 10 for dilution of standards and samples. The basic dilution formula is $C_1V_1 = C_2V_2$. The dilution factor is given by DF = V_2/V_1.
- ***Calibration checks*** and metrics – perform calibration checks according to directions specified in the method used for instrumental analysis. If problems are encountered, recalibrate as needed and reanalyze samples accordingly. Use a Standard Method of analysis stipulating metrics for evaluation of calibration and adhere to same.
- Precision for replicates – report precision of replicates as relative standard deviation (RSD), where RSD = stdev × 100%/average. Acceptable RSDs for instrument replicate analysis should always be below 10%, with Acceptance Criteria for pollutants analyzed in sediments or tissues being larger and more variable. Poor precision for replicates using this method (>20% RSD) should be investigated. If the cause of the poor precision is not discovered, Corrective Action is to reanalyze samples if enough sample material is available.
- Equipment rinsate and procedural blanks – results for blanks must be calculated and reported in units of absolute concentration (not bioavailable) for all blank samples. The acceptance criterion for blanks is a value below the ***limit of detection*** (LOD). Blanks above

the LOD are indicative of field contamination (***equipment blanks***), no Corrective Action is possible, and all samples must be regarded as potentially contaminated.

- LOD and ***limit of quantification*** (LOQ) – these critical parameters define the lowest concentration of heavy metal that can be reliably detected (LOD) and quantified (LOQ) using a given method of instrumental analysis. LOD is often defined as three times the standard deviation of the calibration blank divided by the slope of the ***calibration curve***. LOQ is often defined as three times the LOD. Standard Methods for instrumental analysis commonly contain guidelines concerning what LOD and LOQ are typical. Substantial deviation from such reported guidelines may typically reflect issues with user competency in performing analysis. LOD and LOQ must be reported, as they reflect upon competence of analysis and interpretation of results (e.g., lower precision typically occurs between the LOD and LOQ).
- Split samples (lab and field) – precision is reported as relative percent difference (RPD) and RPD = $200\% \times \text{abs } (R_1 - R_2)/(R_1 + R_2)$, where subscripts 1 and 2 represent the two duplicate samples. Precision acceptance is no greater than 20% RPD for sediment samples and no greater than 15% RPD for tissues. Corrective Actions include reviewing the sample handling and preparation processes to ensure consistency.
- Recovery spikes and matrix spikes – these samples and their method of preparation are discussed in more detail in Chapter 4. Acceptable recovery spikes for this method should be stated in the Standard Methods for extraction and instrumental analysis used. If recovery spike values fall outside of this, Corrective Action involves reviewing and adjusting the sample preparation and analytical procedures to identify potential sources of error.
- Recovery and matrix spike duplicates – these samples should be evaluated by RPD for the duplicate samples, with results reported and explained. The recovery spike duplicate acceptance criterion is 15% RPD or less. Matrix spike duplicates may be higher if ***matrix effects*** are severe.

8.6 CONCLUSIONS AND OUTLOOK

Studying bioavailability using macrophytes, as compared to other test organisms in use, has some inherent points of interest. Macrophytes perform as a type of environmental filter, absorbing and accumulating contaminants, and with root systems having a large surface area for uptake. Many macrophytes exhibit good tolerance to a range of environmental conditions, including water quality, climate, and pollutant load. The ability to easily separate different tissue types of these organisms lends itself to uptake studies that reveal how test organisms process pollutants internally. Bioavailability studies using macrophytes, while labor-intensive compared to many ***chemoavailability*** approaches, is nonetheless reasonably straightforward to implement by nonexperts. Since macrophytes are relatively easily cultured, in fact often cultured by aquarium hobbyists in home settings, it is therefore surprising that this approach is not as widely used as approaches using invertebrates, for example. As more research is published on macrophytes, the approach of using macrophytes in bioavailability studies will no doubt become more common, as more investigators discover the relative ease, flexibility, and utility of this approach.

REFERENCES

Ammeri, R.W., Souid, F., Hajjeji, F., Eturki, S., Moussa, M., 2022. Potential ofaquatic plants for pesticide removal in wastewater: a case study on pentachlorophenol. In: Shukla, G., A. Bhat, J., Chakravarty, S., W. Almutairi, A., Li, M. (eds), *Floristic Diversity - Biology and Conservation*. pp. 156–176. IntechOpen, Rijeka.

Angerer, J., Bird, M.G., Burke, T.A., Doerrer, N.G., Needham, L., Robison, S.H., Sheldon, L., Zenick, H., 2006. Strategic biomonitoring initiatives: moving the science forward. *Toxicol. Sci.* 93, 3–10. https://doi.org/10.1093/toxsci/kfl042

ANHP (Alaska Natural Heritage Program), 2011. *Canadian Waterweed (Elodea canadensis michx.)*. University of Alaska, Alaska.

Caillat, A., Ciffroy, P., Grote, M., Rigaud, S., Garnier, J.M., 2014. Bioavailability of copper in contaminated sediments assessed by a DGT approach and the uptake of copper by the aquatic plant *Myriophyllum aquaticum*. *Environ. Toxicol. Chem.* 33(2). 278–285. https://doi.org/10.1002/etc.2422

Chi, J., Liu, H., 2016. Effects of biochars derived from different pyrolysis temperatures on growth of *Vallisneria spiralis* and dissipation of polycyclic aromatic hydrocarbons in sediments. *Ecol. Eng.* 93, 199–206. https://doi.org/10.1016/j.ecoleng.2016.05.036

Coimbra, E.C.L., Borges, A.C., 2023. Removing Mn, Cu and Fe from real wastewaters with macrophytes: Reviewing the relationship between environmental factors and plants' uptake capacity. *Toxics* 11(2), 158. https://doi.org/10.3390/toxics11020158

Dhir, B., Sharmila, P., Saradhi, P.P., 2009. Potential of aquatic macrophytes for removing contaminants from the environment. *Crit. Rev. Env. Sci. Technol.* 39(9), 754–781. https://doi.org/10.1080/10643380801977776

Diepens, N.J., Arts, G.H.P., Focks, A., Koelmans, A.A., 2014. Uptake, translocation, and elimination in sediment-rooted macrophytes: A model-supported analysis of whole sediment test data. *Environ. Sci. Technol.* 48(20), 12344–12353. https://doi.org/10.1021/es503121x

Doust, J.L., Schmidt, M., Doust, L.L., 1994. Biological assessment of aquatic pollution: A review, with emphasis on plants as biomonitors. *Biol. Rev.* 69(2), 147–186. https://doi.org/10.1111/j.1469-185x.1994.tb01504.x

Gao, J., Garrison, A.W., Hoehamer, C., Mazur, C.S., Wolfe, N.L., 2000. Uptake and phytotransformation of *o,p′*-DDT and *p,p′*-DDT by axenically cultivated aquatic plants. *J. Agric. Food Chem.* 48(12), 6121–6127. https://doi.org/10.1021/jf990956x

Gecheva, G., Stankova, S., Varbanova, E., Kaynarova, L., Georgieva, D., Stefanova, V., 2023. Macrophyte-based assessment of upland rivers: Bioindicators and biomonitors. *Plants*, 12, 1366. https://doi.org/10.3390/plants12061366

Gerhardt, K.E., Huang, X.-D., Glick, B.R., Greenberg, B.M., 2009. Phytoremediation and rhizoremediation of organic soil contaminants: Potential and challenges. *Plant Sci.* 176(1), 20–30. https://doi.org/10.1016/j.plantsci.2008.09.014

Guittonny-Philippe, A., Petit, M.-E., Masotti, V., Monnier, Y., Malleret, L., Coulomb, B., Combroux, I., Baumberger, T., Viglione, J., Laffont-Schwob, I., 2015. Selection of wild macrophytes for use in constructed wetlands for phytoremediation of contaminant mixtures. *J. Environ. Manage.* 147, 108–123. https://doi.org/10.1016/j.jenvman.2014.09.009

Huesemann, M.H., Hausmann, T.S., Fortman, T.J., Thom, R.M., Cullinan, V., 2009. In situ phytoremediation of PAH- and PCB-contaminated marine sediments with eelgrass (*Zostera marina*). *Ecol. Eng.* 35(10), 1395–1404. https://doi.org/10.1016/j.ecoleng.2009.05.011

International Organization for Standardization (ISO), 1992. Water quality. Measurement of biochemical parameters. Spectrometric determination of the chlorophyll-a concentration. In: Prepared by Technical Committee ISO/TC 147: Water Quality, Sub-committee SC 2: Physical, Chemical, Biochemical Methods. 1. ISO International Standard (ISO), 10260.

Jiang, H.S., Zhang, Y., Yin, L., Li, W., Jin, Q., Fu, W., Zhang, T., Huang, W., 2018. Diurnal changes in photosynthesis by six submerged macrophytes measured using fluorescence. *Aquat. Bot.* 149, 33–39. https://doi.org/10.1016/j.aquabot.2018.05.003

Kilminster, K., 2013. Trace element content of seagrasses in the Leschenault Estuary, Western Australia. *Mar. Pollut. Bull.* 73(1), 381–388. https://doi.org/10.1016/j.marpolbul.2013.05.030

Knauer, K., Mohr, S., Feiler, U., 2008. Comparing growth development of *Myriophyllum* spp. in laboratory and field experiments for ecotoxicological testing. *Environ. Sci. Pollut. Res. Int.* 15(4), 322–331. https://doi.org/10.1007/s11356-008-0008-1

Lee, G., Suonan, Z., Kim, S.H., Hwang, D.W., Lee, K.S., 2019. Heavy metal accumulation and phytoremediation potential by transplants of the seagrass *Zostera marina* in the polluted bay systems. *Mar. Pollut. Bull.* 149, 110509. https://doi.org/10.1016/j.marpolbul.2019.110509

Li, J.H., Guo, H.Y., Mu, J.L., Wang, X.R., Yin, D.Q., 2006. Physiological responses of submerged macrophytes to dibutyl phthalate (DBP) exposure. *Aquat. Ecosyst. Health. Manag.* 9(1), 43–47. https://doi.org/10.1080/14634980600561052

Li, X.Q., Hua, Z.L., Zhang, J.Y., Gu, L., 2022. Ecotoxicological responses and removal of submerged macrophyte *Hydrilla verticillate* to multiple perfluoroalkyl acid (PFAA) pollutants in aquatic environments. *Sci. Total Environ.* 825, 153919. https://doi.org/10.1016/j.scitotenv.2022.153919

Liu, H., Bu, H., Liu, G., Wang, Z., Liu, W., 2015. Effects of surrounding land use on metal accumulation in environments and submerged plants in subtropical ponds. *Environ. Sci. Pollut.* 22(23), 18750–18758. https://doi.org/10.1007/s11356-015-5067-5

Liu, J.K., 1999. *Advanced Aquatic Biology*. Science Publishers, Beijing.

Łoboda, A.M., Bialik, R.J., Karpiński, M., Przyborowski, Ł., 2018. Seasonal changes in the biomechanical properties of *Elodea canadensis michx*. *Aquat. Bot.* 147, 43–51. https://doi.org/10.1016/j.aquabot.2018.03.006

Loonen, H., Muir, D., Parsons, J., Govers, H., 1997. Bioaccumulation of polychlorinated dibenzo-*p*-dioxins in sediment by oligochaetes: Influence of exposure pathway and contact time. *Environ. Toxicol. Chem.* 16(7), 1518–1525. https://doi.org/10.1002/etc.5620160728

Maleva, M.G., Malec, P., Prasad, M.N.V., Strzałka, K., 2016. Kinetics of nickel bioaccumulation and its relevance to selected cellular processes in leaves of *Elodea canadensis* during short-term exposure. *Protoplasma* 253(2), 543–551. https://doi.org/10.1007/s00709-015-0832-3

Mascaró, O., Valdemarsen, T., Holmer, M., Pérez, M., Romero, J., 2009. Experimental manipulation of sediment organic content and water column aeration reduces *Zostera marina* (eelgrass) growth and survival. *J. Exp. Mar. Biol. Ecol.* 373(1), 26–34. https://doi.org/10.1016/j.jembe.2009.03.001

Meng, F., Chi, J., 2017. Effect of *Potamogeton crispus* L. on bioavailability and biodegradation activity of pyrene in aged and unaged sediments. *J. Hazard. Mater.* 324, 391–397. https://doi.org/10.1016/j.jhazmat.2016.11.001

Meng, F., Huang, J., Liu, H., Chi, J., 2015. Remedial effects of *Potamogeton crispus* L. on PAH-contaminated sediments. *Environ. Sci. Pollut.* 22(10), 7547–7556. https://doi.org/10.1007/s11356-015-4280-6

Mishra, V.K., Tripathi, B.D., 2008. Concurrent removal and accumulation of heavy metals by the three aquatic macrophytes. *Bioresour. Technol.* 99(15), 7091–7097. https://doi.org/10.1016/j.biortech.2008.01.002

Moore, M.T., Kröger, R., Cooper, C.M., Smith, S., 2009. Ability of four emergent macrophytes to remediate permethrin in mesocosm experiments. *Arch. Environ. Contam. Toxicol.* 57(2), 282–288. https://doi.org/10.1007/s00244-009-9334-7

Nazneen, S., Mishra, A.K., Raju, N.J., Mehmood, G., 2022. Coastal macrophytes as bioindicators of trace metals in the Asia's largest lagoon ecosystem. *Mar. Pollut. Bull.* 178, 113576. https://doi.org/10.1016/j.marpolbul.2022.113576

Nguyen, T.Q., Sesin, V., Kisiala, A., Emery, R.J.N., 2021. Phytohormonal roles in plant responses to heavy metal stress: Implications for using macrophytes in phytoremediation of aquatic ecosystems. *Toxicol. Environ. Chem.* 40(1), 7–22. https://doi.org/10.1002/etc.4909

Norouznia, H., Hamidian, A.H., 2014. Phytoremediation efficiency of pondweed (*Potamogeton crispus*) in removing heavy metals (Cu, Cr, Pb, As and Cd) from water of Anzali wetland. *Int. J. Aquat. Biol.* 2(4), 206–214. https://doi.org/10.22034/ijab.v2i4.87

Nyquist, J., Greger, M., 2007. Uptake of Zn, Cu, and Cd in metal loaded *Elodea canadensis*. *Environ. Exp. Bot.* 60(2), 219–226. https://doi.org/10.1016/j.envexpbot.2006.10.009

Patel, D., Verma, R., & Hait, M. (2024). Macrophytes as a sustainable biomonitoring tool of aquatic pollution. In S. Kumar & A. Kumar (Eds.), *Biomonitoring of pollutants in the Global South* (pp. 211–226). Springer. https://doi.org/10.1007/978-981-97-1658-6_11

Polechońska, L., Klink, A., Dambiec, M., Rudecki, A., 2018. Evaluation of *Ceratophyllum demersum* as the accumulative bioindicator for trace metals. *Ecol. Indic.* 93, 274–281. https://doi.org/10.1016/j.ecolind.2018.05.020

Pujar, A., Jaiswal, P., Kellogg, E.A., Ilic, K., Vincent, L., Avraham, S., Stevens, P., Zapata, F., Reiser, L., Rhee, S.Y., 2006. Whole-plant growth stage ontology for angiosperms and its application in plant biology. *Plant Physiology* 142(2), 414–428. https://doi.org/10.1104/pp.106.085720

Qian, Y., Cheng, C., Drouillard, K., Zhu, Q., Feng, H., He, S., Fang, Y., Qiao, S., Kolenčíka, M., Chang, X., 2019. Bioaccumulation and growth characteristics of *Vallisneria natans* (Lour.) hara after chronic exposure to metal-contaminated sediments. *Environ. Sci. Pollut.* 26(20), 20510–20519. https://doi.org/10.1007/s11356-019-05347-z

Raskin, I., Kumar, P.N., Dushenkov, S., Salt, D.E., 1994. Bioconcentration of heavy metals by plants. *Curr. Opin. Biotechnol.* 5(3), 285–290. https://doi.org/10.1016/0958-1669(94)90030-2

Regier, N., Larras, F., Bravo, A.G., Ungureanu, V.G., Amouroux, D., Cosio, C., 2013. Mercury bioaccumulation in the aquatic plant *Elodea nuttallii* in the field and in microcosm: Accumulation in shoots from the water might involve copper transporters. *Chemosphere* 90(2), 595–602. https://doi.org/10.1016/j.chemosphere.2012.08.043

Sapci, Z., Ustun, E.B., 2015. Heavy metal uptakes by *Myriophyllum verticillatum* from two environmental matrices: The water and the sediment. *Int. J. Phytoremediation* 17(3), 290–297. https://doi.org/10.1080/15226514.2014.898022

Sculthorpe, C.D. (1967). The Biology of Aquatic Vascular Plants. London: Edward Arnold.

Smart, M.R., Barko, J.W., 1985. Laboratory culture of submersed freshwater macrophytes on natural sediments. *Aquat. Bot.* 21(3), 251–263. https://doi.org/10.1016/0304-3770(85)90053-1

Song, Y., Zhang, L.L., Li, J., He, X.J., Chen, M., Deng, Y., 2018. High-potential accumulation and tolerance in the submerged hydrophyte *Hydrilla verticillata* (L.f.) Royle for nickel-contaminated water. *Ecotoxicol. Environ. Saf.* 161, 553–562. https://doi.org/10.1016/j.ecoenv.2018.06.032

Srivastava, S., Mishra, S., Dwivedi, S., Tripathi, R.D., Tandon, P.K., Gupta, D.K., 2009. Evaluation of zinc accumulation potential of *Hydrilla verticillata*. *Biol. Plant.* 53(4), 789–792. https://doi.org/10.1007/s10535-009-0145-5

Suman, J., Uhlik, O., Viktorova, J., Macek, T. 2018. Phytoextraction of heavy metals: A promising tool for clean-up of polluted environment? *Front. Plant Sci.* 9, 1476. https://doi.org/10.3389/fpls.2018.01476

Tarish, M., Ali, R.T., Shan, M., Amjad, Z., Rui, Q., Akher, S.A., Al Mutery, A., 2024. Plant tissues as biomonitoring tools for environmental contaminants. *Int. J. Plant Biol.* 15(2):375–396. https://doi.org/10.3390/ijpb15020030

Van Beek, J.M., Zheng, T., Wang, Z., Kovach, K.R., Townsend, P.A,. 2024. Evaluation of the reliability of the CCM-300 chlorophyll content meter in measuring chlorophyll content for various plant functional types. *Sensors* 24(15), 4784. https://doi.org/10.3390/s24154784

Wang, C., Zheng, S.S., Wang, P.F., Qian, J., 2014. Effects of vegetations on the removal of contaminants in aquatic environments: A review. *J. Hydrodyn.* 26(4), 497–511. https://doi.org/10.1016/S1001-6058(14)60057-3

Windisch, J., Plessl, C., Christian, C., Zechmeister, T., Jirsa, F., 2022. Unexpected pathways of mercury in an alkaline, biologically productive, saline lake: A mesocosm approach. *J. Hazard. Mater.* 427, 128163. https://doi.org/10.1016/j.jhazmat.2021.128163

Xing, W., Wu, H., Hao, B., Liu, G., 2013. Metal accumulation by submerged macrophytes in eutrophic lakes at the watershed scale. *Environ. Sci. Pollut.* 20(10), 6999–7008. https://doi.org/10.1007/s11356-013-1854-z

Yin, Y., Wang, X., Sun, Y., Guo, H., Yin, D., 2008. Bioaccumulation and oxidative stress in submerged macrophyte *Ceratophyllum demersum* L. upon exposure to pyrene. *Environ. Toxicol.* 23(3), 328–336. https://doi.org/10.1002/tox.20330

Yu, G., Wang, J., Cheng, H., Luo, X., Wang, S., Zheng, Z., 2024. Toxicological effects of bisphenol f on growth, antioxidant response, ultrastructure, and microbial properties of the submerged macrophyte *Ceratophyllum demersum* L. *Sci Total Environ.* 957, 177481. https://doi.org/10.1016/j.scitotenv.2024.177481

Yuan, Q., Qiang, J., Zhang, Z.Q., Chen, H., Zhang, S.H., 2023. Physiological changes and the effect on water during the decay of *Myriophyllum verticillatum* in winter. *J. Anhui Univ. Nat. Sci.* 47(6), 93–101. https://doi.org/10.3969/j.issn.1000-2162.2023.06.013

Zhang, B., Wu, Z., Cheng, S., He, F., Wang, Y., Gao, Y., 2007. Primary study on phytodegradation of bisphenol a by *Elodea nuttallii*. *Wuhan Univ. J. Nat. Sci.* 12(6), 1118–1124. https://doi.org/10.1007/s11859-007-0110-0

Zhao, J., Liu, Y., 2009. *Illustrated Catalogue of Aquatic Plants*. Huazhong University of Science and Technology Press, Hubei.

Zhao, W., 2005. *Aquatic Biology*. China Agricultural Publishing House, Beijing.

Zhou, J., Gao, J., Liu, Y., Ba, K., Chen, S., Zhang, R., 2012. Removal of fluoride from water by five submerged plants. *Bull. Environ. Contam. Toxicol.* 89(2), 395–399. https://doi.org/10.1007/s00128-012-0710-z

9 Measuring Bioavailability Using Whole-Cell Bioreporters

9.1 INTRODUCTION AND SCOPE

In recent years, as a rapid, cost-efficient, and effective technology, the use of whole-cell ***bioreporters*** to assess the ***bioavailability*** of pollutants is becoming increasingly attractive (Belkin, 2003; Zhang et al., 2017). Whole-cell bioreporters are living organisms that have been engineered to respond to physical or chemical elements in their environment by creating electrochemical or optical signals as a "report" that can be quickly and quantitatively identified due to the action of ***reporter genes*** (Van Der Meer et al., 2004). These bioreporters can be categorized into three groups, the mechanism for action of which is illustrated in Figure 9.1; two of these are "lights-on", and one is "lights-off".

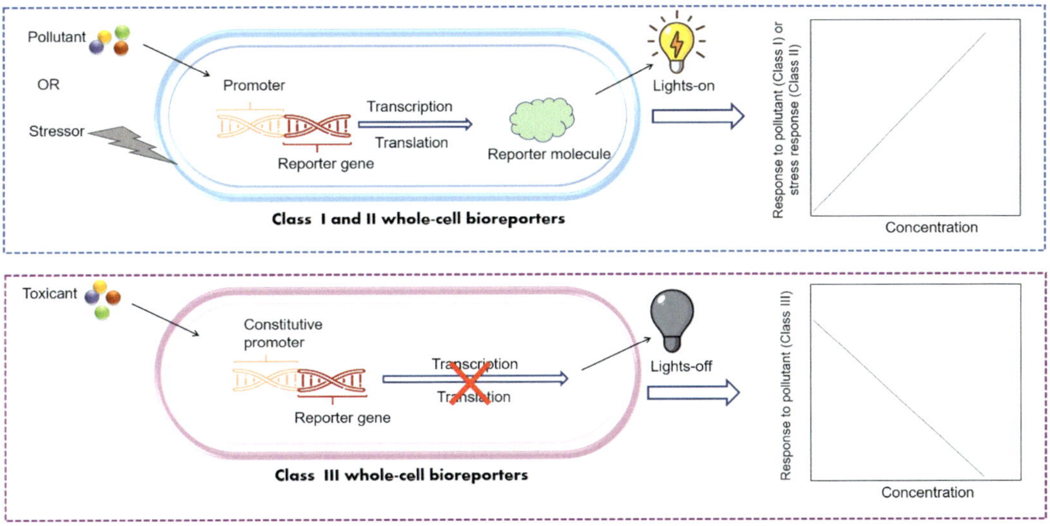

FIGURE 9.1 Schematic illustrating the response mechanisms of Class I, Class II, and Class III, bioreporters, responding to, respectively, specific pollutants (top panel), pollutants as stressors according to stressor type (top panel), and pollutants as toxicants (bottom panel). These bioreporters are engineered organisms (often bacteria) that can detect the presence of pollutants by generating a measurable response in the form of a reporter molecule. Class I bioreporters have specific DNA sequences that control cellular machinery that confers a certain resistance to the target pollutant upon ***exposure***. This quality makes the Class I bioreporter highly selective in detection. Class II bioreporters have similar machinery that is activated by global stress response, i.e., may be caused by a number of pollutants, but is specific to a particular type of stress such as heat shock or DNA damage. Class III bioreporters are based on a constitutive part of the cellular machinery that is always active and thus drives continuous expression of a reporter molecule under normal circumstances; when a toxicant disrupts normal cellular processes, signal decreases (Zhu et al., 2022). In the context of whole-cell bioreporters, transcription and translation are the two essential processes that allow the cell to respond to specific pollutants and produce a detectable signal. Transcription is the process in which the DNA sequence of a gene is copied into messenger RNA (mRNA). Translation is the next step after transcription, where the mRNA produced is used as a template to synthesize proteins. These proteins are responsible for carrying out various functions and include ***reporter molecules***. ***Promoters*** and reporter genes are discussed further in Section 9.2.

DOI: 10.1201/9781003355229-11

Class I bioreporters are those that output a signal based on the bioavailability of a particular pollutant or class of pollutant in a dose-dependent manner; for instance, bioreporter zntA produces dose-dependent signaling in response to Pb exposure (Zhang et al., 2017). Class II bioreporters output a signal based on the type of molecular or biological stress that pollutant stressors cause. For instance, Hartono et al. (2018) found that Class II bioreporters could be used to measure the oxidative stress and cellular damage caused by multi-walled carbon nanotubes (Hartono et al., 2018). Class III bioreporters exhibit a non-specific reduction in signal output as a result of the toxicity of the chemical, i.e., because Class III bioreporters measure toxicity, they only provide an indirect reflection of bioavailability (Abd-El-Haleem et al., 2006; Zhu et al., 2022).

Many studies have shown that Class I whole-cell bioreporters can be used to evaluate the bioavailability of various pollutants, including both heavy metal and organic pollutants (Kang et al., 2018; Magrisso et al., 2009; Sun et al., 2017). Class I bioreporters have been used to evaluate the bioavailability of heavy metals – for instance, lead (Zhang et al., 2017), mercury (Ndu et al., 2012; Xu et al., 2015), arsenic (As) (Trang et al., 2005), nickel (Ni) (Cayron et al., 2017), zinc (Zn) (Liu et al., 2012), cadmium (Cd) (Elcin and Öktem, 2020), and copper (Cu) (Pang et al., 2020), among others. They can also be constructed to measure the bioavailability of organic pollutants such as endocrine disruptors (Wang et al., 2015) and priority pollutants such as polycyclic aromatic hydrocarbons (Sun et al., 2017; Tecon et al., 2006; Wells et al., 2005). Class I bioreporters are usually constructed based on sensing elements from pollutant resistance genes (van der Meer and Belkin, 2010). This ensures that organisms engineered with such sensing elements remain active at relatively high pollutant concentrations without experiencing toxic effects. In addition, compared to other methods that have been used to evaluate pollutant bioavailability, including all the other methods for estimating ***chemoavailability*** and true bioavailability in this book, the whole-cell bioreporter method is sensitive, cost-effective, and high throughput. Instead of lengthy manual preparations of samples, followed by the need for extraction or digestion and instrumental analysis using complicated and costly equipment, bioreporter analyses can be completed in 1 day, using a microplate reader to measure signal output, such that hundreds of samples may be processed in a single day by one user. This number would increase to thousands of samples in an efficient laboratory using high-density microplates and automated plate movers and stackers.

This chapter will introduce and provide a generalized method for using Class I whole-cell bacterial bioreporters to detect the bioavailability of heavy metals and organic pollutants in water and soil/sediment. It can also be used to test the bioavailability of pollutants in seawater if the host organism (i.e., ***host cell***) is adapted to it. While the method focuses on Class I bacterial bioreporters, it should be generally applicable and adaptable for use with Class II bioreporters as well. Because bioreporter methods vary according to the different pollutants detected and the individual bioreporter strains used, this chapter provides guidance on how users tailor methods for their particular applications. Information is provided for the selection and handling of materials, how to culture bioreporters, and how to make measurements. Chapters containing terms and definitions, on avoiding contamination of samples in field and lab work, and guidance on Quality Assurance (QA)/Quality Control (QC) are also within the scope of this chapter.

9.2 PRINCIPLE

Bioreporters contain a ***promoter*** fused to a reporter gene. The promoter is a region of DNA that functions in a manner to initiate and regulate the transcription of one or more genes downstream. The reporter gene is a non-endogenous gene that encodes and is able to express a reporter molecule. Commonly used reporter genes produce ***reporter molecules*** that are fluorescent proteins (such as green fluorescent protein [GFP]) or enzymes (such as luciferase, which catalyzes a reaction producing luminescence, or β-galactosidase, which catalyzes a reaction leading to a color change). In a bioreporter, the promoter functions as the sensor, detecting the presence of a substance and producing a response, and the reporter gene a transducer, converting the sensor's response into a measurable output, usually optical (fluorescence, luminescence, color). To construct a bioreporter, a promoter is fused to a reporter gene, which ensures that whenever the promoter is active, the reporter gene is continuously expressed. The promoter:reporter gene fusion is then ***transformed***, i.e., introduced into a suitable host (micro)organism or host cell, either on a plasmid or as a stable chromosomal insertion, to create genetically distinct bacterial bioreporter strains with reporting capabilities tailored to a pollutant or class of pollutants of interest. The host organism or bacterial species into which the promoter:reporter gene fusion is transformed is chosen based on its compatibility with hosting the fusion, ease of genetic manipulation, growth rate and ease of culture, ability to function well in environmental sample media, and other relevant characteristics.

Class I bacterial bioreporters derive the specificity of their response from regulatory proteins that bind to, or influence access to, the promoter, thereby governing its behavior. The exact mechanism of how this regulation occurs can vary. For instance, a pollutant may bind directly to a receptor on the regulatory protein causing the regulatory protein to change shape (i.e., change confirmation), facilitating access to the promoter and initiation of the downstream gene transcription process. In some cases, a regulatory protein may block transcription, and pollutant binding causes a confirmation change that causes the regulatory protein to dissociate from the promoter, enabling transcription to proceed. The overall process by which the pollutant interacts with the regulatory system to affect a change in the status of the promoter is ***induction***; induction results in direct ***activation*** of the promoter, leading to the expression of the reporter and other downstream genes. Expression of the reporter gene results in a signal from reporter molecules that, for fluorescent, luminescent, and colored reporter molecules produced by most bioreporters, is measured by a microplate reader, providing high-throughput data. Some bioreporters produce reporter proteins that need substrates for particular chemical reactions to emit signal. For example, the reporter protein β-galactosidase requires the substrate X-Gal (5-bromo-4-chloro-3-indolyl β-D-galactopyranoside) to produce a colored signal.

9.3 AVOIDING CONTAMINATION AND PRECAUTIONARY NOTES

When conducting bioreporter experiments, it is particularly important to prevent contamination, and simultaneously difficult to do so since, for these experiments, contamination may arise from two ***sources***, both chemical ***contaminants*** and biological contamination of the ***test organisms***. Chapter 3 provides guidelines on avoiding chemical contamination in field and laboratory work.

Regarding biological contamination, in the process of cultivating strains, many routes may introduce bacterial contamination; for example, if the cultivation flask is not scrupulously sterile, or a cap on the bottle of cultivation media is not secure. It is important to ensure all instruments (e.g., petri dishes, pipettes, tubes, inoculation loops) are properly sterilized before use. This can be done via autoclaving (loops may be flamed). Ensure that all media used are sterile, either by autoclaving for media that can withstand autoclaving without alteration or by sterile filtering for media that cannot be autoclaved. Ensure to wear gloves and lab coats; change gloves and launder lab coats frequently to avoid cross-contamination. All procedures must be performed under sterile conditions to prevent biological contamination that may affect the results of the experiment, i.e., in a biosafety cabinet. Regularly clean work surfaces and equipment using a disinfectant solution (e.g., 70% ethanol, bleach or other suitable disinfectants). This helps minimize the risk of biological contamination from ambient microorganisms. Procedures given herein provide means of ensuring that bioreporter stock is not contaminated. Maintain sufficient quantities of uncontaminated cryopreserved bioreporter stock to ensure that reculturing is possible if a contamination event does occur.

9.4 TEST ORGANISMS, EQUIPMENT, CONSUMABLES, AND REAGENTS

9.4.1 Guidance Notes on Obtaining Test Organisms

For bioreporter experiments, the most important thing is to obtain a suitable bioreporter. There have been few companies that provide bioreporters or bioreporter-based sensing technology commercially, and the lifespan of companies that have tried has not been long. One of the main factors that has limited the success of bioreporter commercialization is likely that, even if the technology is available, there is a lack of generalized familiarity with using bioreporters for Environmental Monitoring. Table 9.1 provides a representative list of culture collections, some of which have bioreporters. Finding bioreporters in the vast and sometimes incomplete catalogues of these organizations is difficult; however, they offer a first point of contact to discover regional resources to pursue. Regional resources are important in some countries due to restrictions on importing such strains. Alternatively, a good strategy is to directly approach researchers who have expertise in bioreporter construction to request the use of existing strains. A list of representative highly cited papers authored by such investigators is given in the footnote to Table 9.1.

9.4.2 Equipment and Consumables

The list of equipment, consumables, and reagents given in this section is intended to be reasonably comprehensive; users should, however, read the entire method before finalizing needs and modify this indicative list accordingly. Note: All equipment are presumed appropriately clean and contaminant-free. Refer to Chapter 3 for guidance on avoiding contamination. Additionally, all materials that come into contact with bioreporters must be sterile.

Aluminum foil – used to cover the openings of vessels when autoclaving and for wrapping organic-contaminant-free implements.

TABLE 9.1
Culture Collections That May Have Bioreporters Or Be Able to Provide Information on Sourcing Bioreporters[a]

Abbreviation of Organization	Culture Collection	Website
ATCC	American Type Culture Collection	https://www.atcc.org/
BCCM	Belgian Coordinated Collections of Microorganisms	https://bccm.belspo.be/
CABRI	Common Access to Biological Resources and Information	http://www.cabri.org/
CBE[b]	Chemical and Biomolecular Engineering department at the University of Tennessee	https://www.utk.edu/
CCUG	Culture Collection, University of Gothenburg	https://www.gu.se/en
CGMCC	China General Microbiological Culture Collection Center	https://cgmcc.net/english/
EBPI[b]	Environmental Bio-Detection Products Inc.	https://biotoxicity.com/
IMI	CABI - Centre for Agriculture and Bioscience International	https://www.cabi.org/
JCM	Japan Collection of Microorganisms	https://jcm.brc.riken.jp/en/?doing_wp_cron=1735105592.4306008815765380859375
JSCC	Japan Society for Culture Collections online database	https://www.nite.go.jp/nbrc/information/20190325.html
NCTC	National Collection of Type Cultures (UK)	https://www.culturecollections.org.uk/about-us/nctc/
SCK CEN[b]	Belgian Nuclear Research Centre	https://www.sckcen.be/en/node/13
VKM	All Russian Collection of Microorganisms	https://vkm.ru/

[a] More information can be found from https://wfcc.info/sites/.

[b] Represents organizations that are confirmed to have bioreporters. Bioreporters may also be acquired from College of Water Sciences at Beijing Normal University (https://cws1.bnu.edu.cn/Default.aspx); alternately, investigators developing bioreporters are sometimes amenable to sharing their strains and may be approached directly, e.g., Al-Anizi et al., (2014), Belkin et al., (2017), Din et al., (2019), Gao et al., (2017), Hui et al., (2021), Lopes et al., (2012), Moratti et al., (2024), Van Der Meer (2006), Wu et al., (2023), Xu et al., (2018), and Zhang et al., (2012).

Auger – used to sample soils at depth.
Autoclave – to sterilize utensils and solutions used during experiments.
Balance – to measure masses of chemicals needed for media and reagent solutions (accurate to 0.001 mg).
Beakers, volumetric flasks, volumetric pipettes – for preparing and handling reagent solutions.
Biosafety cabinet(s) – to perform experimental procedures under sterile conditions and avoid microbiological contamination of test organism cultures.
Cryopreservation tubes (~0.25–1 mL) – for cryopreservation of bioreporter stocks.
Coring sampler – used for collecting sediments at depths.
Cultivation flasks (e.g., Erlenmeyer type with sterile vented closures) or sterile tubes – to incubate bioreporter test organisms in liquid media; must have enough capacity to produce the amount of cell culture needed and to ensure adequate aeration during cultivation.
Cuvette(s) – used in experiments for spectrophotometric measurements to determine growth via optical density.
Fiber-reinforced tape – used to seal thermo-insulated boxes used for transport of field samples.
Field logbook – in addition to a sample collection log sheet, field personnel should record information about sample collection activities, field conditions, pH, temperature, and any unexpected events that may affect the results of experiments, which will help to assist in the subsequent analysis of the results of the experiment.
Filtration equipment – when needed for filtration of field collected water samples; consists of

A. A vacuum filtration device, vacuum, and ***filter membranes***;

OR

B. syringes with syringe filter cartridges.

The filter membranes associated with either option above must be 0.22 or 0.45 μm pore size to remove large particles and/or microorganisms from samples. Filters composed of polytetrafluoroethylene (PTFE) are often used to avoid loss of heavy metals and organic pollutants from samples via adhesion to the filter. In practice, the user should solicit technical specifications for products used. For option A, vacuum pumps used must also be of a suitable design to avoid sample contamination. For option B, it is important to choose materials that will not contaminate the samples. See Chapter 3 for more details on avoiding sample contamination.

Glass or PTFE containers with PTFE-lined lids (depending on pollutant) – for collection of field soil, sediment, or water samples.
Glass rod – to stir/dissolve chemicals in preparation of media and reagent solutions.
GPS/electronic or physical devices – to ensure unambiguous location of sample ***sites*** (e.g., GPS, beacons) and depths (e.g., depth gauge).
Grab sampler – to collect surficial sediment samples.
High-speed centrifuge – for harvesting cells; should be able to accommodate 1–50 mL tubes.
Ice/dry ice – ice is used to ensure that the sample temperature is around 4°C during transportation. Ice packs are preferred if they can be kept well frozen prior to use; alternatives include dry ice (subject to certain shipping requirements) or wet ice (which increases the chance of cross-contamination of samples as it melts). When using dry ice, the following points should be noted: dry ice in contact with water sublimates rapidly and generates large amounts of gas, which may result in package breakage or sample damage. Therefore, avoid direct contact between dry ice and water during transportation and ensure that thermo-insulated boxes are dry and free of moisture. During transportation, ventilation should be maintained inside the packing box to prevent carbon dioxide from accumulating. At the same time, excessive pressure or squeezing of the box should be avoided to avoid damaging the packaging material or affecting the insulation effect of dry ice. When handling dry ice, gloves, goggles, and other protective measures should be worn to avoid direct skin contact and eye injury.

Inoculating loop – used to streak plates/petri dishes with solid media and to inoculate liquid media.

Indelible pens – to record relevant data without losing fidelity in wet field environments and for marking labels.

Lab coat – to provide protection of skin and personal clothing from incidental contact, preventing the spread of chemical and microbial contamination outside the lab.

Lab gloves – clean/contaminant-free, of a material suitable for avoidance of contamination.

Lab freezer – −80°C, used for cryopreservation.

LIVE/DEAD® BacLight™ Bacterial Viability Kit, or similar – used in bioreporter cell viability assays.

Microplate reader and microtiter plates – to detect the optical signal output by bioreporter (visible, fluorescent, or luminescent, depending on bioreporter construction); note: plates come in a variety of formats, however, 96-well microplates are a standard that is often used. For colorimetric detection, a clear, flat-bottom plate is needed. Black plates are usually used for fluorescent signals and white plates are used for luminescent signals.

Muffle furnace – for high-temperature decontamination of glassware (organic pollutants).

Multichannel pipette (8 and 12 channels, 10–200 µL) and tips – for pipetting cell cultures and reagents into microtiter plates.

Multichannel pipette solution basin/trough – to hold solution for rapid loading of microtiter plates.

Multisensor sonde – e.g., YSI or similar; used for on-site measurements during water sample collection. Calibration is required before use to ensure the ***precision*** of the instrument. Multiple sensors are desirable, including those capable of measuring temperature, pH, conductivity, and dissolved oxygen (DO).

Niskin-type water collection bottle(s), or similar, and line – for the collection of water samples, needs to seal well, be corrosion-resistant, and the bottle itself needs to be clean and free of contamination. Specialized PTFE-lined bottles should be used for heavy metals (e.g., sold by General Oceanics).

Paper towels – used to wipe down tools and surfaces to prevent external contaminants from being introduced into the sample during the sampling process, as well as to effectively remove soil from the surface of tools as part of decontamination/to avoid cross-contamination.

Petri dishes – for holding solid media and culturing bioreporters.

pH meter – to test the pH of solutions as needed.

Pipettes (single channel, 100–1,000 µL) and tips – general use for pipetting in experiments.

Reagent bottles – assortment of sizes, to store reagents.

Refrigerators – 4°C, used for storing samples and media.

Sample identification labels – completed in indelible ink, labels are used for unique identification and shall contain (1) a unique sample identifier (also recorded in the field logbook), (2) location of the sample sampling point, (3) date of collection, and (4) type of sample collected.

Scraper/spoon (stainless steel) – used to scrape soil samples from an auger head.

Secchi disk – typically half black, half white, with a diameter commonly around 30 cm; used for gauging the transparency of waters.

Several sizes of resealable watertight (press-seal/ziplock type, polyethylene) plastic bags – contaminant-free, for secondary packaging of field implements and samples for analysis of ***bioavailable*** pollutants. Note: Plastic may be a source of contamination for organic pollutants; if used as secondary packaging for implements or samples in studies of organic pollutants, care must be taken to ensure there is no direct contact between the secondary packaging and the implement or sample. For instance, implements should be wrapped in clean aluminum foil as primary packaging.

Soil/sediment mixing tool – to homogenize soil or sediment samples; e.g., rolling mill, feed mixer, or gardener's soil mixer (e.g., DaRos MX1000 or similar).
Solvents and/or cleaners – as needed for decontamination of field equipment.
Stainless-steel shovels/trowels– used when collecting soil and sediment samples; must be appropriately decontaminated before each use.
Stainless-steel buckets – used when collecting/field processing larger volume soil/sediment samples, or for collecting surface water samples.
Sterile syringes and syringe filters – used to sterile filter antibiotic and carbon source solution into media.
Test organisms – bioreporters for sample testing; must be specific to target pollutant(s) or stress response.
Thermo-insulated box with ice packs, ice, or dry ice – clean, free of contaminants, and made of an appropriate material for the target pollutant(s); must be able to keep samples and/or test organisms at ambient temperature while being handled and transported.
Thermostatic shaker – for the cultivation of test organisms.
Trays – sterile, for transferring items used during experiments.
Tweezers – used to remove weeds or extraneous material from soil/sediments.
Type I water and solvents – for rinsing/decontamination, as needed.
UV/visible photometer and cuvettes – to determine the optical density of test organism cultures at 600 nm and monitor growth.

9.4.3 Guidance Notes on Reagents

Target pollutant(s) bioavailability to be measured using a bioreporter are determined according to the purpose of each study. In bioreporter experiments, different bioreporters will require different media compositions and therefore will require different reagents. These need to be tailored to the specific experiment. This section provides generalized reagents for culturing bioreporters as well as some guidance on selecting reagents. All reagents must be of guaranteed reagent quality or better to ensure that the reagents used are of the highest purity to avoid contamination.

All solutions use ultrapure water – also known as Type I water, has a resistance of 18 MΩ·cm.

Reagents for preparing, storing, and maintaining stocks:

Antibiotics – e.g., ampicillin, kanamycin, or tetracycline are often used in stock solutions – assisting in reduction of biological contamination when added to bioreporter cultivation media.
Biocompatible solvents – e.g., methanol, acetone, dimethyl sulfoxide (DMSO) – used as ***carrier*** to retain organic pollutants in standards in solubilized form.
Complex or rich media – rich in nutrients and containing ingredients whose exact chemical composition is unknown; usually used for bioreporter cultivation. Examples of commonly used media are lysogeny broth (LB), or tryptic soy broth (TSB). LB is by far the most commonly used rich media.
Ethanol (75%) and/or disinfectants –for disinfection, as needed.
Glycerol (cell culture grade) – for cryopreservation of bioreporter, it ensures that the bacteria remain active during long-term preservation by lowering the freezing point to prevent the formation of ice crystals that can damage the cellular structure at low temperatures.
Minimal media (MM) – a cultural media with minimal necessities to support the growth of a wide type of microorganisms (e.g., bioreporter host cells). Usually used in bioavailability (bioreporter induction/activation) experiments, minimal media is a defined media (exact composition is known, composed of pure ingredients in carefully measured concentrations) with the composition varying according to the microorganism being cultured.

Typically, MM contains the following: (1) a carbon source: glucose, glycerol, gluconate, acetate, and succinate are commonly used carbon sources. (2) Nitrogen and phosphorous sources – usually obtained from ammonium and phosphate, though nitrate, urea, and β-glycero-phosphate are also used. (3) Basal salts: sodium, potassium, calcium, and magnesium in forms such as NaCl, Na_2HPO_4, NaH_2PO_4, KCl, K_2HPO_4, KH_2PO_4, KNO_3, $CaCl_2$, $MgSO_4$, and $MgCl_2$. (4) Buffer: The choice of buffer depends on the environment required by the bioreporter and in addition needs to be based on whether it will react with the target pollutant and thus affect pollutant bioavailability. Commonly used buffers include HEPES (2-(4-(2-hydroxyethyl) piperazin-1-yl)ethanesulfonic acid; pH range: 6.8–8.2), MES (2-(*N*-morpholino)ethanesulfonic acid; pH range: 5.5–6.7), MOPS (3-(*N*-morpholino)propanesulfonic acid; pH range: 6.5–7.9), phosphate buffer (pH range: 6.0–8.0), and Tris (tris(hydroxymethyl)aminomethane; pH range: 7.0–9.0). (5) Micronutrients – optional in some MM and are decided based on the growth requirements of different bioreporters, determined by prior knowledge from the bioreporter source or literature on the bioreporter host organism.

Solid media – usually agar.

Standards and calibration solutions – for instrument calibration and quantification. Certified stock standards with highly accurate known concentrations should be purchased, from which standards for calibration and/or spiking need to be prepared.

9.5 PROCEDURE

9.5.1 Preparing Bioreporter Stock Solutions

Prior to testing pollutant bioavailability using a bioreporter, the acquired strain first needs to be processed to produce bioreporter stocks for routine use. Stocks in individual vials are kept in a freezer at −80°C. In this manner, a new vial of bioreporter stock can be taken out of the freezer and used each time an experiment is performed. The basic process of preparing bioreporter stock solutions is to (a) prepare needed liquid media, solid media, and equipment; (b) prepare a liquid inoculum; (c) inoculate a sample of the acquired strain into solid ***medium***; (d) incubate/grow new culture; (e) sample single colonies to regrow as purified stock; (f) collect a large number of samples of purified stock, prepare for freezing, and freeze at −80°C; and (g) test stock. If, at the end of this process, the cryopreserved stocks respond suitably to induction and activation, this stock is fit to be used in experiments. If not, the process needs to be repeated using more new single colonies to assess whether the entire sample of the acquired bioreporter is unsuitable for use, or if an individual colony can be identified that can produce stock suitable for use. Part of the process of preparing a bioreporter strain up to freezing stocks is graphically illustrated in Figure 9.2.

For preparation of a cultivation medium, the most common approach is to use a rich medium in liquid form and in agar plates (petri dishes). LB is typically used, though rarely there may be bioreporters based on organisms that do not do well on LB, in which case an organism-specific medium should be used (refer to Section 9.5.3 for more details on media). Bioreporters for use in environmental applications contain antibiotic resistance genes for strain selection and maintenance during the engineering and cultivation processes. As such, an appropriate antibiotic, corresponding to the antibiotic resistance gene used in strain engineering, must be in the cultivation media. Providers of strains will advise on the antibiotic and dosage. A representative method to prepare antibiotic LB agar culture plates is to dissolve 12.5 g of LB powder in ~400 mL of Type I water. If the ready-made powder is not available, 12.5 g can be made by combining 5 g of tryptone, 2.5 g yeast extract, and 5 g of sodium chloride. Next, add 7.5 g of agar powder into the 400 mL of LB, mix well to fully dissolve the agar, then make up to 500 mL total volume and autoclave. After autoclaving, add additional sterile water to replace any lost in the autoclaving process. The weights of LB and agar may vary – consult supplier recommendations. Also, in actual use conditions, the

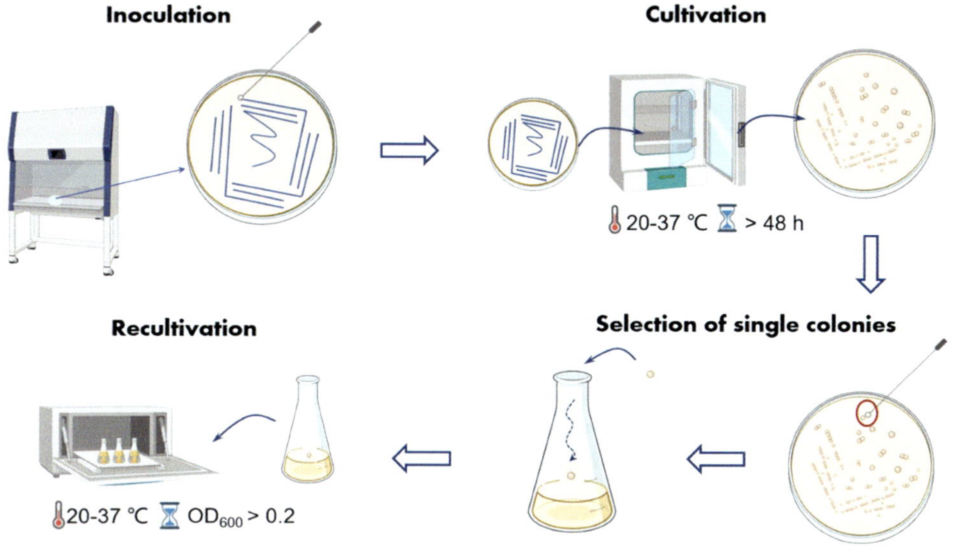

FIGURE 9.2 Schematic diagram illustrating part of the experimental procedure for preparing whole-cell bioreporters strains for making stocks. The processes shown are (clockwise from top left) streaking liquid inoculum onto a petri dish containing solid medium (top left, step c in text), incubating/growing new culture on the solid medium (top right, step d in text), sampling single colonies and inoculating into liquid media to regrow as purified stock (bottom panel, step e in text).

degree to which agar firms up in plates may vary. If the agar is too soft to streak, i.e., if a loop cannot be run over it without cutting the surface, using a greater amount of agar is indicated. After autoclaving this solution, cool the LB/agar medium to 40°C–50°C (i.e., cool, but not so cool that the agar starts to solidify), then add antibiotic to the medium by sterile filtration. If a given bioreporter is to be cultured in 40 µg/mL ampicillin, for example, make a stock antibiotic solution of 40 mg/mL; at this concentration, 1 mL stock antibiotic solution in 1 L total medium volume will result in a final concentration of 40 µg/mL. For 500 mL of LB agar, 0.5 mL of stock antibiotic solution is needed. When the autoclaved LB/agar solution has cooled to 40°C–50°C, in a biosafety cabinet, use a sterile syringe and syringe filter to slowly inject the 0.5 mL of 40 mg/mL ampicillin solution into the LB/agar solution. Store the stock antibiotic solution in a refrigerator at 4°C for future use. While still warm enough to pour, pour the antibiotic LB/agar medium into 10 mm petri dishes in a sterile biosafety cabinet; generally, 15–20 mL can be poured into each plate. Allow the plates to solidify at room temperature (this typically takes 15–30 minutes). Bioreporter strains are generally provided on plates, as stab cultures, or freeze-dried. In all cases, liquid LB medium will be required as well. Prepare this exactly as for agar plates, including the required amount of antibiotic, only without adding the agar.

If the bioreporter is received as freeze-dried, first, hydrate the freeze-dried inoculum. To do this, transfer the freeze-dried inoculum into a sterile container (such as a sterile test tube or small sterile vial) and add sterile liquid medium (usually LB as above). Gently swirl or mix to ensure the freeze-dried inoculum is fully dissolved into the liquid medium. Shake the culture at ~90 rpm at an optimal temperature until the optical density[1] at 600 nm (OD_{600}) increases by ~0.2, normally on the order of 4–24 hours depending on the strain's growth characteristics. The temperature for cultivation should be between 20°C and 37°C, with most bacteria used as hosts preferring 30°C–37°C. If the provider of the bioreporter strain has not indicated cultivation conditions, a mid-range low and a higher temperature within this range should be tested. In some cases, more cultivation time is

[1] Accounting for both absorption and scattering of light.

needed for fully dormant strains; however, if this growth is not achieved in a few days, there is likely a problem with the strain. Note: OD measurements should be in reference to a ***blank***, which in this case would be rich medium/LB; OD_{600} increases with cell growth as cells impede light transmittance by both ***absorption*** and scattering. Once the culture has grown to an appropriate OD_{600}, it is ready for use for streaking plates.

If the bioreporter is received on plates or as a stab culture, take a sample using a flame-sterilized loop, rubbing the loop gently into the culture growing in the stab culture tube or on the agar plate, and inoculate it into a liquid medium (e.g., LB broth). Shake at ~90 rpm at the optimum growth temperature until an increase in OD_{600} of ~0.2 is reached.

Once the bioreporter strain is growing well in the liquid medium, it must be plated on solid media in order to purify the strain. Working in a biosafety cabinet and using a sterile loop, dip the loop into the growing bioreporter liquid culture, then streak onto the surface of an agar plate (normally LB/agar, as above). Use a streaking technique to spread the sample in a way that isolates individual cells. This is usually done by performing a "streak plate" method. First, streak the inoculum in three parallel rows across the outer edge of one quadrant of the plate; each streak will dilute the bacteria. Then, flame sterilize the loop and streak the sterile loop through the most dilute streak in first section, again making three parallel streaks to dilute the culture further. Continue this process across the plate until a pattern such as shown in the top right of Figure 9.2 is achieved. This creates areas where single cells will be spread out enough to grow into isolated colonies. Normally two or three plates are made at the same time.

After inoculation, incubate plates until growth is sufficient for single colonies to form in the areas of the plate where the inoculum was most diluted during streaking (indicatively 2 days, but some strains could take significantly longer). For purification, there must be single colonies present. As growth occurs and colonies begin to appear on the plates, ensure to examine the plates frequently. Make sure there is just one kind of colony. If there are multiple types of colonies (e.g., different colors, appearance of texture, sheen), it suggests contamination from another bacterial species. This is problematic and means that expanded testing and recultivation/purification is needed to try and recover a pure bioreporter strain. Contamination of the strain is most likely due to poor laboratory practice and poor maintenance of sterile conditions.

After single colonies are apparent on the plate and enough growth has occurred to be confident of which growth does represent single colonies, recultivate a pure sample of bioreporter strain. To do this, in a biosafety cabinet, and using a sterile loop, select a single colony from a plate and inoculate and incubate it in liquid LB (with antibiotic) at 20°C–37°C (recommendations on temperature as above), shaking at ~90 rpm for 16 hours. It is best to choose several single colonies to recultivate and test (i.e., in separate cultivation tubes). After 16 hours of incubation with shaking at ~90 rpm, the OD_{600} value should be >0.2. This indicates that the culture has reached a reasonable density of cells, suggesting that the bacteria are growing well in the liquid LB medium. A low OD_{600} (<0.2) suggests that the bioreporter is not growing properly. Possible reasons could include insufficient inoculum, contamination, poor growth conditions (wrong medium, wrong temperature), and incorrect antibiotic concentration. Tests should be made to vary conditions in order to obtain adequate growth. For fast-growing strains (e.g., *Escherichia coli*), the OD_{600} may reach 1.0 or higher after 16 hours of incubation. If the OD_{600} is very high (>1.5), the culture has likely overgrown, resulting in nutrient depletion of the culture and cell dormancy or senescence. The goal at this point is to have actively growing cells. Overnight cultures with an OD_{600} >0.2 should be subcultured to an OD_{600} ~ 0.02. To do this, take a volume of the overnight culture, V_{ON_cult}, according to $V_{ON\_cult} = 0.02 \times V_{subcult}/OD_{600,ON\_cult}$, where $V_{subcult}$ is the total volume of medium that will be used in the subculture, and OD_{600,ON_cult} is the OD_{600} of the overnight culture. This subculture should reach an OD_{600} ~ 0.2 in 2–5 hours, unless the cells in the overnight culture were senescent, in which case more time may be needed.

After the subculture reaches an OD_{600} of ~0.2+, it is ready for cryopreservation. A sterile glycerol solution is needed for this. Use molecular biology-grade glycerol, which is sterile or can be autoclaved before use. Typically, a 50% (v/v) glycerol:Type I water sterile solution is used. Mix the 50% sterile glycerol solution with the bioreporter subculture at a 1:1 volume ratio (e.g., 0.5 mL bacterial culture + 0.5 mL 50% glycerol solution) mixing well into small volume (e.g., 1 mL) cryotubes. Freeze the tubes at −80°C for long-term storage. Ensure that both glycerol and bacterial cultures are kept sterile throughout the process to prevent contamination.

All purified cryopreserved bioreporters need to be evaluated for performance prior to use to determine if purification was successful in isolating viable strains for bioavailability studies. Typically, several single colonies are selected simultaneously for parallel recultivation and then cryopreserved. This results in several batches of purified bioreporter being cryopreserved at a time, with each batch representing a single clone. Hundreds of stock vials can be cryopreserved in each batch, which ensures that if the purification is successful, the performance of the strains will remain consistent over a long period of time. To test the stocks, follow the procedure in Section 9.5.5 and construct a bioavailability ***calibration curve*** (Section 9.5.6). For each stock that is able to produce a dose-dependent signal as a function of the concentration of target pollutant, evaluate the ***sensitivity*** (slope of calibration curve), range of linearity, and limits of detection (LODs). All of the different batches should exhibit the same performance according to these metrics, within experimental uncertainty; the performance of stocks should agree with existing data on strain performance if available. If these performance metrics are in agreement, then the recultivation, purification, and preservation of stocks have been successful. In some cases, one of more of the cryopreserved stocks may have better performance than others – in this case, these stocks should be used and the remainder discarded. If the dose-dependent response pattern as a function of target pollutant concentration in the bioreporter signals produced is weak or absent, there is either a problem with strain or with the purification process. The purification process should be repeated to troubleshoot. In addition, when first using a strain, the performance of the bioreporter should be regularly assessed for consistency over a regular period of time, e.g., 2 or 3 months. It is also required to evaluate the performance of stocks if they are not used for a long period of time (e.g., 1 year or more). Stocks need to be recultivated at long-term intervals to remain viable, with the recommended interval being strain-dependent, but usually on the order of 1–5 years. It is very important to recultivate stocks before they begin to lose viability or run low. It is important to note that all labware used during the above experiments must be sterilized. Examples include beakers, centrifuge tubes for storing bioreporter, pipette tips, cylinders, flasks, and reservoir bottles. For vessels with lids, before autoclaving, it is necessary to add a small amount of Type I water to the vessel and loosely screw the lid on to sterilize it. For vessels without lids, add a small amount of water and cover the mouth of the bottle with foil. Centrifuge tubes are sterilized by placing them in a large beaker, adding a small amount of water, and then covering the beaker with foil.

9.5.2 Sample Collection, Transport, and Handling

Sampling containers and sampling equipment should be made of materials that do not interact with pollutants in samples (e.g., stainless steel or plastic, see Chapter 3). The potential for light to affect the samples in some manner that affects bioavailability of the target pollutants must also be taken into account when selecting sample container materials. Sample containers and equipment must be contaminant-free. There are also some differences in cleaning methods for containers for heavy metal-contaminated samples and samples for organic pollutants, as described in Chapter 3. Equipment must be decontaminated between each use for sampling and an ***equipment rinsate*** blank must be collected after each decontamination. ***Field duplicates***, whether co-located (soils and sediments only) or split samples, must be collected to evaluate field variability in soil, sediment, and water (see Chapter 4). A split sample is created by dividing a single sample into two or more parts for simultaneous testing or analysis, serving to measure precision as affected by field heterogeneity.

9.5.2.1 Water Samples

Once appropriate sampling sites have been determined, a Niskin bottle sampler on a line is used to take water samples at each site from a depth of 0.2 m or more below the water's surface. The bioreporter procedure described here requires 0.1 mL of water per sample, or less; however, collecting 25–50 mL is conservative and usually convenient. At each sampling site, water transparency is tested using a Secchi disk, and the temperature, pH, conductivity, and DO levels are measured on-site using a multisensor sonde. Although more determinands may be added if necessary to meet experimental goals, these field assessments of water quality are routine. Water samples are transferred from the Niskin bottle into sterile sample containers that are clearly and sequentially labeled to ensure unambiguous identification.

9.5.2.2 Soil Samples

In soil sampling, the typical depth of the soil horizon is between 0 and 20 cm, collected with a digging tool, preferably stainless steel. However, for certain studies, deeper soil layers may be targeted, requiring the use of a suitable auger. When selecting the sampling location and depth, it is important to consider the site's background and history, including soil properties, previous environmental assessments, and pollution events. The soil sample should be transferred into a clean container using a trowel, shovel, or material scraped from the auger head.

9.5.2.3 Sediment Samples

Grab samplers are effective for collecting surface layer samples of sediments, while sediment corers are used to obtain profiles. Coring or using a benthic grab is generally preferred over sampling dredged material. Sediment samples should be gathered in clean, pre-prepared containers, noting the depth range from which they were taken.

9.5.2.4 Transport and Handling

To prevent exposure to direct sunlight during collection, especially for samples that may contain compounds reactive to light, it is important to minimize sunlight exposure. Samples containing light-sensitive pollutants should be stored in light-blocking containers. After collection, samples (water, soil, and sediment) are then transported to the laboratory within 6 hours at a temperature of 4°C. When field samples arrive at the lab, they are all kept at 4°C in a dark setting (Zhang et al., 2020b). The US EPA advises that water samples should not be held for longer than 96 hours for biologically significant measurements (US EPA, 2016). Soil and sediment samples should not be stored for more than 2 weeks, as even brief storage can alter concentrations of labile substances (ASTM, 1999). These recommended hold times are likely too long for some samples, given changes that can occur in bioavailability, and in particular reactions of organic pollutants. Ideally, samples are processed as soon as possible, within 24 hours.

Before testing for bioavailability, water samples may be filtered using a 0.22 or 0.45 μm filter to obtain the operationally defined dissolved fraction. If filtering is required, it should be completed as soon as possible, and no later than 5 hours following the collection of the sample (Reed et al., 2023). Either a suction filtration device or a large syringe with a syringe filter can be utilized. It is critical to find filters that are compatible with the target pollutant, meaning the filters will not absorb the target pollutants in the water samples, and free of contaminants. If analyses of both heavy metals and organics are performed, separate containers and filtration materials will be used for each. If users are unsure about which filter to use, consultation with the supplier is recommended.

Soil and sediment samples need to be processed in the laboratory before use. The soil or sediment should be thoroughly mixed using techniques such as stirring, a rolling mill, feed mixer, or other suitable equipment. Careful mixing is essential to prevent significant alterations to its physicochemical properties. Large soil/sediment materials may need to be removed manually or through wet sieving.

9.5.3 Preparation of Media Used in Bioavailability Tests

After bioreporter stocks and samples for testing are secured, it is necessary to prepare media for use in testing. Unlike culturing, bioreporter performance in sample testing is not about growth. Performance may be significantly impacted by the liquid medium in which the bioreporter is cultivated prior to use for testing and in most cases is directly affected by the liquid medium used for testing. Investigation must be conducted prior to bioreporter use to ascertain what optimized growth and testing conditions are. Factors to consider include growth, as bioavailability testing needs to occur using active cells, and sensitivity to pollutants – i.e., some media interfere with response either via direct interactions with pollutants or indirectly through effects on bioreporter response characteristics.

The most common approach for routine cultivation and testing of bioreporters is to use two media: the first is usually a nutrient-rich medium to accelerate the growth of the bioreporter (Section 9.5.1), while the second is a MM to flatten growth while satisfying basic needs for cellular maintenance. MM provides only the essential nutrients required for the bioreporter's growth; hence, there are fewer ***components*** of media to interfere with the bioavailability of pollutants in tested samples. By supporting the bioreporter's differential response to the environmental stimuli of pollutants rather than excess growth, the signal-to-noise ratios/detection capability is enhanced. For instance, AE1433 and zntA bioreporters used in heavy metals analysis are used with two media; one is LB, for growth/cultivation, and the second is MM, for testing (Magrisso et al., 2009; Zhang et al., 2017). In contrast, some bioreporter experiments may be conducted with a single medium for growth and testing, in which case either rich or MM media have been used. Jiang et al. (2017) used LB medium to both culture and test bioreporter strains to evaluate the bioavailability of phenol. Using only rich medium for both growth and cultivation may function when (1) the medium does not interfere with bioavailability of the pollutant and (2) the medium does not adversely impact the response of the bioreporter. Commonly used rich media are LB and TSB. None of the rich media in common use are selective, i.e., supporting the growth of a specific organism, and all are suitable for the cultivation of a wide range of non-fastidious microorganisms. LB is primarily used to support the growth of *E. coli* for a variety of purposes. Since many bioreporters are constructed using *E. coli* as a host, LB is most widely used among bioreporter researchers. Preparation and use of a rich medium, *viz*., LB, for cultivation of bioreporters, have been discussed above and are relatively simple. In practice, it is very difficult to test for the interference of the medium with bioavailability of the pollutant; hence, it is safer to use MM in testing. How the medium affects bioavailability of non-ionic organic pollutants is unclear, whereas the role of rich medium in interfering with the bioavailability of ionic species lies clearly in ***complexation***, among other reactions. Some rich media contain several undefined complex organic constituents in the media that may affect the bioavailability of pollutants. Therefore, some studies only use MM for both cultivation and testing (Elcin and Öktem, 2020; Hansen et al., 2019). For more details on the use of rich media and MM in bioreporter work, see Hynninen and Virta (2010).

The medium in which the bioreporter is cultivated immediately prior to use in sample testing is usually supplemented with the same antibiotic and dose used in preparing plates and stocks, whereas it is a general recommendation not to add antibiotic to the medium used for sample testing. By the time a culture is prepared for testing, its use is time-limited, antibiotic has no functional purpose, and minimizing the amount of components in sample testing media is desirable.

Understanding how the different components of MM may be varied and why is somewhat nuanced. MM contains a minimum amount of nutrients necessary for cell growth and ensures that bioreporters can maintain a modicum of growth and basic function. There are many different formulations of MM, components of which usually include a carbon source, a nitrogen source, a phosphorous source, salts, micronutrients (optional in some MM formulations), and a buffer. Glucose, glycerol, gluconate, acetate, and succinate are examples of carbon sources used, each of which has specific advantages and disadvantages in terms of energy yield, utilization rate, specific metabolic pathways required for utilization, oxygen requirement, potential inhibitory by-products, and potential to inhibit growth at high concentrations. Glucose is frequently used, particularly for *E. coli* cultivation, due to its fast utilization and ability to support rapid bacterial growth (Bren et al., 2016). Nitrogen and phosphorous are usually obtained from ammonium and phosphate, though nitrate, urea, and β-glycero-phosphate, as examples, are also used. The basal salts include sodium, potassium, magnesium, and sometimes calcium. Commonly used buffers and the pH intervals in which they work are listed in Section 9.4.3.

A typical salts solution that can support *E. coli* and other laboratory strains commonly used as hosts is called M9 minimal salts solution. Formulations for this vary, but often contain KH_2PO_4, NaCl, Na_2HPO_4, and NH_4Cl. The KH_2PO_4 serves as both a source of phosphate and, in combination with the HPO_4^{2-}, a buffer. K and Na enable cells to balance osmotic pressure, and NH_4Cl is a nitrogen source. In many formulations, $MgSO_4$ and $CaCl_2$ are added, which provide (variously) (1) Mg as a cofactor essential for enzyme function, (2) sulfur necessary for the synthesis of sulfur-containing amino acids and cofactors, and (3) calcium, which is used in some physiological processes (Wallace et al., 2020). In addition to buffering the pH of the solution, the buffer should be selected to minimize interference with the bioavailability of target pollutants, which is notably a difficulty with heavy metals. Table 9.2 provides information on the strength of some buffers in terms of their ability to complex different heavy metals.

TABLE 9.2
An Overview of the Intensity of Complexation Between the Various Metal–Buffer Pairs

Buffer	Cr(III)	Mn(II)	Fe(III)	Co(II)	Ni(II)	Cu(II)	Zn(II)	Cd(II)	Pb(II)	Final
ACES		LC		LC	LC	SC	LC			(–)
ADA		SC		SC	SC	SC	SC	SC	SC	(–)
AMPSO				SC	SC	SC	LC	LC	LC	(–)
BES				LC		LC				(–)
Bicine		LC	SC	SC	SC	SC	SC	SC		(–)
Bis-Tris		LC		LC	LC	SC	LC	LC	SC	(–)
BTP				SC	SC	SC	SC	SC	SC	(–)
CABS										(+)
CAPS										(+)
CAPSO										(+)
CHES		SC		SC					SC	(+)
DIPSO		LC		LC	LC	SC	LC	LC	LC	(–)
EPPS						LC	NC	NC	LC	(+)
HEPBS										(+)
HEPES						–	–	–	–	(+)
HEPPSO				LC	LC	–	–	–	LC	(+)
MES		LC	SC	LC	LC	–	–	NC	NC	(+)
MOBS			SC							(+)
MOPS		LC	SC	LC	LC	–	–	NC	NC	(+)
MOPSO			SC		LC	–	NC	NC	–	(+)
PIPES				LC	LC	–	–	NC	–	(+)
POPSO										(+)
TABS			SC							(–)

(Continued)

TABLE 9.2 (*Continued*)
An Overview of the Intensity of Complexation Between the Various Metal–Buffer Pairs

Buffer	Cr(III)	Mn(II)	Fe(III)	Co(II)	Ni(II)	Cu(II)	Zn(II)	Cd(II)	Pb(II)	Final
TAPS	SC		SC			SC	LC	LC	LC	(–)
TAPSO		LC	SC	LC	LC	SC	LC	LC	SC	(–)
TEA				LC	LC	SC	LC	LC	LC	(–)
TES	SC		SC	LC	LC	LC	LC			(–)
Tricine				SC	SC	SC	SC			(–)
Tris	SC		SC	SC	SC	SC	LC	LC	LC	(–)

Source: Ferreira et al. (2015).
Note: Final evaluation regarding the overall suitability of the buffer: (+) represents suitable and (–) represents not suitable.
Abbreviations: "–", data not in agreement; LC, light complexation; NC, no complexation; SC, strong complexation. ACES, N-(2-Acetamido)-2-aminoethanesulfonic acid; ADA, N-(2-Acetamido)iminodiacetic acid; AMPSO, 3-(N-Morpholino)-2-hydroxypropanesulfonic acid; BES, N,N-Bis (2-hydroxyethyl)-2-aminoethanesulfonic acid; Bicine, N,N-Bis (2-hydroxyethyl)glycine; Bis-Tris, Bis (2-hydroxyethyl)iminotris(hydroxymethyl)methane; BTP, 1,3-Bis [tris(hydroxymethyl)methyl]amino propane; CABS, N-Cyclohexyl-3-aminopropanesulfonic acid; CAPS, N-Cyclohexyl-3-aminopropanesulfonic acid; CAPSO, N-Cyclohexyl-2-hydroxyl-3-aminopropanesulfonic acid; CHES, 2-(N-Cyclohexylamino)ethanesulfonic acid; DIPSO, 3-[N,N-Bis (2-hydroxyethyl)amino]-2-hydroxypropanesulfonic acid; EPPS, N-(2-Hydroxyethyl)piperazine-N′-(3-propanesulfonic acid); HEPBS, N-(2-Hydroxyethyl)piperazine-N′-(4-butanesulfonic acid); HEPES, N-(2-Hydroxyethyl)piperazine-N′-(2-ethanesulfonic acid); HEPPSO, N-(2-Hydroxyethyl)piperazine-N′-(2-hydroxypropanesulfonic acid); MES, 2-(N-Morpholino)ethanesulfonic acid; MOBS, 3-(N-Morpholino) butanesulfonic acid; MOPS, 3-(N-Morpholino)propanesulfonic acid; MOPSO, 3-(N-Morpholino)-2-hydroxypropanesulfonic acid; PIPES, Piperazine-N,N′-bis (2-ethanesulfonic acid); POPSO, Piperazine-N,N′-bis (2-hydroxypropanesulfonic acid); TABS, is N-Tris(hydroxymethyl)methyl-4-aminobutanesulfonic acid; TAPS, N-Tris(hydroxymethyl)methyl-3-aminopropanesulfonic acid; TAPSO, N-Tris(hydroxymethyl)methyl-3-aminopropanesulfonic acid; TEA, Triethanolamine; TES, N-[Tris(hydroxymethyl)methyl]-2-aminoethanesulfonic acid; Tricine, N-[Tris(hydroxymethyl)methyl]glycine; Tris, Tris(hydroxymethyl)aminomethane.

Before conducting sample testing, it is first necessary to determine the suitability of the prepared MM. The determination is based on whether the bioreporter produces a dose-dependent signal for the pollutant and how sensitive the bioreporter is to the pollutant. The carbon source chosen and the pH/buffer used are most likely to have an effect on response, with the formulation of the salts solution being secondary. Accordingly, different carbon sources and buffers/pHs (salts if necessary) should be tested and MM formulation adjusted to achieve the best response performance. In addition to the basic components of MM, some bioreporter host strains will require additional trace nutrients, amino acids, or vitamins. It is often possible to formulate bioreporter MM on the basis of literature on host–strain requirements. Moreover, the range of pollutant concentrations at which the dose-dependent signal is generated should be as large as possible. Generally, the provider of the bioreporter strain will have protocols on how to prepare the MM; however, even so, some testing and optimization are in order to confirm full optimization. For reference, in this section, the formulation of American Type Culture Collection (ATCC) M9 will be compared and contrasted with two other protocols for MM from the literature. These examples serve to illustrate the whys and wherefores of how MM formulations are varied for bioreporter work.

9.5.3.1 ATCC Protocol for Preparing M9

For the M9 medium, prepare an M9 salts solution consisting of 12.8 g of Na_2HPO_4, 3.0 g of KH_2PO_4, 0.5 g of NaCl, and 1.0 g of NH_4Cl dissolved in 500 mL of Type I water, autoclaved at 121°C, and cooled. Additionally, prepare a 20% glucose solution (20 g glucose in a total volume of 100 mL Type I water), a 1 M $MgSO_4$ solution (1.204 g $MgSO_4$ in a total volume of 10 mL Type I water), a 1 M $CaCl_2$ solution (1.11 g $CaCl_2$ in a total volume of 10 mL Type I water), and a 0.5% w/v thiamine solution (0.05 g thiamine in a total volume of 10 mL Type I water). Sterile filter 20 mL glucose, 2 mL $MgSO_4$, and 0.1 mL each of $CaCl_2$ and thiamine solutions into the M9 salts solution. Dilute the resulting solution to a total volume of 1,000 mL with sterile Type I water (Wallace et al., 2020). In this formulation, the $MgSO_4$ and $CaCl_2$ are not added to the salts solution in order to prevent precipitation during autoclaving, and the thiamine is considered optional, depending on the host organism bioreporter requirements. The pH of M9 is adjusted to 7.4. The final concentrations of the essential components in MM9 are summarized as 0.4% glucose, 0.0187 M NH_4^+, 0.11 M PO_4^{3-}, 0.19 M Na^+, 0.022 M K^+, 0.002 M $MgSO_4$, and 0.0001 M $CaCl_2$.

9.5.3.2 MM for Bioreporter zntA

The *E. coli* PHL268-based bioreporter strain zntA has been used for the evaluation of heavy metal bioavailability using a MM composed of 0.03 g of $CaCl_2{\cdot}2H_2O$, 1.49 g of KCl, 0.2 g of $MgCl_2{\cdot}6H_2O$, 4.68 g of NaCl, 0.43 g of Na_2SO_4, 1.07 g of NH_4Cl, 6.06 g of MOPS, and 2 g of sodium gluconate, dissolved in 1,000 mL of distilled water adjusted to pH 7.0 ± 0.1 (Magrisso et al., 2009; Zhang et al., 2020a, 2024). To prepare this MM, the required masses of $CaCl_2{\cdot}2H_2O$, KCl, $MgCl_2{\cdot}6H_2O$, NaCl, Na_2SO_4, and NH_4Cl are weighed, then dissolved and mixed in distilled water (less than 1,000 mL). This salts solution, along with 200 mL of 1 M NaOH solution, is autoclaved. Sodium gluconate and MOPS are weighed, then dissolved in ~50 mL water; without autoclaving, then this solution is added to the salts solution by sterile filtration. The pH of the resulting solution is checked, then adjusted to a final pH of 7.0 ± 0.1 using the sterile NaOH. After pH adjustment, the final volume is made up to 1,000 mL with sterile Type I water.

On first examination, the composition of this MM contains many of the same ions as the ATCC M9. On closer examination, the differences for salts are in reasonably close alignment:

- Carbon source, 0.4% glucose, in M9 versus ~0.2% as gluconate for zntA MM;
- NH_4^+ ~0.02 M in both (from NH_4Cl);
- PO_4^{3-} 0.11 M in M9 (from Na_2HPO_4, KH_2PO_4) versus none for zntA;
- Na^+ 0.19 M (from Na_2HPO_4, NaCl) in M9 versus 0.096 M (from NaCl, Na_2SO_4, and sodium gluconate) for zntA;
- K^+ ~0.02 M in both (from KH_2PO_4 and from KCl for zntA);
- Mg^{2+} 0.002 M (from $MgSO_4$) in M9 versus 0.001 M (from $MgCl_2{\cdot}6H_2O$) for zntA;
- Ca^{2+} 0.0001 M (from $CaCl_2$) in M9 versus 0.0002 M (from $CaCl_2{\cdot}2H_2O$) for zntA;
- SO_4^{2-} 0.002 M (from $MgSO_4$) in M9 versus 0.003 M (from Na_2SO_4) for zntA; and
- Cl^- 0.028 M in M9 (from NaCl, NH_4Cl, and $CaCl_2$) versus 0.12 M (from NaCl, KCl, $MgCl_2{\cdot}6H_2O$, NH_4Cl, and $CaCl_2{\cdot}2H_2O$) for zntA.

The central points of difference in the ATCC M9 and the MM used with zntA is in the carbon source used, the difference in phosphorous amendment, and the difference in buffer. The zntA medium reflects the use of this bioreporter for Pb work. MOPS is used as a buffer because phosphate is fully unsuitable for use with Pb. MOPS buffer firstly reduces the risk of precipitation of Pb compounds, and secondly the bioreporter at pH 7 has the highest sensitivity and shortest time to maximum bioluminescence response (Zhang et al., 2022a). The initial formulation of MM for zntA utilized 0.00132 M β-glycero-phosphate in lieu of phosphate due to incompatibility of free phosphate with Pb, however, this was later omitted by Zhang et al. (2017) when further testing demonstrated that omitting β-glycero-phosphate altogether did not alter results. The sensitivity of Pb in particular to any trace of phosphate entails that it is better to omit sources of phosphate when possible. When eliminating the components of the phosphate buffer, it is necessary to reformulate the salts, resulting in a final formulation that is different, but only marginally, from the ATCC M9 formulation. The other main difference between M9 and the zntA MM buffer is that the former uses glucose as a carbon source, whereas the latter uses gluconate. Much has been written about the effect of these two carbon sources on different host strains; however, practically speaking, gluconate was chosen for zntA because response characteristics were better, and a lower concentration of gluconate was preferred to modulate bioreporter growth during sample testing. The zntA MM has a much higher concentration of Cl^-, due to the high prevalence of this anion as a counter ion in the absence of phosphate species; this does not have a functional purpose, and despite the high Cl^-, the zntA MM has a lower ***ionic strength*** than the M9 medium.

9.5.3.3 MM for ADPWH_Nah Bioreporter

The ADPWH_Nah whole-cell bioreporter demonstrates a positive response to naphthalene and does not respond to other polycyclic aromatic hydrocarbons, such as pyrene, anthracene, and phenanthrene (Sun et al., 2017). The MM used for the induction of this bioreporter contains 20 mM sodium succinate, 2.5 g Na_2HPO_4, 1.0 g NH_4Cl, 2.5 g KH_2PO_4, 0.1 g $MgSO_4{\cdot}7H_2O$, 10 μL saturated $FeSO_4$ (44.8%), 1 mL Bauchop and Elsden solution, and 10 μL saturated $CaCl_2$ (20.8%), in 1.0 L Type I water. Bauchop and Elsden solution is a stock salt solution that contains 10.75 g MgO, 2.0 g $CaCO_3$, 4.5 g $FeSO_4{\cdot}7H_2O$, 1.44 g $ZnSO_4{\cdot}7H_2O$, 1.12 g $MnSO_4{\cdot}4H_2O$, 0.25 g $CuSO_4{\cdot}5H_2O$, 0.28 g $CoSO_4{\cdot}7H_2O$, 0.06 g H_3BO_3, and 51.3 mL HCl (~12 mol/L) (Bauchop and Elsden, 1960). To prepare this MM, Na_2HPO_4, NH_4Cl, KH_2PO_4, and $MgSO_4{\cdot}7H_2O$ are weighed, then dissolved and mixed in Type I water (200 mL). Then the $FeSO_4$ and $CaCl_2$ are added to it. This salt solution needs to be autoclaved. Sodium succinate is weighed, then dissolved in ~ 10 mL Type I water, without autoclaving, then this solution, together with 1 mL Bauchop and Elsden solution, are added to the salts solution by sterile filtration. The final volume is made up to 1,000 mL with sterile Type I water, with an unadjusted pH in the vicinity of neutral.

The composition of this MM for ADPWH_Nah bioreporter contains many of the same ions as the ATCC M9; however, it is different in many respects. On closer examination, the differences for salts are in reasonably close alignment:

- Carbon source – 0.4% glucose in M9 versus ~0.3% as sodium succinate for ADPWH_Nah MM;
- NH_4^+ ~0.02 M from NH_4Cl in both;
- PO_4^{3-} (from Na_2HPO_4 and KH_2PO_4) 0.11 M in M9 versus 0.036 M (from Na_2HPO_4 and KH_2PO_4) for ADPWH_Nah MM;
- Na^+ (from Na_2HPO_4 and NaCl) 0.19 M in ATCC M9 versus 0.0752 M (from Na_2HPO_4 and sodium succinate) for ADPWH_Nah MM;
- K^+ 0.002 M in M9 versus 0.0184 M or ADPWH_Nah MM both from KH_2PO_4;
- Mg^{2+} (from $MgSO_4$) 0.002 M in M9 versus 0.00065 M (from $MgSO_4{\cdot}7H_2O$ and MgO) for ADPWH_Nah;
- Ca^{2+} (from $CaCl_2$) 0.0001 M in M9 versus 0.000039 M (from $CaCl_2$ and $CaCO_3$) for ADPWH_Nah MM;
- SO_4^{2-} (from $MgSO_4$) 0.002 M in M9 versus 0.00046 M (from $MgSO_4{\cdot}7H_2O$, $FeSO_4$, $FeSO_4{\cdot}7H_2O$, $ZnSO_4{\cdot}7H_2O$, $MnSO_4{\cdot}4H_2O$, $CuSO_4{\cdot}5H_2O$, $CoSO_4{\cdot}7H_2O$) for ADPWH_Nah MM;
- Cl^- (from NH_4Cl) 0.0275 M in M9 versus 0.0011 M (from NH_4Cl, $CaCl_2$, and HCl in ADPWH_Nah);
- Micronutrients: 0.97 μM borate, 1 μM Co, 1 μM Cu, 46 μM Fe, 5 μM Mn, and 5 μM Zn in ADPWH_Nah MM.

The ATCC M9 medium and ADPWH_Nah MM differ significantly in their composition, particularly in carbon source (ADPWH_Nah MM uses succinate instead of glucose), amount of phosphorous (ADPWH_Nah MM has approximately one-third of the phosphorous), ion concentrations (much lower in ADPWH_Nah MM for all major ions except K^+ and Cl^-), and micronutrients (ADPWH_Nah MM includes the micronutrients borate, Co, Cu, Fe, Mn, and Zn at trace concentrations). Unlike the ATCC M9, no adjustment is made to the ADPWH_Nah MM pH, as a result of which results may be more variable as the pH is neither fully controlled nor verified.

9.5.4 Preparation of Bioreporter Cultures for Induction

To date, there are no ***Standard Methods*** for using Class I bioreporter for measuring pollutant bioavailability. Although there are some differences in the media required for different bioreporters, as described above, the process of preparing bioreporters for induction is generally similar. The first step is the cultivation of the bioreporter. This may be done by recultivating a cryopreserved stock in liquid medium. Alternately, the cryopreserved stock may be thawed and samples plated (per Section 9.5.1), then single colonies from plates may be used.[2] If using a cryopreserved stock, the cryopreserved stock is first thawed and then ~0.1–1 mL is pipetted directly into liquid medium, usually a rich medium. Whether using a single colony from a plate or cryopreserved stock, the liquid medium contains antibiotics (as described in Section 9.5.1), recultivation should be undertaken under sterile conditions, and the dilution rate should ensure transparent medium. The liquid culture made using either a single colony from a plate or from direct inoculation using a thawed stock will be harvested when it reaches a certain stage of growth for subsequent induction experiments. Since there may be a long lag time after freezing, growth may require overnight culture (or even longer). Depending on the length of the lag-phase, overnight culturing may result in overgrowth/high turbidity, in which case a new subculture will be required the next day. Overgrown cultures are usually diluted/subcultured in order to keep the bioreporter in a highly active state.

The dilution ratio of subculture in fresh media needs to be determined according to the requirements of each specific experiment. Typically, the dilution ratio is 100:1 (i.e., fresh medium to overgrown culture). The optimal dilution ratio may be adjusted according to the growth of the subsequent subculture. Usually, the subculture reaches the early log-stage rapidly.

Alternately, bioreporters may be cultivated continuously using a chemostat. A chemostat is a bioreactor system used for continuous culture of microorganisms or cells that maintains a stable environment by continuously adding fresh nutrient medium while simultaneously removing an equal volume of culture. The rate of inflow and outflow is adjusted to maintain the culture at the growth stage required. Many investigators do not use chemostats, according to the adage "continuous culture, continuous problems", and therefore, this method is not described further here.

Ideally, a bioreporter culture is harvested for induction at the phase of growth that provides the best response, and this should be tested before routine use of any strain. The measurement of OD_{600} should start at <~0.05 (transparent medium), which ensures that increases in OD_{600} correspond to active growth. An OD_{600} ~ 0.2 corresponds to early log-stage for many strains. This early exponential growth stage of the bioreporter is usually better, because using an active culture provides a better measured response. The cultivation time differs from one bioreporter to another. For some strains, optimal response may instead occur in mid- or late log-phase. OD_{600}s at harvest are reported to range from 0.1 to 0.6 (Augustiniene et al., 2023; Li et al., 2022, 2023). As to how to determine the growth phase of a strain, OD_{600} is measured at different time points, and then a growth curve, OD_{600} versus time, is plotted, which reveals the ODs associated with lag, exponential or logarithmic, stationary, and senescence phases. Once the growth curve is established, the output signals of the bioreporter are measured at OD_{600}s corresponding to different growth phases to determine the optimal harvest OD for the strain. The optimum OD_{600} for harvest is considered to result in the output that results in the lowest LOD.

[2] If using the plating method, new plates should be made approximately every week or two.

When the bioreporter culture grows to the most suitable OD_{600} value determined for optimum response, then the strain is harvested. The culture is transferred to a centrifuge tube and centrifuged at speeds usually around 4,000–10,000×*g*, depending on the strain and cell size. High speeds may damage some cells, or otherwise affect response, so it may be necessary to test and optimize the centrifugation conditions and, if necessary, perform a low-speed pre-centrifugation. Check if a clean pellet can be obtained. After centrifugation, the supernatant is removed and the remaining pellet may or may not be washed, i.e., resuspended in the medium that will be used for induction, then centrifuged again, discarding supernatant. The final pellet is resuspended in the medium that will be used for induction, typically MM, for subsequent experiments. Best practice is to resuspend harvested cells to a specific OD_{600}. The OD_{600} of the cell suspension used for induction experiments should produce a final solution with sufficient optical transparency for reliable and sensitive measurement of optical signals during testing, yet the suspension should contain sufficient cells as to be able to produce good signal. For most bioreporters, an OD_{600} of 0.2 of the cell suspension after being resuspended is suitable. For strains with an OD_{600} of 0.2 at the time of harvest, this requires adding MM for resuspension of the pellet at approximately a 1:1 ratio with the original media; usually slightly less MM is added since some cells are lost with the supernatant – the final OD should be verified. For strains that are harvested with an OD_{600} greater than 0.2, it is necessary to add a larger volume of MM than that of the original media in order to ensure that the final OD_{600} of the harvested bioreporter cell suspension is 0.2.

Once bioreporter cell suspensions are prepared in this way, they should be used in induction experiments immediately. Sterile conditions are no longer as important as working quickly, and likewise antibiotic is usually not added to media used in induction experiments. There is still a need to avoid contamination and ensure that the work area, equipment, and materials are clean/free of contaminants as this may affect the results of your experiment. In particular, before conducting experiments, the user should be aware of the levels of contaminants in the media required for the bioreporter to be used.

9.5.4.1 Taking zntA and ADPWH_Nah Bioreporters as Examples

Two literature examples show the generalizable aspects of the process of preparing bioreporter suspensions for use in testing. For the zntA bioreporter, a 100 μL portion of thawed stock (from cryopreserved stock at −80°C in 1:1 glycerin) was grown overnight (~16 hours) in liquid LB amended with 40 μg/mL ampicillin with shaking at 90 rpm and at 30°C. The overnight culture was subcultured and regrown at 26°C with shaking at 200 rpm after being diluted by a factor of 100 in fresh LB+ ampicillin. Cells were collected by centrifugation (10,625×*g*) after the subculture arrived at an OD_{600} of 0.2. The supernatant was discarded, and cells were then resuspended in enough zntA MM to reach a final OD_{600} of 0.2 in order to create final bioreporter cell suspensions for measurements.

For the ADPWH_Nah bioreporter, the strain was incubated in LB media supplemented with 300 μg/mL ampicillin at 30°C overnight. After overnight incubation, 1 mL of the ADPWH_Nah overnight culture was centrifuged at 3,000 rpm for 10 minutes. Subsequently, the supernatant was discarded, and the cell pellet was resuspended in 1 mL ADPWH_Nah MM. The bioreporter cell suspension for ADPWH_Nah MM prepared in this way was used for experiments without regard to OD, suggesting that the procedure is either not optimized or that this particular bioreporter is highly sensitive across much of the growth curve.

There are similarities in the method of preparation of cell suspensions for use in induction experiments for these two examples of zntA and ADPWH_Nah. The approach to producing the zntA suspension, in contrast to the ADPWH_Nah approach, is based on the determination, using OD, of the exact growth phase corresponding to the optimum bioreporter response, whereas the less precise approach to producing ADPWH_Nah cell suspensions is likely to lead to less day-to-day reproducibility in response.

9.5.5 Bioreporter Induction and Response Measurement

After harvesting and preparing the bioreporter cell suspension, bioreporter induction and response measurement is performed. A suitable microtiter plate should first be selected before the experiment. While microtiter 96-well plates are most commonly used, 384-well plates (112 μL typical volume) have also been used. For colorimetric detection, a clear, flat-bottom plate is needed. Black plates are usually used for fluorescent signals, and white plates for luminescent signals. The volume of a standard 96-well microtiter plate is 100–300 μL. Standard solutions of the target pollutant at a range of concentrations are pipetted (e.g., 50–100 μL) into individual wells for use in preparing a bioavailability calibration curve (described in the next section). The same volume of each sample is then plated in the same manner. A minimum of three laboratory replicates are plated for each standard or sample; four provides better precision. During experimental procedures, standards/samples and bioreporter-MM suspension are often used in a 1:1 ratio to ensure enough bioreporter to provide optimal signal strength, while also ensuring enough solution transparency to avoid loss of signal from scattering. If the OD_{600} of the bioreporter cell suspension is 0.2 and the ratio of cell suspension to standard/sample is 1:1, the final OD_{600} in the plate reader will be 0.1, which is suitably transparent (refer to Section 9.5.7 concerning optical effects). This dilution rate also ensures that the bioreporter has ample MM. Conversely, during harvesting the bioreporter, a much more concentrated suspension could be made. For instance, if the bioreporter cell suspension has an OD_{600} of 1.0,[3] and 20 μL of cell suspension is plated to every 180 μL of standard or sample, as for ADPWH_Nah induction described above, the final measurement OD_{600} would still be 0.1, but there would be much less media for the bioreporter. This approach would be more environmentally realistic, but is also problematic in being less controlled (less MM to support the bioreporter). For this reason, the approach of using a suspension of $OD_{600} = 0.2$, and plating suspension 1:1 with standard or sample, as for the example of zntA described above, is far more common.

When utilizing a bioreporter system that requires a substrate for signal generation, the appropriate substrate must be added to the assay to facilitate the reaction needed for subsequent detection. Common examples of such substrates include n-decanal, which is used for luminescent bioreporters expressing bacterial luciferase (*lux*) to produce bioluminescence; X-gal, which is hydrolyzed by β-galactosidase (*lacZ*) to yield a blue chromogenic product; and *p*-nitrophenyl phosphate, which is cleaved by alkaline phosphatase to produce a yellow-colored *p*-nitrophenolate anion. The optimal concentration of these substrates is critical to ensure accurate and sensitive detection of the bioreporter ***activity***, and it is typically recommended by the bioreporter provider based on established protocols. If such recommendations are unavailable, however, the substrate concentration can be experimentally optimized by performing a titration series to determine the minimal amount required to achieve maximal signal intensity without causing substrate saturation or background interference. This optimization process should also consider factors such as incubation time, temperature, and the specific assay conditions to ensure reproducibility and reliability of the results. The substrate, if needed, should be added before the bioreporter cell suspension is added to the plate.

Subsequent to plating standards, samples, and substrate (if needed), the bioreporter-MM suspension is added to the microtiter plate wells having standards or samples. The bioreporter suspension is added last so that the plate will go into the plate reader immediately after induction begins. Adding the bioreporter cell suspension should be as fast as possible. This is because bioreporters begin to respond as soon as they are exposed to a pollutant, and some bioreporters respond very quickly. To achieve rapid addition, the bioreporter cell suspension is poured into a disposable sterile basin or trough. Then, the suspension is transferred to the microtiter plate by using a multichannel pipette (e.g., 12-channels) and mixed thoroughly with the standards/samples using the pipette. To

[3] Note: The relationship between cell numbers in a bacterial/bioreporter suspension and OD_{600} is nonlinear; if measured OD_{600} is above ~0.2, the suspension should be diluted for measurement, and the OD_{600} from the diluted suspension corrected with the appropriate DF (e.g., for a measured OD_{600} of 0.3, the suspension is diluted with media by a factor of 2, the OD_{600} of the dilute suspension is measured, and this dilute OD_{600} is multiplied by 2).

maximize signal output, each well in a microplate should be filled to its maximum capacity, with standard or sample plus bioreporter suspension as much as possible, as using less than the full volume of wells will result in diminishing the signal strength. Although the mixing ratio of 1:1 is standard, this might be adjusted; however, the basic principle remains to ensure that the wells are filled to their maximum capacity with sufficient concentration of bioreporter to produce a strong signal and sufficient transparency to avoid signal blocking. The recommended volume for a standard 96-well microplate is generally 300 μL; however, under shaking, each well is ideally filled with no less than 100 μL and no more than 200 μL.

Immediately after mixing the standards or samples with the bioreporter cell suspension, the microtiter plate is placed in the plate reader and the bioluminescence, visible light, or fluorescence signal that is produced by the bioreporter is recorded. The temperature for the plate reader is usually set to 24°C–30°C. The optimal temperature should be determined by testing. The time needed for detection of maximum signal should also be determined, which is easily accomplished by letting the plate reader take measurements over a few hours. The decision to use a plate cover depends on the specific requirements of the bioreporter assay, including the need to prevent evaporation, allow gas exchange, and ensure compatibility with the optical requirements of the plate reader. For longer measurement times, plate covers are required because otherwise the entire sample volume might evaporate, even for 96-well plates. For shorter measurement times, it may be better to omit plate covers. Plate covers must be optically transparent with respect to the type and wavelength of light signal measured. Users should evaluate these factors and choose the appropriate approach to maintain assay ***accuracy*** and reliability. The plate needs to be shaken during each measurement (typically ~180 rpm). The most common protocol is to measure bioluminescence, visible light, or fluorescence every 10 minutes. It takes the plate reader a certain time to perform a measurement over a whole plate (longer for larger format plates), during which shaking ceases. The 10-minute interval ensures that, for most of the total measurement period, the plate is not stagnant, while also affording enough measurement frequency to observe the pattern of changes in the signal produced by the bioreporter over time after exposure to pollutants and to ascertain with confidence the time when the response of the bioreporter reaches its maximum. All measurement parameters discussed herein, and any other applicable parameters (e.g., optical parameters associated with fluorescence measurements) should be optimized for specific experiments. For every plate, standards for bioavailability calibration curves, as described in the next section, must be included.

In addition to the foregoing, in bioreporter experiments, two critical issues must be addressed with regard to induction and response measurements: First, a nearly zero signal may indicate that the bioreporter is dead, potentially due to high pollutant concentrations/toxicity. Because living bioreporters exhibit non-zero blank signals for bioavailability calibration curve blanks (raw signal, e.g., absorbance, relative luminescence units [RLU]), if a sample has a signal lower than the blank, it is a strong indication of a toxicity problem. To confirm this, it is advisable to run dilution series of the samples or rerun samples with low signals, especially if high pollutant levels are suspected. Additionally, the well-known LIVE/DEAD toxicity bioassay has been used as a benchmark for toxicity studies, including bioreporter studies. This assay distinguishes live and dead cells (viable and inviable cells) through fluorescent staining with SYTO 9 ("live") and propidium iodide (PI, "dead"), respectively. Li et al. (2022) assessed toxicity-related signal inhibition by conducting the LIVE/DEAD toxicity assay immediately after bioreporter signal measurements. The kit comes with full directions for use; refer to Li et al. (2022) for supplemental guidance. If toxic effects are confirmed, the only way to know the cause of toxic effects is further experimentation, including instrumental analysis. In practice, bioreporters are, by nature, generally robust, and the literature is full of reports of use without incident from toxicity.

A second issue of concern with bioreporters is potential interferences – i.e., when a bioreporter responds to pollutants other than the target pollutant(s)/class or is affected by other toxic components in the sample, in addition to any toxic effects that may be conveyed by high concentrations of the target pollutant (Xu et al., 2013; Yoon et al., 2016). To address the interference issues from non-target pollutants, these should be assessed during the testing phase, i.e., during the experimental design and method development phases, by systematically testing the bioreporter's response to multiple potential interferents to ensure its specificity. Specific steps include (1) selecting pollutants known to be potentially cross-reactive for testing and (2) evaluating the response of the bioreporter to these pollutants to assess whether the bioreporter selected can be viably used in planned bioavailability studies based on interferences and likely sample composition.

9.5.5.1 Taking zntA and ADPWH_Nah Bioreporters as Examples

Continuing with the comparison of the general method above to literature examples, for the bioluminescent zntA bioreporter, individual wells in a white 96-well microtiter plate were filled with 50 μL of standard or sample. Next, the bioreporter suspension in MM was poured into a disposable sterile trough. A 50 μL portion of bioreporter suspension was added using a multichannel pipette and mixed thoroughly with the bioreporter suspension. Four replicates for each concentration of standard and for each sample were used. Using a microtiter plate luminometer, bioluminescence was measured every 10 minutes for up to 7 hours at a temperature of 24°C and shaking at 180 rpm.

For the ADPWH_Nah bioreporter, since it produces a very strong bioluminescence signal, a 96-well black flat and clear-bottomed microplate was filled with 180 μL of standard/sample solution, followed by 20 μL of bioreporter suspension in each well with thorough mixing. The black microplate reduces scattering and cross-talk of light signals, minimizing background noise, thereby increasing detection sensitivity and accuracy for very strong signals (while white plates reflect light signals to enhance the detected bioluminescence). Bioluminescence was measured every 10 minutes over the course of 5 hours, and 5 minutes of shaking was used before each reading, with the plate reader set to a temperature of 30°C. For each standard or sample, three replicates were measured.

Both the zntA and ADPWH_Nah bioreporter methods measured bioluminescence at regular intervals using a plate reader under controlled temperature and shaking conditions. The shaking speed and the temperature at which the samples were tested varied slightly. Both methods use replicates for characterizing measurement precision and to increase accuracy through averaging of random errors. The ADPWH_Nah method is tailored for strong bioluminescence signals, and further adaption could have been to reduce the well volumes to 100 μL. The zntA method represents common practice, and, having produced enough signal for highly sensitive detection of the target pollutant, 100 μL well volumes were used. These differences highlight adaptations in plate type, volume ratios, well volume, and shaking protocols to optimize detection based on the bioreporter's signal strength and experimental requirements. The ADPWH_Nah method uses a 20 μL cell suspension to 180 μL MM ratio, i.e., presumably the cells are overgrown, and the method does not measure the OD at harvest, which prevents precise determination of bioreporter growth phase. This lack of precision may lead to less reproducibility of the bioavailability calibration curves. Additionally, if the suspensions are optically thick, which was not reported and presumably not measured, it is not possible to know how much loss of signal occurred through reduced solution transmittance, thus affecting the experimental results.

According to Zhang et al. (2017), bioavailability calibration curve slopes obtained from independent experiments using zntA had a relative standard deviation (RSD) of less than 10%, i.e., demonstrating repeatable performance on induction day after day in a manner approaching that for methods of instrumental analysis. The RSDs for calibration curve intercepts and calculated LODs, respectively, were 40% and 30%, reflecting the higher variability that occurs at very low signal-to-noise ratios. Such results were not reported in the ADPWH_Nah method studies. As a result, the ADPWH_Nah method is less suitable for collecting (QA)/(QC) metrics, particularly for applications relating to, for instance, Water Quality Criteria where a regulatory aspect is pertinent.

Also consistent with a robust QA/QC approach, the work on zntA included evaluation of other ***target analyte*** responses, as well as characterizing non-linear ranges of the bioavailability calibration curve and target pollutant concentrations giving rise to toxic response (Li et al., 2022; Zhang et al., 2024). The study on ADPWH_Nah method did not address whether other target analytes interfere with the bioreporter, and the standard calibration curve established in the ADPWH_Nah study was confined to the linear range of the bioreporter and did not address studies of toxicity to the bioreporter at high pollutant concentrations.

9.5.6 Preparation of Standard Solutions for Bioavailability Calibration Curves

For this method, a bioavailability calibration curve is constructed to determine the bioavailability of heavy metals in samples. While an instrumental standard calibration curve quantifies the total concentration of target pollutants in a sample, the bioavailability calibration curve represents ~100% bioavailability under conditions free from environmental ***matrix effects***; however, this 100% bioavailable pollutant may be less than the amount that would be measured by an instrumental standard calibration curve. For instance, instrumental standard calibration curves for heavy metals are based on standards that are acidified to ensure the stability of the metal ions in solution, such that heavy metals are in their most labile (i.e., biologically speaking, bioavailable) form, whereas bioavailability calibration curves use standards at environmentally relevant pHs, at which some portion of heavy metal may be in a form that is not available (e.g., oxyhydroxide form). The bioavailability calibration curve plots bioreporter optical signal versus either the concentration or ***activity*** of pollutant. Due to the challenge of preparing standard solutions having highly accurate concentrations, commercially available standards with certified concentrations should be purchased and should be sourced at high concentrations (e.g., 1,000 μg/g, stock). Heavy metal solutions should contain as little acid as possible (often this will be 2% HNO_3), and organic pollutant standards should contain as little strong solvent (e.g., hexane) as possible. These conditions maximize the potential for dilution to biocompatible pH or biocompatible solvent. As to how to many standards to have, typically five to seven standards are used to prepare a bioavailability calibration curve. In addition, a blank (i.e., no target pollutant, but containing whatever the other standards contain, e.g., a dilute HNO_3 solution for heavy metals or dilute biocompatible solvent for organics) is required. The concentrations of standards should follow a dilution series by factor of 2 (i.e. 10, 5, 2.5, 1.25, 0.625, etc.) OR "tailored" dilution which the user may decide (for instance, 10, 8, 6, 4, 2, … or 10, 5, 1, 0.5, 0.1, …). Dilutions for standards require best accuracy, normally using volumetric flasks and Class A pipettes.

To prepare a bioavailability calibration curve for heavy metals, the following steps are representative. A series of standard solutions need to be prepared by diluting certified heavy metal stock. While the stock is usually invariably in an acidic environment in order to maintain the stability of the heavy metal, when the pH is acidic, it affects the viability of the bioreporter. The stock must then be diluted with water to reduce its acidity (and according to the general guideline that no dilution should exceed a factor of 10). When the acidity is low enough to be buffered by the bioreporter induction media, any further dilutions are made at low acidity. As an example, if the stock solution has a concentration of 1,000 μg/g heavy metal in 2% HNO_3, i.e., approximately 2 g of HNO_3 per 1,000 g of water or ~0.32 M HNO_3, to create standards for the bioavailability calibration curve, the user performs a three-step dilution of the stock solution (to 100, then 10, and finally 1 μg/g) using water for each dilution step. The resulting 1.00 μg/g heavy metal solution is then used to prepare the standards for the bioavailability calibration curve. Through this initial dilution process with water, the final concentration of HNO_3 in the 1.00 μg/g metal solution will be approximately 0.00032 M, resulting in a pH of around 3.5. In order to ensure a constant pH and that the heavy metal remains solubilized in acid for a series of concentrations of standard solutions diluted from the 1.00 μg/g, a solution of 0.00032 M HNO_3 is used for any further dilutions Obviously, this is far less acidic than the directly purchased certified heavy metal stock solution. Due to the lower acidity, in order to maintain the stability of the metal, it is necessary to prepare standard solutions in this manner on a regular basis. According to Zhang et al. (2017), such solutions are stored at 4°C, and the concentration of the metal in the prepared standard solution can remain stable for 1 month; users should verify calibration standard stability when using this approach of acid dilution. For the study of Zhang et al. (2017), the buffer in the bioreporter medium was sufficient to neutralize this dilute acid upon mixing (1:1, bioreporter cell suspension in buffered MM to standard), ensuring the viability or function of the whole-cell bioreporter. Buffers used to neutralize acid in heavy metal standards should be chosen not only to buffer the pH of the solution but also to reduce interference with the target pollutants' bioavailability, which is particularly problematic for heavy metals. The selection of buffers is discussed in Section 9.5.3. Details of how to calculate the required amount of buffer based on the acidity of the solution are given in Chapter 13. Also see Sections 13.5.1, 13.5.6, and 13.6.1 for further details of the role of buffers in bioavailability measurements and construction of bioavailability calibration curves.

To prepare bioavailability calibration curve standards for organic chemical pollutants, start by selecting an appropriate solvent based on the solubility of the target pollutants. Common potentially biocompatible solvents include methanol, acetone, DMSO, or water, depending on the polarity of the target compound. If aqueous solubility is very low, it may be that the concentrated stock solution is purchased and dissolved in a more non-polar solvent, in which case it is necessary to gradually serially dilute the stock into a more biocompatible solvent. Always use a certified stock solution of the target pollutant, when available. The user must perform sequential tenfold dilutions, using either the biocompatible solvent alone or a combination of biocompatible solvent and water, while using the minimum amount of solvent that ensures the target pollutant remains in solution. Ultimately, the standards used in constructing a bioavailability calibration curve should be in a biocompatible solvent such as DMSO or a mixture of DMSO and water. If solvents must be used, i.e., if it is not possible to prepare organic pollutant standards that are fully dissolved in water, then the potential impact of the solvent on bioreporter growth must be independently tested. If the organic target pollutant is volatile or photoreactive, ensure to use sealed vials or amber containers to minimize degradation.

After bioreporter measurements are made on bioavailability calibration curve standards, to plot a bioavailability calibration curve, the user must select either concentration or activity for the x-axis and an appropriate data processing method for plotting response on the y-axis. The activity of bioavailable pollutant in solution reflects the effective concentration of pollutant, i.e., its ability to actually participate in chemical reactions in solution. As such, the activity of pollutant is a much better indicator for assessing environmental pollution and ecological risks; however, it is often not used because it is difficult to quantify and not relatable to regulatory standards, which are always in units of concentrations. For heavy metals, the primary bioavailable species in most cases is the free ion; hence, activity of the free ion is typically calculated based on the total concentration of the standard using a chemical speciation software such as Visual MINTEQ (see Chapter 15 for an introduction to the use of this software). For organic pollutants, if the pollutant is ionizing, activity may also be estimated using chemical speciation software. CHEAQS Next is preferable to Visual MINTEQ for organics and can also handle reduction of activity that occurs with degradation reactions, if applicable. The activity of non-ionizing organic pollutants will primarily be a function of the concentration of organic pollutant – if degradation reactions apply, this concentration may be determined by taking the kinetics of degradation into account (as is possible in CHEAQS Next). Activity is calculated from concentration using an ***activity coefficient***, which for non-ionizing organics is most commonly determined either from using literature values or calculated using Universal Quasi-Chemical Functional-Group activity coefficients (Bell, 2016; Fredenslund et al., 1977; Kheirati, 2025).

Regarding plotting data on the y-axis for the bioavailability calibration curve, a bioreporter's response to target pollutant in a standard or sample will be a raw signal based on (1) RLU (for luminescent or fluorescent signals) or (2) absorbance or transmittance (for colored bioreporter signals). The output signal varies with time as increasing levels of reporter protein are produced in response to induction and activation. Common metrics for processing the time-dependent raw signal used on the y-axis include the following:

- At a given measurement time, t, response is quantified by taking the difference in signal intensity (e.g., ΔRLU) between two conditions, usually the standard or sample minus the blank. The time t is typically chosen as being closest to the measurement time reflecting the average maximum signal output for all samples on the plate.

- A response ratio (RR) of the signal in the presence of the inducer divided by the signal in its absence (blank) for a given value of t. For the RR, t is again the measurement time reflecting the average maximum signal output for all samples on the plate.
- Alternatively, the maximum response ratio (MRR) is the ratio of an induced standard or sample's highest signal, i.e., at any measurement time, to that of the non-induced control (Zhang et al., 2017). For the MRR, the time to maximum signal output may be different for a standard/sample versus the blank.
- The slope of the linear part of the response-versus-time curve – this refers to the rate of change of the response signal with respect to time (e.g., $\Delta RLU/\Delta t$, or $\Delta A/\Delta t$) during a period where the relationship between the response and time is linear.

When using a bioreporter for the first time, it is useful to plot a variety of metrics and test which one provides the best calibration curve (e.g., slope/sensitivity, precision, LOD) to meet application needs. For instance, the method of initial response slopes is highly variable in use; for some reporters, the metric is too insensitive. For others, this metric is useful because it allows for signal collection at much shorter measurement times, i.e., well before maximum signal intensity is attained. Taking a bioluminescent bioreporter as an example, Figure 9.3a shows the bioluminescence response produced by the bioreporter, in RLU, versus time in response to different concentrations of a target pollutant. Figure 9.3b and c shows the resulting bioavailability calibration curves based upon the MRR calculated from data in Figure 9.3a and plotted versus concentration (Figure 9.3b) or activity (Figure 9.3c, determined using Visual MINTEQ). The slope of a bioavailability calibration curve reflects the sensitivity of the method to concentration changes of the target pollutant. Too small a slope may result in inadequate sensitivity. Users should be able to reproduce bioavailability calibration curves from 1 day's measurement to the next.

9.5.7 Measurements of Correction Factors for Bioreporter Signal Attenuation (Soils and Sediments)

Evaluating the bioavailability of pollutants in soil, sediment, and other environmental matrices through direct-contact assays with whole-cell bioreporters is more challenging compared to aqueous-phase assessments, as solid particles can significantly interfere with the transmittance of the optical signals generated by bioreporters, leading to a reduction or attenuation in the measured signal intensity. Some investigators neglect the effect on the bioreporter signal by solid particles in the

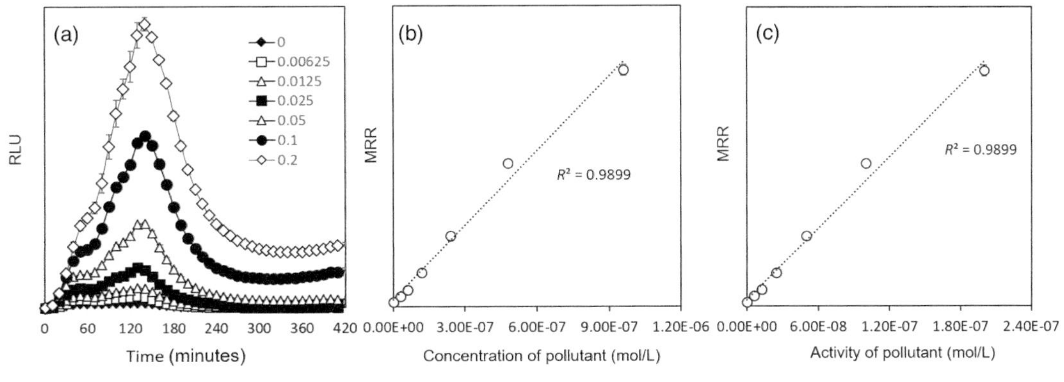

FIGURE 9.3 (a) Bioreporter response (in RLU) kinetics as a function of time, (b) bioavailability calibration curve plotting MRR versus concentration, and (c) bioavailability calibration curve plotting MRR versus activity.

sample. This, however, can result in an underestimation of the concentration of bioavailable pollutants in samples. The assay described herein involves the preparation of samples comprising soil or sediment slurries and the correction of the resulting signal to compensate for reduced signal transmittance.

The main mechanisms explaining the transmittance of the bioreporter optical signal as it passes through a slurry are depicted in Figure 9.4. They are as follows: (1) transmittance of ballistic and quasi-ballistic or snake photons, (2) absorption, (3) scattering (discussed below, subject to the glow-ball effect), and (4) reflectance (Meretska et al., 2017; Myers et al., 2013). The observed response is a function of ballistic/quasi-ballistic photons, scattered light, and reflectance because the microtiter plate reader detector cannot monitor absorbed photons. All observed photons that do not encounter scattering/reflectance interactions while moving through a material are referred to as ballistic photons (Vasefi, 2010); photons that are only marginally impacted are known as quasi-ballistic or snake photons (Martelli and Binzoni, 2018).

The selection of slurry concentration directly affects the accuracy of the results. Figure 9.5 provides a representative example of the relationship between percent transmittance ($\%T$) and OD[4] versus increasing soil/sediment slurry concentration. The reduction in $\%T$ from low to high slurry concentration exhibits strong nonlinearity (Figure 9.5a), which is anticipated due to the differential optical effects on the polychromatic nature of the signal. When plotted as OD (Figure 9.5b), a logarithmic pattern becomes evident, also aligning with expected behavior. OD is mathematically related to the logarithm of transmittance according to

$$\text{OD} = 2 - \log_{10}(\%T).$$

The points in Figure 9.5 represent experimental measurements and the lines are a non-linear curve fit.

When measuring luminescence with a plate reader, the plate reader is operated in luminescence mode, also sometimes referred to as open aperture or broad spectrum, collecting all light across wavelengths that the reader can detect. Therefore, the OD for characterizing the effect of soil/sediment particles reflects a measurement across the same spectral range. Alternately, for GFP-based bioreporters, for instance, the optical settings might be set at an excitation wavelength of ~485 nm and an emission wavelength of ~510 nm, with a bandwidth of ~10–20 nm, in which case OD would correspond to the diminution of signal under these settings. Similarly, when detecting the β-galactosidase via reaction with X-gal, the colored reaction product is measured at ~615 nm, to within some bandwidth, and OD would refer to these settings. While the underlying principle – relating attenuation of signal intensity to target slurry concentration – remains the same, the specific optical properties, interference factors (e.g., background fluorescence, light scattering), and dynamic range vary between systems. Therefore, users should construct their own calibration curves, as shown in Figure 9.5, for each bioreporter and associated optical settings, to account for these differences and ensure accurate quantification. This step is necessary because signal responses are influenced by the unique characteristics of the bioreporter, detection wavelengths, and sample ***matrix***.

The ballistic and snake photon-dominated region (the high-slope region below 6 mg/L for the work shown in Figure 9.5) is a response region wherein the $\%T$ and OD change rapidly as a function of slurry concentration, however, this region has been found to be associated with the lowest average relative error (8%) for results after signal correction. The ballistic/snake photon-dominated signal also represents a measurement domain wherein the signal correction factor, CF, is not theoretically anticipated to be a function of bioreporter signal intensity. These considerations argue for measurement at low slurry concentrations. It is not the case, however, that the lower the slurry concentration, the better. Primary problems with measurement at low slurry concentration relate

[4] Note: Here, OD is used to quantify the turbidity or particle concentration of the slurry, rather than microbial cell density.

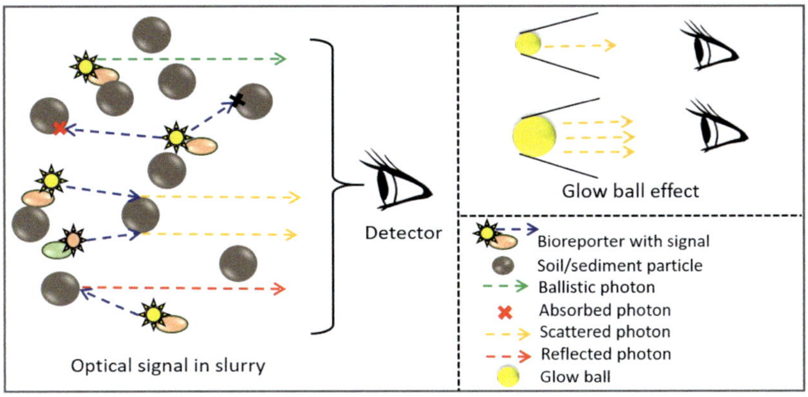

FIGURE 9.4 A schematic showing the primary optical effects governing detection of bioreporter response in the presence of soil or sediment particles. These effects include (1) ballistic photon and quasi-ballistic or snake photon transmittance, (2) absorption, (3) scattering (subject to glow-ball effect), and (4) reflectance.

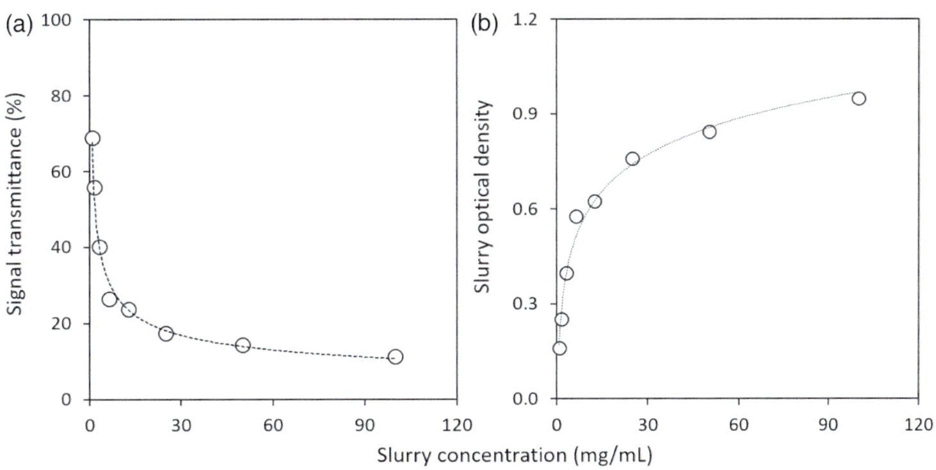

FIGURE 9.5 Representative graphs showing the effect of slurry concentration on (a) %T and (b) OD for bioreporter optical signal; an example using a bioluminescent bioreporter is shown (Zhang et al., 2022b).

to issues with precision and LOD. This is because a smaller slurry concentration means a smaller amount of soil- or sediment-associated pollutant. If the concentration of pollutant in the slurry is close to the LOD, the reduced transmittance of the bioreporter optical signal due to soil/sediment particles in the slurry may result in a signal below the LOD. Therefore, an evaluation of the pollutant concentration in the sample combined with the LOD of the bioreporter is needed. Usually, some pre-experimentation is required.

As the slurry concentration increases, there is a transition from ballistic/snake photon-dominated signal to glow-ball signal, and the highest curvature in the %T/OD versus slurry concentration response marks this transition. This area of curvature in the signal is associated with a large error. Zhang et al. (2022b) found that the greatest error, just over 30% relative, occurs at this transition point at a slurry concentration of ~10–20 mg/mL (for those experimental conditions, Figure 9.5). As this level of relative error would be unacceptable for most quantitative applications, working with slurries having %T/OD in this region of curvature should be avoided. For the data shown in Figure 9.5, ballistic photons are effectively absent for slurry concentrations higher than 20 mg/mL, which is the glow-ball-dominated spectral region.

At higher ODs, where scattering predominates (e.g., in solutions with increasing opacity due to higher suspended particle concentrations), light introduced into the particle field creates a glow ball or a nearly spherical region of scattered light. This phenomenon occurs because the increased particle concentration enhances light scattering, reflection, and absorption, overall leading to a higher intensity of the glow ball as the signal increases. Hence, in this glow-ball region, the CF will be a function of the bioreporter signal intensity. The relative error from using a single CF in the glow-ball region has been reported to be 16%. Most problematically, working in the glow-ball-dominated part of the curve is fraught due to the amount of optical signal that is lost; at some point, if too much signal is lost, it is not tenable to work in the glow-ball region and obtain accurate results, though there may be circumstances where it is possible and practical. For certain experiments (*vide infra*, multipoint/curve fitting), working in the glow-ball region is suitable; however, when the signal is blocked to a greater extent (e.g., less than 20% transmittance), the signal-to-noise ratio decreases significantly, making it difficult to accurately determine the CF. This is particularly problematic because, in this region, the calibration curve tends to flatten. Therefore, it is advisable to avoid working in regions where %*T* is not less than 15%–20%.

There are two approaches for bioreporter measurements using soil or sediment slurries: single-point and multipoint/curve fitting. For samples that are relatively homogenous, that do not have a large particle size range, and that are of reasonably uniform color, it may be possible to use a single-point CF. If best accuracy is needed to quantify bioavailability, or, e.g., for studying bioavailability under varying solid-phase:water ratios (***sorption***/desorption), multipoint measurements should be made. In either case, to ensure accurate and representative measurements, the sediment or soil samples must be thoroughly mixed well before preparing the slurry concentrations, as uneven ***distribution*** of particles can lead to inconsistent optical signal transmittance and compromise CF reliability.

In the single-point correction method, a microtiter plate with soil/sediments only is prepared in advance. This is referred to as a signal correction plate. To prepare this plate, it is first necessary to prepare soil or sediment slurries. When preparing a sample slurry, weigh an easily measurable amount of soil or sediment into a container, then dilute or serially dilute in Type I water to obtain the desired slurry concentration (for example, 5 g diluted to a total volume of 100 mL produces a slurry concentration of 50 mg/mL; 16 mL of this solution diluted to a total volume of 100 mL produces a final concentration of 8 mg/mL [in the range of ballistic photons for the results in Figure 9.5]). Prepare one slurry per sample and ensure that all slurries are of the same concentration. Next, plate the slurries using the same volume of slurry for each well as will be used for bioreporter induction experiments. To obtain a representative sample and consistent findings, slurries must be pipetted immediately after vigorous agitation, because settling happens as soon as agitation stops. After plating the slurries, place the plate in an oven at around 30°C (there is no specific requirement for this temperature) until dry.

After the signal correction plate is ready, a second plate, referred to as the pre-induced bioreporter plate, is prepared. Making this plate involves inducing bioreporter with standards, as described in Section 9.5.5. When working in the ballistic/snake-photon region, to prevent excessive errors due to weak signals, the concentration of the bioreporter with the largest response signal in the linear range of the standard solution is usually chosen as the induced bioreporter concentration. The total volume of standard plus bioreporter cell suspension in the pre-induced bioreporter plate must be the same as will be used in analysis of unknown samples. After the signal induction plate is prepared with induction standards plus bioreporter cell suspension, it is placed in the plate reader for measurement. When the response signal indicates that this plate has reached the time point normally used for signal detection (e.g., maximum response, or end of the linear part of the response-versus-time curve), the plate is immediately removed from the plate reader, and contents of this pre-induced bioreporter plate are immediately transferred using a multichannel pipette to the signal correction plate with soil/sediments and mixed vigorously – the dried soil/sediment on the signal correction plate must be forcefully mixed into solution. In this approach, soil or sediment is not "seen" by the bioreporter until the bioreporter is completely activated, and this avoids having the bioavailability of target pollutant in the standard from being affected by the presence of soil/sediment during induction and activation. It is very important to thoroughly mix sediment from the signal correction plate with induced bioreporter from the signal induction plate as rapidly as possible, and then to immediately replace this plate in the plate reader and measure the resulting attenuation of signal that the soil/sediment particles effect on the pre-induced bioreporter. This measurement of signal attenuation serves as the single-point CF (Zhang et al., 2022b). Experiments conducted to obtain single-point CFs should be performed with four replicates, i.e., four replicates of each slurry sample in the first plate and four identical wells with pre-induced bioreporter in the second plate. The equation used to calculate a single-point CF is

$$\mathrm{CF} = \frac{R_\mathrm{s}}{R_\mathrm{c}},$$

where R_c is the "control" pre-induced bioreporter signal, i.e., in the absence of solid material, and R_s is the signal after mixing with the soil/sediment on the signal correction plate. Replicate CFs are averaged.

After determining average single-point CFs for each sample, it remains to measure bioreporter response to the samples. Prepare and induce bioreporter as described in Sections 9.5.4 and 9.5.5; instead of aqueous samples, use slurries at the slurry concentration used to obtain the CF. Note: The bioavailability calibration curve will not have slurry, and hence, the responses to standards will not require correction. Calculate a response metric for the slurry samples, as described in Section 9.5.5 (e.g., ΔRLU or MRR), then multiply the actual sample response by the CF value to obtain the corrected response. Finally, use the corrected response with the bioavailability calibration curve to calculate the bioavailability of target pollutant(s) in the slurry.

When multipoint correction is required, then experiments need to be carried out using detailed measurements for each sample on a series of slurry samples with different solid-phase:water ratios. The series are prepared to cover an expected range of slurry concentrations that must all be either in the ballistic photon or glow-ball region; normally four to five slurry concentrations are needed. The CF is measured for each slurry concentration in the same manner as described for the single-point method above. A calibration curve is generated plotting CF versus slurry concentration, then a best fit to the trend is obtained (potentially linear, but probably nonlinear; choose the non-linear function that gives the best fit). Corrections that are made using the fitting function are more accurate than single-point measurements because random errors for the four to five measurements cancel in the fitting process. As with the single-point method, after obtaining the CF fitting function, bioavailability measurement experiments to detect the bioavailability of pollutants in soil/sediment samples are conducted, and the CF-versus-slurry concentration curve is used to correct for the attenuation of signal transmittance by soil/sediment particles. Two approaches are possible: first, a single slurry concentration of the soil/sediment sample is used as a sample for bioavailability testing. The corresponding CF value for the same slurry concentration is then obtained from the fitting function discussed above, and this CF value is applied to correct the quantified bioreporter signal prior to determining bioavailable pollutant using the bioavailability calibration curve. Alternately, if the goal is to conduct experiments investigating the bioavailable target pollutant desorbed, bioavailability induction experiments are conducted across the range of slurry concentrations used for CF measurement. The resulting calculated bioreporter response metrics are each corrected with the CF associated with the slurry concentration at which the bioavailability measurement is made, and then bioavailable pollutant concentrations are determined from the bioavailability calibration curve using the corrected responses.

9.6 VALIDITY OF RESULTS, DATA ANALYSIS, INTERPRETATION, AND REPORTING

9.6.1 Notes on QA/QC Planning Requirements

Both laboratory and field activities require a QA/QC plan to ensure the validity of measured results. Chapter 4 provides detailed QA/QC guidance that is applicable to this chapter, and a full understanding of the concepts and practices described in Chapter 4 is required prior to implementing QA/QC in this method. Specific QA/QC information pertinent to this chapter, and not already discussed above, is as follows:

- All analysts performing this method must undergo an Initial Demonstration of Proficiency (IDP), as described below.
- Samples that are analyzed as having concentrations at or near the upper limit of the bioavailability calibration curve must be reprocessed and reanalyzed. This involves diluting the field sample.
- For every microplate, a series of standard solutions need to be analyzed; an independent bioavailability calibration curve must be prepared for each plate.
- Select the appropriate bioreporter strain – ensure that the bioreporter has a well-characterized, measurable response to the target analyte (e.g., fluorescent protein expression, enzyme activity). This response may be somewhat variable due to the biological nature of the measurement, however, response characteristics for each calibration curve should lie within an envelope consistent with findings from optimization work done to assess bioreporter response (Section 9.5 – optimized assay conditions include time to maximum signal response during testing of standards and samples, temperature, components of growth and assay media, pH of testing medium, and substrate concentrations. These factors can affect the bioreporter's response and should be investigated before running experiments).
- The bioreporter needs some time to develop full signal in response to the target pollutant. When a bioreporter strain is used for the first time, the response of the bioreporter to the target pollutant should be observed for a sufficient length of time to determine the appropriate detection time and to characterize response characteristics, i.e., time to maximum response.
- The quality and reproducibility of the bioavailability calibration curve need to be evaluated. Metrics may include R^2, slope, and LOD. Response is often linear, with R^2s of >0.95. Slope and LOD are both indicators of sensitivity and should be reproducible across experiments from day to day. The degree of reproducibility is variable, depending upon the bioreporter (see comments in Section 9.6.2), however, a lack of reproducibility is indicative of poor adherence to systematic control of experimental factors. Additionally, accuracy should be evaluated when possible. For heavy metals, using speciation modeling, which predicts bioavailable fractions of pollutants based on their chemical forms, is a good approach (see, for instance, Li et al., 2022; Zhang et al., 2017). This approach works well for heavy metals and ionizing organic species, but is more challenging for non-polar organics; in the case of such organics, it may be possible to design controlled ***partitioning*** experiments to assess the accuracy of the bioavailability calibration curve approach. In bioreporter experiments, calibration is indeed critical and must be reproducible, but it is not limited to being consistent for IDP. Reproducibility should be maintained throughout the entire lifecycle of method use to ensure reliable and accurate results. Regular verification of the calibration curve, i.e., periodic checks using known standards or Data Validation samples, is essential to confirm that the curve remains consistent over time.

- If samples containing pollutants are stored for any appreciable amount of time, this may result in changes in bioavailability. It is important to make bioavailability measurements as soon as possible, and in no case should hold times exceed those specified herein.
- A field duplicate sample must be collected and analyzed for every 20 samples to assess the precision of natural field sample variability. A minimum of three replicate samples (laboratory replicates) are required for sample.
- For samples where toxicity is suspected (notably if sample response is less than that of the bioavailability calibration curve blank), serial dilutions and/or a LIVE/DEAD test toxicity bioassay should be conducted (Zhang et al., 2024).
- To exclude the issue of other pollutants interfering with the bioreporter, pollutants known to be potentially cross-reactive should be selected for testing and the response of the bioreporter to these pollutants should be evaluated (Xu et al., 2013; Trang et al., 2005).

9.6.1.1 Initial Demonstration of Proficiency

To demonstrate initial proficiency in a bioreporter experiment, first, verify the whole response range, testing bioreporter response to the target pollutant at concentrations from below the LOD, i.e., below the point that signal greater than the blank can be detected, up to the point where cells become inviable due to toxicity or saturation. Verify cell viability at each concentration using a LIVE/DEAD assay (e.g., staining with PI or a metabolic dye [Li et al., 2022]) to confirm that the signal reflects either active bioreporter cells or fractions of inviable cells. From these results, determine if there is a linear response range. Based on whether a suitable linear response range can be identified (preferred) or not, choose a concentration range to use for constructing bioavailability calibration curves. Construct a minimum of three bioavailability calibration curves, characterizing response metrics for each (e.g., if linear, slope, and intercept, if nonlinear, fitting parameters) and calculating LOD. IDR is demonstrated when response characteristics are the same to within statistical uncertainty for three bioavailability calibration curves performed independently of each other.

9.6.2 Data Analysis, Interpretation of QA/QC Results (Acceptance Criteria), Corrective Action, and Reporting

Data needs to be analyzed, interpreted in the context of QA/QC requirements, and if needed subjected to Corrective Action prior to reporting. This includes, at minimum, the following:

- Dilution factors (DFs) – unless specified by an instrumental method, DFs are never greater than a factor of 10 for dilution of standards and samples. The basic dilution formula is $C_1V_1 = C_2V_2$. The dilution factor is given by DF = V_2/V_1.
- ***Calibration checks*** and metrics – for the bioavailability calibration curve, the sensitivity (slope of calibration curve), range of linearity, and LOD need to be assessed for stability and should be repeatable. If the curve is unreliable (e.g., high variability in slopes or LOD), results may lack precision. If instability in the normally repeatable bioavailability calibration curve occurs, Corrective Action should be taken. First, verify the accuracy of standard solution preparation, ensuring proper dilution and pH adjustment. Next, check bioreporter viability and induction conditions. Then, assess plate reader stability to confirm consistent optical signal measurements. After double-checking all influencing factors, the bioavailability calibration curve should return to being repeatable to within statistical uncertainty of metrics. *NB*: Some changes may occur in factors not fully controllable, such as the chemical reagents supplied (e.g., LB is sometimes variable between batches and suppliers). Nonetheless, even if some factor causes a change in the bioavailability calibration curve, it should be possible to re-establish a repeatable result. If a repeatable bioavailability

calibration curve cannot be obtained, this raises concerns about the accuracy of results that should be discussed in reporting.

- Hold times – because of the potential extreme sensitivity to changes in bioavailability, hold times and their effect on bioavailability must always be reported and their effects discussed.
- Equipment rinsate blanks – these blanks ensure the absence of contamination during cleaning processes that could interfere with results. The acceptance criterion for blanks is a value below the LOD. Blanks above the LOD are indicative of field contamination (***equipment blanks***), no Corrective Action is possible, and all samples must be regarded as potentially contaminated.
- Reagent and ***procedural blanks*** – in bioreporter experiments, reagent and procedural blanks are necessary to identify potential contamination, but since the bioreporter itself will usually have a non-zero ***calibration blank***, these blanks cannot be assessed with the bioreporter itself. Instrumental analysis is required to confirm the absence of target pollutants in these blanks. If contamination is found, Corrective Action is to investigate and address the source (e.g., reagents, equipment, or environmental exposure) before proceeding. This step ensures that any signal from the bioreporter is solely due to the intended pollutant and not external contamination, maintaining the integrity of the results.
- Dilutions – for samples with pollutant concentrations at the upper calibration limit (but not toxic), dilute to fall within the linear range, ensuring accuracy. First, confirm the signal is above the upper calibration limit but not erratic or declining (which would suggest toxicity). Dilute the sample incrementally (e.g., two- or tenfold) until the signal falls within the linear range. Then, reanalyze the diluted sample.
- Precision for field duplicates/replicates – poor precision for field duplicates may be a measure of high environmental heterogeneity of sampled media; this is more likely to occur with soil and sediment samples. RSDs for field duplicates of aqueous samples should normally be on the order of 10% or below as an acceptance criterion. Based on Zhang et al. (2022b), the RSDs should be lower than 20% for soil and sediment work. To investigate issues of poor precision for field sample duplicates/replicates, compare metrics for field replicates to equivalent metrics for the bioavailability calibration curve standard in the middle of the bioavailability calibration curve. Report RSD values and any discrepancies and provide context on how environmental heterogeneity impacts the results. Poor precision suggests inconsistencies due to sampling errors, environmental factors, or lack of standardization in performing bioreporter experiments.
- Precision for laboratory replicates – when detecting the optical signals of the bioreporter using a plate reader, a metric for precision must be calculated and reported, and the RSD of replicate measurements is used. Report precision of replicates as RSD, where RSD = stdev × 100%/average. Acceptable RSDs for instrument replicate analysis should be below 10% for pollutants analyzed in water samples. However, for soil/sediment samples, the RSD values are more variable, occasionally exceeding 10% and, in some cases, approaching 20%, which aligns with the expectation that ***Acceptance Criteria*** for these matrices are broader. As the measurement times increase, RSDs increase. The degree to which this is problematic can be evaluated in part by scrutiny of the bioavailability calibration curve. For work that uses soil/sediment slurries, the heterogeneity of the samples due to small soil/sediment volumes, varying particle sizes, and differences in colloidal minerals within the samples may also lead to large RSDs. At this point, it is necessary to adjust the concentration of the slurry and select the most suitable slurry concentration. Poor precision and how this affects interpretation of results must always be reported.
- LOD and LOQ (***limit of quantification***) – these critical parameters define the lowest concentration of pollutant that can be reliably detected (LOD) and quantified (LOQ) using a given method of instrumental analysis. For this method, LOD is calculated as three times

the standard deviation of the calibration blank divided by the slope of the bioavailability calibration curve. LOQ is defined as three times the LOD. Since whole-cell bioreporters are single-celled organisms, they are very sensitive to environmental factors. This may result in slight differences in the LOD and LOQ calculated for each microplate bioavailability calibration curve, as compared to Standard Methods of instrumental analysis. In general, as long as the user is proficient enough in the method, these small variations do not have a significant impact on the results of the experiment or the variability of LODs/LOQs. Indicatively, the RSD of LOD obtained from multiple independent experiments should be less than 50%. LOD and LOQ must be reported as they reflect upon competence of analysis and interpretation of results (e.g., lower precision typically occurs between the LOD and LOQ). If deviations from an established, repeatable LOD/LOQ are observed, Corrective Action requires investigation into potential causes (e.g., changes in bioreporter performance, instrument drift, or matrix effects). See comments above on corrective measures for bioavailability calibration curves.

- Toxic response – if the concentration of pollutants in a sample is too high, it can be toxic to the bioreporter, causing a decrease in output signal and the calculated bioavailable pollutant concentration lower than actually is. Because bioreporters do not have a non-zero calibration blank, observing signals lower than the calibration blank is indicative of the possible presence of toxic effects. If one sample exhibits toxic effects or if toxic effects are suspected, it is then necessary to dilute the samples. To confirm, perform a dilution series: dilute the sample incrementally and measure the bioreporter signal for each dilution. If the signal increases with dilution and stabilizes (final concentration after correction for DFs) within the linear range of the calibration curve, the original sample was likely too concentrated/toxic. Alternately, a method such as LIVE/DEAD can also be used to determine cell viability. The results of the LIVE/DEAD toxicity bioassay should clearly report the ratio of live to dead cells to quantify the toxic effects of the sample on the cells. These results should be combined with bioreporter bioluminescence inhibition data to illustrate the effect of toxicity on the bioreporter signal. If the problem of toxicity cannot be fixed via dilution, the sample should be flagged as potentially toxic (the matrix may be toxic), and results should be reported and interpreted with caution.
- Interferences – when a bioreporter responds to non-target pollutants or is affected by other toxic components in the sample, data analysis and interpretation must account for potential interferences and toxicity. During data analysis, compare the bioreporter response to the target pollutant(s) against responses to known interferents, and apply Acceptance Criteria such as a minimum specificity threshold (e.g., <10% cross-reactivity) or a signal-to-noise ratio to ensure reliable results. If interferences are identified, Corrective Action might include using a more specific bioreporter strain or employing mathematical techniques to isolate the signal of interest (which might also involve instrumental analysis to obtain additional information on the relative amounts of total pollutant versus interferents). In the case of interferences, results should be interpreted with caution, clearly distinguishing between target and non-target responses, and reporting should include detailed descriptions of the bioreporter's specificity, interference testing results, and any steps taken to mitigate issues.
- Data reporting – users should specify the bioreporter response units used, as described in directions for constructing bioavailability calibration curves, and fully explain how response is characterized (e.g., absolute units such as RLU, calculated response units such as MRR). If concentration units are used, a discussion of the difference between concentration and activity should also be included. Specifically, if the bioavailability calibration curve is calibrated in units of concentration, this reflects "100% bioavailability" in the absence of matrix, but likely not 100% bioavailability overall, since some species may not be bioavailable or may have activity coefficients greater than 1.0 (more common for organic pollutants).

9.7 CONCLUSIONS AND OUTLOOK

Whole-cell bioreporter technology has existed and evolved for almost 30 years. During this time, a large number of whole-cell bioreporters have been constructed for the detection of different pollutants. As discussed in this chapter and demonstrated by many investigators, bioreporter technology provides a faster, easier, and less expensive way to detect pollutant bioavailability than any other method in common use (e.g., other methods for measuring bioavailability and chemoavailability in this book). To date, however, there is no readily available source of bioreporter strains (e.g., from a centralized provider) and no guidance approaching a standardized methodology for best practice in use. This has arguably greatly limited the dissemination and uptake of the technology. This chapter is a first attempt to generalize how bioreporter technology can be used to measure the bioavailability of pollutants. Given the increasing prevalence of environmental pollution and accompanying need to assess and manage such pollution, it is both inefficient and fiscally imprudent to not harness this powerful technology in reducing the cost and complexity of ***Environmental Risk Assessment***.

REFERENCES

Abd-El-Haleem, D., Zaki, S., Abulhamd, A., Elbery, H., Abu-Elreesh, G. 2006. Acinetobacter bioreporter assessing heavy metals toxicity. *J. Basic Microbiol.* 46, 339–347. https://doi.org/10.1002/jobm.200510122

Al-Anizi, A.A., Hellyer, M.T., Zhang, D., 2014. Toxicity assessment and modelling of *Moringa oleifera* seeds in water purification by whole cell bioreporter. *Water Res.* 56, 77–87. https://doi.org/10.1016/j.watres.2014.02.045

ASTM (American Society for Testing and Materials), 1999. Standard guide for collection, storage, characterization, and manipulation of sediments for toxicological testing. E1391-94. In: *Annual Book of ASTM Standards, Vol. 11.05.* ASTM, Philadelphia, PA.

Augustiniene, E., Kutraite, I., Valanciene, E., Matulis, P., Jonuskiene, I., Malys, N., 2023. Transcription factor-based biosensors for detection of naturally occurring phenolic acids. *N. Biotechnol.* 78, 1–12. https://doi.org/10.1016/j.nbt.2023.09.004

Bauchop T., Elsden S.R., 1960. The growth of micro-organisms in relation to their energy supply. *J Gen Microbiol.* 23, 457–469. https://doi.org/10.1099/00221287-23-3-457

Belkin, S., 2003. Microbial whole-cell sensing systems of environmental pollutants. *Curr. Opin. Microbiol.* 6(3), 206–212. https://doi.org/10.1016/S1369-5274(03)00059-6

Belkin, S., Yagur-Kroll, S., Kabessa, Y., Korouma, V., Septon, T., Anati, Y., Zohar-Perez, C., Rabinovitz, Z., Nussinovitch, A., Agranat, A.J., 2017. Remote detection of buried landmines using a bacterial sensor. *Nat. Biotechnol.* 35(4), 308–310. https://doi.org/10.1038/nbt.3791

Bell, C., 2016. Chemical properties component of Chemical Engineering Design Library (ChEDL). https://github.com/CalebBell/chemicals

Bren, A., Park, J.O., Towbin, B.D., Dekel, E., Rabinowitz, J.D., Alon, U., 2016. Glucose becomes one of the worst carbon sources for *E. coli* on poor nitrogen sources due to suboptimal levels of cAMP. *Sci. Rep.* 6(1), 24834. https://doi.org/ 10.1038/srep24834

Cayron, J., Prudent, E., Escoffier, C., Gueguen, E., Mandrand-Berthelot, M.-A., Pignol, D., Garcia, D., Rodrigue, A. 2017. Pushing the limits of nickel detection to nanomolar range using a set of engineered bioluminescent *Escherichia coli*. *Environ. Sci. Pollut. Res.* 24, 4–14. https://doi.org/10.1007/s11356-015-5580-6

Din, G., Hasan, F., Conway, M., Denney, B., Ripp, S., Shah, A.A., 2019. Engineering a bioluminescent bioreporter from an environmentally sourced mercury-resistant *Enterobacter cloacae* strain for the detection of bioavailable mercury. *J. Appl. Microbiol.* 127(4), 1125–1134. https://doi.org/10.1111/jam.14399

Elcin, E., Öktem, H.A., 2020. Immobilization of fluorescent bacterial bioreporter for arsenic detection. *J. Environ. Health Sci. Eng.* 18, 137–148. https://doi.org/10.1007/s40201-020-00447-2

Ferreira, C.M.H., Pinto, I.S.S., Soares, E.V., Soares, H.M.V.M., 2015. (Un)suitability of the use of pH buffers in biological, biochemical and environmental studies and their interaction with metal ions – A review. *RSC Adv.* 5, 30989–31003. https://doi.org/10.1039/C4RA15453C

Fredenslund, Å., Gmehling, J., Michelsen, M.L., Rasmussen, P., Prausnitz, J.M., 1977. Group-contribution estimation of activity coefficients in nonideal liquid mixtures. *Ind. Eng. Chem. Res.* 16(3), 450–462. https://doi.org/10.1016/0378-3812(79)80015-1

Gao, G., Fang, D., Yu, Y., Wu, L., Wang, Y., Zhi, J., 2017. A double-mediator based whole cell electrochemical biosensor for acute biotoxicity assessment of wastewater. *Talanta* 167, 208–216. https://doi.org/10.1016/j.talanta.2017.01.081

Hansen, M.G., Marcussen, H., Holm, P.E., Brandt, K.K., 2019. Evaluation of pH-Dependent Metal Speciation Artifacts in Whole-Cell Bioreporter Analysis. *J. Environ. Qual.* 48(1), 199–204. https://doi.org/10.2134/jeq2018.10.0352

Hartono, M.R., Kushmaro, A., Chen, X., Marks, R.S., 2018. Probing the toxicity mechanism of multi-walled carbon nanotubes on bacteria. *Environ. Sci. Pollut.* 25(5), 5003–5012. https://doi.org/10.1007/s11356-017-0782-8

Hui, C.Y., Guo, Y., Wu, J., Liu, L., Yang, X.Q., Guo, X., Xie, Y., Yi, J., 2021. Detection of bioavailable cadmium by double-color fluorescence based on a dual-sensing bioreporter system. *Front. Microbiol.* 12, 696195. https://doi.org/10.3389/fmicb.2021.696195

Hynninen, A., Virta, M., 2010. Whole-cell bioreporters for the detection of bioavailable metals. In: Belkin, S., Gu, M.B. (eds) *Whole Cell Sensing System II: Applications*, pp. 31–63. Springer Berlin Heidelberg, Berlin, Heidelberg. https://doi.org/10.1007/10_2009_9

Jiang, B., Li, G., Xing, Y., Zhang, D., Jia, J., Cui, Z., Luan, X., Tang, H., 2017. A whole-cell bioreporter assay for quantitative genotoxicity evaluation of environmental samples. *Chemosphere* 184, 384-392.https://doi.org/10.1016/j.chemosphere.2017.05.159

Kang, Y., Lee, W., Jang, G., Kim, B.G., Yoon, Y., 2018. Modulating the sensing properties of *Escherichia coli*-based bioreporters for cadmium and mercury. *Appl. Microbiol. Biotechnol.* 102(11), 4863–4872. https://doi.org/10.1007/s00253-018-8960-2

Kheirati, S., 2025. UNIFAC group contribution method activity calculator function MATLAB Central File Exchange. Retrieved March 18, 2025. https://www.mathworks.com/matlabcentral/fileexchange/64885-unifac-group-contribution-method-activity-calculator-function

Li, B., Zhang, X., Tefsen, B., Wells, M., 2022. From speciation to toxicity: Using a "two-in-one" whole-cell bioreporter approach to assess harmful effects of Cd and Pb. *Water Res.* 217, 118384. https://doi.org/10.1016/j.watres.2022.118384

Li, S., Chen, D., Liu, Z., Tao, S., Zhang, T., Chen, Y., Bao, L., Ma, J., Huang, Y., Xu, S., Wu, L., Chen, S., 2023. Directed evolution of TetR for constructing sensitive and broad-spectrum tetracycline antibiotics whole-cell biosensor. *J. Hazard. Mater.* 460, 132311. https://doi.org/10.1016/j.jhazmat.2023.132311

Liu, P., Huang, Q., Chen, W., 2012. Construction and application of a zinc-specific biosensor for assessing the immobilization and bioavailability of zinc in different soils. *Environ. Pollut.* 164, 66–72. https://doi.org/10.1016/j.envpol.2012.01.023

Lopes, N., Hawkins, S.A., Jegier, P., Menn, F.M., Sayler, G.S., Ripp, S., 2012. Detection of dichloromethane with a bioluminescent (*lux*) bacterial bioreporter. *J. Ind. Microbiol. Biot.* 39(1), 45–53. https://doi.org/10.1007/s10295-011-0997-5

Magrisso, S., Belkin, S., Erel, Y., 2009. Lead bioavailability in soil and soil components. *Water Air Soil Pollut.* 202(1), 315–323. https://doi.org/10.1007/s11270-009-9978-y

Martelli, F., Binzoni, T., 2018. Heuristic model for ballistic photon detection in collimated transmittance measurements. *Opt. Express* 26, 744–761. https://doi.org/10.1364/OE.26.000744

Meretska, M., Uppu, R., Vissenberg, G., Lagendijk, A., IJzerman, W.L., Vos, W.L., 2017. Analytical modeling of light transport in scattering materials with strong absorption. *Opt. Express* 25(20), A906–A921. https://doi.org/10.1364/OE.25.00A906

Moratti, C.F., Yang, S.N.N., Scott, C., Coleman, N.V., 2024. Development of a whole-cell biosensor for ethylene oxide and ethylene. *Microb. Biotechnol.* 17(6), e14511. https://doi.org/10.1111/1751-7915.14511

Myers, J.A., Curtis, B.S., Curtis, W.R., 2013. Improving accuracy of cell and chromophore concentration measurements using optical density. *BMC Biophys.* 6(1), 1–16. https://doi.org/10.1186/2046-1682-6-4

Ndu, U., Mason, R.P., Zhang, H., Lin, S., Visscher, P.T., 2012. Effect of inorganic and organic ligands on the bioavailability of methylmercury as determined by using a mer-lux bioreporter. *Appl. Environ. Microbiol.* 78(20), 7276–7282. https://doi.org/10.1128/AEM.00362-12

Pang, Y., Ren, X., Li, J., Liang, F., Rao, X., Gao, Y., Wu, W., Li, D., Wang, J., Zhao, J., 2020. Development of a sensitive *Escherichia coli* bioreporter without antibiotic markers for detecting bioavailable copper in water environments. *Front. Microbiol.* 10, 3031. https://doi.org/10.3389/fmicb.2019.03031

Reed, M.H., Strope, E.K., Cremona, F., Myers, J.A., Newell, S.E., McCarthy, M.J., 2023. Effects of filtration timing and pore size on measured nutrient concentrations in environmental water samples. *Limnol. Oceanogr-Meth.* 21(1), 1–12. https://doi.org/10.1002/lom3.10529

Sun, Y., Zhao, X., Zhang, D., Ding, A., Chen, C., Huang, W.E., Zhang, H., 2017. New naphthalene whole-cell bioreporter for measuring and assessing naphthalene in polycyclic aromatic hydrocarbons contaminated site. *Chemosphere* 186, 510–518. https://doi.org/10.1016/j.chemosphere.2017.08.027

Tecon, R., Wells, M., Van Der Meer, J.R., 2006. A new green fluorescent protein-based bacterial biosensor for analysing phenanthrene fluxes. *Environ. Microbiol.* 8(4), 697–708. https://doi.org/10.1111/j.1462-2920.2005.00948.x

Trang, P.T.K., Berg, M., Viet, P.H., Mui, N.V., Van Der Meer, J.R., 2005. Bacterial bioassay for rapid and accurate analysis of arsenic in highly variable groundwater samples. *Environ. Sci. Technol.* 39(19), 7625–7630. https://doi.org/10.1021/es050992e

US EPA (United States Environmental Protection Agency), 2016. *Quick Guide to Drinking Water Sample Collection*. United States Environmental Protection Agency, CO 80403, Washington, DC.

Vasefi, F., 2010. Advancements in angular domain optical imaging in biological tissues. PhD Thesis, Simon Fraser University.

van der Meer, J., Belkin, S. 2010. Where microbiology meets microengineering: design and applications of reporter bacteria. *Nat. Rev. Microbiol.* 8, 511–522. https://doi.org/10.1038/nrmicro2392

van der Meer, J.R., 2006. Analytics with engineered bacterial bioreporter strains and systems. *Curr. Opin. Biotech.* 17, 1–3. https://doi.org/10.1016/j.copbio.2006.01.005

van der Meer, J.R., Tropel, D., Jaspers, M., 2004. Illuminating the detection chain of bacterial bioreporters. *Environ. Microbiol.* 6(10), 1005–1020. https://doi.org/10.1111/j.1462-2920.2004.00655.x

Wallace, M.J., Dharuman, S., Fernando, D.M., Reeve, S.M., Gee, C.T., Yao, J., Griffith, E.C., Phelps, G.A., Wright, W.C., Elmore, J.M., Lee, R.B., Chen, T., Lee, R.E., 2020. Discovery and characterization of the antimetabolite action of thioacetamide-linked 1,2,3-triazoles as disruptors of cysteine biosynthesis in gram-negative bacteria. *ACS Infect. Dis.* 6(3), 467–478. https://doi.org/10.1021/acsinfecdis.9b00406

Wang, J., Eldridge, M., Menn, F.-m., Dykes, T., Sayler, G., 2015. Standardized application of yeast bioluminescent reporters as endocrine disruptor screen for comparative analysis of wastewater effluents from membrane bioreactor and traditional activated sludge. *Ecotoxicology* 24, 2088–2099. https://doi.org/10.1007/s10646-015-1556-z

Wells, M., Wick, L.Y., Harms, H., 2005. Model polymer release system study of PAH bioaccessibility: The relationship between "rapid" release and bioaccessibility. *Environ. Sci. Technol.* 39(4), 1055–1063. https://doi.org/10.1021/es035067b

Wu, S., Zheng, H., Wang, Y., Wang, L., Chen, W., 2023. Cyanobacterial bioreporter of nitrate bioavailability in aquatic ecosystems. *Water Res.* 247, 120749. https://doi.org/10.1016/j.watres.2023.120749

Xu, X., Oliff, K., Xu, T., Ripp, S., Sayler, G. and Zhuang, J., 2015. Microbial availability of mercury: effective detection and organic ligand effect using a whole-cell bioluminescent bioreporter. *Ecotoxicology* 2, 2200–2206. https://doi.org/10.1007/s10646-015-1553-2

Xu, T., Close, D.M., Sayler, G.S., Ripp, S., 2013. Genetically modified whole-cell bioreporters for environmental assessment. *Ecol. Indic.* 28, 125–141. https://doi.org/10.1016/j.ecolind.2012.01.020

Xu, T., Young, A., Marr, E., Sayler, G., Ripp, S., Close, D., 2018. A rapid and reagent-free bioassay for the detection of dioxin-like compounds and other aryl hydrocarbon receptor (AhR) agonists using autobioluminescent yeast. *Anal. Bioanal. Chem.* 410(4), 1247–1256. https://doi.org/10.1007/s00216-017-0780-1

Yoon, Y., Kang, Y., Chae, Y., Kim, S., Lee, Y., Jeong, S.W., An, Y.J., 2016. Arsenic bioavailability in soils before and after soil washing: The use of *Escherichia coli* whole-cell bioreporters. *Environ. Sci. Pollut. Res. Int.* 23, 2353–2361. https://doi.org/10.1007/s11356-015-5457-8

Zhang, D., He, Y., Wang, Y., Wang, H., Wu, L., Aries, E., Huang, W.E., 2012. Whole-cell bacterial bioreporter for actively searching and sensing of alkanes and oil spills. *Microb. Biotechnol.* 5(1), 87–97. https://doi.org/10.1111/j.1751-7915.2011.00301.x

Zhang, X., Li, B., Deng, J., Qin, B., Wells, M., Tefsen, B. 2020a. Advances in freshwater risk assessment: improved accuracy of dissolved organic matter-metal speciation prediction and rapid biological validation. *Ecotoxicol. Environ. Saf.* 202, 110848. https://doi.org/10.1016/j.ecoenv.2020.110848

Zhang, X., Li, B., Deng, J., Qin, B., Wells, M., Tefsen, B., 2020b. Regional-scale investigation of dissolved organic matter and lead binding in a large impacted lake with a focus on environmental risk assessment. *Water Res.* 172, 115478. https://doi.org/10.1016/j.watres.2020.115478

Zhang, X., Li, B., Schillereff, D.N., Chiverrell, R.C., Tefsen, B., Wells, M., 2022b. Whole-cell biosensors for determination of bioavailable pollutants in soils and sediments: Theory and practice. *Sci. Total Environ.* 811, 152178. https://doi.org/10.1016/j.scitotenv.2021.152178

Zhang, X., Qin, B., Deng, J., Wells, M., 2017. Whole-cell bioreporters and risk assessment of environmental pollution: A proof-of-concept study using lead. *Environ. Pollut.* 229, 902–910. https://doi.org/10.1016/j.envpol.2017.07.068

Zhang, X., Zhu, Y., Elçin, E., He, L., Li, B., Jiang, M., Yang, X., Yan, X.P., Zhao, X., Wang, Z., Wang, F., Shaheen, S.M., Rinklebe, J., Wells, M., 2024. Whole-cell bioreporter application for rapid evaluation of hazardous metal bioavailability and toxicity in bioprocess. *J. Hazard. Mater.* 461, 132556. https://doi.org/10.1016/j.jhazmat.2023.132556

Zhang, X., Zhu, Y., Wells, M., 2022a. A case study comparing lead-response in different bioreporters: What constitutes a good biosensor? *Case Stud. Chem. Environ. Eng.* 5, 100192. https://doi.org/10.1016/j.cscee.2022.100192

Zhu, Y., Elcin, E., Jiang, M., Li, B., Wang, H., Zhang, X., Wang, Z., 2022. Use of whole-cell bioreporters to assess bioavailability of contaminants in aquatic systems. *Front. Chem.* 10, 1018124. https://doi.org/10.3389/fchem.2022.1018124

Section 3

Abiotic Methods

Measurement of Chemoavailability using Chemical Proxies

10 Chemoavailability of Pollutants Determined by Diffusive Gradients in Thin-Films

10.1 INTRODUCTION AND SCOPE

The use of aquatic organisms to predict the ***bioavailability*** of pollutants in the aquatic environment is desirable on the a fortiori basis that biological methods reflect actual bioavailability, however, often methods based on ***chemoavailability***, i.e., pollutants available to a chemical proxy, are more convenient. While convenient, results from chemovailability measurements often do not correlate well with bioavailability measurements. ***Diffusive Gradients in Thin-Films*** (DGT) is a method of measuring chemoavailability that has become popular in recent years as having a higher performance than many other chemovailability methods with regard to obtaining results that correlate with bioavailability methods (Diviš et al., 2007; Senila and Kovacs, 2024). DGT results have also been demonstrated to correlate well with expectations of bioavailability based on modeling of chemical speciation in complex environmental media. For instance, Yapici et al. (2008) used DGT to measure cadmium (Cd), nickel (Ni), copper (Cu), lead (Pb), cobalt (Co), and zinc (Zn) in municipal wastewater and mining wastewater using DGT, and the measured values from DGT were in good agreement with the values of these pollutants expected to be ***bioavailable*** based on calculations using the Windermere Humic Aqueous Model (WHAM) Model IV (see Chapter 14).

Another feature of the DGT approach that is appealing is that it can be used to detect pollutants *in situ* in a process referred to as *in situ* ***passive sampling***. *In situ* passive sampling is an alternative to traditional sampling techniques that differ in a number of respects from traditional active sampling. This approach provides time-averaged concentrations of ***chemoavailable*** pollutants over sampler deployment times of hours to weeks. While this may be disadvantageous if understanding a time series evolution of pollutant chemoavailability is the goal, it is an advantage in situations wherein pollutant levels are below the limits of detection (LODs) for instrumental analysis as the technique ***preconcentrates*** pollutants. In some situations, *in situ* ***preconcentration*** can also have the effect of protecting pollutant analytes from degradation during field deployment of *in situ* passive samplers and subsequent transportation to the laboratory (Kot et al., 2000). In this context, the DGT technique has been shown to be a versatile strategy for better understanding bioavailable concentrations of some pollutants.

The DGT technique is based on the idea that the thin film in the sampler is biomimetic to membranes in aquatic biota that contact natural waters, pore waters, etc., and the principle of diffusion (***Fick's first law***). Thus, the mechanism of action of DGT could or should be able to mimic the environmental content of the "bioeffective state" of pollutants, including binding strength and kinetics of complexes, solid–liquid exchange kinetics, and other reactions that control the actual ***bioavailable*** pollutant in the ambient ***medium***. The simplicity, ease of use in the field, applicability to different environmental matrices, and ability to preconcentrate a wide range of pollutants have made the DGT technique a frequent choice for environmental research (Chen et al., 2017; Diviš et al., 2007). DGT has been applied to the Environmental Monitoring of the chemoavailability of a variety of pollutants, with particular development and use for heavy metals studies in aquatic environments, soils, and sediments, though recently focus increasingly shifts to organic pollutants (refer to the case study in Chapter 16, and also see Chen et al., 2012, 2015).

DOI: 10.1201/9781003355229-13

This chapter provides an implementable methodology for DGT measurement of the bioavailability of heavy metals and organic pollutants in water, soils, and sediments. This includes a description of the selection use of reagents, equipment and materials used in the technique, procedures, and the precautions to be taken during operation. Chapters containing Terms and Definitions, guidance on avoiding contamination of samples in field and lab work, guidance on Quality Assurance (QA)/Quality Control (QC), and ***Standard Methods*** for instrumental analysis are also within the scope of this chapter.

10.2 PRINCIPLE

Figure 10.1 illustrates the mechanism of action of the DGT device. Bioavailable pollutants in an environmental medium (e.g., water, soil, sediment) in which the DGT device is suspended, pass from the environmental medium through the DGT device's ***filter membrane***, and then diffuse through a thin layer of ***diffusive gel*** to a layer of ***binding gel***. The filter membrane prevents fouling of the diffusive gel, and the diffusive gel layer sets up the ***concentration gradient*** that drives diffusion. In strongly binding the pollutant, the binding gel removes pollutants from the diffusion process such that at this point, the concentration of pollutant in the binding gel may be approximated as zero, hence creating and maintaining the concentration gradient between the pollutant in the environmental medium and the DGT device's internal binding gel. The diffusive flux of the to-be-measured pollutant during the DGT deployment time, i.e., the time during which preconcentration occurs, is calculated according to an equation based on Fickian diffusion.

10.3 EQUIPMENT

All field equipment and laboratory materials must be properly cleaned and sanitized prior to use, per Chapter 3. In addition, to avoid contamination, cleaned field gear must be transported in clean, sealed containers. Depending on experimental design, some equipment listed may not be needed; conversely, some users may find it useful to add additional items. The list is indicative.

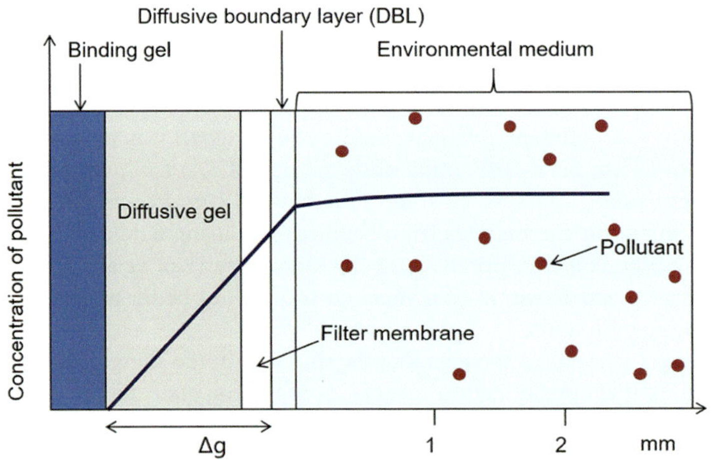

FIGURE 10.1 Schematic drawing showing the mechanism of action of the DGT device. The abscissa shows a conceptual cross-section, distance, through the DGT device and window of ***exposure*** to the environmental medium. Pollutant (solid circles) diffuses through the medium toward and through the DGT device's filter membrane and diffusive gel to the binding gel, driven by a concentration gradient, as shown by the solid blue line and label on the ordinate axis. The concentration at the binding gel is effectively zero since, once bound, the pollutant is no longer free to diffuse. Thus, the concentration gradient is approximately over the distance Δg, being the thickness of the diffusive gel and filter membrane, with the highest concentration in the environmental medium. The ***diffusive boundary layer*** (DBL) denotes the transition between advective transport and diffusion-limited transport, i.e., in aqueous settings with current. In case of fluid flow across the DGT device, the thickness of this layer approaches a constant.

10.3.1 Sample Collection, Deployment, Packaging, and Transport-Related

Anchors and anchor ropes – to ensure that buoy systems are stabilized.
Boat – for transportation of laboratory personnel and equipment to sampling ***sites***.
Buoys – surface buoys and/or subsurface buoys to help locate the deployment of DGT devices if deployment is not in very shallow water.
Containers – for, e.g., samples.
Dead weight – to stabilize the line(s) on which DGT devices are deployed; ensures a stationary location in currents (aqueous deployments).
Decontamination equipment – as needed to decontaminate field gear during operations.
DGT devices – bagged and in storage cases.
DGT unit storage cases – to store and protect DGT devices from physical damage (additionally, each DGT device should have its own bag).
Filter membranes, 0.45 or 0.2 μm pore size, 25 mm diameter – polyethersulfone filter membranes for heavy metals, oxyanion, and glyphosate studies, and polytetrafluoroethylene (PTFE) membranes for organic pollutant studies. This filter protects the diffusion and binding layers inside the DGT device from possible contamination caused by large particles and contains the gels to prevent damage or loss of these during the deployment of the DGT devices.
Fixing clamps (clips or clamps) – on a case-by-case basis, depending on how DGT devices are attached, string or net (aqueous deployments).
Frame as needed – provides additional support to secure DGT devices, ensuring stability; must not interfere with current across DGT devices (aqueous deployments).
Gloves – ***contaminant***-free, powder-free gloves – for wearing when placing and accessing DGT devices.
GPS/electronic or physical devices – to ensure the exact siting of deployment locations and depths (e.g., depth gauge).
Ice – ice packets are preferred if it is possible to keep them well frozen until used; alternatives include dry ice (subject to certain shipping requirements) or wet ice (which increases chances of sample cross-contamination when melting). When using dry ice, thermo-insulated boxes must be drilled in places to allow pressure regulation as the dry ice evaporates.
Line (heavy duty) or netting – to suspend DGT devices and ensure their correct ***distribution*** in the water for aqueous deployments.
Moorings or other fixed structures – to which the DGT device assembly is attached with a rope to ensure stationary deployment.
Muffle furnace – for high-temperature decontamination of glassware (organic pollutants).
Multiparameter water quality sonde – to measure pH, temperature, conductivity, and other water quality parameters at test sites.
Plastic bags – watertight, resealable (zip-type), several sizes; packaging for devices before and after deployment period, samples, DGT devices, and items of equipment that need to be kept decontaminated during transport. 0-DGT devices must not contact plastic bags directly; in this case primary packaging could be contaminant-free aluminum foil.
String/twine – for securing packaging and/or labels to packaging.
Thermo-insulated boxes – to transport samples and DGT devices at 4°C.
Tool kits – including screwdrivers, pliers, etc., for assembling and adjusting buoy devices or other fixed structural equipment.

10.3.2 Field Data Records-Related

Digital camera – to take photographs of the sampling sites to document the site environment and setup as ancillary data support.
Field logbook – in addition to sample tracking forms, the field collection team should document in a field logbook any additional information on sample collection activities,

hydrologic conditions, weather conditions, boat or equipment operations, or any unusual activities observed or problems encountered that would be useful in evaluating the quality of data collected using this method.

Field record ledger – a type of sample tracking form that enables an at-a-glance summary of details of all samples. Each line of the form usually includes (1) project number and/or name, and/or short description; (2) unique sample identifier; (3) sampling date and time (24-hour clock format); (4) sampling location latitude/longitude and potentially including site name, geographic description, waterbody name/description; (5) sampling depth (with units); (6) collection method; (7) if Chain-of-Custody is applicable, collectors' names and signatures; and (8) contact for responsible person or entity.

Indelible pens – to record field observations without loss of fidelity in wet environment.

Labels – clear, durable labels for each sample, DGT device, and device bag.

10.3.3 Laboratory/Device Preparation-Related

(Amber) vials/containers – for storage of DGT pollutant eluates; vials used for storage of eluates containing organic contaminants require amber vials.

Aseptic cutting plates – for cutting gels to prevent tearing of the gel during the cutting process (available from purveyors of electrophoresis supplies).

Balance – for weighing laboratory samples.

Beakers – various preparatory uses.

Conductivity meter – used to estimate ***ionic strength***, per guidelines in this method.

Filtration unit/centrifuge – to remove particles from binding gel eluate.

Laminar flow hood, positive pressure – for processing of DGT devices to avoid contamination.

Gel cutter, razor, scalpel, or cutter coated with PTFE – used to cut gels.

Glass plates and inert plastic/PTFE spacers – used for casting gels (available from purveyors of electrophoresis supplies).

Incubator – for preparation of gels.

Magnetic stirrer and stir bar – for gel and other preparations.

Nitrogen evaporator – to reduce solvent in DGT extracts to ultralow volumes under a nitrogen stream, e.g., the N-EVAP model from Organomation, or similar.

pH meter – for determining the acidity or ***alkalinity*** of samples and prepared reagents.

Pipettes (10–1,000 μL) and micropipettes (50–300 μL) – for precise dosing/spiking of samples and reagents.

Refrigerators – for storage of collected samples and devices.

Sonicator – to prepare titanium dioxide binding gel.

Tweezers/Blades – used to remove filter membranes and diffusion gels when collecting binding gels from DGT devices.

Vacuum pump – degassing of gel solutions during gel preparation.

Water bath – used for heating when preparing gels.

10.4 REAGENTS

Ultrapure water – also known as Type I water (Milli-Q, resistivity >18.2 MΩ·cm).

Note: All reagents must be of guaranteed reagent (GR) quality or better to ensure that the reagents used are of the highest purity to avoid contamination.

Acetonitrile – eluent to extract some organic pollutants from DGT binding gels.

Acrylamide – for making diffusive gels.

Agarose-derived crosslinker – to cross-link polyacrylamide in making gels (from, e.g., DGT Research, or similar).

Agarose – for making diffusive gels.

Ammonia – 10% solution used with methanol to extract some organic pollutants from DGT binding gels.
Ammonium persulfate – reacts with tetramethyl ethylenediamine to form free radicals for initiation of polymerization in acrylamide and bisacrylamide-based gels.
Chelex-100 – used to prepare hydrophilic–lipophilic balanced (HLB) binding gels.
HLB binding agents – used to prepare HLB-based binding gels.
Methanol – eluent to extract some organic pollutants from DGT binding gels.
Nanocrystalline titanium dioxide – MetSorb (Graver Technologies) or equivalent – to prepare titanium dioxide binding gels.
Nitric acid, ultra-trace purity (Optima® grade) – used for sample acidification and gel elution.
Sodium chloride (NaCl) – used to make solutions for gel storage.
Sodium hydroxide (NaOH) – eluent for titanium dioxide resins.
Sodium nitrate – used to make solutions for gel storage.
N,*N*,*N*′,*N*′-tetramethylethylenediamine (TEMED) – a catalyst commonly used in polymerization reactions, especially when preparing polyacrylamide gels.

10.5 PRECAUTIONS AND RECOMMENDATIONS

Field deployment:

- For watercourse deployment, favor watercourse accesses that are practical, easy to reach, and safe throughout the year.
- Anticipate fluctuations in river levels (flood/recession) depending on the weather conditions and snow melt.
- Anticipate activities upstream of measurement stations that can impact water quality during the deployment time; for example, the release of water from dams, cattle fording, or any other ***activity*** that might induce sporadic high increases in total suspended solids in the water column.
- Place DGT devices as far away as possible from areas where human or animal activity might interfere with them (e.g., nearby construction, or sporting activities such as fishing, swimming, and canoeing).
- For DGT devices deployed in the water column over longer deployment times, biofilms may develop, and these may interfere with the measurements. It is preferable to shorten the DGT device deployment time in cases where biofilms might develop in order to avoid same.

Per Chapter 3, any field gear that has not been properly cleaned and decontaminated must not be used, i.e., all materials used (ice chests, processing tools, plastic bags, etc.) must be subjected to appropriate cleaning and decontamination in the laboratory prior to use. Further, after cleaning, field gear must be transported and stowed in a manner to remain contaminant-free. Stowing typically involves carrying cleaned gear in clean, sealed, containers.

Additional considerations are as follows:

- For water column deployment, particularly for shorter deployment times, ideally current speed is greater than 2 cm/s during DGT device exposure to limit the influence of the ***diffusive boundary layer*** (DBL) on diffusive transport (Gimpel et al., 2001).
- For pollutants that are charged or ionic species (e.g., heavy metals), the targeted water pH should ideally range from 5 to 9 as the pH can impact the ***adsorption*** capacities of the selected binding phases. Proton competition at pH <5 in particular limits the capacity of DGT devices using Chelex as a binding phase (Wang et al., 2017).

- At a low ionic strength (electrical conductivity of ~120 μS/cm or less), the diffusion coefficients, which determine the inherent rate that a pollutant may diffuse through a medium, may be highly variable. In this case, laboratory determination of diffusion coefficients in synthetic waters with the same or similar composition may be in order (Scally et al., 2006).
- All processing of DGT devices should take place in a contaminant-free laminar flow hood to avoid contamination. For various reasons, DGT devices are extremely easy to contaminate.

10.6 OBTAINING LOADED DGT DEVICES

Gels may either be prepared or purchased in pre-assembled, "loaded", devices that are ready for deployment. Purchasing is preferable for reliable and consistent results; however, depending on the number of devices deployed, it may be cost-effective to prepare them. Purchased loaded devices have the advantage of being prepared in a clean laboratory, by experienced technical staff and with high standards of QC. It is also possible to purchase the filter membranes, diffusive gels, and binding gels separately for self-assembly in DGT devices. Self-fabrication of diffusive and binding gels has great potential for error due to contamination and incorrect process control in the multistep process. Tables 10.1 and 10.2 show the variety of gels available as loaded devices from DGT Research, by pollutant, for, respectively, water and soil/sediment deployment.

TABLE 10.1
Loaded DGT Devices Available for Deployment in Aqueous Media

Target Pollutants	DGT Research Specification	Filter Membrane	Diffusive Gel	Binding Layer
Up to 30 metals, including Cd, Co, Cu, Fe, Mn, Ni, Pb, Zn, etc.	LSNM-NP-loaded DGT device for metals (A)	PES	0.8 mm APA	Chelex-100
P and metals present as oxy-ions, including As, Mo, Sb, Se, U, V, W, etc.	LSNP-NP-loaded DGT device for P and metals (B)	PES	0.8 mm APA	Iron oxide
P and metals present as oxy-ions or strongly hydrolyzed, including Al, As, Mo, Sb, Se, U, V, W. Also, glyphosate	LSNT-NP-loaded DGT device for P, metals (B), and glyphosate	PES	0.8 mm APA	Titanium oxide (MetSorb)
P and metals present as oxyanions, including As, Mo, Sb, Se, V, etc.	LSNZ-NP-loaded DGT device for P and metals (B)	PES	0.8 mm APA	Precipitated zirconium oxide
Phosphate and polyphosphates and up to 40 metals, including Al, As, Cd, Co, Cu, Fe, Mn, Ni, Pb, Sb, Se, U, V, W, Zn, etc.	LSNX-NP-loaded DGT device for metals (A and B) and P	PES	0.8 mm APA	Mixed binding layer of Chelex-100 and MetSorb
Phosphate and polyphosphates and up to 40 metals, including As, Cd, Co, Cu, Fe, Mn, Ni, Pb, Sb, Se, U, V, W, Zn, etc.	LSNX-NP-loaded DGT device for metals (A and B) and P	PES	0.8 mm APA	Mixed binding layer of Chelex-100 and iron oxide (ferrihydrite)
Hg^{2+}, methyl mercury, and As(III)	LSNB-AP-loaded DGT device	PES	0.8 mm agarose	3-Mercapto-propyl-functionalized silica gel
Cr(VI) in the presence of Cr(III)	LSNE-NP-loaded DGT device	PES	0.8 mm APA	N-methyl-D-glucamine functional resin
Technetium (Tc)	LSNV-NP-loaded DGT device	PES	0.8 mm APA	Dispersed TEVA resin

(*Continued*)

TABLE 10.1 (*Continued*)
Loaded DGT Devices Available for Deployment in Aqueous Media

Target Pollutants	DGT Research Specification	Filter Membrane	Diffusive Gel	Binding Layer
Nitrate in freshwater	LSNN-AP-loaded DGT device	PES	0.8 mm agarose	SIR-100-HP resin
Sulfide	LSNH-NP-loaded DGT device	PES	0.8 mm APA	Silver iodide
A range of polar compounds, including antibiotics and some illicit drugs	LSNC-AP-loaded DGT device for antibiotics and illicit drugs	PES	0.8 mm agarose	XAD-18 resin
A range of polar compounds, including pharmaceuticals other than antibiotics	LSND-AP-loaded DGT device for pharmaceuticals other than antibiotics	PES	0.8 mm agarose	HLB
Bisphenols	LSNR-AT-loaded DGT device	PTFE	0.8 mm agarose	Activated charcoal
Household and personal care products	LSND-AN-loaded DGT device	Nucleopore polycarbonate	0.8 mm agarose	HLB
Pesticides and herbicides	LSND-AG-loaded DGT device	GHP	0.8 mm agarose	HLB

Abbreviations: APA, agarose cross-linked polyacrylamide; GHP, hydrophilic polypropylene; HLB, hydrophilic–lipophilic balanced binding agent; PES, polyethersulfone; PTFE, polytetrafluoroethylene; TEVA, Technetium extraction via anion exchange.

TABLE 10.2
Loaded DGT Devices Available for Deployment in Soil/Sediment

Target Pollutants	DGT Research Specification	Filter Membrane	Diffusive Gel	Binding Layer
Up to 30 metals, including Cd, Co, Cu, Fe, Mn, Ni, Pb, Zn	LSPM-NP-loaded DGT device for metals	PES	0.8 mm APA	Chelex BL
P and metals present oxyanions, including As, Mo, Sb, Se, U, V, W	LSPP-NP-loaded DGT device for P and metals	PES	0.8 mm APA	Iron oxide
P and metals present as oxyanions or strongly hydrolyzed, including As, Mo, Sb, Se, U, V, W	LSPT-NP-loaded DGT device for P and metals	PES	0.8 mm APA	MetSorb
P and metals present as oxyanions, including As, Mo, Sb, Se, V	LSPZ-NP-loaded DGT device for P and metals (B)	PES	0.8 mm APA	Precipitated zirconium oxide
P present as phosphate and polyphosphates and up to 40 metals, including Al, As, Cd, Co, Cu, Fe, Mn, Ni, Pb, Sb, Se, U, V, W, Zn	LSPX-NP-loaded DGT device for P and metals	PES	0.8 mm APA	Mixed binding layer of Chelex and MetSorb
P present as phosphate and polyphosphates and up to 40 metals, including As, Cd, Co, Cu, Fe, Mn, Ni, Pb, Sb, Se, U, V, W, Zn	LSPY-NP-loaded DGT device for P and metals	PES	0.8 mm APA	Mixed binding layer of Chelex and iron oxide (ferrihydrite)
Selectively measures Cr(VI) in the presence of Cr(III)	LSPE-NP-loaded DGT device for chromium (VI)	PES	0.8 mm APA	N-methyl-D-glucamine functional resin

(*Continued*)

TABLE 10.2 (*Continued*)
Loaded DGT Devices Available for Deployment in Soil/Sediment

Target Pollutants	DGT Research Specification	Filter Membrane	Diffusive Gel	Binding Layer
Hg^{2+}, methyl mercury, and As(III)	LSPB-AP-loaded DGT device for mercury and arsenic	PES	0.8 mm agarose	3-Mercaptopropyl-functionalized silica gel
Tc	LSPV-NP-loaded DGT device for technetium	PES	0.8 mm APA	Dispersed TEVA resin
Nitrate in sediments where the ionic strength of the pore water is <0.02 mmol/L (freshwater)	LSPN-AP-loaded DGT device for nitrate	PES	0.8 mm agarose	SIR-100-HP resin

Abbreviations: APA, agarose cross-linked polyacrylamide; PES, polyethersulfone; TEVA, Technetium extraction via anion exchange.

10.6.1 Notes on DGT Devices

Early DGT devices were prepared in different ways according to the fabrication capabilities of specific research labs (Davison and Zhang, 1994, Zhang et al., 1995). Currently, most devices are purchased from a supplier. For water column work, the most common design is a so-called piston design, whereas for soil and sediment work, a probe-type design is used to ensure intimate contact on insertion. Soil/sediment probe designs also differ from the piston design used in water column work in having a wider lip to better seal and prevent fine soil particles from entering the device. Figure 10.2 shows examples of these designs.

The piston design offered by DGT Research is such that there is a 1.2 mm gap between the piston (a in Figure 10.2) and the covering sleeve (e in Figure 10.2), accommodating 25 mm diameter filter membrane/gels with a 20 mm diameter exposure window when assembled. The commercially available filter membranes, diffusive gels, and binding gels for such a device are, respectively, 0.14, 0.8, and 0.4 mm thick for a total thickness of 1.34 mm, which allows for a slight amount of compressibility when the device is assembled, which ensures a good seal. Similar comments apply to

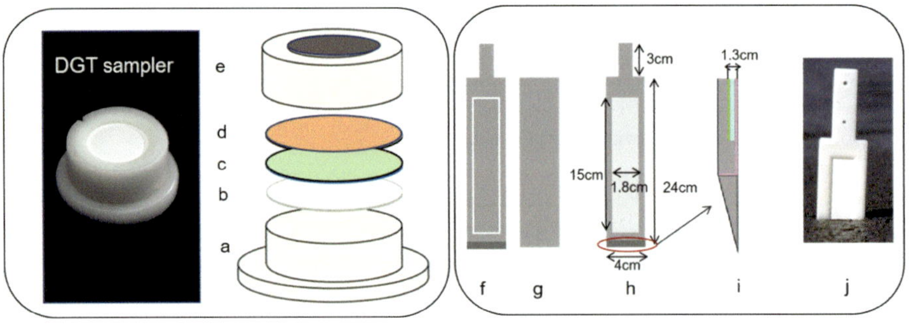

FIGURE 10.2 Left panel shows a loaded piston-type DGT device used in aqueous environments, along with a schematic indicating ***components*** of the device (a – piston base; b – binding gel; c – diffusive gel; d – filter membrane; e – exposure window) (Marrugo-Madrid et al., 2021). Right panel shows a schematic of a soil/sediment probe (f and g – window and base of device, respectively; h – loaded device and representative dimensions; i – a cutaway view of the bottom of h loaded with binding gel, diffusive gel, and a filter membrane, a representative total thickness of which at the exposed device window is typically on the order of ~1.1–1.5 cm [1.3 cm shown]; j – picture of a soil/sediment probe).

commercially available soil/sediment probes that are in common use, although more thicknesses are available to accommodate different field conditions/levels of bioavailable pollutant and the need to use different diffusive gel thicknesses to determine DBL thickness. The important point is that the combined filter membrane/gel thicknesses must be such as to ensure a good seal once loaded in a DGT device.

10.6.2 Preparation of Diffusive and Binding Gels and Choice of Filter Membranes

Often both diffusive and binding gels are prepared with the same gel stock solution. The most common diffusive gel types are agarose and agarose cross-linked polyacrylamide (APA) gels, with APA being somewhat more common. To prepare an APA gel, add 47.5 mL of Type I water to 37.5 mL of 40% acrylamide solution (e.g., EM Science/BDH Inc.; OmniPur) and mix thoroughly. Then add 15 g of a 2% agarose-derived crosslinker solution (DGT Research Ltd, Lancaster, UK) and mix again. This results in a final crosslinker concentration of 0.3% in the gel solution. This solution should be prepared fresh each time DGT gels are made. If temporary storage is necessary, keep at 4°C in the dark, though fresh preparation is recommended for consistency. Diffusive gels with smaller pore sizes can be made by increasing the degree of cross-linking during polymerization. Such smaller pore sizes are useful to exclude high-molecular weight compounds such as humic and fulvic acids; hence, these types of gels are often used for heavy metal pollutant studies. Prepare such a gel in the same manner as the APA gel, however, instead of using the agarose-derived crosslinker at a final concentration of 0.3% in the stock gel solution, add bis (acrylamide) to a final concentration of 0.8% (Scally et al., 2006).

To initiate gel polymerization, add 70 μL of 10% ammonium persulfate and 20 μL TEMED to 10 mL of APA gel stock solution and mix well. It is very important to avoid air bubbles in mixing; on one hand, any bubbles or foam formed during the mixing must be allowed to collapse before the solution is cast. On the other hand, the TEMED will initiate polymerization of the gel solution, so it is important to cast the gel in the shortest possible time after mixing. In some conditions, degassing the gel solution (i.e., under vacuum or suction) before initiating polymerization may help. Cast the APA gel between two glass plates separated by a plastic spacer with an appropriate thickness and set the gel at a temperature of 42°C ± 2°C for at least 45 minutes. In practice, it is difficult to know if the thickness of the gel is appropriate until assembling devices. The set gel is delicate and hence, must be carefully removed from the plates and left to hydrate in Type I water for a day or more, with at least four changes of water (until the rinse water is pH 7 or less). This allows the gel to hydrate, swell, and release excess TEMED and ammonium persulfate. After rinsing, the gel is then stored in 0.01 M NaCl or $NaNO_3$ solution at room temperature.

Ideally, make diffusive and binding gels to target thicknesses and then test whether the final actual thicknesses are appropriate to the DGT device in use. If the combined thickness of filter membrane and gels is too thick, too much compression occurs on assembly, and this damages the gels and their performance. If the combined thickness on assembly is too thin, there will not be a seal at the DGT device window, and the results will not be meaningful. Other gel thicknesses can be used as long as fitting well/sealing in the DGT device. If the total thickness of filter membrane, diffusive, and binding layers is too thin to achieve a good seal, it may be possible to add an appropriate spacer at the base of the DGT receiving device (piston or probe), between the base of the device and the binding gel layer.

To make agarose diffusive gels, dissolve 0.15 g agarose in 10 mL of water and heat to 100°C to create a 1.5% agarose stock solution. Cast the hot solution between two pre-warmed glass sheets separated by a PTFE spacer, then let cool to room temperature. Carefully remove the gels and store in 0.01 M NaCl solution.

Binding gels differ from diffusive gels mainly due to the presence of a binding agent. Many different binding agents have been researched for different types of pollutants and field deployment conditions. Binding gels are designed based on the ***selectivity***, rate of binding, and performance of binding agent with respect to the target pollutant(s) in typical environmental field conditions.

By far the most common binding agent used in DGT devices is Chelex-100 resin. Binding gels incorporating this resin have been used for studies on Al, Ca, Cd, Co, Cr(III), Cu, Fe, Mg, Mn, Ni, Pb, U, Zn, and rare earth elements. To prepare this binding gel, add 4 g (wet weight, i.e., after cleaning and washing following manufacturer's directions) of Chelex-100 resin to 10 mL of the stock APA gel solution. Initiate gel polymerization as for the APA diffusive gel, however, use only 50 μL of ammonium persulfate and 15 μL TEMED. Cast and set as for the diffusive gels described above. This gel may take longer to set.

Titanium dioxide binding resins are another popular choice as they enable detection of P and oxyanion forms of Al, As, Mo, Sb, Se, U, V, and W. Glyphosate has also been studied using DGT with this type of binding gel. To prepare this gel, combine 1 g of nanocrystalline titanium dioxide powder (MetSorb) with 10 mL of the APA gel stock solution, sonicate for 5 minutes, mix well on a magnetic stirrer for 1 minute, and then cast between two glass plates separated by an inert plastic spacer. Set as for diffusive gels, then hydrate the set gel in Type I with at least three to four changes of water over 24 hours.

For organic pollutants, a gel that has come into popular use is based on HLB binding agents. HLB-based binding gels have been used to study household and personal care products, pesticides and herbicides, microcystin, and many other organic pollutants (Chen et al., 2017; Li et al., 2019; Yao et al., 2019). Because of their unique balance of binding agents, gels made with HLBs are suitable for use with a wide range of organic pollutants. This binding gel is in agarose and paired with an agarose diffusive gel. To make the binding gel, mix 400 mg of HLB (Oasis HLB, 30 μm, Waters) in 10 mL of 1.5% warm agarose solution. Then cast as for the agarose diffusive gel. Store in Type I water.

When casting binding gels, the binding agent tends to settle on one side of the gel; this side should be placed facing up, i.e., toward the diffusive gel, when assembling in DGT devices. Keep track of what side of the binding gel is down when casting – the side that is facing downward during casting should face toward the diffusive gel during DGT device assembly. For both diffusive and binding gels, the stock gel solution may be made at weaker gel concentrations. Lower gel concentrations produce a weaker gel, which may be difficult to work with, but the coefficient of diffusion will be higher, which may be desirable for higher molecular weight pollutants.

Filter membranes in common use are made either of polyethersulfone or PTFE, with either 0.45 or 0.2 μm pore size. Polyethersulfone is used with Chelex-100 and titanium dioxide/APA gels, and PTFE is used with HLB/agarose gels.

10.6.3 DGT Device Assembly

An advantage of purchasing loaded DGT devices commercially is that they come assembled. In any use scenario, processing of DGT device components and assembly of DGT devices is performed under clean room or clean-room equivalent conditions. Additionally, it is extremely important to the function of the final DGT devices that the gels be kept fully hydrated with Type I water at all times. It is also quite important when assembling devices to prevent entrainment of air, which is straightforward if hydration is sufficient. In addition to buying loaded DGT devices, it is possible to purchase gel sheets from suppliers, in which case the gels must be cut in-house for assembly in DGT devices.

To cut gels for piston-type DGT devices, remove a gel sheet from the storage container or packaging (if purchasing gels, in which case follow manufacturer's directions for preparing for use) and place the gel sheet on a contaminant-free cutting board. Care is needed in this step since unsupported gel sheets tear easily. Check that the gel is in good condition, with no bubbles, and for binding gels that there is a consistent distribution of the binding resin across the gel. Discard any gels that do not meet these criteria. Press the gel cutter directly and firmly onto the gel and twist while pressing downward (for 25 mm diameter gels, it is possible to purchase stainless-steel cookie cutters of this size). For probe-type DGT devices, gels must be cut in rectangular shapes with an appropriately cleaned razor, scalpel, or PTFE-coated cutting blade.

To assemble DGT devices, place a cut piece of binding gel on the base of the DGT device holder, binder side up, followed by a piece of diffusive gel on top and a single filter membrane after that. The layers must be correctly aligned with the base and with each other. When this is done, the sleeve of the DGT device is placed to seal the membrane and gel layers. These usually snap into place, and an even force must be used. A good seal must be obtained. If the membrane cannot be moved when "nudged" lightly with a clean, gloved finger, then the seal is adequate. Otherwise, thicker layers or spacers are needed.

Once prepared the DGT devices must be kept wet. They may be stored at 4°C in small containers or clean watertight plastic bags (zip type). While commercially available devices may have longer shelf-lives, for in-house constructed DGT devices, due to the absence of QC in manufacturing that is possible in commercial settings, prolonged storage is inadvisable to ensure that biological growth is minimized.

10.6.4 Number of Devices Needed

A minimum of five DGT devices are required per sampling site. This includes one device for a DGT field blank and four DGT devices for sampling. Of these latter four, three are for deployment and the fourth is kept as a reserve. If water column sampling is conducted at multiple depths per site, an additional four devices are needed for each extra depth. Laboratory blanks are additionally required; however, these are never exposed per se. These blanks are for the detection of anomalous contamination of the devices in the manufacturing process (be it commercial or in-house) of the devices. Three laboratory blanks are required for every batch of DGT devices. Batches should be designated by the manufacturer, or are known by individual labs who fabricate their own gels, i.e., each time gels are made, devices made from such gels require at least one laboratory blank (three is best). Handling of DGT field blanks is discussed further in the next section.

For flows less than ~2 cm/s or for stagnant conditions, it is necessary to estimate the DBL thickness. This is done by deploying multiple sets of DGT devices with different diffusive gel thicknesses – a minimum of four thicknesses, with a minimum of three replicates each. As above, an extra DGT device is needed for each three replicates. No additional field blank is needed.

10.6.5 DGT Device Deployment

DGT devices are transported to sites stored in clean, sealed bags. Ideally, if clusters of devices, i.e., on lines or nets, are to be deployed, these are assembled and bagged before transport so that deployment in the field is rapid. Bags containing DGT devices are then placed entirely in foam boxes/coolers at or near 4°C. It is important that devices are never allowed to freeze.

10.6.5.1 Deployment in Natural Waters

The location for DGT-based sampling should minimize chances for disturbance of the devices or vandalism, should have a dependable depth to ensure that devices will not dry out during deployment, and should have flowing, but ideally not turbulent flow. For heavy metals analysis, the site should be away from submerged metallic structures. Ideally, sites are accessible by land, boat, and/or with a diver. In the case of ports, sites should be without risk of entanglement of the system on motors of ships that circulate near the exposed DGT devices.

The equation used to calculate bioavailable concentration for DGT assumes a constant DBL, an assumption that is only true for flows of ~2 cm/s or greater. The effect of the DBL on DGT calculations will be most problematic when currents are low and the thickness of diffusive gel layer is thin. When currents or flow across DGT devices deployed in water is anticipated to be low, the general approach is to use multiple DGT devices with different thicknesses of diffusive gel layers (as much as 2 mm thick). This enables the DBL to be compensated for in calculating final results (Section 10.7).

Even in cases of sufficient flow, a minimum of three DGT devices are required at any deployment site and depth. It is important that the manner in which devices are deployed does not interfere with currents. Many investigators use plastic plates that house DGT devices for deployment; however, this practice is undesirable since plates are very likely to disrupt localized flow across devices. The simplest deployment scheme that does not disrupt flow is to string DGT devices on a line or mount them on netting.

The easiest method to deploy DGT devices in water is to have some fixed mount to which to attach strings or nests of devices; this could involve moorings and buoys, or other fixed structures such as a picket, to which the DGT assembly is attached with a rope. Examples of deployment using moorings/buoys are shown in Figure 10.3. If deploying from a boat, all deployment preparation and activities must be done with the engine switched off.

When deploying in water, ensure that the sampling windows of the DGT devices are fully immersed during the deployment period. Deploy DGT devices ideally at a depth of 1–1.5 m below the surface and at least 1 m above the bottom of a river, lake, or sea bed. When work in shallow areas (depth <1 m) is required, ensure that the DGT devices are fully immersed and at least 0.3–0.5 m.

10.6.5.2 Deployment in Soils and Sediments

Deployment of DGT devices in soils in field conditions is, for the most part, impractical. In actual field conditions, soil moisture conditions vary a great deal both spatially and temporally, and *in situ* soils are subject to evaporation. DGT devices must be fully wet over the course of use for bioavailability studies, which is incompatible with natural soil conditions in many cases. In some situations, e.g., rice paddies, where the soil is fully wet through the period of deployment, DGT devices may be of use, in which case the procedure would be as for sediments. Otherwise, soil studies with DGT are limited to laboratory work. In this case, field soil samples are collected, homogenized (in some studies dried, ground, and sieved), placed in beakers, and wetted, and DGT planar probes are inserted into the beakers with a covering to prevent evaporation.

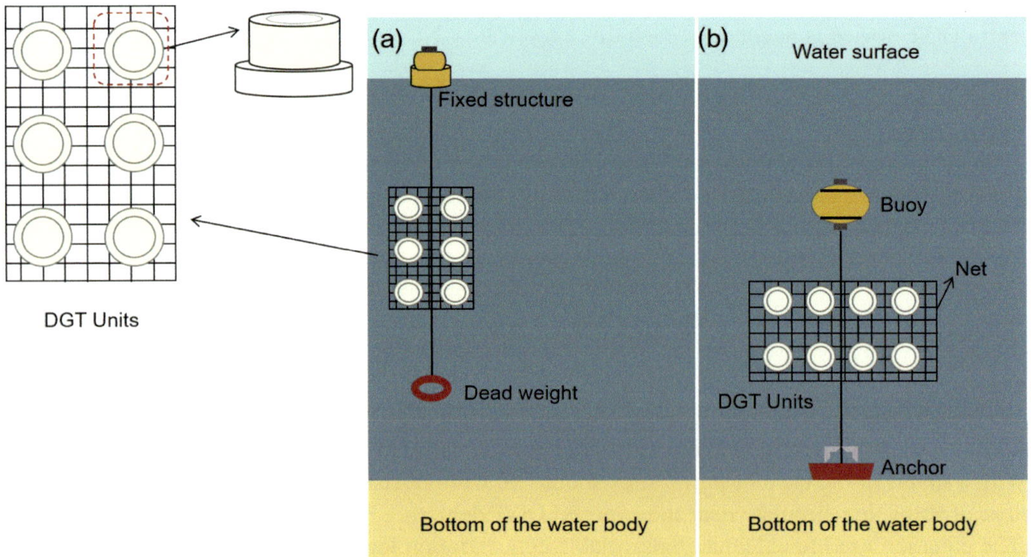

FIGURE 10.3 Examples of options for DGT device deployment using moorings/buoys showing deployment in water (a) from the surface and (b) from the bottom.

In sediments, probes are deoxygenated before deployment if it is suspected that sediments (or the depth layer to which probes are inserted) are suboxic or anoxic. Deoxygenating reduces the disruption of sediment chemistry in such conditions (e.g., oxidative precipitation of Fe(II) to Fe(III) and dissolution of FeS). It is also useful to consider conditioning the probe to a similar ionic strength to deployment conditions. Deoxygenation is accomplished by bubbling nitrogen through the solution in which the DGT device is stored, in a closed container with a single nitrogen inlet and outlet that can be tightly sealed after deoxygenation. Even very tightly sealed containers will only keep oxygen out for a few hours at most. Deoxygenation of the DGT probes is difficult in practice because after deoxygenation, the DGT probes must be transferred from their deoxygenation solution to the sediment essentially immediately; otherwise, oxygen is rapidly replenished. Because the DGT sediment probe assembly is thin, however, it does not contain a lot of oxygen and, within a short period of time, any small amounts of oxygen will be consumed in the anoxic environment. While this will compromise the DGT measurement of redox-sensitive and redox-affected components for DGT devices during the first day or more after deployment, this initial period where the presence of oxygen can have an effect on the DGT device accumulation represents a small portion of the total accumulation time. Nonetheless, in an abundance of caution, deoxygenation should be performed when appropriate to the deployment environment because the exact effect of oxygen on DGT probes has not been systematically evaluated.

As sediments are stratified, achieving a representative assessment of sediment environments is challenging, and this point should be taken into consideration in sampling plans. To test deeper sediments, DGT device deployments may be made in sediment cores in laboratory settings, which also admits the possibility to control temperature. For deployment in field sediments (overlain by water), it is important not to stand on or disturb sediments around the deployment site, as a result of which snorkeling or diving is preferred. It is also possible to fix an extension device to the handle of the DGT probe. In all cases, DGT probes should be deployed as close to vertical as possible during insertion. Insertion must be done in a manner that ensures close contact between the face of the probe and the sediment. Insertion must avoid tearing the DGT membrane and gels during deployment and retrieval. Some users self-fabricate a kind of sheath to protect the face of the probe during insertion, which is withdrawn immediately after insertion. Probes should be marked in a manner to facilitate ease of retrieval.

10.6.5.3 Other Considerations for Deployment

Deployment should be well planned ahead of reaching deployment sites because of the need to achieve deployment in a rapid manner. The time that DGT devices should be exposed to the air or ambient environment other than the deployment medium must be strictly minimized. Once the DGT devices/device assemblies are removed from bags, they should be immediately placed in the deployment medium.

DGT field blanks identify potential contamination that might occur during handling, and should be treated as follows: (a) field blanks are exposed to air in the laboratory during the time that other DGT devices are exposed to air during handling and assembly of DGT devices that will be deployed for sampling; (b) field blanks are exposed to air at the sampling site, one field blank per site; to do this, while deploying each DGT assemblage, expose the DGT field blank device on a clean surface until the assemblage of DGT devices is submerged in water (or sediment probes inserted) for both deployment and retrieved (after field deployment, place the DGT field blank device back in its original container and store refrigerated until the DGT retrieval day); and (c) field blanks are also exposed to air in the laboratory after retrieval and during the dismantling/processing of field-deployed DGT devices. One field blank is needed for each site.

DGT device windows should not be touched during deployment, and contaminant-free, powder-free gloves should be worn. In the field, remove the DGT assemblages (i.e., replicate DGT devices on a net or line) from their containers and attach onto the attachment point (e.g., mooring if a mooring is used). Deploy the DGT devices immediately. The overall process and procedure is the same for sediment probe deployments, aside from the mode of instatement of deployed devices. Record accurately the deployment time to the nearest minute. Keep the original container for each device or assemblage of devices (clearly labeled and sealed) to store the DGT assemblages at the retrieval stage. Record the depth of deployment and temperature of the water. If the temperature variation during the deployment period is within ±2°C, a mean (or start and end temperature) will suffice. If the variation is greater, ideally the mean temperature should be obtained from a temperature logger that is deployed with the DGT devices (one per site and depth).

Best practice is to also monitor physicochemical parameters at each site and depth. Typically, determinands measured include pH, specific conductivity, and dissolved oxygen. Measuring these over the deployment time by data logger is preferable to one-off measurements at deployment and retrieval. For sediment deployments, measuring these parameters in both sediments and overlying water is desirable.

10.6.6 Length of Deployment, Retrieval, and Post-retrieval Processing of DGT Devices

Deployment times typically vary between 3 and 21 days, though as little as one day (24 hours) may be used for some sediment work. Longer deployment periods allow a higher concentration of pollutant to be sampled, i.e., resulting in lower detection limits. Since the DGT devices provide a time-weighted average, sampling for longer provides a more representative picture of the integrated bioavailable pollutant. Long deployment periods entail that devices are more likely to be subject to biofouling (compromising results) and/or reaching the maximum capacity of DGT devices. Deployment time/pollutant accumulation also depends on temperature, ionic strength, and diffusive layer thickness. Thicker diffusive layers extend the time it takes for accumulation. Generally, shorter times are appropriate for sediments, and longer times for aqueous environments where pollutant concentrations are high.

DGT device retrieval is much the reverse of deployment. If a boat is used for retrieval, the engine must be off during times that DGT devices are being handled. At each site, the field blank must be exposed during retrieval as it was for deployment. On removal from environmental media, devices should be rinsed well in ambient site water, or alternately in generous amounts of Type I water, after which excess water is shaken off. Individual DGT devices and/or assemblages should be replaced in the clean containers they were in immediately prior to deployment (which were stored, refrigerated, and sealed during deployment). After placement of DGT devices in containers, they should be placed in zip-type bags sealed in a manner to minimize air. Time of retrieval is recorded to the minute, and depth is measured. If water quality physicochemical parameters were not collected by data logging, these measurements should be repeated at the time of retrieval. Store all bagged DGT devices in a cool box with ice packs for transport to the laboratory (ensure devices do not dehydrate or freeze).

DGT devices should be stored in the laboratory at 4°C until processing. Reasonable hold times for heavy metals is one month; for organic pollutants, hold times should not exceed three days unless prior knowledge ensures that no degradation/reaction will occur. To process DGT devices, use a clean, positive-pressure, laminar flow hood. Disassemble each device, popping the sleeve with window off the top, then carefully remove the filter membrane and diffusive gel with a properly cleaned tweezer or blade of an appropriate material (see Chapter 3). Place the binding gel in a capped tube of appropriate material and add eluent. For Chelex-100 resins, the approach to elution, the most commonly used, is to fully submerge the binding gel in 1 mL of 1.0 M HNO_3 and leave it for 24 hours. After 24 hours, the binding gel is carefully removed from the eluent (i.e., there should not be particles or pieces of the gel in the eluent), and the resulting solution is filtered or centrifuged as needed to remove particulates and analyzed by instrumental analysis. Inductively coupled plasma–mass spectrometry (ICP–MS) is preferred. For titanium dioxide resins, the same procedure is followed; only the eluent is 1.0 M NaOH; post-processing acidification is needed; eluent solutions may be diluted tenfold with 2% HNO_3.

Organic compounds studied using HLB-based DGT devices are more variable in behavior and less studied. For household and personal care products, pharmaceuticals (other than psychiatric pharmaceuticals), and endocrine disrupting compounds, the procedure in recognized use is to fully submerge the binding gel in 5 mL of acetonitrile and sonicate for 30 minutes (Chen et al., 2017, 2018), after which the binding gel is carefully removed from the eluate. For organophosphorus flame retardants and some pesticides (Zou et al., 2018), the procedure is to add 10 mL of methanol and sonicate for 30 minutes as above. For psychiatric pharmaceuticals, including alprazolam, bupropion hydrochloride, caffeine, carbamazepine, diazepam, estazolam, mirtazapine, naproxen, oxazepam, paroxetine, risperidone, and temazepam, add 10 mL of methanol amended with concentrated ammonia to have a 5% ammonia content, sonicate for 3 hours (Fang et al., 2019), and carefully remove binding gel as above. For all extracts having organic pollutants, extracts should be evaporated to near dryness (important! do not evaporate to total dryness) under a gentle stream of nitrogen, then 1 mL of methanol is added. Samples may be filtered at this time if gel residue remains in the extraction liquid; alternately, samples may be centrifuged. Extracts with organic pollutants should be stored in amber vials prior to instrumental analysis by a suitable method for the specific pollutants under study. Standard Methods for instrumental analysis are given in Chapter 5.

It should be noted that the extraction efficiencies of organics for the above protocols are generally high, above 80%; however, for selected compounds, they may be as low as 60%. As such, for organics, it is recommended to test the efficiency of elution. This involves running replicate samples (minimum three) of DGT device extraction and elution in a laboratory-controlled environment.

10.7 VALIDITY OF RESULTS, DATA ANALYSIS, INTERPRETATION, AND REPORTING

10.7.1 Notes on QA/QC Planning Requirements

Chapter 4 provides guidance that is applicable to this chapter. For DGT QA/QC specifically –

- Users performing this method must perform an ***Initial Demonstration of Proficiency*** (IDP), including laboratory tests as described below. This IDP includes procedures for instrumental methods provided in individual ***Standard Operating Procedures*** for instrumental analysis. The IDP process described below is designed to additionally provide information useful for field operation planning.
- Routine instrumental calibration and ***calibration blank*** checks are required; guidelines for ***Acceptance Criteria*** for these are stipulated in appropriate Standard Methods used for instrumental analysis. Frequent calibration blank and standard checks are required to ensure the absence of instrument drift. Drift affects both ***precision*** and ***accuracy***.
- A minimum of one field blank is required for each sampling site.
- Batch sizes for instrumental analysis shall not exceed 20 samples; samples batched include sediment and tissue digests/extracts and replicate samples; QA/QC samples are additional and should not exceed 5 per batch. Instrument check samples assess calibration stability and are not part of batching.
- Sample order in batches is to be randomized over all samples and sample types.
- For every four batches, one ***procedural blank*** is required, unless different batches of reagents were used, in which case more procedural blanks may be needed – each manufacturer's batch of reagents (e.g., solvents, acid) must be subject to procedural blank testing. A procedural blank consists of a binding gel that has not been used in a DGT device and has not been exposed in the field (i.e., is not a field blank), but is processed/eluted in the laboratory according to the same procedure used for field sample DGT gels, i.e., store the procedural blank in a clean container filled with Type I water or a blank ***matrix*** solution (e.g., filtered synthetic water) that mimics the ionic strength of samples, using the same

soaking time as for samples, the same temperature conditions, the same transport and storage duration, the same elution procedure. This enables evaluation of whether contamination from reagents used in the procedure occurred.

- One ***recovery spike*** is required per batch. This enables evaluation of whether artifactual contamination (recoveries significantly above 100%) or analyte loss (recoveries significantly below 100%) is occurring.
- Isotope-labeled ***internal standards*** should be used as appropriate for organic pollutants. Standard Methods of instrumental analysis will provide guidance on this.

For this method, IDP consists of successfully performing a series of laboratory experiments. For each experiment, pH is measured before and after the experiment for experiments of duration less than a day, and daily otherwise. If pH changes by more than 0.2 pH units, adjustment is required (using sodium acetate, hydrochloric acid, sodium carbonate, or sodium hydroxide). Temperature should be kept constant to within 2°C. All experiments should be undertaken in containers that are covered by a watch glass or aluminum foil to prevent possible evaporation, and dark conditions may be appropriate for some photoreactive organic pollutants. Blanks should be included for all laboratory samples, and all tests must be conducted in triplicate.

Adsorption testing – materials used for making DGT devices should be assessed for possible adsorption of target pollutants. Devices consisting of DGT holder (piston and cap), diffusive gel, and a filter membrane are used (i.e., devices without binding gel). These are immersed in a solution containing a level of target pollutant (or mixture thereof) and left on a shaker for 24 hours. Triplicate samples of pollutant solutions used for the experiment should be taken at the beginning of the experiment. Type I water is used as a blank and containers with pollutant but no device serve as a positive control. After 24 hours, the devices are removed from the water, and the water is analyzed for pollutant. The amount of pollutant adsorbed by devices is calculated according to mass balance based on concentrations in the solution before and after the experiment. The blank should be below the LOD, and the positive control should show no loss of pollutant to ***sorption*** on the container. Sorption of pollutant to devices should be undetectable to negligible (<10%).

Elution recovery testing – elution recovery is the ratio of the measured amount of pollutant in binding gel eluent to the amount adsorbed by the binding gel. To measure this in the laboratory, binding gels are immersed in a solution containing a level of target pollutant and left on a shaker for 24 hours as above for adsorption testing (a single binding gel in each container, minimum triplicate sampling). Samples of the initial solution are also taken as for adsorption testing. After shaking, gels are removed from solution and pollutant is eluted from the binding gels. At the same time, samples of the solutions in which the gels were shaken are taken. The amount of pollutant adsorbed by binding gels is obtained by mass balance using the concentration difference in the solution before and after the experiment, and this amount should correspond to the amount eluted by the binding gels. It is important to note that the results from this experiment can depend on the concentration of the pollutant used for the test solution. In some cases, elution recovery may be a function of concentration, and in all cases, if the concentration of pollutant in the solution is too high, the mass of pollutant will exceed the capacity of the gel. Stable recoveries may still be obtained. Ideally, this experiment is performed across a range of pollutant concentrations in order to understand uptake behavior and how this might affect field planning.

Recovery for this experiment is calculated as

$$R = \left(m_{\text{soln, lost}} \times 100\ \%\right) / m_{\text{gel}},$$

where $m_{\text{soln, lost}}$ is the mass of pollutant lost from solution over the course of the experiment (i.e., if the solution volume is kept constant as it should be, the difference between the pollutant concentration at the beginning and end of experiment) and m_{gel} is the mass of pollutant eluted from the gel (the concentration in the eluent multiplied by the volume of the eluent). For heavy metals, recoveries should be 90%–110%. A range of as much as 80%–120% may be acceptable for organics. If recoveries within these ranges are not obtained, then the elution conditions and gel properties need to be reviewed and revised until acceptable recoveries are obtained. If recoveries are variable with concentration, this presents another type of problem that may relate to deployment time (to ensure gels are not overloaded) or choice of gel (if gel layers do not efficiently capture pollutants across the range of concentrations anticipated to be needed in the field).

Uptake capacity of binding gels – this experiment is performed similarly to those above, using DGT devices in containers of solution and measurements of solution pollutant concentration being made before and after 24 hours of shaking. Multiple concentrations (typically ten or more) of each target pollutant must be used, ranging from low to high (usually ranging from low ppb to high ppm). Results from this type of experiment are often pH-dependent, so it is important to ensure a steady pH, and near conditions foreseen for field deployment. This experiment should be run with Type I water blanks and DGT devices without binding gels as a positive control.

After 24 hours, DGT binding gels are eluted, and, for each concentration of pollutant tested, mass of each target pollutant, *M*, is plotted versus the mass in solution, $M_{soln,t=0}$, at the beginning of the experiment (Note: The mass taken up should agree with the mass determined by difference between the start and end of the experiment). The upper limit of linearity for the low-loading portion of this curve illustrates the maximum uptake capacity for the gel in question. This is shown conceptually in Figure 10.4.

Deployment time – if uncertainties concerning deployment time exist, it is useful to perform laboratory experiments prior to field planning. In this case, conduct DGT experiments as above (DGT devices in containers with solutions of pollutant of interest) using one to three concentrations below the DGT devices' binding capacity. These experiments should be run over the possible range of deployment times (i.e., up to one month), sampling at least ten times during the total time of the experiment. The amount of target pollutant uptake is determined by difference, i.e., diminution in solution is an estimate of uptake, and uptake, expressed as mass, is plotted versus time. The slope of the linear part of this curve is the uptake rate, from which deployment times can be gauged according to expectations of bioavailable pollutant concentrations in field conditions.

Effect of pH, ionic strength, and dissolved organic matter – unless a user knows the expected performance of a given loaded DGT device with respect to pH, ionic strength, or dissolved organic matter, it is best to test the performance according to the experiments described above. In particular,

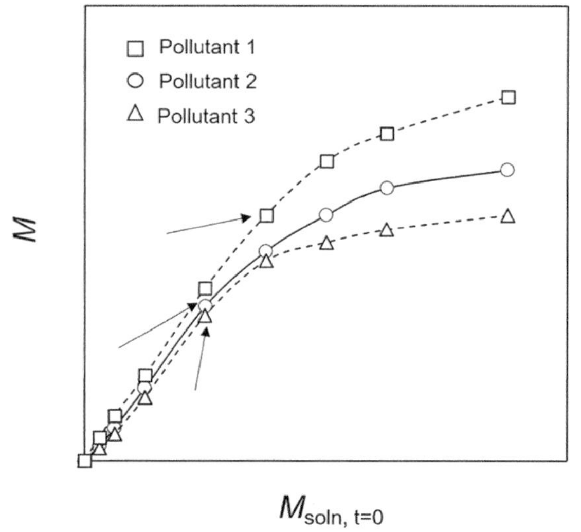

FIGURE 10.4 Figure illustrating what typical plots of *M* versus $M_{soln,\ t=0}$, look like. Arrows indicate approximate upper limit of linearity.

many DGT gels are sensitive to the high ionic strength in seawater. For laboratory testing, high-purity artificial seawater and commercially available humic acid can be useful for this type of testing.

Diffusion coefficient measurements – diffusion coefficients that are applicable to the target pollutants of interest and temperature during deployment are needed for final calculations. For many pollutants, these exist in literature or may be obtained from the website of DGT Research. In cases wherein diffusion coefficients are not available, these must be determined. This requires a two-compartment diffusion cell with a window between the two compartments that is loaded with the diffusion gel to be used in DGT devices (such cells are available through DGT Research and other suppliers). Both compartments are filled with a solution having an ionic strength and pH at or near that for which the diffusion coefficient is to be measured. Test pollutant is spiked into the diffusion cell ***source*** compartment (spiking involves adding a high concentration in low volume, typically 5% of the volume of the spiked solution, such that the final concentration in the receiving solution is as desired, with a negligible change in volume, see Section 4.3). The final concentration of the spiked source compartment solution needs to be high enough to ensure that the amount of spiked pollutant diffusing into the receiving compartment is small enough that there is no substantive change in the concentration of pollutant in the source compartment (typically, final concentration of spiked source compartment solution is in excess of 1,000 µg/L). Both compartments must be under adequate stirring for the duration of the experiment to minimize the DBL across the diffusive gel. Negligibly small subsamples, e.g., 0.1 mL out of 100 mL, are collected from both compartments at the intervals of 30–60 minutes over the course of the experiment (~8–24 hours are typically adequate). The solution in the source compartment must be sampled as well, to ensure that the concentration of pollutant in that compartment remains effectively constant. After instrumental analysis, the mass of pollutant diffused into the receiving compartment at different times is calculated (concentration of subsample multiplied by volume of solution in receiving compartment), and then this mass is plotted versus subsampling time in seconds. The slope of this plot is k, the rate at which pollutant is diffused. The diffusion coefficient of the pollutant in the gel, D, in cm^2/s, is then calculated according to

$$D = \frac{k\Delta g_{\text{eff}}}{C_s A_s},$$

where Δg_{eff} is the effective thickness of the diffusive layer (in cm), C_s is the pollutant concentration in the source solution (numerator of concentration unit must be the same as for k, denominator in cm^3), and A_s is the effective sampling area of the DGT devices (in cm^2). Many investigators take $\Delta g_{\text{eff}} \sim \Delta g$, where Δg is the sum of the gel(s) and filter membrane thicknesses, and $A_s \sim A_g$, where A_g is geometric area equal to the physical area defined by the window of the DGT device's plastic molding. The $\Delta g_{\text{eff}} \sim \Delta g$ approximation considers the DBL to be effectively zero, whereas 0.2 cm is a typical value (for experiments that involve flows greater than ~2 cm/s, the thickness of the DBL has been shown to be more or less constant at ~0.2 cm; Garmo et al., 2006); hence, $\Delta g_{\text{eff}} \sim \Delta g + 0.2$ cm. It has also been shown that A_s is on the order of 20% larger than A_g, i.e., $A_s \sim 1.2 \times A_g$ (Garmo et al., 2006). If the experiment is performed at 25°C (D_{25}), values of D_T at other temperatures T can be estimated according to

$$\log D_T = \frac{1.37023(T-25) + 8.36 \times 10^{-4}(T-25)^2}{109+T} + \log\frac{D_{25}(273+T)}{298}$$

Initial Demonstration of Proficiency – successful determination of DGT device performance according to the laboratory experiments described above, which includes meeting performance specifications of the applicable methods of instrumental analysis, constitutes IDP for this method.

10.7.2 Data Analysis, Interpretation of QA/QC Results (Acceptance Criteria), Corrective Action, and Reporting

Generalized guidelines for the QA/QC requirements of this method are as follows:

- Dilution factors (DFs) should never be more than a factor of 10 and must always be accounted for in calculations. The basic dilution formula is $C_1V_1 = C_2V_2$. The dilution factor is given by DF = V_2/V_1.
- ***Calibration checks*** and metrics shall be performed as specified in a Standard Method of instrumental analysis applicable to the pollutant under study; ***Corrective Action*** must be taken in response to any irregularities of the standard conditions of calibration.
- Similar remarks apply to internal standards and ***surrogates***.
- Precision for all instrumental ***duplicate/replicates*** – the instrumental method chosen will specify the needs for analysis of duplicate or replicate instrumental analysis of processed samples. A metric for precision must be calculated and reported. Generally, the relative percent difference (RPD) or relative standard deviation (RSD) of measurements, respectively, is used to evaluate the precision for sample duplicates and replicates. The formula for RPD is as follows: RPD = 200% × abs $(R_1 - R_2)/(R_1 + R_2)$, where subscripts 1 and 2 represent the two duplicate samples. The formula for RSD is as follows: RSD = stdev × 100%/average. If metrics such as RPD or RSD are too high, it indicates that there is a significant amount of variability or inconsistency between the measurements of the same sample. For example, typically, an RSD value greater than 2% (acceptance criterion) for analysis of sample replicates using ICP–MS at concentration ranges in the middle of the ***calibration curve*** indicates poor precision. Poor precision for replicates of lab samples is generally indicative of an operator or instrument error, and Corrective Action involves rectifying the problem prior to analysis and reporting of results from bioavailability studies.
- Precision for field replicates – this is calculated as RSD – a minimum of three replicates are required. Due to the integrative nature of the DGT method, a well-planned and executed DGT study should be able to achieve RSDs <15% for aqueous deployment unless concentrations are very low (e.g., below the ***limit of quantification*** or LOQ). Precision may be similar for soils/sediments in cases where laboratory studies on samples that have been mixed are performed. For field deployment, soils and sediments will normally have worse performance. Ideally, having at least one aqueous deployment with good precision helps rule out poor method execution when precision of field replicates is low for soil/sediment studies. Report RSD values and any discrepancies and provide context on how environmental heterogeneity impacts the results.
- Field and procedural blanks – procedural blanks must have a concentration below the LOD for pollutants; otherwise, this is indicative of contamination by reagents/materials. Contaminated samples cannot be used in reporting; thus, it is better to establish that proper laboratory conditions are achievable to ensure this criterion for procedural blanks before major work is undertaken. If procedural blanks are acceptable, then having field blanks above the LOD is indicative of field contamination. If the contamination is effectively constant around an average value that is low, compared to field samples, then this should be reported with results and considered in the interpretation of results. For highly varying contamination, in the case of field sample contamination, reporting is not possible and field work must be repeated, taking stronger measures for avoiding field contamination.
- LOD and LOQ – these critical parameters define the lowest concentration of pollutant that can be reliably detected and quantified using a given method of instrumental analysis. LOD is often defined as three times the standard deviation of the calibration blank divided by the slope of the calibration curve. LOQ is often defined as three times the LOD. Standard Methods for instrumental analysis commonly contain guidelines concerning

what LOD and LOQ are typical. Substantial deviation from such reported guidelines may typically reflect issues with user competency in performing analysis. LOD and LOQ must be reported as they reflect upon competence of analysis and interpretation of results (e.g., lower precision typically occurs between the LOD and LOQ).

- Recovery spikes – these samples and their method of preparation are discussed in more detail in Chapter 4. For both types of spikes, calculate recovery as follows: Recovery (%) = $100\% \times (C_{\text{Total}} - C_{\text{bkg}})/C_{\text{spike added}}$, where C_{Total} is the total concentration of pollutant in the spiked sample, i.e., the concentration from the sample and the added spike; C_{bkg} is the background concentration of the pollutant in the sample without the ***spike***; and $C_{\text{spike added}}$ is the known amount of pollutant contributed to the sample by addition of the spike. Acceptable recovery for DGT device eluates, i.e., simple solutions, should be on the order of 90%–110%. If recovery spike values fall outside of this, Corrective Action involves reviewing and adjusting the sample preparation and analytical procedures to identify potential sources of error.
- Reporting DGT results – in DGT, results are reported in terms of C_b, the concentration of pollutant in bulk solution (pore water in the case of soil/sediments). This is calculated in one of two ways:
  1. For experiments that involve flows greater than ~2 cm/s, C_b is calculated as

     $$C_b = (M \times \Delta g_{\text{eff}})/(D \times t \times A),$$

     where M is the mass of pollutant bound to the DGT device over deployment time, t.
  2. For conditions where flow is low or absent, deployment of devices with multiple diffusive gel thicknesses is required. Subsequent to deployment, M, for each thickness of diffusive gel used, is plotted versus Δg_{eff} of the devices. This will produce a curve such as shown in Figure 10.5. This experimental data is then fit, with non-linear curve fitting, to the equation (Garmo et al., 2006)

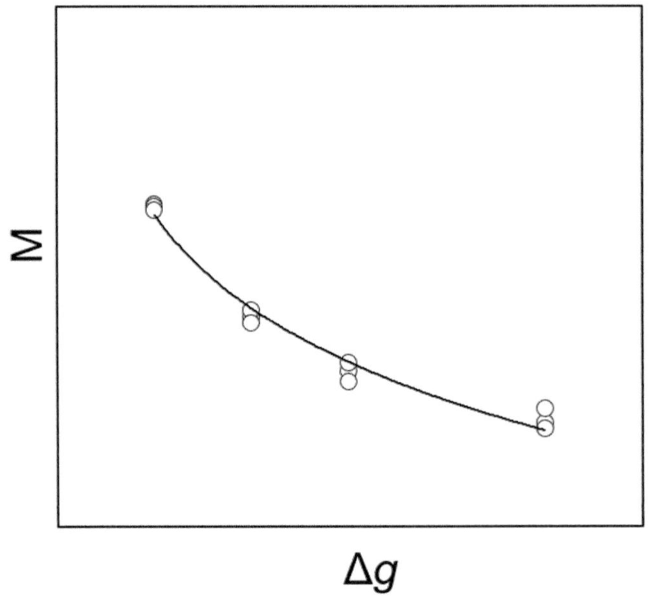

FIGURE 10.5 Figure illustrating what the curve resulting from plotting M versus Δg looks like. A minimum of four values of Δg, with three replicates/points for each value, are needed. Parameters C_b and δ are obtained from non-linear curve fitting.

$$M = (A_s \times A_g \times D_w \times D \times t \times C_b)/(A_s \times D \times \delta + A_g \times D_w \times \Delta g)$$

where A_s, A_g, D, t, C_b, and Δg are as above, D_w is the diffusion coefficient of the pollutant in water, δ is the DBL thickness, and the equation is fit to solve for parameters C_b and δ.

The effect (if any) of the DBL on results should be mentioned in reporting. Additional considerations that would normally be discussed include whether biofouling occurred, and, if so, how it might have affected DGT pollutant levels measured, and whether other field conditions (pH, ionic strength, dissolved organic matter) were appropriate to the performance envelope of the DGT devices used. If field conditions are indicative of potential binding gel saturation over the course of deployment, this should be reported and discussed. Finally, the implications of the inherently time-integrated results obtained from DGT, the consequences of which may be either advantageous or disadvantageous, depending on the questions to be answered by a bioavailability study, should be addressed in reporting results.

10.8 CONCLUSIONS AND OUTLOOK

The use of DGT in environmental studies has seen a meteoric rise since its introduction in the 1990s, and it is unquestionably a powerful and important method for use in environmental studies, both for routine monitoring and in impact studies. The outstanding advantages of the DGT approach that have made it popular are no doubt the convenience of the method if executed well and that devices can be deployed directly in the environment, *in situ*. To some extent, DGT devices can address speciation in a limited way as well (e.g., for heavy metals, it is possible to use gels that occlude certain size molecules, enhancing capture of metal free-ion species), and certainly for studies wherein a time-averaged concentration is acceptable or even desired, DGT holds a distinct advantage. Disadvantages include complex interpretation, as devices can be influenced by numerous environmental factors (pH, ionic strength, temperature, dissolved organic matter, water flow). Additionally, while DGT results often correlate with measures of true bioavailability, they also often do not – or not in a useful way (see case study in Chapter 16). Future research in developing DGT as a technique will undoubtedly improve its already substantial standing as a highly useful Environmental Monitoring technique.

REFERENCES

Chen, C.E., Chen, W., Ying, G.G., Jones, K.C., Zhang, H., 2015. In situ measurement of solution concentrations and fluxes of sulfonamides and trimethoprim antibiotics in soils using o–DGT. *Talanta*. 132, 902–908. https://doi.org/10.1016/j.talanta.2014.08.048

Chen, C.E., Zhang, H., Jones, K.C., 2012. A novel passive water sampler for in situ sampling of antibiotics. *J. Environ. Monit*. 14(6), 1523–1530. https://doi.org/10.1039/C2EM30091E

Chen, W., Li, Y., Chen, C.E., Sweetman, A.J., Zhang, H., Jones, K.C., 2017. DGT passive sampling for quantitative in situ measurements of compounds from household and personal care products in waters. *Environ. Sci. Technol*. 51(22), 13274–13281. https://doi.org/10.1021/acs.est.7b03940

Chen, W., Pan, S., Cheng, H., Sweetman, A.J., Zhang, H., Jones, K.C., 2018. Diffusive gradients in thin-films (DGT) for in situ sampling of selected endocrine disrupting chemicals (EDCs) in waters. *Water Res*. 137, 211–219. https://doi.org/10.1016/j.watres.2018.03.029

Davison, W. and Zhang, H., 1994. In situ speciation measurements of trace components in natural waters using thin–film gels. *Nature*. 367(6463), 546–548. https://doi.org/10.1038/367546a0

Diviš, P., Dočekalová, H., Brulík, L., Pavliš, M., Hekera, P., 2007. Use of the diffusive gradients in thin films technique to evaluate (bio)available trace metal concentrations in river water. *Anal. Bioanal. Chem*. 387(6), 2239–2244. https://doi.org/10.1007/s00216-006-0996-y

Fang, Z., Li, K., Li, Y., Zhang, H., Jones, K.C., Liu, X., Liu, S., Ma, L.Q., Luo, J., 2019. Development and application of the diffusive gradients in thin–films technique for measuring psychiatric pharmaceuticals in natural waters. *Environ. Sci. Technol.* 53(19), 11223–11231. https://doi.org/10.1021/acs.est.9b03166

Garmo, ØA., Razi, N.K., Røyset, O.E., Steinnes, E., 2006. Estimation of diffusive boundary layer thickness in studies involving diffusive gradients in thin films (DGT). *Anal. Bioanal. Chem.* 386(7–8), 2233–2237. https://doi.org/10.1007/s00216-006-0885-4

Gimpel, J., Zhang, H., Hutchinson, W., Davison, W., 2001. Effect of solution composition, flow and deployment time on the measurement of trace metals by the diffusive gradient in thin films technique. *Anal. Chim. Acta.* 448(1), 93–103. https://doi.org/10.1016/S0003-2670(01)01323-X

Kot, A., Zabiegała, B., Namieśnik, J. 2000. Passive sampling for long-term monitoring of organic pollutants in water. TrAC Trend. Anal. Chem. 19(7), 446–459. https://doi.org/10.1016/S0165-9936(99)00223-X

Li, Y., Chen, C.E.L., Chen, W., Chen, J., Cai, X., Jones, K.C., Zhang, H., 2019. Development of a passive sampling technique for measuring pesticides in waters and soils. *J. Agric. Food Chem.* 67(22), 6397–6406. https://doi.org/10.1021/acs.jafc.9b00040

Marrugo-Madrid, S., Turull, M., Zhang, H., Díez, S., 2021. Diffusive gradients in thin films for the measurement of labile metal species in water and soils: A review. *Environ. Chem. Lett.* 19(5), 3761–3788. https://doi.org/10.1007/s10311-021-01246-3

Scally, S., Davison, W., Zhang, H., 2006. Diffusion coefficients of metals and metal complexes in hydrogels used in diffusive gradients in thin films. *Anal. Chim. Acta.* 558(1), 222–229. https://doi.org/10.1016/j.aca.2005.11.020

Senila, M. and Kovacs, E., 2024. Use of diffusive gradients in thin–film technique to predict the mobility and transfer of nutrients and toxic elements from agricultural soil to crops – An overview of recent studies. *Environ. Sci. Pollut. Res.* 31(24), 1–22. https://doi.org/10.1007/s11356-024-33602-5

Wang, Y., Ding, S., Shi, L., Gong, M., Xu, S., Zhang, C., 2017. Simultaneous measurements of cations and anions using diffusive gradients in thin films with a ZrO–Chelex mixed binding layer. *Anal. Chim. Acta.* 972, 1–11. https://doi.org/10.1016/j.aca.2017.04.007

Yao, L., Steinman, A.D., Wan, X., Shu, X., Xie, L., 2019. A new method based on diffusive gradients in thin films for in situ monitoring microcystin–LR in waters. *Sci. Rep.* 9(1), 17528. https://doi.org/10.1038/s41598-019-53835-6

Yapici, T., Fasfous, I.I., Murimboh, J., Chakrabarti, C.L., 2008. Investigation of DGT as a metal speciation technique for municipal wastes and aqueous mine effluents. *Anal. Chim. Acta.* 622(1), 70–76. https://doi.org/10.1016/j.aca.2008.05.061

Zhang, H., Davison, W., Miller, S., Tych, W., 1995. In situ high resolution measurements of fluxes of Ni, Cu, Fe, and Mn and concentrations of Zn and Cd in porewaters by DGT. *Geochim. Cosmochim. Acta.* 59(20), 4181–4192. https://doi.org/10.1016/0016-7037(95)00293-9

Zou, Y.T., Fang, Z., Li, Y., Wang, R., Zhang, H., Jones, K.C., Cui, X.Y., Shi, X.Y., Yin, D., Li, C., 2018. Novel method for in situ monitoring of organophosphorus flame retardants in waters. *Anal. Chem.* 90(16), 10016–10023. https://doi.org/10.1021/acs.analchem.8b02480

11 Tenax Extraction as a Simple Method to Determine the Chemoavailability of Organic Pollutants

11.1 INTRODUCTION AND SCOPE

With the development of industrial and agricultural production, large quantities of organic chemicals are released into the environment. Soil and sediments, as sinks for pollutants, have been polluted to varying degrees. Organic pollution in soil and sediments may be found at highly increased concentrations in localized areas (Dai et al., 2022); at the same time, the rising level of pollution globally is staggering. For example, some per- and polyfluoroalkyl substances (PFAS) have been found in rainwater at concentrations that exceed US EPA Health Advisory limits for drinking water by a factor of up to 100,000, with the least polluted areas of Tibet and Antarctica still having levels over ten times that recommended (Cousins et al., 2022). Various organic additives used in industrial production, excessive application of fertilizers and pesticides in agricultural production, and irrigation effluents carry organic pollutants that result in enrichment of pollutants in soil and sediments (Rashmi et al., 2020). In addition to new and emerging organic pollutants such as PFAS, polybrominated diphenyl ethers (flame retardants), pharmaceuticals and personal care products, and others, there is a longstanding catalog of "traditional" pesticide and chemical organic pollutants such as Organochlorine pesticides, polycyclic aromatic hydrocarbons (PAHs), and polychlorinated biphenyls (PCBs). These pollutants pose a serious threat to environmental quality and human health. In the face of so much pollution, the modern paradigm is to base ***Environmental Risk Assessment*** on ***bioavailability***; thus, the detection of ***bioavailable*** organic pollutants in soil/sediment is essential to guide environmental management of pollution (Chen et al., 2023).

Currently, two main methods for the study of organic pollutant bioavailability include the biological ***exposure*** method and the chemical solvent extraction method (Chen et al., 2023). The former determines true bioavailability in one form or another (e.g., organism-based, short term, or via ***bioaccumulation***). The latter utilizes mild solvents or adsorbents to extract or adsorb the pollutant in the environmental media and tries to estimate pollutant bioavailability in terms of the amount extracted or adsorbed by a chemical proxy, i.e., ***chemoavailability***. Among the chemical solvent extraction methods, the Tenax extraction method has been widely used for, not in environmental media such as soil, sediment, and sludge. Tenax (commercial name) is a porous polymer, 2,6-diphenylene oxide, which is a superior proxy because of its excellent ***sorption*** of organic substances and high-temperature resistance (375°C). In particular, Tenax has a high affinity for ***hydrophobic organic pollutants*** (HOCs), i.e., non-polar compounds with low water insolubility and a relatively high ***octanol–water partition coefficient***. This property allows Tenax to effectively capture HOC pollutants from the environment (Lydy et al., 2015). Since the 1970s, Tenax has been widely used as a more reliable adsorbent than activated carbon for the capture of volatile organic compounds (Wallace, 2001). The technique was then gradually developed to extract organic pollutants from soil/sediment. Tenax's initial application in estimating chemoavailability in soils and sediments focused on the desorption of chlorinated chemicals (Pignatello, 1990a, 1990b, 1991). The original approach involved characterization of the full desorption curve. Subsequently, the Tenax method

DOI: 10.1201/9781003355229-14

was developed to measure a single, fixed-time desorption metric to gauge bioavailability of a range of organic pollutants including pesticides, PCBs, and PAHs (Cornelissen et al., 1997a, 2001; Hua et al., 2023); this approach has been referred to as the simple method. Tenax extractions, whether single-point or sequential extractions to assess desorption, are straightforward to perform and adhere to similar general procedures (Kraaij et al., 2001; Landrum et al., 2007).

The simple method for Tenax extraction was based on prior work that characterized desorption curves for organic pollutants in continuous extraction by Tenax. Such curves were fit to a two-compartment or three-compartment desorption model that included fast desorption, slow desorption, and, for the three-component model, a very slow desorption fraction. Experimental data for desorption were fit with such two- or three-component models using non-linear curve fitting, and extracted parameters (fractions of each of the two or three ***components*** and desorption rate constants for each component) were scrutinized for correlation with true bioavailability data. The best correlate with true bioavailability is the rapidly desorbing fraction, which is logical as it contains not only rapidly desorbing pollutants, but also freely dissolving pollutants (Cornelissen et al., 2001). Since the original work describing the simple method, subsequent work has shown that results from single-point extractions with Tenax (e.g., 6 or 24 hours) often show significant correlation with measures of true bioavailability. In addition to its wide range of applications in soils and sediments, Tenax has also demonstrated good extraction results in water. As research confirms, Tenax-TA is an effective adsorbent for aromatic hydrocarbons in freshwater and seawater, as well as in water bodies containing large amounts of dissolved organic matter (Zhao and Pignatello, 2004).

It has been reported that many organic pollutants can be adsorbed onto Tenax, and there is a general range of octanol–water partition coefficients, K_{ow}, associated with them (Jantunen, 2010). Generally, organic pollutants with $\log K_{ow}$ values between approximately 3 and 8 are more likely to be adsorbed effectively by Tenax (Jantunen, 2010). Examples of pollutants that fall within this range include PCBs, PAHs, and some pesticides. PCBs are a class of Persistent Organic Pollutants that vary in the number and position of chlorine atoms on the biphenyl ring; PCBs have been extensively studied in environmental samples using Tenax-based methods. Compounds like naphthalene, phenanthrene, and pyrene are common PAHs. Naphthalene has a $\log K_{ow}$ value of about 3.3, phenanthrene about 4.5, and pyrene about 5.1 (Van Hattum et al., 1998). These values put them in the range where Tenax can effectively adsorb them from environmental matrices. Certain pesticides, such as dichlorodiphenyltrichloroethane (DDT), have a $\log K_{ow}$ of approximately 6.2. It has been reported that Tenax can be used to extract DDT from environmental samples, such as soil, sediment and water (Biziuk et al., 1996; De la Cal et al., 2008; Raffy et al., 2018; Yang et al., 2010).

This chapter describes a method for how to use a simple Tenax extraction method to determine the chemoavailability of organic pollutant bioavailability in soil/sediment, as well as freshwater and seawater samples with high concentrations of bioavailable organic pollutants. The principle of the method, the procedure, and the influencing factors are described in detail. Performed correctly, the method is capable of providing high reproducibility, with short test periods. As the method is not labor-intensive and Tenax is easily regenerated for reuse, the method is cost-effective as well. Chapters containing Terms and Definitions, guidance on avoiding contamination of samples in field and laboratory work, guidance on Quality Assurance (QA)/Quality Control (QC), and methods of environmental sample extraction and instrumental analysis are also within the scope of this chapter.

11.2 PRINCIPLE

The Tenax extraction principle is based on the idea that the bioavailability of organic pollutants can be predicted according to the amount of organic pollutants that are rapidly desorbed from soil, sediment, water, or seawater and extracted by Tenax. The process of organic pollutant extraction from soil/sediment by Tenax occurs by desorption of organic pollutant from soil/sediment into the aqueous phase and subsequent sorption of organic pollutant in the aqueous phase by Tenax. Tenax present in the aqueous phase acts as a quasi-infinite sink, which disrupts the equilibrium ***distribution*** of organic pollutant between water and soil/sediment. This leads to the continuous desorption of organic pollutant from soil/sediment into the aqueous phase, enabling uptake from solution by Tenax. The function of Tenax relies on its ability to quickly adsorb the desorbed pollutants from soil/sediment. Tenax may also be used directly for the extraction of organic ***contaminants*** in water or seawater (Zhao and Pignatello, 2004). In practice, when constructing a full desorption curve, Tenax must be refreshed periodically to maintain quasi-infinite sink conditions; however, for relatively short periods of time, as given in the method here, this is not necessary. After exposure to an environmental sample, Tenax is collected, eluted, and then target pollutants in eluates are subject to instrumental analysis.

Since HOCs desorb from soil/sediment at different rates from pools of pollutant that are more or less bioavailable, i.e., different fractions with varied apparent desorption rates/bioavailability, Tenax extractable concentrations provide a more accurate assessment of bioavailability than exhaustive extractions (Cornelissen et al., 1997a, 1997b, 1998). With a two- or three-compartment desorption model, compound desorption from soil/sediment is commonly characterized as bi- or triphasic (Cornelissen et al., 1997a), as seen in this equation for the three-compartment or triphasic desorption:

$$\frac{S_t}{S_0} = f_{\text{rap}}\left(e^{-k_{\text{rap}} \times t}\right) + f_{\text{slow}}\left(e^{-k_{\text{slow}} \times t}\right) + f_{\text{veryslow}}\left(e^{-k_{\text{veryslow}} \times t}\right),$$

where f_{rap}, f_{slow}, and f_{veryslow} are the decimal percent pollutant fractions present in the rapidly, slowly, and very slowly desorbing soil/sediment compartments; k_{rap}, k_{slow}, and k_{veryslow} are the rate constants (inverse time units) for the corresponding desorbing compartments at some time t during desorption; and S_t and S_0 represent the amount of pollutant sorbed to soil/sediment at time t and at the beginning of Tenax desorption ($t = 0$), respectively. In the process of development of the Tenax method, non-linear curve fitting of an experimental Tenax desorption curve (experimental parameters of S_t/S_0 plotted versus t) was used to solve for f and k parameters, and in some cases, the parameter f_{rap} was found to correlate well with measures of true bioavailability. Subsequent work discovered that taking the fraction desorbed at a single time "cut" (of desorption curves) of 6 hours extraction, $f_{t=6}$, is also suitable and much more convenient, hence the basis for this method. Figure 11.1 shows an example of the difference in f_{rap} and $f_{t=6}$, and the time needed to reach f_{rap}, for experimental data (Hua et al., 2023). In this example, $f_{t=6}$ differs from f_{rap} by as much as 5%, or 3.7% on average, which corresponds to an average relative error of 18%. Since f_{rap} has been used to approximate the bioavailable fraction and 6-hour Tenax extraction, i.e., $f_{t=6}$ is both a reasonable estimation of f_{rap} and much more convenient to measure, many researchers opt for a 6-hour Tenax extraction time as a measure of chemoavailability (Lydy et al., 2015; Moermond et al., 2004). Using $f_{t=6}$, as in this method, trades convenience for the time and labor needed to calculate f_{rap} from full desorption curves. Figure 11.2 shows a graphical depiction of the process behind the single-point Tenax extraction method presented herein.

11.3 EQUIPMENT AND CONSUMABLES

All field equipment and laboratory materials must be properly cleaned and sanitized prior to use, per Chapter 3, and decontaminated in between each use. In addition, to avoid contamination, contaminant-free field gear must be transported in clean, sealed containers. Depending on experimental design, some equipment listed may not be needed; conversely, some users may find it useful to add additional items. The list is indicative.

Sample collection, handling, packaging, and transport-related:

Auger – to collect soil samples at depth.
Aluminum foil – for wrapping organic contaminant-free implements.
Coring sampler – used for collecting sediments at depths.

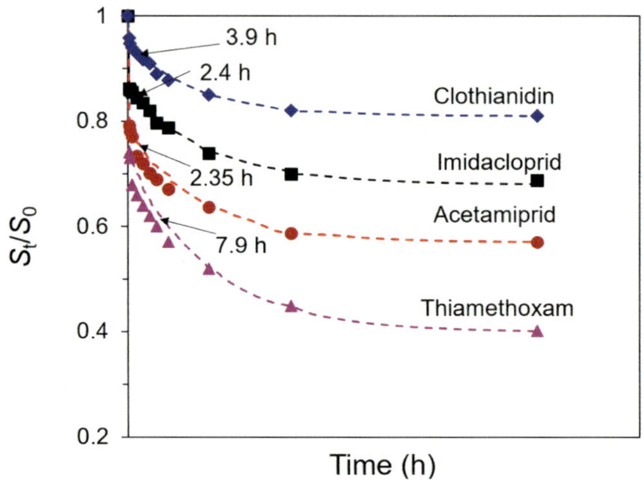

FIGURE 11.1 An example of Tenax desorption curves for four neonicotinoid insecticides and corresponding non-linear curve fits, taken from the literature. For the four curves shown (top to bottom), f_{rap} is 0.07, 0.15, 0.25, and 0.3, respectively, whereas $f_{t=6}$ is 0.08, 0.19, 0.30, and 0.25, respectively. While the desorption time to reach $f_{t=6}$ is constant at 6 hours, the time at which the desorption corresponding to f_{rap} is reached (taken from formulas of non-linear fits for these four curves) is 3.9, 2.4, 2.35, and 7.9 hours, respectively, marked by black arrows in the plot. The time to f_{rap} is clearly not 6 hours and instead varies considerably. Nonetheless, values of $f_{t=6}$ differ from values of f_{rap} by only 3.7%, on average, making $f_{t=6}$ a reasonable approximation of f_{rap} (Hua et al., 2023).

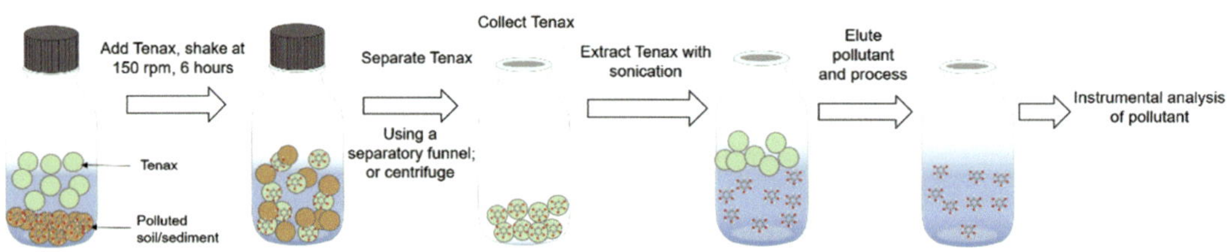

FIGURE 11.2 Graphic illustrating the single-point Tenax extraction method for the determination of ***chemoavailable*** organic pollutants presented in this chapter.

Field logbook – along with a sample collection log, to note details such as collection activities, field conditions, temperature, and any unexpected events that may influence the experiment's outcomes, aiding in the analysis of the results later, and to itemize collected samples and sample information.

Glass (standard) or PTFE (polytetrafluoroethylene) containers, with PTFE-lined lids – for collection of field soil, sediment, or water samples.

GPS/electronic or physical devices – to ensure unambiguous location of sample ***sites*** (e.g., GPS, beacons) and depths (e.g., depth gauge).

Grab sampler – to collect surficial sediment samples.

Ice packs or dry ice – used to keep samples at a low temperature and inhibit biological ***activity***.

Indelible pens – used to ensure data are accurately recorded in wet field conditions without loss of clarity.

Niskin-type bottle – for the collection of water samples, must seal well, be corrosion-resistant, and the bottle itself needs to be clean and free of contamination; specialized Teflon-lined samplers are available for use with organic pollutants.

Paper towels – used to wipe down tools and surfaces to prevent external contaminants from being introduced into the sample during the sampling process, as well as to effectively remove soil from the surface of tools as part of decontamination/to avoid cross-contamination.
Plastic bags – for secondary containment of the soil/sediment samples, plastic bags are usually used; however, plastic bags should only be used as secondary containment due to the risk of contaminating samples via contact with plastic.
Sample identification labels – written in indelible ink, these labels are for unique identification and must include (1) a unique sample identifier (also logged in the field book), (2) sampling point location, (3) collection date, and (4) sample type.
Scraper/spoon (stainless steel) – used to scrape soil samples from an auger head.
Stainless-steel buckets – used when collecting/field processing larger volume soil/sediment samples, or for collecting surface water samples.
Stainless-steel digging tool – to collect soil samples.
Thermo-insulated boxes – to transport samples.
Tweezers – used to remove weeds or extraneous material from soil/sediments.
Type I water and solvents – for rinsing/decontamination, as needed.

Tenax extraction and handling-related:

Amber vials – chromatography-type or similar, with solid caps, 1–2 mL, capacity; used to store final eluates awaiting instrumental analysis.
Balance – for weighing laboratory samples.
Beakers – various sizes for preparatory uses.
Benchtop centrifuge – used to separate soil/sediment/Tenax and liquids. To operate with 30 mL centrifuge tubes at 2,400 rpm.
Dessicator – to retain dryness of anhydrous sodium sulfate after preparation.
Drying oven – used to heat at 60°C–275°C for preparation of Tenax and anhydrous sodium sulfate.
Glass centrifuge tubes – used to separate Tenax from sample.
Glass flasks/containers – used to mix the Tenax with soil/sediment/water samples.
Graduated glass test tubes – used to hold the eluate and for nitrogen evaporation to concentrate eluates/reduce eluate volumes.
Muffle furnace – for high-temperature decontamination of glassware (organic pollutants).
Nitrogen evaporator – to reduce solvent in extracts to ultralow volume under a nitrogen stream, e.g., the N-EVAP model from Organomation, or similar.
Refrigerators – for storage of collected samples.
Separatory funnel – used to manually separate Tenax after extraction.
Soil/sediment mixing tool – to homogenize soil or sediment samples; e.g., rolling mill, feed mixer, or gardener's soil mixer (e.g., DaRos MX1000 or similar).
Sonicator – to elute pollutants from Tenax.
Syringes and syringe filters – polypropylene syringes with PTFE ***filter membranes*** of 0.2 μm pore size and 25 mm diameter (or similar other appropriate solvent resistant material compatible with organic pollutant studies).
Tenax TA – a porous polymer based on 2,6-diphenyl-p-phenylene oxide. It can be purchased from Alltech International and other providers (mesh 60–80, Deerfield, USA).
Thermostatic shaker – used in Tenax extraction.

Personnel protection-related:

Lab gloves – contaminant-free, with good acid, alkali, and corrosion resistance, and effective protection against the hazards of organic solvents.
Lab goggles – to prevent eye injuries during experiments.

11.4 REAGENTS

Note: All reagents must be of ultrapure quality to avoid contamination.

Acetone – to clean Tenax before use and to elute pollutants from Tenax after extraction.

Benzalkonium chloride – used to prepare a 2,500 mg/L solution in water to ***spike*** samples for inhibition of microbial activity during Tenax extraction.

Hydrated sodium sulfate, $Na_2SO_4 \cdot 10H_2O$ – to prepare anhydrous Na_2SO_4 – used to remove traces of water from eluates.

Hexane – to clean Tenax before use and to elute the pollutants from Tenax after extraction.

Mercuric chloride – used to prepare a 5,000 mg/L solution in water to spike samples for inhibition of microbial activity during Tenax extraction. Users may choose to omit use of this reagent or use an alternate biocide – see procedure notes for details.

Nitrogen gas – used to dry Tenax after cleaning and for evaporation of eluates.

Standards and calibration solutions – for instrument calibration and quantification (recommended in chosen method of instrumental analysis).

Surrogate compounds – for spiking samples to assess surrogate recovery efficiencies; recommended in chosen method of instrumental analysis or chosen by user according to chemical similarity with target organic pollutants. Surrogates are often deuterated versions of the target pollutant.

Ultrapure water – also known as Type I water (Milli-Q, resistivity >18.2 MΩ·cm).

11.5 PROCEDURE

11.5.1 Sample Collection, Transport, and Handling

In the case of soil sampling, the sampling depth of the soil horizon is typically between 0 and 20 cm, sampled using a spade or digging tool, ideally stainless steel; however, for some studies deeper soils might be the target, in which case an appropriate auger is needed. The site background and history monitoring requirements (e.g., soil characteristics, prior environmental assessments, pollution events) should be considered when choosing the sampling point and depth for sample collection. The soil sample should be loaded into a clean sample container using a trowel, shovel, or scrapings from an auger head.

For sediments, grab samplers are useful for gathering samples of the surface layer. Alternately, sediment corers are used to collect profiles (see Chapter 7, Section 7.4.3). Coring or using a benthic grab is preferred over sampling dredged material. Sediment samples are collected in clean, pre-prepared containers recording the depth range from which sediments were collected. For organic pollutants in both soils and sediments, sampling equipment made of high-strength, wear-resistant stainless steel that has appropriately decontaminated prior to and between each use is needed (refer to Chapter 3 regarding contamination).

As with soil/sediments, when collecting water samples in the field, it is important to use well-cleaned, contaminant-free materials appropriate for organic pollutants analysis. Additionally, care must be taken to prevent the suspension of sediments, which can be achieved by carefully selecting the sampling time to avoid resuspension due to natural factors such as wind or weather. After identifying suitable sampling locations, samples of water are collected at each sampling location from depths of 0.2 m below the water surface (for surface grab samples), or from deeper locations using Niskin-type water collection bottles. Water samples are then transferred from the Niskin bottle into suitable containers, free of contamination, labeled clearly and sequentially for precise identification.

Field duplicate (co-located or split samples) must be taken in all cases to assess field variability in soil, sediment, and water at the rate of 1 per every 15 field samples. A split sample involves dividing a single sample into two or more portions for parallel testing or analysis. Its function is to assess ***precision*** as influenced by field heterogeneity. Equipment must be decontaminated between each sample, and ***equipment rinsate blanks*** taken for soil/sediments. While efforts should be made to minimize disturbance during collection, subsampling, compositing, or homogenizing samples will be required for many study designs. For all samples, direct sunlight exposure should be minimized during collection, especially if the sample might contain photolytically reactive compounds. Samples with pollutants that are susceptible to light should be placed in light-proof containers. Samples must be cooled to 4°C with ice, ice packs, or dry ice before transportation, avoiding freezing (ASTM, 1999; US EPA, 1999). Samples should be transported to the laboratory within 6 hours of collection. On arrival at the laboratory, soil, sediment, and water samples should be stored at 4°C in the dark and used promptly for bioavailability tests. Soil and sediment samples should not be stored for more than two weeks, as even brief storage can alter concentrations of labile substances (ASTM, 1999). The US EPA recommends that hold times of filtered water samples for biologically relevant measurements be no more than 96 hours (US EPA, 2016). In both cases, the recommended hold times are likely too long for some samples, given changes that can occur in bioavailability, and in particular reactions of organic pollutants. Ideally, samples are processed as soon as possible.

Soil and sediment samples must be prepared in the laboratory prior to use. The soil or sediment should be thoroughly mixed using methods such as stirring, a rolling mill, feed mixer, or other appropriate equipment. It is important to mix the soil or sediment carefully to avoid significantly changing its physicochemical properties; structural changes in soil/sediment might cause changes in bioavailability.

11.5.2 Preparation of Tenax

Prior to use, Tenax must be cleaned. Tenax should be pretreated to remove possible contaminants from the manufacturing process and to ensure that the Tenax is contaminant-free and ready to adsorb the target compound. To achieve this, combine Tenax with 10 mL of acetone per gram of Tenax, then sonicate for one hour in a sonicating water bath. Once the Tenax has been sonicated, remove the acetone and replace it with 10 mL of a 1:1 vol/vol acetone/hexane solution per gram of Tenax. Then, sonicate again. Repeat the acetone followed by 1:1 acetone:hexane wash twice more for a total of three cleaning cycles. After the last 1:1 acetone:hexane wash, perform a final wash/sonication with hexane only. Following the last wash, the Tenax is heated to 60°C to eliminate any remaining solvent. When the mass of Tenax reaches a constant weight, the Tenax is dry and ready for use. A good way to check for constant weight is to wash the Tenax in large quantities, then dry in a large container, such as a large, wide beaker, and then weigh the total mass of the container and Tenax at intervals during drying. Each time the container is weighed, it should be at room temperature. During drying, the Tenax should be covered with some suitable material to prevent contamination, but the container must not be fully sealed, which would impede drying.

11.5.3 Single-Point Tenax Extraction

This method is based on a single-point Tenax extraction, which has been found to provide concentrations of organic pollutant that are proportional to f_{rap} and, in some cases, correspond well to measures of true bioavailability (Cornelissen et al., 2001; Kraaij et al., 2001; Landrum et al., 2007; Liu et al., 2011). To perform a single-point desorption/Tenax extraction, 1.0 g of wet soil/sediment is added to a glass container and then 75 mL of Type I water is added. Spike 1 mL of a 5,000 mg/L solution of mercuric chloride ($HgCl_2$) into each sample (i.e., 5.0 mg $HgCl_2$ or 0.5% of the soil/sediment wet weight) as a biocide to inhibit microbial activity.[1] Ensure the $HgCl_2$ mixes well with the soil/sediment slurry.

[1] Note: Other biocides might be used, notably sodium azide. Tests to determine the efficcacy of biocide, such as glucose utilization with and without biocide, should be performed if an alternate biocide is used in order to confirm that the alternate biocide is suitably inhibitory. Mercuric chloride is known to be highly effective. If $HgCl_2$ is used, ensure to collect all waste containing Hg in a labeled, leak-proof, and chemically resistant container and that it is handled by a licensed hazardous waste disposal facility. Never pour Hg-containing liquids or slurries down the drain or dispose of Hg-contaminated material in trash, as it is highly toxic and poses significant environmental and health risks.

Next, spike the slurry with an appropriate amount of surrogate compound,[2] mix well, and add 1.5 g Tenax to the container and seal it (Ten Hulscher et al., 2003; Van Noort et al., 2002). Note that the 6-hour extraction is a short-term test. Because of this, some investigators perform this extraction without the use of Hg_2Cl, to avoid dealing with the hazards and waste associated with the use of this material. Bacterial doubling times and stimulation from shaking and oxygenation are relevant on this time scale, and it is more conservative to add Hg_2Cl, or some alternate efficacious biocide, to avoid artifactual changes in bioavailable pollutant from microbial action during the extraction.

To perform a single-point desorption/Tenax extraction for organic pollutants in water samples, add 0.3 g Tenax to a 100 mL portion of water sample in a glass container (Zhao and Pignatello, 2004). Then spike 0.6 mL of a 5,000 mg/L solution of mercuric chloride into each sample (i.e., 3 mg $HgCl_2$) to inhibit microbial activity. As an alternative to $HgCl_2$, spike 2.0 mL of a 2,500 mg/L solution of benzalkonium chloride into each sample. As with testing soil sediment samples, this step is a recommended precaution. Finally, spike sample with an appropriate amount of surrogate compound, and seal the container.

After Tenax, biocide, and surrogate compounds have been added to the containers with soil, sediment, or water, containers are placed on a shaker at 150 rpm, preferably with a thermostatic control set to 25°C. At the end of 6 hours, the containers are removed from the shaker and Tenax must be separated from each container. This can be done manually using a separatory funnel, or contents may be centrifuged. The Tenax beads will float, facilitating separation from the soil/sediment slurry in either method.

Collect Tenax in a 30 mL glass centrifuge tube. Subsequently, add 10 mL acetone/hexane (1:1, v/v) elution solvent and mix well with Tenax. The centrifuge tube containing Tenax and the elution solvent is sonicated using a sonicating water bath for 15 minutes at room temperature. After sonication, the centrifuge tubes are centrifuged to separate the Tenax beads, which float, and then the eluate is collected. The process of adding solvent, sonicating, centrifuging, and collecting the eluate is repeated three times in total, and the three portions of eluate are combined into one sample.

The use of Tenax in large quantities of water often results in water in the Tenax eluate. When collecting the eluates into a single sample, ensure not to collect small amounts of water that may have coalesced in the bottom of the centrifuge tube. To dry further, which is important for some methods of instrumental analysis, add 1.5 g of anhydrous Na_2SO_4 to the eluate. To prepare anhydrous Na_2SO_4, heat $Na_2SO_4 \cdot 10H_2O$ in an oven at 275°C for a minimum of 24 hours. Ensure the material is covered loosely during heating to prevent contamination and still allow the hydrated material to lose its waters of hydration. Anhydrous sodium sulfate will appear as a white, crystalline powder or solid. If used regularly, it may be left in the oven. Immediately prior to use, transfer the needed quantity of anhydrous sodium sulfate to a dessicator and allow to cool to room temperature. Remove from dessicator immediately before use.

After this drying step, filter samples using a syringe and syringe filter having 0.2 μm organic-pollutant compatible membranes. Syringes made of polypropylene are chemically stable and resistant to most organic solvents, and should be suitable for this application since the target pollutant is in solvent. Using a nitrogen evaporator, dry the processed eluate samples to near dryness under a gentle stream of nitrogen, then add 3 mL of hexane to each and continue to nitrogen evaporate to a volume of 1 mL. It is very important not to let samples evaporate to complete dryness during this process. Eluates prepared in this manner are typically suitable for direct use in instrumental analysis. In some cases, cleanup of the eluate may be needed, in which case the unprocessed eluate may be suitable for direct use in cleanup. Consult the method of instrumental analysis for the target pollutants in question for guidance. For a representative list of ***Standard Methods*** for instrumental analysis of pollutants, refer to Chapter 5.

---

[2] The surrogate compounds to be used must not be target pollutant(s), but must behave chemically identically, or nearly so, to the target pollutant(s). Most Standard Methods of instrumental analysis for organic pollutants have guidance about the use of surrogate compounds. This method assumes that users will consult such methods for a determination of the exact compounds and spiking levels to be used for surrogates (see Chapter 5).

The samples used in the Tenax process are wet samples; however, the reporting of pollutant concentrations in the samples is based on dry weight. When testing soil/sediment samples, it is necessary to take a separate subsample to determine the dry weight content of the sample. For measuring soil/sediment dry weight, gravimetric analysis is thought to be the most accurate technique. Using this technique, the water content of a soil or sediment is measured by weighing a portion of sample before and after it has been dried in an oven. The moist soil/sediment sample should be placed in an oven that is set to a certain temperature (usually 105°C) until all of the water has evaporated, at which point its weight should remain constant. Usually, this process takes 24–48 hours. Dry weight content (*D* or *D*%, for decimal or percent dry weight, respectively) is D = Dry Weight/Wet Weight, or $D\%$ = Dry Weight/Wet Weight) * 100%.

Field rinsate blank (soils and sediments only) samples, ***procedural blank*** samples, ***laboratory duplicates***, and ***field duplicates*** should be prepared and extracted at the same time as field samples. The ***field blank*** sample is prepared as for soil or sediment samples, only using a soil/sediment that does not contain the target pollutants. The soil or sediment used for the field blank should ideally originate from a control site with similar characteristics, except that field blank soil or sediment crucially must not be polluted. The procedural blank is a sample that does not contain the matrix (i.e., water, soil, or sediment sample), which is brought through the entire measurement procedure and analyzed in the same manner as other samples. When preparing procedural blanks, water is often used in place of the matrix; though sometimes solvent is used if the final samples subject to instrumental analysis are in solvent. Laboratory duplicates consist of two aliquots of a field sample that are independently extracted. These QA/QC samples are spiked with biocide and surrogate in the same manner as for field samples. Additionally, eluate blank samples should be prepared at a rate of at least three per sample processing run, i.e., each time a set of field samples is processed. To make an eluate blank sample, combine the total amount of eluate that is used to elute Tenax samples (30 mL), then spike with surrogate compound.

Surrogate compound(s) ideally behave in an identical chemical way as the target pollutant(s), however, are also chemically unique, such that when added to field samples prior to any substantive sample handling or preparation, surrogates provide a kind of marker by which mass conservation can be tracked. If surrogate is lost in processing, it is a fair assumption that target pollutant has been lost as well. Problematically, however, surrogate compounds may experience ***matrix effects***, i.e., in some sense, a surrogate also bears some similarity to a ***matrix spike*** (though a matrix spike involves spiking with the target pollutant, and surrogate compounds are unique/not target pollutants). This circumstance entails that, as used in methods such as the single-point Tenax extraction, if surrogate recovery is low, it is difficult to know if this occurs due to surrogate loss in processing or due to matrix effects on the surrogate. If surrogate recovery is low, by analyzing surrogate content in procedural blanks and eluate blanks, some insight may be gained as to the causes. If eluate blank surrogate recovery is low, there is a problem with target pollutant/analyte loss during the processing of eluates, which can occur if care is not taken, particularly in nitrogen evaporation. To whatever extent procedural blank surrogate recoveries are lower than those for eluate blanks provides an indication of surrogate loss prior to eluate processing. If surrogate recoveries for field samples are low, but both of these types of blanks exhibit reasonable surrogate recoveries, this suggests that the surrogate recovery for field blanks is reflecting short-term matrix effects that occur in the short time between surrogate addition to samples and sample processing.

11.5.4 Variation of Method: Factors Affecting the Tenax Extraction Method

A number of factors affect the success of this method, importantly, the sample-to-Tenax ratio, the amount of water used relative to sample/Tenax, the extraction time, and the solvent used for elution. The sample-to-Tenax ratio directly impacts the interaction between them. Tenax needs to be present at an amount that supports its function as a quasi-infinite sink. An appropriate sample-to-Tenax ratio enables the efficient transfer of target pollutants from the sample to Tenax. If the amount of sample is too high, the sorption capacity of Tenax may be insufficient to produce quasi-infinite sink conditions, resulting in incomplete extraction (Nutile et al., 2017). Assuming a fixed amount of sample and sample pollutant loading, an increase in the amount of Tenax provides more sorption capacity, enabling more thorough capture of target substances and enhancing the extraction efficiency (Cornelissen et al., 2001). While a larger quantity of Tenax can adsorb more pollutant, an excessive amount of Tenax not only raises costs, but also has implications for the subsequent elution process. Not only the amount of sample is important, but also its pollutant load. For a fixed amount of sample, if the concentration of target pollutants in the sample is high, the amount of Tenax used should be increased. If the concentration of target substances is low, the proportion of Tenax can be appropriately decreased (Cornelissen et al., 2001).

Pollutants are extracted from sample to Tenax via the aqueous phase, and the amount of water, relative to Tenax and sample, affects the distribution of target pollutants between the aqueous phase and Tenax. A suitable amount of water helps maintain good mass transfer efficiency, facilitating the transfer of target pollutants from the aqueous phase to Tenax. If the relative proportion of water is too high, although the target substances are easily desorbed, they may be overly diluted, reducing the ***concentration gradients*** that are useful to effect desorption and binding with Tenax (Cornelissen et al., 2001). If the proportion of water is too low, this reduces mass transfer, which reduces extraction efficiency. Substances with strong hydrophobicity may require less water for effective extraction and vice versa, since desorption is favored for less hydrophobic HOCs. For more hydrophobic pollutants, it is important to achieve higher concentration gradients to drive desorption (Yeom et al., 1996).

Extraction time refers to the time at which the single-point extraction is terminated and the Tenax is collected. The extraction time determines whether the extraction is sufficient (Kong et al., 2023) to adequately represent f_{rap}. If the extraction time is too short, the target pollutant extracted under-represents bioavailability and vice versa. Additionally, if the extraction time is too long, Tenax may begin to saturate, causing a deviation from quasi-infinite sink conditions. While some investigators use longer single-point extraction times to represent f_{rap}, the time used here, 6 hours, has been shown to be a better reflection of f_{rap}, and, being a shorter time during which much less pollutant will desorb, is also more likely to avoid the pitfall of Tenax saturation.

The choice of elution solvent determines whether target pollutant(s) can be effectively eluted from Tenax. A suitable elution solvent should have a strong affinity for the target pollutant(s), capable of disrupting pollutant sorption to Tenax (Sinche et al., 2017). Organic pollutants that are less hydrophobic might be eluted with more polar solvents and vice versa. For strongly non-polar target pollutants, non-polar organic solvents should be used (Yang et al., 2010). If the solvent is not properly selected, it may result in incomplete elution.

The method provided here is suitable for most studies. The sample-to-Tenax ratio is conservative and should not require adjustment in most cases. If the concentrations of pollutant in samples are known to be very low, the 1.5 g Tenax called for here could be halved. Similarly, if it is known that pollutants are present in samples at very high concentrations (e.g., 10s–100s of ppm), then a user might choose to double the amount of Tenax used. Another factor that might argue for increasing the amount of Tenax used is the organic matter content of samples. Organic matter may compete with Tenax for sorption of organic pollutants, reducing the extraction efficiency. The amount of water used in this method should not require adjustment; however, it can be reduced to as little as 40 mL in case of concerns for samples with very low concentrations of very hydrophobic pollutants. The extraction time should not be adjusted without testing the effect, which would involve constructing full desorption curves, at which point f_{rap} can be calculated from the full desorption curve. For this method, the elution solvent is the main factor that a user might wish to alter. The mixture of hexane and acetone used here is effective and generally applicable for elution of a wide range of organic pollutants; however, for very hydrophobic pollutants, using hexane alone or using a first elution with hexane:acetone and then using hexane only would be a better choice.

Table 11.1 shows how factors affecting Tenax extraction have varied in some representative studies that used the Tenax extraction method to determine HOC bioavailability in soils and sediments. Many of these studies did not systematically analyze the effect of varying the factors affecting

TABLE 11.1
Representative Summary of the Literature Illustrating How Factors Affecting Tenax Extraction Have Varied[a]

Organic Pollutant	Tenax Mass (g)	Sediment Mass (g)	Water Volume (mL)	Elution Solvent	Extraction Times (d)	Reference
Benzo[a]pyrene, hexachlorobiphenyl, pyrene, and tetrachlorobiphenyl	0.2	2.3–3.7 g[b]	48	2:1 H:A	2, 4, 8, 14, 21, 28 for the benzo[a]pyrene and hexachlorobiphenyl. 2, 4, 7, 13, 20, 28 for pyrene and tetrachlorobiphenyl.	Kukkonen et al. (2004)
Bifenthrin and permethrin	0.05	0.2	14	1:1 A:H	0.5, 1, 4, 7, 10, 14	You et al. (2009)
Brominated diphenyl ethers, DDD, DDE, and DDT	2	2	40	1:1 A:P	0.125, 0.25, 0.79, 1.25, 3, 6, 12, 18	De la Cal et al. (2008)
Halogenated aliphatic hydrocarbons	0.25	1	10	A	4	Pignatello (1990a)
13 priority PAHs, 7 PCBs, 3 chlorinated aromatic hydrocarbons, β-hexachloro cyclohexane, and DDD	1.5	1	70	H	–	Van Noort et al. (2002)
1,2,3,4-tetrachlorobenzene, Pentachlorobenzene, hexachlorobenzene, fluoranthene, 2,3,5,6-tetrachlorobiphenyl, and 2,3',4,4',5-pentachlorobiphenyl	0.2	0.6	100	H	34	Cornelissen et al. (1997a)
Chlorpyrifos, DDE, hexachlorobiphenyl, permethrin, phenanthrene	0.05	0.25	45	1:1 A:H	14	You et al. (2007)
DDD, DDE, DDT, and dieldrin	0.1 or 0.2	0.2 or 0.5	25	1:1 A:H	–	Morrison et al. (2000)
12 priority PAHs, 6 PCBs, pentachlorobenzene, hexachloro benzene, DDD, and DDE	1.5	1	70	1:9 A:H	–	Ten Hulscher et al. (2003)
DDT	0.2	1	100	1:1 A:H	0.006, 0.013, 0.02, 0.04, 0.08, 0.16, 0.33, 0.5, 1, 2, 3, 8, 16	Yang et al. (2010)
Nonylphenol	1.5	2	40	1:1 A:P	0.04, 0.08, 0.12, 0.2, 1, 3, 12, 18	de Weert et al. (2008)
10 priority PAHs, 4 PCBs, and 9 chlorobenzenes	0.6	1	70	H	0.25, 1.25	Cornelissen et al. (2001)
Polychlorinated dibenzo-p-dioxins, dibenzofurans, diphenyl ethers, and hydroxydiphenyl ethers	0.2	0.57	48	T	0.08, 0.16, 0.5, 1, 2.12, 4, 7, 12, 19, 26, 33, 40	Sormunen et al. (2008)
Phenanthrene	0.1	0.15–0.25 g	20	A or H	–	White et al. (1999)

[a] Most of the studies reported in this table collected data for full desorption curves, as reflected in multiple extraction times; – signifies full desorption curve with multiple and/or unspecified time points. Abbreviations: A, acetone; DDD, dichlorodiphenyldichloroethane; DDE, dichlorodiphenyldichloroethylene; DDT, dichlorodiphenyltrichloroethane; H, hexane; P, petroleum ether; W, water; T, toluene.

[b] Wet weight.

Tenax extraction, and hence, while this information may be indicative, users are advised caution in varying the procedure based on this data alone. Note that, while there is no particular systematic trend to the variation for most of the factors listed in Table 11.1, overall, it does appear that some investigators use more hydrophobic solvents to elute more hydrophobic pollutants. Hexane was used for hydrophobic pollutants, e.g., chlorinated hydrocarbons (Van Noort et al., 2002) and PAHs (Cornelissen et al., 1997a, 2001), which are known for their low water solubility. An elution solvent of 1:1 acetone:petroleum ether was used for nonylphenol (de Weert et al. 2008), and toluene was used as an elution solvent for polychlorinated dibenzo-p-dioxin (PCDD) (Sormunen et al., 2008), known for its hydrophobic nature.

11.5.5 Regeneration of Tenax

Regenerating Tenax after use is required to ensure it is contaminant-free and to restore its sorption capacity for subsequent analyses. The regeneration process involves cleaning, drying, and reconditioning Tenax to remove adsorbed pollutants and contaminants. Before starting the regeneration process, inspect the used Tenax for any visible signs of contamination or degradation, such as color changes or apparent damage to the porous structure. If the visual indications of contamination are severe, consider replacing the Tenax. If none of the above is the case, then proceed with the following steps. First, wash Tenax with water to remove water-soluble compounds including biocide that may remain in the material. Use an excess amount of solvent to the amount of Tenax to facilitate cleaning (approximately 50 mL/g sample Tenax or more) and leave on a shaker for one hour. Then, wash Tenax with acetone to remove moderately polar organic compounds and/or interstitial water that may have been sorbed during use. Use an excess amount of solvent to the amount of Tenax to facilitate cleaning (approximately 50 mL/g sample Tenax or more) and leave on a shaker for 1 hour. Follow this with a wash using hexane to remove any non-polar pollutants. Again, use approximately 50 mL of hexane per gram of Tenax and leave on a shaker for 1 hour.

Next, Tenax must be dried to remove any residual solvent. Place Tenax, covered to prevent contamination, in a drying oven or use a nitrogen gas flow to gently dry it at 100°C–120°C overnight or longer. Ensure constant weight, as in Section 11.5.2. This step ensures that any remaining solvent is removed, and Tenax is ready for further use. After the regeneration process, inspect Tenax to ensure that it has the visual appearance of its original form. Perform a blank analysis (i.e., as for procedural blanks) and/or sample test to ensure that the Tenax is free of contaminants and performs as expected. If necessary, repeat the regeneration process until Tenax achieves the desired performance. Once Tenax has been regenerated and tested, store it in a clean, dry, and sealed container to prevent contamination before the next use.

Precautions and considerations – always wear appropriate personal protective equipment, such as gloves and goggles, when handling chemicals like acetone, methanol, and hexane. Ensure that the Tenax containers used for regeneration are clean and free from contamination. Monitor Tenax performance regularly to determine when regeneration may no longer be effective and when replacement is necessary. By following these steps, Tenax can be effectively regenerated, ensuring its optimal performance in future analyses. Regular maintenance and monitoring of the condition of Tenax condition are essential to ensure reliable and accurate results.

11.6 VALIDITY OF RESULTS, DATA ANALYSIS, INTERPRETATION, AND REPORTING

11.6.1 Notes on QA/QC Planning Requirements

Both laboratory and field operations require a QA/QC plan to verify the ***accuracy*** of the results obtained. Chapter 4 offers comprehensive guidance relevant to this chapter, and a thorough understanding of the concepts and procedures outlined in Chapter 4 is essential before applying QA/QC in this method. Standard techniques for instrumental analysis provided in Chapter 5 offer further guidance on QA/QC procedures. Specific QA/QC information pertinent to this chapter, and not already discussed above, is as follows:

- All users conducting experiments utilizing the methods described in this chapter need to undergo an Initial Demonstration of Proficiency (IDP). IDP for instrumental methods are provided in individual ***Standard Operating Procedures*** for same; this involves analysis of a Certified or Standard Reference Material and obtaining results that fall within accepted ranges for accuracy and precision. IDP for this method entails performing analysis of three procedural blanks and three eluate blanks and obtaining surrogate recoveries in the range of, respectively, 90%–110% and 85%–115%.
- Routine instrumental calibration and ***calibration blank*** checks are required; guidelines for ***Acceptance Criteria*** for these are stipulated in appropriate Standard Methods used for instrumental analysis.
- Batch sizes for instrumental analysis shall not exceed 20 samples; samples batched include soil/sediment and water extracts and replicate samples; QA/QC samples are additional and should not exceed five per batch. Instrument check samples assess calibration stability and are not part of batching.
- Sample order in batches is to be randomized over all samples and sample types.
- For every batch, one randomly chosen equipment rinsate blank, OR one equipment rinsate blank per ***duplicate/replicate sample***, is required, whichever is smaller. These samples enable evaluation of whether or not field samples may have been contaminated by field equipment. If contamination is found in any ***equipment blank***, then all equipment blanks must be analyzed.
- For every batch, one procedural blank is required.
- For every batch, one eluate blank is required.
- One field duplicate sample is required for every fifteen field samples; this enables estimation of precision of inherent field sample variability.
- One laboratory duplicate sample is required for every fifteen samples, unless otherwise indicated by the Standard Method in use; this enables estimation of precision of duplicate analysis.
- One ***recovery spike*** is required per batch. This enables evaluation of whether artifactual contamination (recoveries significantly above 100%) or analyte loss (recoveries significantly below 100%) is occurring.
- One recovery spike duplicate per instrument run (i.e., assumes multiple batches run sequentially) or one recovery spike duplicate every 12 hours is required, whichever is more frequent.
- Although Tenax beads can be regenerated (recycled), the issue of Tenax aging should also be considered. Tenax sorption properties should be evaluated periodically.

11.6.2 Data Analysis, Interpretation of QA/QC Results (Acceptance Criteria), Corrective Action, and Reporting

Data need to be analyzed, interpreted in the context of QA/QC requirements, and, if needed, subjected to ***Corrective Action*** prior to reporting. These include, at minimum, the following:

- Dilution factors (DFs) – unless specified by an instrumental method, DFs are never greater than a factor of 10 for dilution of standards and samples. The basic dilution formula is $C_1V_1 = C_2V_2$. The dilution factor is given by DF = V_2/V_1. A number of DFs are used in this method, and it is very important to correct for these in calculating results.

- ***Calibration checks*** and metrics – perform calibration checks according to directions specified in the method used for instrumental analysis. This will include guidance on using certified standards and ensuring regular check intervals. Metrics such as linearity, ***sensitivity***, and precision are assessed for stability according to Acceptance Criteria given in methods of instrumental analysis (see Chapter 5). If instability occurs, Corrective Action must be taken. Only data collected under validated instrumental calibration conditions (stability of calibration) should be reported. Should the metrics associated with the instrumental ***calibration curve*** not meet requirements/Acceptance Criteria, or if calibration checks fail, as stipulated in the Standard Method chosen for instrumental analysis, it is mandatory to take Corrective Action to establish or re-establish proper calibration. Any data collected under failed calibration is not reportable.
- Precision for all instrumental duplicate/replicates – a metric for precision must be calculated and reported. For duplicates, the relative percent difference (RPD) is used to evaluate precision. The formula for RPD is as follows: $\text{RPD} = 200\% \times \text{abs}(R_1 - R_2)/(R_1 + R_2)$, where subscripts 1 and 2 represent the two duplicate samples. For more than two replicates, relative standard deviation (RSD) is used; the formula for RSD is as follows: $\text{RSD} = \text{stdev} \times 100\%/\text{average}$. If metrics such as RPD or RSD are too high, it indicates that there is a significant amount of variability or inconsistency between the measurements of the same sample. For example, typically, an RSD value greater than 5% (acceptance criterion) for analysis of sample replicates using gas chromatography–mass spectrometry (GC–MS) (for example) at concentration ranges in the middle of the calibration curve indicates poor precision. Poor precision for replicates of lab samples is generally indicative of an operator or instrument error, and Corrective Action involves rectifying the problem prior to analysis and reporting of results from bioavailability studies.
- Field duplicate sample precision – acceptable precision is typically no greater than 35% RPD for soil/sediment samples and no greater than 20% RPD for aqueous samples. If results are outside these criteria, the ***source*** and implications of field heterogeneity must be discussed in reporting.
- Equipment rinsate blanks – results for blanks must be calculated and reported in units of absolute concentration for all blank samples. The acceptance criterion for blanks is a value below the target pollutant ***limit of detection*** (LOD). Blanks above the LOD are indicative of field contamination (equipment blanks), no Corrective Action is possible, and all samples must be regarded as potentially contaminated. If routine analysis of a small sampling of equipment blanks reveals contamination, then all equipment blanks must be analyzed and descriptive statistics performed to assess degree of contamination. If the contamination is effectively constant around an average value that is low, compared to field samples, then this should be reported with results and considered in interpretation of results. For highly varying contamination, in the case of field sample contamination, reporting is not possible and field work must be repeated, taking stronger measures for cleaning of field equipment and/or replacement of equipment parts responsible for contamination.
- Procedural blanks – as for equipment blanks, calculated and reported in units of absolute concentration for all blank samples. The acceptance criterion is a value below the LOD. Blanks above the LOD are indicative of laboratory contamination. If enough field sample remains, it is potentially possible to reprocess and reanalyze samples. First, investigate potential contamination sources by reviewing all procedures, equipment, and reagents used. After confirming the issue, determine whether it could have resulted in contamination of samples in the lab. If so, new field samples are required and reporting is not possible. Otherwise, if field samples are deemed uncontaminated, repeat the analysis with fresh reagents and adjusted protocols to minimize contamination and ensure accurate results.

The acceptance criterion for surrogate recovery of procedural blanks is 85%–115%; if results are outside this range, Corrective Action involves investigation of the cause, rectification, and reprocessing of samples associated with the procedural blank.

- Eluate blanks – these blanks are primarily used to assess surrogate recovery. The Acceptance Criteria range for surrogate recovery of eluate blanks is 90%–110%; if results are outside this range, Corrective Action involves investigation of the cause, rectification, and reprocessing of samples associated with the procedural blank.
- LOD and LOQ (***limit of quantification***) – these critical parameters define the lowest concentration of heavy metal that can be reliably detected (LOD) and quantified (LOQ) using a given method of instrumental analysis. LOD is often defined as three times the standard deviation of the calibration blank divided by the slope of the calibration curve. LOQ is often defined as three times the LOD. Standard Methods for instrumental analysis commonly contain guidelines concerning what LOD and LOQ are typical. Substantial deviation from such reported guidelines may typically reflect issues with user competency in performing analysis. LOD and LOQ must be reported as they reflect upon the competence of analysis and interpretation of results (e.g., lower precision typically occurs between the LOD and LOQ).
- Surrogate spikes – surrogates and the general method spiking are discussed in Chapter 4. Surrogate recovery acceptance for sample is 80%–120%. Recoveries outside this range should be assessed as potentially arising from matrix effects (by comparison to procedural and eluate blank surrogate recoveries), and the implications of this should be discussed in reporting of results. Calculate recovery as Recovery (%) = $100\% \times (C_{\text{spike analyzed}}/C_{\text{spike added}})$, where $C_{\text{spike analyzed}}$ is the amount of surrogate analyzed to be in a sample after processing and $C_{\text{spike added}}$ is the total concentration of surrogate spike added to each sample before processing (as independently confirmed by analysis).
- Recovery spikes – these spikes and the general method of spiking are discussed in Chapter 4. Calculate recovery as Recovery (%) = $100\% \times (C_{\text{Total}} - C_{\text{bkg}})/C_{\text{spike added}}$, where C_{Total} is the total concentration of organic pollutants in the spiked sample, i.e., the concentration from the sample and the added spike; C_{bkg} is the background concentration of the analyte in the sample without the spike; $C_{\text{spike added}}$ is the known amount of organic pollutant contributed to the sample by addition of the spike. For a recovery spike, the Acceptance Criteria range is typically 90%–110%; however, acceptable recovery varies, and is stipulated, for different analytical techniques (e.g., GC–MS, High performance liquid chromatography–mass spectrometry (HPLC–MS)). Consult the Standard Method of instrumental analysis for Corrective Action steps in the event of recovery spike non conformance.
- Recovery spike duplicates – these samples should be evaluated by RPD for the duplicate samples, with results reported and explained. The recovery spike duplicate acceptance criterion is 15% RPD or less.
- Data reporting – According to the experimental procedure in Section 11.5, the pollutant concentration that desorbed from soil/sediment at the time point 6 hours in concentrated eluates for instrumental analysis is C_{eluate} in V_{eluate} volume of eluate solution. Typically, concentrations of pollutants are reported generally on a dry weight basis, and so the dry matter content, D, of the soil/sediment needs to be taken into account. The bioavailable pollutant concentration per gram of dry weight then is $C_{\text{soil/sed bioavail}} = (C_{\text{eluate}} \times V_{\text{eluate}})/(D \times W_{\text{soil/sediment}})$, where $W_{\text{soil/sediment}}$ is the wet weight of soil/sediment used in Tenax extraction (1 g) and C_{bioavail}, C_{eluate}, and V_{eluate}, are in units of mg/L, mg/L, and mL, respectively. The pollutant bioavailable concentration in water is $C_{\text{W bioavail}} = (C_{\text{eluate}} \times V_{\text{eluate}})/V_{\text{sample}}$, where V_{sample} is the volume of the water sample (100 mL) and $C_{\text{W bioavail}}$, C_{eluate}, V_{eluate}, and V_{sample} are in units of mg/L, mg/L, mL, and mL, respectively.

11.7 CONCLUSIONS AND OUTLOOK

The Tenax extraction method has been applied to the study of organic pollutants in water, soil, or sediment as a method to approximate the bioavailability of pollutants through a simple chemoavailability process. As with most chemoavailability techniques, Tenax extraction has its shortcomings. For long-term desorption studies using Tenax tests on soils and sediments, one report cites recoveries of 70%–130% (Cornelissen et al., 1998), but in some cases, it is as low as 38% and as high as 146% (Williamson et al., 1998). Additionally, it is an abiotic process (longer term studies consistently use biocides); most problematically, some investigators have reported large deviations in f_{rap} and measurements of true bioavailability (Akkanen et al., 2007; Wells et al., 2005). For ease, the amount of time the single-point extraction takes, and convenience in studying organic pollutants, this method stands out. Two interesting avenues for future outlook with this technique includes functionalization of Tenax sorbent materials to improve ***selectivity*** for specific pollutants and development of smaller, portable Tenax-based extraction techniques that can be used for in-field real-time monitoring. While the former may be challenging, the method is sufficiently simple that the latter may be readily achieved.

REFERENCES

Akkanen, J., Sormunen, A, Lyytikäinen, M., Leppänen, M., Pehkonen, S., Kukkon, J.V.K., 2007. *Does Tenax Extraction Based Desorption Measure (Bio)Availability) of Sediment-Associated Contaminants?* ICES CM 2007/J:14. International Council for the Exploration of the Sea (ICES), Copenhagen.

ASTM (American Society for Testing and Materials), 1999. Standard guide for collection, storage, characterization, and manipulation of sediments for toxicological testing. E1391–94. In *Annual Book of ASTM Standards, Vol. 11.05*. Philadelphia, PA.

Biziuk, M., Namieśnik, J., Czerwiński, J., Gorlo, D., Makuch, B., Janicki, W., Polkowska, Ż., Wolska, L., 1996. Occurrence and determination of organic pollutants in tap and surface waters of the Gdańsk district. *J. Chromatogr. A* 733(1), 171–183. https://doi.org/10.1016/0021-9673(95)00905-1

Chen, H.Y., Tian, Y.X., Cai, Y.X., Liu, Q.Y., Ma, J., Wei, Y., Yang, A.F., 2023. A 50-year systemic review of bioavailability application in soil environmental criteria and risk assessment. *Environ. Pollut.* 335, 122272. https://doi.org/10.1016/j.envpol.2023.122272

Chikushi, H., Toda, K., Yoshida, N., 2011. *Adsorption Properties of Sediments for Pesticides: Investigation with Supercritical Fluid Extraction and Gas Chromatograph Mass Spectrometry*. INTECH Open Access Publisher. https://doi.org/10.5772/14014

Cornelissen, G., Rigterink, H., Ferdinandy, M.M.A., Van, N.P.C.M., 1998. Rapidly desorbing fractions of PAHs in contaminated sediments as a predictor of the extent of bioremediation. *Environ. Sci. Technol.* 32(7), 966–970. https://doi.org/10.1021/es9704038

Cornelissen, G., Rigterink, H., Ten, Hulscher., Vrind, D.E.M., Van, N.B.A., 2001. A simple Tenax extraction method to determine the availability of sediment-sorbed organic compounds. *Environ. Toxicol. Chem.* 20(4), 706–711. https://doi.org/10.1002/etc.5620200403

Cornelissen, G., Van, N.P.C.M., Govers, A.J., 1997a. Desorption kinetics of chlorobenzenes, polycyclic aromatic hydrocarbons, and polychlorinated biphenyls: Sediment extraction with Tenax and effects of contact time and solute hydrophobicity. *Environ. Toxicol. Chem.* 16(7), 1351–1357. https://doi.org/10.1002/etc.5620160703

Cornelissen, G., Van, N.P.C.M., Parsons, I.R., Govers, H.A.I., 1997b. Temperature dependence of slow adsorption and desorption kinetics of organic compounds in sediments. *Environ. Sci. Technol.* 31(2), 454–460. https://doi.org/10.1021/es960300+

Cousins, I.T., Johansson, J.H., Salter, M.E., Sha, B., Scheringer, M. 2022. Outside the safe operating space of a new planetary boundary for per-and polyfluoroalkyl substances (PFAS). *Environ. Sci. Technol.* 56(16), 11172–11179. https://doi.org/10.1021/acs.est.2c02765

Dai, C., Han, Y., Duan, Y., Lai, X., Fu, R., Liu, S., Leong, K.H., Tu, Y., Zhou, L., 2022. Review on the contamination and remediation of polycyclic aromatic hydrocarbons (PAHs) in coastal soil and sediments. *Environ. Res.* 205, 112423. https://doi.org/10.1016/j.envres.2021.112423

De la Cal, A., Eljarrat, E., Grotenhuis, T., Barceló, D., 2008. Tenax extraction as a tool to evaluate the availability of polybrominated diphenyl ethers, DDT, and DDT metabolites in sediments. *Environ. Toxicol. Chem.* 27(6), 1250–1256. https://doi.org/10.1897/07-513.1

De Weert, J., de la Cal, A., van den Berg, H., Murk, A., Lagenhoff, A., Rijnaarts, H., Grotenhuis, T., 2008. Bioavailability and biodegradation of nonylphenol in sediment determined with chemical and bioanalysis. *Environ. Toxicol. Chem.* 27(4), 778–785. https://doi.org/10.1897/07-367.1

Hua, L., Zhao, D., Wang, H., Wei, T., 2023. Residues and bioavailability of neonicotinoid pesticide in Shaanxi agricultural soil. *Water Air Soil Pollut.* 234(2), 129. https://doi.org/10.1007/s11270-023-06159-1

Jantunen, P., 2010. The role of sorption in the ecological risk assessment of xenobiotics PhD Dissertation, University of Eastern Finland, Joensuu, Finland.

Kong, J., Cao, X., Huang, W., Li, C., Xian, Q., Yang, S., Li, S., Sun, C., He, H., 2023. Predicting the bioavailability of nitro polycyclic aromatic hydrocarbons in sediments: ZIF-8/h-BN solid-phase microextraction versus Tenax extraction. *Environ. Pollut.* 318, 120896. https://doi.org/10.1016/j.envpol.2022.120896

Kraaij, R.H., Ciarelli, S., Tolls, J., Kater, B.I., Belfroid, A., 2001. Bioavailability of lab-contaminated and native polycyclic aromatic hydrocarbons to the amphipod *Corophium volutator* relates to chemical desorption. *Environ. Toxicol. Chem.* 20(8), 1716–1774. https://doi.org/10.1002/etc.5620200814

Kukkonen, J.V.K., Landrum, P.F., Mitra, S., Gossiaux, D.C., Hunnarsson, J., Weston, D., 2004. The role of desorption for describing the bioavailability of select polycyclic aromatic hydrocarbon and polychlorinated biphenyl congeners for seven laboratory-spiked sediments. *Environ. Toxicol. Chem.* 23(8), 1842–1851. https://doi.org/10.1897/03-474

Landrum, P.F., Robinson, S.D., Gossiaux, D.C., You, J., Lydy, M.J., Mitra, S., ten Hulscher, T.E.M., 2007. Predicting bioavailability of sediment-associated contaminants for *Diporeia* sp. and oligochaetes. *Environ. Sci. Technol.* 41(18), 6442–6447. https://doi.org/10.1021/es0706807

Liu, M., Tian, S.Y., Chen, P., Zhu, L.Y., 2011. Predicting the bioavailability of sediment-associated polybrominated diphenyl ethers using a 45-d sequential Tenax extraction. *Chemosphere.* 85(3), 424–431. https://doi.org/10.1016/j.chemosphere.2011.07.069

Lydy, M.J., Harwood, A.D., Nutile, S., Landrum, P.F., 2015. Tenax extraction of sediments to estimate desorption and bioavailability of hydrophobic contaminants: A literature review. *Integr. Environ. Assess. Manag.* 11(2), 208–220. https://doi.org/10.1002/ieam.1603

Moermond, C.T.A., Roozen, F.C.J.M., Zwolsman, J.J.G., Koelmans, A.A., 2004. Uptake of sediment-bound bioavailable polychlorobiphenyls by benthivorous carp (*Cyprinus carpio*). *Environ. Sci. Technol* 38:4503–4509. https://doi.org/10.1021/e

Morrison, D.E., Robertson, B.K., Alexander, M., 2000. Bioavailability to earthworms of aged DDT, DDE, DDD and dieldrin in soil. *Environ. Sci. Technol.* 34(4), 709–713. https://doi.org/10.1021/es9909879

Nutile, S.A., Harwood, A.D., Sinche, F.L., Huff Hartz, K.E., Landrum, P.F., Lydy, M.J., 2017. The robustness of single-point Tenax extractions of pyrethroids: Effects of the Tenax to organic carbon mass ratio on exposure estimates. *Chemosphere.* 171, 308–317. https://doi.org/10.1016/j.chemosphere.2016.12.045

Pignatello, J.J., 1990a. Slowly reversible sorption of aliphatic halocarbons in soils. I. Formation of residual fractions. *Environ. Toxicol. Chem.* 9(9), 1107–1115. https://doi.org/10.1002/etc.5620090901

Pignatello, J.J., 1990b. Slowly reversible sorption of aliphatic halocarbons in soils. II. Mechanistic aspects. *Environ. Toxicol. Chem.* 9(9), 1117–1126. https://doi.org/10.1002/etc.5620090902

Pignatello, J.J., 1991. Desorption of tetrachloroethene and 1,2-dibromo-3-chloropropane from aquifer sediments. *Environ. Toxicol. Chem.* 10(11), 1399–1404. https://doi.org/10.1002/etc.5620101104

Raffy, G., Mercier, F., Glorennec, P., Mandin, C., Le Bot, B., 2018. Oral bioaccessibility of semi-volatile organic compounds (SVOCs) in settled dust: A review of measurement methods, data and influencing factors. *J. Hazard. Mater.* 352, 215–227. https://doi.org/10.1016/j.jhazmat.2018.03.035

Rashmi, I., Roy, T., Kartika, K.S., Pal, R., Coumar, V., Kala, S., Shinoji, K.C., 2020. Organic and inorganic fertilizer contaminants in agriculture: Impact on soil and water resources. In: Naeem, M., Ansari, A., Gill, S. (eds.) *Contaminants in Agriculture*. pp. 3-41. Springer, Cham. https://doi.org/10.1007/978-3-030-41552-5_1

Sinche, F.L., Nutile, S.A., Landrum, P., Lydy, M.J., 2017. Optimization of Tenax extraction parameters for polychlorinated biphenyls in contaminated sediments. *Talanta* 164, 386–395. https://doi.org/10.1016/j.talanta.2016.11.061

Sormunen, A.J., Koistinen, J., Leppänen, M.T., Kukkonen, J.V.K., 2008. Desorption of sediment-associated polychlorinated dibenzo-p-dioxins, dibenzofurans, diphenyl ethers and hydroxydiphenyl ethers from contaminated sediment. *Chemosphere.* 72(1), 1–7. https://doi.org/10.1016/j.chemosphere.2008.02.057

Ten Hulscher, T.E.M., Postma, J., Den Besten, P.J., Stroomberg, G.J., Belfroid, A., Wegener, J.W., Faber, J.H., Van Der Pol, J.J.C., Hendriks, A.J., Van Noort, P.C.M., 2003. Tenax extraction mimics benthic and terrestrial bioavailability of organic compounds. *Environ. Toxicol. Chem.* 22(10), 2258–2265. https://doi.org/10.1897/02-488

US EPA, 1999. *Methods for Collection, Storage, and Manipulation of Sediments for Chemical and Toxicological Analyses*. United States Environmental Protection Agency, Washington DC.

US EPA, 2016. *Quick Guide to Drinking Water Sample Collection*. United States Environmental Protection Agency, Washington DC.

Van Hattum, B., Pons, M.J.C., Montañés, J.F.C., 1998. Polycyclic aromatic hydrocarbons in freshwater isopods and field-partitioning between abiotic phases. *Arch. Environ. Con. Tox.* 35(2), 257–267. https://doi.org/10.1007/s002449900374

Van Noort, P.C.M., Cornelissen, G., Ten Hulscher, T.E.M., Belfoid, A., 2002. Influence of sorbate planarity on the magnitude of rapidly desorbing fractions of organic compounds in sediments. *Environ. Toxicol. Chem.* 21(11), 2326–2330. https://doi.org/10.1002/etc.5620211110

Wallace, L.A., 2001. Human exposure to volatile organic pollutants: Implications for indoor air studies. *Annu. Rev. Environ. Resour.* 26(1), 269–301. https://doi.org/10.1146/annurev.energy.26.1.269

Wells, M., Wick, L.Y., Harms, H., 2005. Model polymer release system study of PAH bioaccessibility: The relationship between "rapid" release and bioaccessibility. *Environ. Sci. Technol.* 39, 1055–1063. https://doi.org/10.1021/es035067b

White, J.C., Hunter, M., Nam, K., Pignatello, J.J., Alexander, M., 1999. Correlation between biological and physical availabilities of phenanthrene in soils and soil humin in aging experiments. *Environ. Toxicol. Chem.* 18(8), 1720–1737. https://doi.org/10.1002/etc.5620180816

Williamson, D.G., Loehr, R.C., Kimura, Y., 1998. Release of chemicals from contaminated soils. *J. Soil Contam.* 7(5), 543–558. https://doi.org/10.1080/10588339891334492

Yang, X., Wang, F., Gu, C., Jiang, X., 2010. Tenax TA extraction to assess the bioavailability of DDTs in cotton field soils. *J. Hazard. Mater.* 179(1–3), 676–683. https://doi.org/10.1016/j.jhazmat.2010.03.055

Yeom, I.T., Ghosh, M.M., Cox, C.D., 1996. Kinetic Aspects of Surfactant Solubilization of Soil-Bound Polycyclic Aromatic Hydrocarbons. *Environ. Sci. Technol.* 30(5), 1589–1595. https://doi.org/10.1021/es950567t

You, J., Brennan, A., Lydy, M.J., 2009. Bioavailability and biotransformation of sediment-associated pyrethroid insecticides in *Lumbriculus variegatus*. *Chemosphere*. 75(11), 1477–1482. https://doi.org/10.1016/j.chemosphere.2009.02.022

You, J., Pehkonen, S., Landrum. P.F., Lydy, M.J., 2007. Desorption of hydrophobic compounds from laboratory-spiked sediments measured by Tenax absorbent and matrix solid-phase microextraction. *Environ. Sci. Technol.* 41(16), 5672–5678. https://doi.org/10.1021/es0700395

Zhao, D., Pignatello, J.J., 2004. Model-aided characterization of Tenax®-TA for aromatic compound uptake from water. *Environ. Toxicol. Chem.* 23(7), 1592–1599. https://doi.org/10.1897/03-161

12 Using Chelating Resins for the Measurement of Heavy Metal Chemoavailability

12.1 INTRODUCTION AND SCOPE

Early ***chelating resins*** appeared in the 1940s, when the Norwegian scientist Anders Skogseid developed resins that exhibited ***selectivity*** for potassium ions, i.e., resins that showed discrimination in binding potassium as compared to other cations, allowing potassium to be extracted from seawater (Skogseid, 1947). Since then, chelating resins have developed rapidly. A ***chelating resin*** is a kind of functional material with a cross-linked polymer skeleton to which are bound functional groups. The latter selectively chelate specific metal ions from aqueous solution, forming a multi-cyclic complex through ionic and covalent bonding (Chu et al., 2022). ***Chelation*** is a form of ***complexation*** reaction between a metal and a chelating agent that has multiple binding ***sites*** allowing the agent to "grab" onto the metal ion. As the polymer skeleton of chelating resins is insoluble in acid and alkali aqueous phases as well as organic solvents, enrichment and separation of heavy metals from a wide variety of liquid matrices are very convenient. Chelating resins have been shown to bind with ***labile metals***, i.e., heavy metals that are in a chemically active or easily exchangeable form and thus more likely to be ***bioavailable*** (Bowles et al., 2006). Chelating resins have been used to study ***chemoavailable*** metal ions in various kinds of wastewater, and other complex matrix samples of environmental relevance (Chu et al., 2022).

Analysis of chemoavailable heavy metal ions, especially at trace levels, is a difficult and complicated task since it necessitates the rigorous requirements of versatility, specificity, ***sensitivity***, and ***accuracy*** in the analysis. Chelating resins are attractive for measuring the ***chemoavailability*** of heavy metals, achieving two goals at once: (1) ***preconcentration*** (a term used in chemical analysis that refers to augmenting the concentration to enhance analysis) – to concentrate very low levels of target heavy metal pollutant, and (2) only chemoavailable heavy metal free ions are extracted (Sud, 2012). Advantages of chelating resins include convenience, rapidity, and being economical. Additionally, specific chelating resins have been synthesized as per the need of specific applications, making selective determination possible by using a chelating group (immobilized on the resin) having a high affinity for a given pollutant of interest (Saçmacı et al., 2012). This high affinity results in enrichment of the chemoavailable heavy metal on the resin solid phase.

This chapter provides a detailed description of the chelating resin method for the determination of chemoavailable heavy metals in natural waters, including both freshwater and seawater. An adaptation of this method has been used for sediments and is suitable for water-logged or wetted soils, however, the method of using chelating resins to determine chemoavailable heavy metals in sediments/soils has been superseded by the ***Diffusive Gradients in Thin-Films*** (DGT) method (Chapter 10). Resins prepared for use under this method are suitable for use in DGT ***binding gels***. The procedure as given is applicable to the measurement of chemoavailable Ag, As, Cd, Co, Cr(III), Cu, Ni, Pb, and Zn. Two approaches are given: one using batch experiments and one using column experiments. Chapters containing terms and definitions, guidance on avoiding contamination of samples in lab work, guidance on Quality Assurance (QA)/Quality Control (QC), and ***Standard Methods*** for instrumental analysis are also within the scope of this chapter.

DOI: 10.1201/9781003355229-15

12.2 PRINCIPLE

The mechanism of metal ion ***adsorption*** by chelating resin is that functional groups on the resin react with ***labile*** (heavy) ***metals*** (hereafter assumed to include As, a metalloid) to form a stable structure. Chelating resins can have stronger binding with metal ions due to multidentate binding, and the potential combination of ion exchange with coordinate bonding. These resins may also have higher selectivity due to steric effects in binding resulting from these properties.

In this method, chelating resins having large ***distribution*** coefficients, K_D, of about 10^3–10^5 for the heavy metal(loid) ions Ag, As, Cd, Co, Cr(III), Cu, Ni, Pb, and Zn are used. Although this approach offers instructions for performing both batch and column-type tests with aqueous samples, the batch method is highly recommended due to the extended contact durations it guarantees. The methods for determining chemoavailable heavy metals using the chelating resin method by both batch and column experiments are shown schematically in Figure 12.1. Both batch and column approaches involve exposing a sample to the chelating resin to sorb heavy metals. The chemoavailable heavy metals, after being complexed to and immobilized on the chelating resin, are then eluted off the resin using a suitable eluent. A sample is prepared from the eluent for instrumental quantification of chemoavailable heavy metals (Chu et al., 2022).

12.3 EQUIPMENT, CONSUMABLES, AND REAGENTS

The lists of equipment, consumables, and reagents given in this section is intended to be reasonably comprehensive; users should, however, read the entire method before finalizing needs and modify this indicative list accordingly.

12.3.1 Equipment and Consumables

Note: All equipment and consumables are presumed appropriately clean and ***contaminant***-free. See Chapter 3 for further guidance on avoiding contamination.

Sample collection/preparation-related:

Acid-washed/heavy metal clean plastic (high-density polyethylene or polypropylene) or PTFE (polytetrafluoroethylene) bottles, 250 mL to 1.0 L, with good impact resistance – used for storage of water samples collected in the field. A minimum of 250 mL of water is needed for this method.

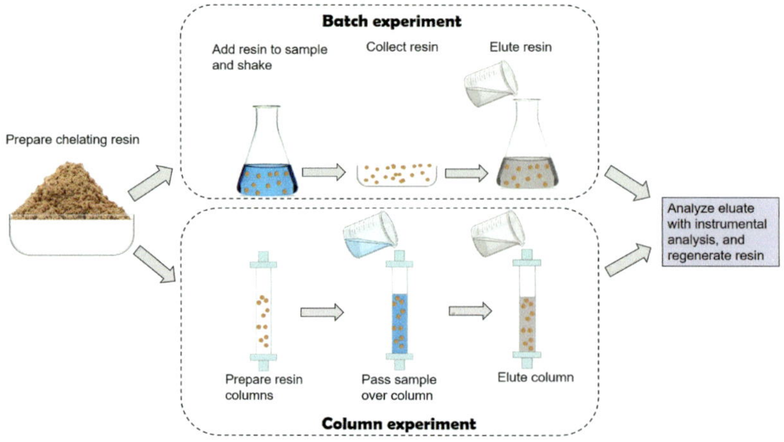

FIGURE 12.1 Schematic diagram of the experimental setup for analyzing chemoavailable heavy metals using chelating resins.

Fiber-reinforced tape – used to seal thermo-insulated boxes used for transport of field samples.

Field logbook – in addition to a sample collection log sheet, field personnel should record information about sample collection activities, field conditions, pH, temperature, and other events that may affect the results of experiments, which will help to assist in the subsequent analysis of the results of the experiment.

Filtration equipment consists of

A. a vacuum filtration device, vacuum, and ***filter membranes***;

OR

B. syringes with syringe filter cartridges.

The filter membranes associated with either option above must be 0.45 μm pore size to remove large particles and/or microorganisms from the sample. Filters composed of polypropylene are often used to avoid loss of heavy metals from samples via adhesion to the filter. In practice, the user should solicit technical specifications for products used. For option A, plastic or PTFE is suitable for use with heavy metals, in order to avoid sample contamination, and vacuum pumps used must also be of a suitable design to avoid sample contamination. For option B, it is important to choose materials that will not contaminate the samples. Generally, high-density polyethylene and polypropylene that offer low reactivity with metals are proper materials for syringes. See Chapter 3 for more details on avoiding sample contamination.

GPS/electronic or physical devices – to ensure unambiguous location of sample sites (e.g., GPS, beacons) and depths (e.g., depth gauge).

Ice/dry ice – ice is used to ensure that the sample temperature is around 4°C during transportation. Ice packs are preferred if they can be kept well frozen prior to use; alternatives include dry ice (subject to certain shipping requirements) or wet ice (which increases the chance of cross-contamination of samples as it melts). When using dry ice, the following points should be noted: dry ice in contact with water sublimates rapidly and generates large amounts of gas, which may result in package breakage or sample damage. Therefore, avoid direct contact between dry ice and water during transportation and ensure that thermo-insulated boxes are dry and free of moisture. During transportation, ventilation should be maintained inside the packing box to prevent carbon dioxide from accumulating. At the same time, excessive pressure or squeezing of the box should be avoided to avoid damaging the packaging material or affecting the insulation effect of dry ice. When handling dry ice, gloves, goggles, and other protective measures should be worn to avoid direct skin contact and eye injury.

Indelible pens – to record relevant data without losing fidelity in wet field environments.

Multisensor sonde – e.g., YSI or similar; used for on-site measurements during water sample collection. Calibration is required before use to ensure the ***precision*** of the instrument. Multiple sensors are desirable, including those capable of measuring temperature, pH, conductivity, and dissolved oxygen (DO).

Niskin-type water collection bottle(s) and line – for the collection of water samples, needs to seal well, be corrosion-resistant, and the bottle itself needs to be clean and free of contamination. Specialized PTFE-lined bottles should be used for heavy metals (e.g., sold by General Oceanics).

Sample identification labels – completed in indelible ink, labels are used for unique identification and shall contain (1) a unique sample identifier (also recorded in the field logbook), (2) location of the sample sampling point, (3) date of collection, and (4) type of sample collected.

Thermo-insulated boxes – clean/uncontaminated and constructed of appropriate material for the pollutants to be analyzed; must have sufficient insulation properties to keep the samples collected at a reasonably constant temperature during handling and transportation.

Laboratory chelating resin system-related:

Aluminum foil (optional) – for some photosensitive metal ions present in samples to be tested, direct sunlight should be avoided to minimize the effects of photochemical reactions. During sample storage, the container may be wrapped with aluminum foil to prevent metal ions in the sample from reacting when exposed to sunlight.

Analytical balance – to accurately measure the mass of chelating resin and other reagents.

Beakers, volumetric flasks, volumetric pipettes – for preparing and handling reagent solutions.

Chelating resin – used for preconcentration of labile heavy metals from water samples; Chelex® 100 (Bio-Rad, for Ag, Cd, Co, Cr(III), Cu, Ni, Pb, and Zn) and/or AmberSep™ M4195 (Dupont, for As).

Columns (when using column method) – gravity columns or low- to medium-pressure sealed columns – for packing chelating resin. A normal column is approximately 0.5–0.7 cm (diameter) × 10–12 cm. Column height after packing should be maximized to ensure good contact times between the sample and the resin packing. Note: The Ca form of Chelex has approximately half the volume of the Na form; columns used with Chelex need to be able to accommodate this level of shrinkage/swelling. Gravity columns need to be longer and have a reservoir in order to load sample.

Glass rod – to stir solutions/dissolve reagents.

Magnetic stirrers and stir bars – used to mix chelating resin with aqueous media; a shaker is preferred for the batch method.

pH meter – to test the pH of solutions.

Plastic or PTFE containers – for storing filtered, and/or preconcentrated and acidified samples.

Refrigerator – used for storing samples.

Peristaltic pump, with sample loading reservoir – for column method – controls the flow of the sample solution into the column for processing; a multichannel pump is preferred to increase throughput.

Sealable flasks or bottles (plastic or PTFE) – for batch experiments, 500 mL or larger, depending on sample size.

Shakers – used to mix samples in batch method.

Timer – used to verify flow rates or batch contact times.

12.3.2 Reagents

Ultrapure reagents are needed for this method. Reagent purity will be reflected in ***procedural blanks*** (*vide infra*).

Ammonium acetate – used for the pre-elution when preconcentrating heavy metals from seawaters or waters with high concentrations of, e.g., Na, Ca, Mg, and Cl.

Heavy metal(loid) standards – for water spiking experiments. Standards need to be buffered for ***bioavailability calibration curves***.

Hydrochloric acid (HCl) – when using AmberSep for As determinations; to clean the resin from potential contaminants.

4-(2-Hydroxyethyl)-1-piperazineethanesulfonic acid (HEPES) – used to buffer samples containing As.

Iron chloride hexahydrate ($FeCl_3{\cdot}6H_2O$) – used to prepare Fe-4195 from AmberSep M4195 resin.

2-(*N*-morpholino)-ethanesulfonic acid (MES) – used to buffer samples containing Ag, Cd, Co, Cr(III), Cu, or Ni.

3-(*N*-morpholino)-2-hydroxypropanesulfonic acid (MOPSO) used to buffer samples containing Zn.

3-(*N*-morpholino)propanesulfonic acid (MOPS) – used to buffer samples containing Pb.
Nitric acid (HNO_3) – used in cleaning, to acidify samples, and to elute chelating resins.
Ultrapure water – also known as Type I water, has a resistance of 18 MΩ·cm.
Sodium hydroxide (NaOH) – used to adjust the pH of chelating resins and elute As from AmberSep resin.

12.4 FACTORS TO CONSIDER IN CHOOSING A CHELATING RESIN

There are different types of chelating resins, which are frequently categorized based on their functional groups, i.e., functional group containing atoms such as O, N, S, and P, with lone pair electrons. Different functional groups have different chelating effects that confer specificity for certain heavy metal ions. The functional groups commonly used for coordinate bonding are incorporated into the polymer matrix as multi-coordinating ***ligands***, i.e., chelating ligands, to form the chelating resin. For a chelating resin to have good performance, the formulation of the resin must also satisfy that the spatial configuration of the chelating groups is free from steric hindrance with regard to metal binding by the chelating groups.

There are many types of chelating resin now available based on different functional groups, classic examples of which include iminodiacetate-based chelating resins (Dragan, 1998), aminophosphonate-based chelating resins (Nesterenko et al., 1999), and picolylamine chelating resins (Liu et al., 2021). Apart from less frequently used chelating resins that incorporate functional groups such as polyamine and others, there exist several selective resin types occasionally employed for heavy metal preconcentration in chemoavailability studies. Selective resins are designed to have a high affinity for specific ions or molecules, but may not necessarily form multiple coordinate bonds. These selective resins include, for example, thiol functional groups and naturally occurring zeolites, among others. Table 12.1 shows some examples of commercially available chelating and selective resins (Kagaya et al., 2014).

Iminodiacetate-, aminophosphonate-, and picolylamine-type chelating resins are preferred over other types of resins because of their high selectivity and capability of binding heavy metals through multiple functional groups attached to the resin support. This property of multiple coordination of heavy metals on the chelating resin makes such a resin particularly suitable for chemoavailable preconcentration of polyvalent heavy metals. Two important additional properties are required for a selective chelating resin: (1) the resin must be usable over a wide range of pH (in practice, usually not attainable), and (2) the cross-linked polymer skeleton of the resin must have high chemical stability and a compact structure. Each chelating resin type has its operational pH range, but for resins' common use, the resin itself is stable in pH ranges from 0 to 14.

12.4.1 What Makes a Chelating Resin Selective?

The following are the factors of primary importance in controlling selective chelation:

Ionic charge (valence) – chelating resins selectively bind to specific heavy metal ions in part based on the charge of the chelating resin's functional groups and the heavy metal ion charge.
Hydrated ionic radius – when the size of the hydrated radius of a heavy metal ion closely fits the space between the functional/binding groups in the chelating resin, the resin is very selective for that particular heavy metal ion.
Ligand bonding/coordination chemistry – nitrogen and oxygen atoms possess exposed electron pairs, rendering their behavior that of Lewis bases (electron pair donors). Metals that are Lewis acids bond with ligands that are Lewis bases. Chelating resins, therefore, typically have functional groups containing nitrogen and/or oxygen, and the structural arrangement of these atoms within the functional group strongly affects ligand bonding/

coordination chemistry, and thus selectivity to a given heavy metal ion. Table 12.2 gives examples of functional groups used in chelating and selective resins, including the functional groups from resins listed in Table 12.1.

TABLE 12.1
Examples of Commercially Available Chelating and Selective Resins

Chelating/Functional Group	Product Name	Supplier/Manufacturer
	True Chelating Resins	
Iminodiacetic acid	Chelex® 100	Bio-Rad Laboratories
	Purolite® S-930	Purolite International
Aminomethylphosphonic acid	Purolite® S-950	Purolite International
	Purolite® S-940	Purolite International
	Duolite C467	DuPont de Nemours Inc.
Bis-picolylamine	AmberSep™ M4195	DuPont
	Puromet® MTS9600	Purolite International
Hydroxypropylpicolylamine	SEPLITE® LSC485	Sunresin New Materials Co.
Polyamine	Diaion CR20	Mitsubishi Chemical
	Puromet™ MTS9850	Purolite International
Pyridine	REILLEX™ 425 Polymer	Vertellus Specialties
	Selective Resins	
Thiol	AmberSep™ GT74	Dupont
Thiourea	Lewatit® MonoPlus TP 214	Lanxess
Thiouronium	AmberSep™ 43600	Dupont
	Purolite® S-920	Purolite International
Natural zeolite	Clinoptilolite	Zeotech Corporation
	Mordenite	BASF SE
Carboxylic acid	Lewatit® CNP	Lanxess

TABLE 12.2
Examples of Resin Functional Groups

Chelating Resin	Functional Group
Amide type	$-CONH_2$
Aminocarboxylic acid type	$-CH(NH_2)(COOH)$
Aminomethylphosphonic	$-NHCH_2PO(OH)_2$
Carboxylic acid type	$-COOH$
Dithiocarboxylic acid type	$-C(=S)SH$
Iminodiacetic acid type	$-N(CH_2COOH)_2$
Phosphate type	$-PO(OH)_2$
Picolylamine	$-NH\ CH_2C_5H_4N$
Polyamine type	$-[(CH_2)_nNH_2]_m$
Pyridine	$-C_5H_4N$
Schiff base	$-C=N-$
Thiol type	$-SH$
Thiourea type	$-NHC(=S)NH_2$
Thiouronium	$-SC(=NH)NH_2$

12.4.2 Specifics of Commonly Used Chelating Resins

12.4.2.1 Iminodiacetate-Based

Aminocarboxylic acid-based chelating resins have become the most widely used chelating resins due to their strong bonding force with metal ions, large exchange and adsorption capacity, and high selectivity for divalent ions. Among them, the iminodiacetate chelating resin, which is a typical chelating resin with O and N as the mixed coordination atoms, has been studied the most (Biesuz et al., 1998); the resin is formulated with a cross-linked polymer skeleton (e.g., polystyrene or polyacrylic acid) and the iminodiacetate functional group, $-N(CH_2COOH)_2$ and is selective for a wide array of divalent metal ions with high selectivity and good adsorption performance (Botelho et al., 2019). This type of chelating resin is used widely in heavy metal ion separation, analysis, enrichment, and removal from industrial wastewater. Figure 12.2 shows how an iminodiacetate functional group chelates with calcium (Ca) (Chen et al., 2006).

The main factors affecting the adsorption performance of iminodiacetate chelating resins toward heavy metal ions are pH, temperature, reaction time, and the initial concentration of heavy metal ions. The pH value of the solution has the greatest influence on the amount of metal ions adsorbed. With the increase of pH value, adsorption first increases and then decreases. At low pH, the acetate groups are protonated, leading to poorer chelation. At higher pHs, metal hydroxides form, and chelation will be reduced (e.g., see Leinonen and Lehto, 2000), The commonly used Chelex-100 iminodiacetate resin can nominally work properly in a pH range from 4 to 14, however, at higher pHs (pH ~8, depending on the sample), the association of metal hydroxides with the resin can produce artifactual results. Chelation increases with temperature, however, the high temperature damages the mechanical properties and thermal stability of the resin. This is an issue for industrial applications, as environmental studies are usually conducted close to ~25°C. The resin's adsorption rate of heavy metals is highest on initial contact, and as the resin gradually reaches saturation, adsorption becomes gradually slower until a maximum adsorption is reached. The larger the initial concentration of metal, the more metal ions chelate to adsorption sites, due to a higher ***concentration gradient***-induced diffusion rate. In addition to these primary factors governing ***sorption***, some other factors have an influence; these factors include, but are not limited to, the particle size distribution of the resin, degree of cross-linking, porosity, pore size, and the density of iminodiacetate groups in the resin.

Chelex-100 is the most commonly used iminodiacetate chelating resin for environmental applications. With reference to bioavailability studies, it has been used to distinguish "labile" versus "inert" metal species in aqueous settings according to operational definitions. Heavy metals that are retained by the resin on ***exposure*** to a sample represent labile/chemoavailable forms.

12.4.2.2 Aminophosphonate-Based

The most prevalent phosphorous groups found in commercial resins are phosphonic and aminophosphonic groups. The availability of polydentate ligand sites to improve the resin's selectivity is one of the primary benefits of aminophosphonic resins. Some results indicate that even at low solution pH, metal adsorption is made possible by the large number of ligand sites (Silva et al., 2018).

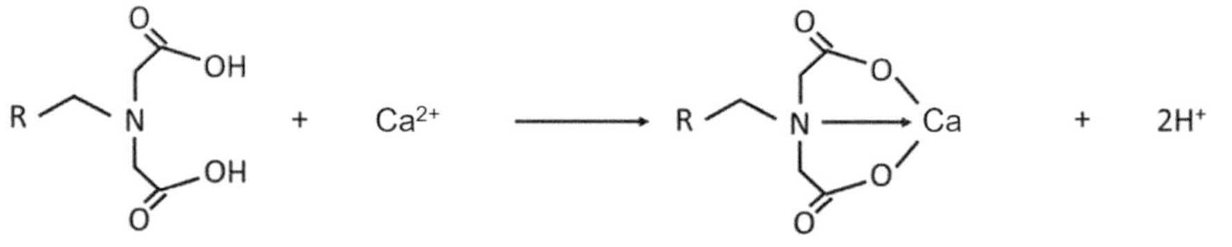

FIGURE 12.2 Reaction showing an iminodiacetate functional group chelating with calcium. Ion exchange occurs between the Ca^{2+} and the two H^+ of the carboxylic acid groups. The imino-N forms a coordinate bond with Ca^{2+}.

In these resins, the two exchangeable ions in the phosphonate groups of aminophosphonate resins, typically Na^+, can be ion-exchanged with a divalent heavy metal ion, and the heavy metal forms a coordinate bond with the nitrogen in the amine, resulting in the formation of a structurally stable ring (Sud, 2012). Figure 12.3 shows how an aminomethylphosphonic functional group chelates with a divalent heavy metal ion.

12.4.2.3 Picolylamine-Based

Pyridine-based chelating resins, including picolylamine and bis-picolylamine, have significantly superior removal ability for heavy metal ions under strongly acidic conditions. This superior performance at lower pHs is primarily because (1) the acid dissociation constant of the pyridine nitrogen is comparatively low, so it deprotonates at lower pHs than many other chelating resin functional groups, and (2) the coordination of heavy metals with the pyridine nitrogen is not only a ligand-to-metal coordination bond, but has π-bond metal-ligand feedback, which increases the stability of the chelate (Pal, 2018; Wołowicz and Hubicki, 2012). As a result of these properties, picolylamine-based resins exhibit greater metal selectivity in acid conditions than do iminodiacetic groups and quaternary amines (Diniz et al., 2000). Commercial resins with picolylamine groups are utilized in detecting metal bioavailability (Van Deventer, 2011). Effective in acid conditions, the selectivity of these resins is reduced at more alkaline solutions above a pH of about 8.0. Figure 12.4 shows how the bis-picolylamine functional group chelates with Cu^{2+}.

12.5 PROCEDURE

12.5.1 Collection, Transport, and Handling of Water Samples

When collecting water samples in the field, it is important to use properly cleaned, contaminant-free materials that are appropriate for heavy metals work (Chapter 3), ensuring that the materials are both clean and free of contaminants.

After identifying suitable sampling locations, water samples are collected at each location from depths of 0.2 m or more below the water surface using a Niskin bottle sampler on a line. The method calls for a volume of 250 mL of water for each sample or replicate. For every ten samples collected, a minimum of two field replicates (duplicates) are required, i.e., requiring collection of 500 mL or more in total, in two aliquots. In practice, less might be needed for more polluted waters, and more might be needed for some environments such as open ocean samples. ***Equipment rinsate*** blanks

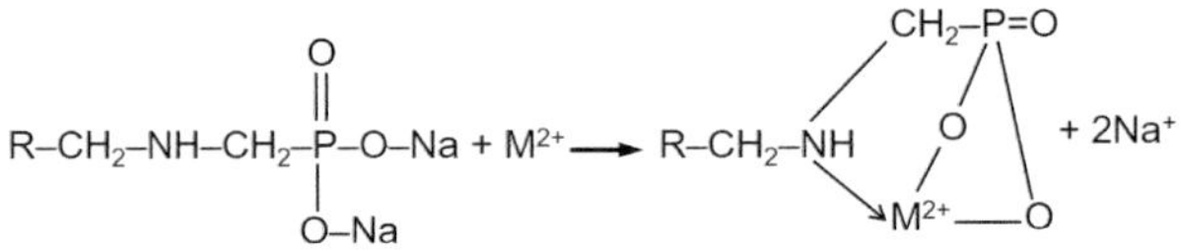

FIGURE 12.3 Reaction showing an aminomethylphosphonic functional group chelating with a bivalent metal. The two monovalent ions on the phosphonate group ion exchange with the heavy metal, while the amino-N forms a coordination bond with the heavy metal.

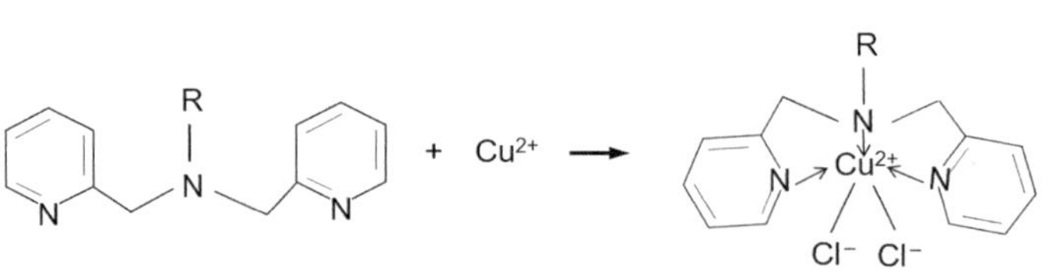

FIGURE 12.4 Reaction showing a bis-picolylamine functional group chelating with Cu^{2+}. For this chelation reaction, all bonds are coordinate. Because this coordinate bonding is not charge-neutral, ion pairs may form with the chelate (chelate-Cl^- on pairs shown).

must also be collected in order to confirm that samples are free of field contamination. Equipment should be rinsed and decontaminated after collection at each sampling locale, at which time the rinsate blank is collected. A Secchi disk is used to test the transparency of the water, while a multisensor sonde is employed for on-site measurement of water temperature, pH, conductivity, and DO levels. These field measurements of water quality are standard, though additional determinands may be added as required by experimental objectives. Water samples are transferred from the Niskin bottle into clean sample containers that are labeled in a clear and sequential manner to ensure unambiguous identification, and then transported to the laboratory at a temperature of 4°C within a 6-hour time frame to minimize post-sampling alterations in bioavailability. To confirm the lack of field contamination in samples, it is essential to decontaminate sampling equipment between samples and to collect equipment rinsate blanks (refer to Chapter 4). All field samples are maintained at 4°C in a light-free environment upon arrival at the laboratory (Zhang et al., 2020).

Water samples may be optionally pretreated prior to preconcentration. The usual procedure to obtain the operationally defined dissolved fraction is to filter the water sample through a 0.2 or 0.45 µm filter. If filtration will be performed, it should be completed no more than 8 hours after sample collection (Reed et al., 2023). Large syringes with syringe filters can be used, or alternately a suction filtration apparatus is used. It is important to ***source*** filters that are both contaminant-free and heavy metal-compatible, i.e., filters that will not sorb the heavy metal ***components***, *or associated metal–organic complexes*, in the water samples. Users should consult the supplier if uncertain. After filtration samples should be stored in a refrigerator at 4°C. The US EPA recommends that filtered samples used for biologically relevant measurements should not be stored for more than 96 hours, which includes time of collection (US EPA, 2016). Some investigators freeze samples at –20°C for longer storage, however, this is not best practice for bioavailability measurements.

12.5.2 Notes on Types of Experiments and Precautions

In lability studies, the contact time of the sample with the resin is an important factor in being able to preconcentrate labile pollutants from the sample. While this method provides details for conducting both batch- and column-type experiments using aqueous samples, the batch method is strongly preferred because it ensures long contact times.

Sample pH is also important because of the way in which chelating resins work. At low pHs, many heavy metals are highly bioavailable, however, the form of functional groups on the chelating resin does not support pollutant retention. At higher pHs, metal hydroxides may form, complicating resin interactions. For Chelex-100, the manufacturer cites a useful pH range of 4–8. This range is reasonable for batch studies with high contact times, however, results of many investigations indicate that for column work a pH of less than 4.5 is untenable, and for some heavy metals a minimum pH of 6.5 is preferred (Bowles et al., 2006). For AmberSep M4195, a pH range of 4–11 is reportedly possible (Chanda et al., 1988). For various reasons, batch experiments are preferred for studies when the detection of labile Ni or As are a goal.

In order to prevent contamination, all sampling containers and experimental vessels should be made of high-purity PTFE, or plastic. Before use, the vessels should be soaked in 20% HNO_3 solution for a minimum of 12 hours, rinsed with Type I water a minimum of three times, and dried under clean conditions. The acids used should be high-purity grade. Specific to this chapter, all sample handling in the laboratory must be performed under clean-room or clean-room equivalent conditions, as assessed by blanks, described in Chapter 4 on QA/QC and in Section 12.6.

12.5.3 Preparing Chelating Resin and Resin Columns

12.5.3.1 Chelex-100 Resin (for Analysis of Ag, Cd, Co, Cr(III), Cu, Ni, Pb, and Zn)

Soak Chelex-100 resin (sodium form) in Type I water overnight. After soaking, decant excess water to obtain a stiff slurry. To clean the resin from potential contaminants, for every gram of resin in the sodium form (~1.5 mL), add an equal volume of 2.5 M ultrapure HNO_3 (diluted in Type I water) and place on a shaker for 5 minutes at 100 rpm (more Type I water can be added to facilitate shaking; alternately, the resin may be processed in a covered beaker on a magnetic stirrer with stir bar if a shaker is not available). Repeat this process twice, decanting the acid between each rinse. After the acid washing is complete, repeat the process of rinsing and shaking three times using Type I water (copious amounts of water may be used) and then repeat three times again using 1.0 M ultrapure NaOH (volume of base as for acid, more water may be added to facilitate shaking). After the last base rinse is decanted, rinse the resin multiple times in Type I water, decanting each time.

The form of the resin is significant, as various changes in the ionic composition of solutions in contact with the resin can cause shrinkage or swelling, and leaving the resin in acid, or H^+, form can be deleterious to the resin. For Chelex-100, using the Ca form has several benefits in terms of its selectivity for heavy metals (the resin has low selectivity for Ca) and stability. To convert the resin to Ca form, add 2 M ultrapure $CaCl_2$ solution to the resin at twice the volume as was used for the acid (as above; note: $Ca(NO_3)_2$ may also be used). Shake at 100 rpm for 5 minutes. Decant solution and repeat the process twice more. Ensure the resin is well drained. The resin is now in Ca form and can be stored moist in a sealed container until use (Bowles et al., 2006).

If the samples to be run are in brackish or seawater, the resin is conditioned using ultrapure artificial seawater. In this case, instead of adding $CaCl_2$ as directed above, add ultrapure artificial seawater to the resin and shake for 20 minutes, then decant. Repeat this once. Ensure the resin is well drained. The resin is now ready to use for seawater experiments and can be stored moist in a sealed container until use.

12.5.3.2 AmberSep M4195 Resin (for Analysis of As)

AmberSep M4195 resin is normally available in 100 g bottles. This method assumes that 100 g is being processed. For larger or smaller quantities, this method should be scaled. Soak AmberSep M4195 resin in 250 mL Type I water overnight. After soaking, decant excess water to obtain a stiff slurry. To clean the resin from potential contaminants, place resin in a flask and add 250 mL 1 M ultrapure HCl (diluted in Type I water) to the flask to facilitate shaking. Place on a shaker for 20 minutes at 100 rpm (or use a covered beaker on a magnetic stirrer with stir bar if a shaker is not available). Decant the acid and add 250 mL of 1 M ultrapure NaOH, shake for 20 minutes at 100 rpm. After this, decant the base and add 250 mL Type I water, then shake for 20 minutes at 100 rpm. Repeat the Type I water rinse twice, shaking for 10 minutes each.

The cleaned resin must now be iron loaded. To iron load, first rinse the resin twice with 250 mL of 2 M HCl for 10 minutes at 100 rpm. Next, rinse multiple times with 250 mL Type I water at 100 rpm until the pH of the effluent after shaking is pH >2. After the pH of the Type I water rinsate rises above 2, decant the water and add 250 mL of a 0.09 M $FeCl_3{\cdot}6H_2O$ solution at pH = 2. Shake this for 10 minutes at 100 rpm, then decant and add another 250 mL of a 0.09 M $FeCl_3{\cdot}6H_2O$ solution (pH = 2). Shake this for 20 hours at 100 rpm. This step will complete the iron loading, however, the resin requires further conditioning and removal of excess iron. After 20 hours, rinse the resin twice with a solution of ultrapure HNO_3 in Type I water; the concentration of HNO_3 should be just enough for the solution to have a pH = 2 (i.e., ~0.01 M). For each rinse, add 250 mL to the resin and shake for 10 minutes at 100 rpm. Next, rinse the resin with a Type I water (250 mL shaking for 10 minutes) until the rinsate pH approaches that of the Type I water. This may require many rinses. Ensure the resin is well drained. The resulting resin is henceforth referred to as Fe-4195 and can be stored moist in a sealed container until use (Lake, 2002).

If the samples to be run are in brackish or seawater, add 250 mL of ultrapure artificial seawater to the resin and shake for 20 minutes, then decant. Repeat this once. Ensure the resin is well drained. The resin is now ready to use for seawater experiments and can be stored moist in a sealed container until use.

12.5.3.3 Amount of Resin to Use, Column Types, and Loading Resin Columns

The amount of resin to use depends on the capacity of the resin and the concentration of bioavailable species in samples. This method recommends the use of 2.0 g of resin for a 250 mL water sample, whether the test approach is batch or column. For many environmental applications, this may be excessive, however, will ensure a suitable margin of safety. Using excess resin in column work is especially important as it increases the contact time between labile pollutants in a sample and the resin.

Columns should be of the type used for low- to medium-pressure chromatography. Gravity columns operate at low pressure, and sealed columns that will be used with a peristaltic pump operate at low to medium pressure. Figure 12.5 shows the two experimental configurations. For use of gravity columns, a piece of PTFE spacer or frit is placed in the bottom of the column, or, alternately, some gravity columns come with a frit at the bottom. The resin is loaded on top of this, and there must be a large head space above the top of the resin in which to load sample. The column has a stopcock to regulate flow, and the weight of sample on top of the resin, along with adjustment of the stopcock, enables regulation of the flow to a desired level, which must be measured using Type

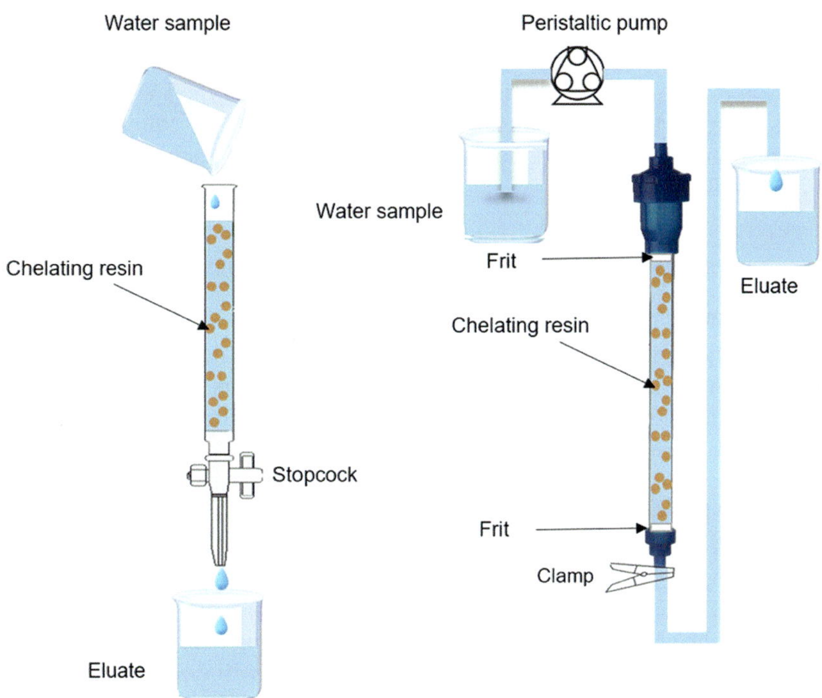

FIGURE 12.5 Schematic diagram showing the two experimental configurations utilized in this method: using a gravity column (left) and low-/medium-pressure column (right). The gravity column is loaded with a beaker; it will have a head of water/sample above the resin bed, which will flow through the column and drip into a second beaker. The flow rate is controlled by a stopcock. For the low-/medium-pressure column, there will be a sealing inlet with a filter or frit on both ends of the column holding the resin bed in place. The column will be full of water/sample, pumped in using a peristaltic pump through/to a tube to the top inlet. The flow is controlled by the pump, which controls the exit of eluate by a tube in the outlet. The container collecting eluate flowing out of the column must be at a higher height than the top of the resin bed; otherwise, water may drain out of the column, causing the column to dry out or channel. Alternately, a flow restrictor can be placed at the outlet tube of the column. In practice, a simple pinch clamp, appropriately adjusted to accommodate the set flow rate, is a way to accomplish this.

I water (alternately artificial seawater) prior to column use. The problem with this approach is that the flow will vary somewhat during use and is difficult to regulate precisely. For this reason, use of low- to medium-pressure columns that have a cap at each end accommodating tubing hooked up to a peristaltic pump is preferable (though more expensive). These columns also have, e.g., PTFE inserts or frits at the column outlet to contain resin. Often, if the amount of resin used does not fill the column, a contaminant-free insert or frit is placed above the top of the resin (inlet side of column), to hold it in place, and the column is operated in a vertical position. Some providers offer columns with adjustable plungers to secure the resin. The size of the column should accommodate the resin. It is better to have a long, thin column with a tall resin bed than vice versa.

The total weight of resin processed in the preparation phase should have been measured, according to the directions above. This weight should be very similar after processing if the resin has been appropriately drained/dewatered. In this case, for batch experiments the amount of resin can be weighed directly. For column experiments, measure the total volume that corresponds to this weight after processing. With the total volume and total weight, an approximate density is calculated. Dividing 2.0 g by this approximate density provides the amount of volume of resin to be used per column. To load the resin into columns, place the needed amount of resin in a beaker and add Type I water (or artificial seawater for experiments on marine or brackish samples). Agitate the slurry and pour the resin slurry directly using a contaminant-free funnel. Allow the excess water to drain out of the column, but do not allow the resin to go dry. Ensure the resin is evenly distributed along the length of the portion of column that it fills. Use a wash bottle with Type I water/artificial seawater to rinse any excess resin particles from the beaker or from the top of the column to the resin bed in the column. The resin bed should have a homogenous distribution of resin throughout, with the top of the resin bed being flat.

12.5.4 Preconcentration of Heavy Metals and Arsenic in Samples

12.5.4.1 Batch Experiments

Add 2.0 g of resin or the equivalent volume to sealable flasks or bottles. Add 250 mL of sample to each flask, seal, and then place on a shaker at 100 rpm for 24 hours (alternately magnetic stirrer with stir bar). Carefully decant resin from each flask/bottle, and then add 250 mL Type I water (or artificial seawater) and shake at 100 rpm for 10 minutes, decanting water afterward.

12.5.4.2 Column Experiments

After columns are prepared, for each column, slowly pass 250 mL of sample through the column using a peristaltic pump or gravity flow. Begin at a very slow flow rate (capabilities for how slow a flow rate can be used vary by equipment). Samples may change the degree of swelling of the resin, and, if the resin shrinks, channeling may occur. If channeling does occur, the integrity of the measurement is compromised. Starting with a very low flow rate will typically prevent channeling. After two column volumes have passed through the column at a low flow rate, increase the flow rate to 4 mL/min. After the sample has passed through the column, wash the column with 25 mL of Type I water (alternately seawater) to remove any unbound sample residues.

12.5.4.3 Pre-elution and Elution of Labile Pollutants

Pre-elution is needed when preconcentrating heavy metals from seawaters or waters with high concentrations of, e.g., Na, Ca, Mg, and Cl, which may in some circumstances be at high enough concentrations in eluent to interfere with instrumental analysis (Stockdale, 2005). Pre-elution is not needed otherwise. If pre-elution is required, add 40 mL of 1 M ammonium acetate to batch Chelex-100 resins, shake for 5 minutes at 100 rpm, decant, and then rinse with 10 mL of Type I water and decant. The rinsing contact time with Type I water should be minimized. Alternately, for Chelex-100 columns, pass 10 mL of 1 M ammonium acetate over the column at a flow of 5 mL/min, followed by a single aliquot of 2 mL Type I water. Pre-elution should not be used for Fe-4195.

For Chelex-100 resin, elute absorbed heavy metals by using 2 M HNO_3. For batch experiments, add 10 mL 2 M HNO_3 and agitate with a stir bar. Allow to stand for 1 hour, agitating occasionally, then extract the eluate from the resin with a contaminant-free pipette or needle and syringe placed through the resin and into the lowest point of the flask. Repeat once, combining the two extracted eluate fractions into one sample. For columns, perform a step-wise elution using seven 1 mL aliquots of 2 M HNO_3, allowing each aliquot to drain from the column before adding the next and collecting all eluent into one sample.

To remove As from Fe-4195, add 10 mL of 1 M NaOH to the resin, cover, and shake for 24 hours at 100 rpm. Extract eluate from the resin as for Chelex-100 above, and repeat, combining both fractions into one sample. For experiments using Fe-4195 to determine labile As species, column experiments are not recommended due to longer contact times being preferable to achieve good elution efficiency (Lake, 2002).

The acid eluates for Chelex-100 are stable for a minimum of 6 months, however, the basic eluates from Fe-4195 must be acidified. To minimize volume changes, 16 M stock HNO_3 should be used. Measure the total volume of eluate, or, alternately, if the density of NaOH at the ambient temperature is known, the volume may be obtained by weighing and converting to volume using density. The moles of HNO_3 needed to neutralize the eluate to ~ pH 2 or less is

$$\text{mol}_{\text{nitric}} = 1\text{ M NaOH } \times V_{\text{eluate}} + 0.0001,$$

where the first term is the amount of acid needed to neutralize the NaOH (in mol), V_{eluate} is the volume of basic eluate collected (in L), and 0.0001 is an added amount of acid needed to make the solution acidic (in mol). The volume of 16 M HNO_3 to add, in mL, is $V_{\text{nitric}} = 1{,}000 \times (\text{mol}_{\text{nitric}}/16)$. For most samples, the amount of concentrated nitric added will be ~0.5–2 mL. The acid must be added slowly and carefully to avoid violent reaction. After acidification, the pH should be checked, and more acid added if needed. Acidified samples are stable for a minimum of 6 months. Sample digestion is not needed, and instrumental analysis is normally conducted using Inductively coupled plasma–mass spectrometry (ICP–MS), however, interference from Na in eluent will need to be addressed, and other instrumental methods are possible. Some Standard Methods for instrumental analysis are given in Chapter 5.

12.5.5 Resin Reuse/Regeneration

For reuse, batch Chelex-100 resin must be rinsed with copious amounts of Type I water after each use until the pH approaches that of the Type I water. If seawater samples are run, the resin needs to be preconditioned with artificial seawater again, as above. Regeneration should be performed after every use if water samples were not filtered and after every ten samples for natural water samples that were filtered. Regeneration for batch resin is the same process as for preparation.

For reuse of Chelex-100 columns, after elution Type I water must be run through the column until the pH reaches a value near that of Type I water. If seawater samples are used, the column must be reconditioned with artificial seawater. To regenerate Chelex columns, pump 5 mL of 2.5 M HNO_3 through the column followed by 20 mL of Type I water, and then 5 mL of 1.0 M NaOH. After the NaOH, the resin is again in sodium form and must be converted to the calcium form by pumping 20 mL of 2 M $CaCl_2$ solution through the column. During this regeneration process, the resin will change volume by as much as a factor of 2. This may cause significant pressure on the column, affecting the flow rate that can be used (4 mL/min or less) and may cause the column to break. Slower flow rates are advantageous to manage these volume changes. If the resin bed develops channels (channeling) or unevenness, this will cause major problems with the analysis. To avoid this, users may prefer to empty the resin out of columns and regenerate it using a batch method. In this case, the method of resin preparation may be used.

Fe-4195 must be regenerated. Regeneration calls for the dissolution of any ferric hydroxide that may have formed on the resin and recharging Fe^{3+} ions into the resin. Treatment with 10% $FeC1_3$ solution acidified to pH 0.8 has been shown to result in 75% removal of ferric hydroxide gel from the resin pores and also in nearly full restoration of the active Fe^{3+} content. The residual ferric hydroxide itself has some affinity for arsenic species; however, this is several-fold less than that of the resin's polymer-chelated Fe in Fe-4195. In practice, the sorption capacity of Fe-4195 regenerated in the manner described above increases over approximately 5–6 regeneration cycles, decreasing thereafter due to a residual ferric hydroxide gel buildup (Chanda et al., 1988). For the "one-step" regeneration, the process is similar to resin preparation. Place 100 g of Fe-4195 in a flask containing 250 mL of Type I water and shake at 100 rpm overnight. Decant excess water, and add 250 mL of 10% $FeCl_3$ solution acidified to pH 0.8 (in Type I water) and shake for 20 hours at 100 rpm (Lake, 2002). After 20 hours, rinse the resin twice with a solution of ultrapure HNO_3 in Type I water. The concentration of HNO_3 should be just enough for the solution to have a pH = 2 (i.e., ~0.01 M). For each rinse, add 250 mL to the resin and shake for 10 minutes at 100 rpm. Next, rinse the resin with a Type I water (250 mL shaking for 10 minutes) until the rinsate pH approaches that of the Type I water. When this criterion is met, ensure the resin is well drained; it can be stored moist in a sealed container until use. If the samples to be run are in brackish or seawater, condition with artificial seawater prior to use as in the preparation step. After five of these one-step regenerations, the resin must be subjected to full regeneration, i.e., it is treated as for initial preparation.

12.5.6 Preparation of Heavy Metal Standards and Bioavailability Calibration Curves

In instrumental analysis, standard solutions with known concentrations of heavy metals are used to establish calibration curves, ensuring precise quantitative measurements of heavy metals in samples. To guarantee accuracy, these standards should be sourced from reputable commercial suppliers. Typically, the standards for heavy metal analysis are prepared in acidic solutions to maintain the stability of the metals, and for optimal performance in this method, they are ideally at high concentrations (e.g. 1,000 mg/L). In general, HNO_3 is used to dilute the high concentration of heavy metal standard solutions into solutions of different concentration gradients. Metal concentrations can be measured using Standard Methods in instrumental analysis (refer to Chapter 5).

In addition to calibration standards for instrumental analysis, in recent years, some investigators have begun to use standards to construct bioavailability calibration curves. For some studies, it may be an interesting option to consider adding this. The bioavailability calibration curve represents approximately 100% bioavailability without the presence of an environmental matrix (including the effect of environmental factors, such as pH), however, this 100% bioavailable pollutant may be less than the amount that would be measured by an instrumental standard calibration curve. For instance, instrumental standard calibration curves for heavy metals are based on standards that are acidified to ensure the stability of the metal ions in solution, such that heavy metals are in their most labile (i.e., biologically speaking, bioavailable) form, whereas bioavailability calibration curves use standards at environmentally relevant pHs, at which some portion of metal may be in a form that is not available (e.g., oxyhydroxide form).

Using a bioavailability calibration curve to quantify bioavailability in samples shows how environmental influences in the sample matrix affect the bioavailability of heavy metals in the sample. More detail on how to prepare the bioavailability calibration curve can be found in Chapter 9.

The bioavailability of heavy metals is strongly dependent on pH. When quantifying bioavailability, one must choose between measuring samples at the ambient sample pH (which may yield a relative result if it differs from the bioavailability calibration curve) and adjusting the sample pH to a fixed value close to the pH used in the bioavailability calibration curve. If a buffer is used to achieve a fixed sample pH, buffering must align with the pH used in the bioavailability calibration curve (see Chapter 9, Section 9.5.3). The impact of pH on speciation varies significantly for different metals. Buffers that are considered acceptable, with "no binding" or interference in heavy metal bioavailability, include, e.g., MES (at pH 6.5) for Ag, Cd, Co, Cr(III), Cu, or Ni; MOPS (at pH 7) for Pb; MOPSO (at pH 6.5) for Zn; and HEPES (at pH 8.0) for As.

After establishing the fixed pH to be utilized, the necessary buffering capacity must be ascertained, per Chapter 9, Section 9.5.6.

12.6 VALIDITY OF RESULTS, DATA ANALYSIS, INTERPRETATION, AND REPORTING

12.6.1 Notes on QA/QC Planning Requirements

Chapter 4 provides guidance that is applicable to this chapter. For chelating resin, QA/QC specific requirements are as follows:

- Users performing this method must perform an Initial Demonstration of Proficiency (IDP), including laboratory tests as described below. This IDP includes procedures for instrumental methods provided in individual ***Standard Operating Procedures*** for instrumental analysis. The IDP process described below is designed to additionally provide information useful for field operation planning.
- Routine instrumental calibration and ***calibration blank*** checks are required; guidelines for ***Acceptance Criteria*** for these are stipulated in appropriate Standard Methods used for instrumental analysis. Frequent calibration blank and standard checks are required to ensure absence of instrument drift. Drift affects both precision and accuracy.
- Samples that are analyzed as having concentrations at or near the upper limit of instrument calibration must be reprocessed and reanalyzed. This involves diluting the field sample, reprocessing through the chelating resin, and reanalysis. This process ensures that the chelating resin preconcentration efficacy was not overloaded by high bioavailability in the original sample.
- In addition to standards necessary for instrumental analysis (described in Standard Methods for same), standards may be prepared to construct a bioavailability calibration curve, as discussed in Section 12.5.6. Absolute concentration is assessed via instrument samples; equivalent bioavailability is assessed using the bioavailability calibration curve. This process is optional.
- Batch sizes for instrumental analysis shall not exceed 20 samples; samples batched include chelating resin processed field samples and replicate samples; QA/QC samples are additional and should not exceed 5 per batch. Instrument check samples assess calibration stability and are not part of batching.
- Sample order in batches is to be randomized over all samples and sample types.
- For every batch, one procedural blank is required. This enables evaluation of whether contamination from the chelation resin or other sources occurred.
- For every batch, one randomly chosen equipment rinsate blank, OR one equipment rinsate blank per ***duplicate/replicate sample***, is required, whichever is smaller. These samples enable evaluation of whether field contamination occurred.
- A field duplicate is required for one in every ten field samples collected.
- After IDP, for every 30 samples processed, three should be laboratory replicates, i.e., three independent extractions of the same sample.
- For every group of samples subject to this method in the same group (e.g., same day if samples are processed once per day), one procedural blank and three independently prepared labile metal samples are required. These samples are analyzed and evaluated as described in Section 12.6.1.1.

- One ***recovery spike*** is required per batch. This enables evaluation of whether artifactual contamination (recoveries significantly above 100%) or analyte loss (recoveries significantly below 100%) is occurring.
- One recovery ***spike*** duplicate per instrument run (i.e., assumes multiple batches run sequentially), OR one recovery spike duplicate every 12 hours, is required, whichever is more frequent.

12.6.1.1 Initial Demonstration of Proficiency and Handling of Standards

IDP for this method follows the approach to QA/QC taken by Bowles et al. (2006). The process consists of three measurements: (1) a procedural blank, (2) a labile metal test, and (3) precision assessment, the latter two of which are repeated under two different buffering conditions. The procedural blank is a 250 mL portion of Type I water that is processed exactly as a sample. The labile metal sample concentration must be determined according to a value at or near the middle of the instrumental standard curve and buffered according to comments in Section 12.5.6. If instrumental analysis will be conducted using ICP–MS, the concentration might be, indicatively, 75 µg/L for Ag, 50 µg/L for As, 25 µg/L for Cd, 250 µg/L for Co, 250 µg/L for Cr(III), 250 µg/L for Cu, 250 µg/L for Ni, 75 µg/L for Pb, or 250 µg/L for Zn. To make the labile metal solutions, sequentially dilute 1,000 ppm standards with buffer at 1:10 using a volumetric pipette and volumetric flask (four dilutions). Dilutions should be made with buffer because heavy metal(loid) standards are acidified; the final pH of the buffered labile metal sample should be checked/verified. Labile heavy metal(loid) solutions may be unstable and should therefore be used within the same day during which prepared. The labile metal sample is also used for precision assessment. For each IDP sample (procedural blank or labile metal), three independent replicates are required, i.e., three independently prepared samples (not three aliquots of one sample).

Results for the procedural blank should be below the limit of detection (LOD). If the procedural blank is above the LOD, then there is a contamination problem. The labile metal sample, once buffered, will give a measured labile metal(loid) concentration of considerably less than if unbuffered/acidified. Results from the labile metal sample should be in good agreement with predictions from speciation modeling, for instance using Visual Minteq. In any case, the results should be consistently repeatable, and the precision for triplicate independent samples of labile metal, as relative standard deviation (RSD), should be 10% or less and certainly no more than 15%. Failure to meet the Acceptance Criteria for the labile metal test and precision assessment is an indication of insufficient operator proficiency; additional practice and up-skilling in the method is needed until IDP can be achieved; otherwise, results cannot be viewed as reliable from a QA/QC standpoint.

12.6.2 Data Analysis, Interpretation of QA/QC Results (Acceptance Criteria), Corrective Action, and Reporting

Generalized guidelines for the QA/QC requirements of this method are as follows:

- Dilution factors (DFs) should never be more than a factor of 10 and must always be accounted for in calculations. The basic dilution formula is $C_1V_1 = C_2V_2$. The dilution factor is given by $DF = V_2/V_1$.
- ***Calibration checks*** and metrics shall be performed as specified in a Standard Method of instrumental analysis applicable to the pollutant under study; ***Corrective Action*** must be taken in response to any irregularities of the standard conditions of calibration.
- Precision for laboratory replicates – the instrumental method chosen will specify needs for analysis of duplicate or replicate instrumental analysis of processed samples. For this method, periodically, independent triplicate extractions on the same sample should be conducted and also analyzed for precision assessment. The results are reported as RSD, where

RSD = stdev × 100%/average. The RSDs for laboratory replicates using this method should be below 15%. Poor precision for replicates of lab samples is generally indicative of an operator or instrument error, and Corrective Action involves rectifying the problem prior to analysis and reporting of results from bioavailability studies.

- Precision for ***field duplicates*** – this should be calculated using relative percent difference (RPD): RPD = $200\% \times \text{abs}(R_1 - R_2)/(R_1 + R_2)$, where R represents a replicate, and subscripts 1 and 2 represent the two duplicate samples. Poor precision for field duplicates may be a measure of high environmental heterogeneity of sampled media, and RPDs for field duplicates of aqueous samples should normally be on the order of 15% or below as an acceptance criterion for heavy metals. If replicate instrumental analysis precision is good and analysis of blanks indicates absence of contamination, then poor precision in field replicates for water samples is likely due to procedural errors. Report RPD values and any discrepancies and provide context on how environmental heterogeneity impacts the results.
- Equipment and procedural blanks – results for blanks must be calculated and reported in units of absolute concentration (not bioavailable) for all blank samples. The acceptance criterion for blanks is a value below the LOD. Blanks above the LOD are indicative of field contamination (***equipment blanks***) or laboratory contamination (procedural blanks). The presence of contamination, particularly random and varying contamination, is indicative of procedural errors and compromises the validity of results. Contamination discovered at this stage cannot be reversed, and correction by "subtracting" the contamination can rarely be performed with confidence. Thus, there is no Corrective Action possible, so efforts should be directed toward ensuring the ability to avoid contamination during the IDP process.
- Recovery spikes and recovery spikes duplicates – These samples and their method of preparation are discussed in more detail in Chapter 4. For both types of spikes, calculate recovery as follows: Recovery (%) = $100\% \times (C_{\text{Total}} - C_{\text{bkg}})/C_{\text{spike added}}$, where C_{Total} is the total concentration of heavy metal in the spiked sample, i.e., the concentration from the sample and the added spike; C_{bkg} is the background concentration of the analyte in the sample without the spike; $C_{\text{spike added}}$ is the known amount of metal contributed to the sample by addition of the spike. Acceptable recovery is in the range stipulated for ICP–MS analysis of aqueous samples according to the Standard Method of analysis used. For a recovery spike, the Acceptance Criteria range is typically 90%–110%. If recovery spike values fall outside of this, Corrective Action involves reviewing and adjusting the sample preparation and analytical procedures to identify potential sources of error. Additionally, recalibrate equipment and verify the accuracy of reagents and ***spiking solutions*** to ensure reliable recovery rates. RPD is calculated for recovery spike duplicates. The acceptable criterion for recovery spike duplicates is an RPD of 15% or lower. Generally, the RPD for duplicate analysis of water samples should be lower than for soils/sediments. Deviations from these criteria and how they affect the interpretation of results should be discussed in reporting.
- LOD and LOQ (***limit of quantification***) – these parameters define the lowest concentration of pollutant that can be reliably detected (LOD) and quantified (LOQ) using a given method of instrumental analysis. LOD is often defined as three times the standard deviation of the calibration blank divided by the slope of the calibration curve. LOQ is often defined as three times the LOD. Standard Methods for instrumental analysis commonly contain guidelines concerning what LOD and LOQ are typical. Substantial deviation from such reported guidelines may typically reflect issues with user competency in performing analysis. LOD and LOQ must be reported as they reflect upon competence of analysis and interpretation of results (e.g., lower precision typically occurs between the LOD and LOQ).

12.7 CONCLUSION AND OUTLOOK

Chelating resins are versatile, easy to use and regenerate, and a relatively inexpensive method for studying the chemoavailability of heavy metals. Because of these qualities, along with their effectiveness in use and the ability to tailor their selectivity toward ionic pollutants of interest, chelating resins have enjoyed a long history of use for studies of labile metals. Two points in particular will likely contribute to the continuing popularity of this approach. First, the ability to tailor selectivity is effectively limitless, and ongoing research continues to innovate in improving the selectivity, capacity, and durability/regenerability of chelating resins, including the development of new resin compositions and functional groups to enhance performance. Second, there is an increasing drive to develop environmentally friendly applications. Not only are resins regeneratable, but some investigators have begun to synthesize resins based on green chemistry principles, e.g., solvent-free techniques and the use of sustainable materials. In the context of the increasing use of tailored chelating resins in DGT binding gels, it seems likely that this method, and variants thereof, will continue to be relevant to labile heavy metals studies in the future.

REFERENCES

Biesuz, R., Pesavento, M., Gonzalo, A., Valiente, M., 1998. Sorption of proton and heavy metal ions on a macroporous chelating resin with an iminodiacetate active group as a function of temperature. *Talanta* 47(1), 127–136. https://doi.org/10.1016/S0039-9140(98)00060-5

Botelho, A.B., Dreisinger, D.B., Espinosa, D.C.R., 2019. A review of nickel, copper, and cobalt recovery by chelating ion exchange resins from mining processes and mining tailings. *Min. Metall. Explor.* 36(1), 199–213. https://doi.org/10.1007/s42461-018-0016-8

Bowles, K.C., Apte, S.C., Batley, G.E., Hales, L.T., Rogers, N.J., 2006. A rapid Chelex column method for the determination of metal speciation in natural waters. *Anal. Chim. Acta.* 558(1), 237–245. https://doi.org/10.1016/j.aca.2005.10.071

Chanda, M., O'Driscoll, K.F., Rempel, G.L., 1988. Ligand exchange sorption of arsenate and arsenite anions by chelating resins in ferric ion form I. Weak-base chelating resin dow XFS-4195. *React. Polym. Ion Exch. Sorb.* 7(2–3), 251–261. https://doi.org/10.1016/0167-6989(88)90147-x

Chen, C.Y., Chiang, C.L., Huang, P.C., 2006. Adsorptions of heavy metal ions by a magnetic chelating resin containing hydroxy and iminodiacetate groups. *Sep. Purif. Technol.* 50(1), 15–21. https://doi.org/10.1016/j.seppur.2005.11.002

Chu, S., Feng, X., Liu, C., Wu, H., Liu, X., 2022. Advances in chelating resins for adsorption of heavy metal ions. *Ind. Eng. Chem. Res.* 61(31), 11309–11328. https://doi.org/10.1021/acs.iecr.2c01353

Diniz, C.V., Doyle, F.M., Martins, A.H., 2000. Uptake of heavy metals by chelating resins from acidic manganese chloride solution. *Min. Metall. Explor.* 17, 217–222. https://doi.org/10.1007/BF03403237

Dragan, E.S., Dinu, M.V., Lisa, G., Trochimczuk, A.W., 2009. Study on metal complexes of chelating resins bearing iminodiacetate groups. *Eur. Polym. J.* 45(7), 2119–2130. https://doi.org/10.1016 /j. eurpolymj.2009.03.012

Kagaya, S., Gemmei-Ide, M., Inoue, Y., 2014. Chelating resins. In: Kobayashi, S., Müllen, K. (eds.) *Encyclopedia of Polymeric Nanomaterials*, pp. 1–10. Springer, Berlin, Heidelberg. https://doi.org/10.1007/978-3-642-36199-9_206-1

Lake, G.E., 2002. *Quantification of Potential Arsenic Bioavailability in Spatially Varying Geologic Environments at the Watershed Scale Using Chelating Resins*. Texas A & M University. College Station. TX. https://calisphere.org/item/83f0626d-db01-41b0-9844-c38da0bfc658

Leinonen, H., Lehto, J., 2000. Ion-exchange of nickel by iminodiacetic acid chelating resin Chelex 100. *React. Funct. Polym.* 43(1–2), 1–6. https://doi.org/10.1016/S1381-5148%2898%2900082-0

Liu, Z., Wang, L., Lv, Y., Xu, X., Zhu, C., Liu, F., Li, A., 2021. Impactful modulation of micro-structures of acid-resistant picolylamine-based chelate resins for efficient separation of heavy metal cations from strongly acidic media. *Chem. Eng. J.* 420(1–3), 129684. https://doi.org/10.1016/j.cej.2021.129684

Nesterenko, P.N., Shaw, M.J., Hill, S.J., Jones, P., 1999. Aminophosphonate-functionalized silica: A versatile chromatographic stationary phase for high-performance chelation ion chromatography. *Microchem. J.* 62(1), 58–69. https://doi.org/10.1006/mchj.1999.1717

Pal, S., 2018. Pyridine: A useful ligand in transition metal complexes. In: Pandey, P. P. (eds.) Pyridine, pp. 57–74, InTech, United Kingdom, London, http://dx.doi.org/10.5772/intechopen.76986.

Reed, M.H., Strope, E.K., Cremona, F., Myers, J.A., Newell, S.E., McCarthy, M.J., 2023. Effects of filtration timing and pore size on measured nutrient concentrations in environmental water samples. *Limnol. Oceanogr-Meth.* 21(1), 1–12. https://doi.org/10.1002/lom3.10529

Saçmacı, Ş., Kartal, Ş., Yılmaz, Y., Saçmacı, M., Soykan, C., 2012. A new chelating resin: Synthesis, characterization and application for speciation of chromium (III)/(VI) species. *Chem. Eng. J.* 181–182, 746–753. https://doi.org/10.1016/j.cej.2011.11.057

Silva, R.A., Hawboldt, K., Zhang, Y., 2018. Application of resins with functional groups in the separation of metal ions/species – a review. *Miner. Process. Extr. Metall. Rev.* 39(6), 395–413. https://doi.org/10.1080/08827508.2018.1459619

Skogseid, A. 1947. *Process for recovering potassium salts from dilute solutions using ion exchange resins.* Norwegian Patent No. 72,583.

Stockdale, A., 2005. The use of cation – exchange resins in natural water trace metals research (Master thesis). Lancaster University, Lancaster, UK.

Sud, D., 2012. Chelating Ion Exchangers: Theory and Applications. In: Inamuddin, Luqman, M. (eds.) Ion Exchange Technology I: Theory and Materials, pp. 373–401, Springer, Netherlands, Dordrecht. https://link.springer.com/book/10.1007/978-94-007-1700-8

US EPA (United States Environmental Protection Agency), 2016. *Quick Guide to Drinking Water Sample Collection.* United States Environmental Protection Agency, Washington DC.

Van Deventer, J., 2011. Selected ion exchange applications in the hydrometallurgical industry. *Solvent Extr. Ion Exch.* 29(5–6), 695–718. https://doi.org/10.1080/07366299.2011.595626

Wołowicz, A., Hubicki, Z., 2012. The use of the chelating resin of a new generation Lewatit MonoPlus TP-220 with the bis-picolylamine functional groups in the removal of selected metal ions from acidic solutions. *Chem. Eng. J.* 197, 493–508. https://doi.org/10.1016/j.cej.2012.05.047

Zhang, X., Li, B., Deng, J., Qin, B., Wells, M., Tefsen, B., 2020. Regional-scale investigation of dissolved organic matter and lead binding in a large impacted lake with a focus on environmental risk assessment. *Water Res.* 172, 115478. https://doi.org/10.1016/j.watres.2020.115478

13 Permeation Liquid Membranes as a Tool for Evaluation of Heavy Metal Chemoavailability

13.1 INTRODUCTION AND SCOPE

Heavy metal pollution in natural waters and seawater has become quite widespread and poses a significant threat to ecosystems on which we rely for much of the world's food. Numerous studies have shown that the ***absorption*** of heavy metals in the environment and their subsequent toxic effects cannot be described in terms of total metal concentrations. Specifically, ***bioavailability*** is a prerequisite to toxic effects on aquatic organisms, which is also strongly controlled by speciation (Bayen et al., 2006; López-López et al., 2010). Therefore, the detection of ***bioavailable*** heavy metals, i.e., metals available to biota, has received increasing attention. Unlike other forms of metals, bioavailable heavy metals in the environment, namely in the form of metal free-ion (refer Chapter 1), can be absorbed by organisms, accumulate, and exhibit toxic effects. These bioavailable heavy metals can enter organisms through their food, water, air, and other pathways, affecting the growth, development, and health of organisms. The bioavailability of these heavy metals varies depending on their chemical speciation in the environment.

A variety of analytical tools have been used to investigate heavy metal bioavailability and speciation in aquatic environments in a framework of ***chemoavailability***. These include, for instance, different forms of voltammetry (Xue et al., 2001), ion exchange (Nielsen et al., 1984), absence of gradient Nernstian equilibrium strip (AGNES) (Galceran et al., 2007), and the Donnan membrane technique (DMT) (Kalis et al., 2006). Nonetheless, many of these techniques are not yet suitable for general environmental use when low detection limits, often required for understanding heavy metal bioavailability in natural samples, are required. Each of these techniques has its own disadvantages. For AGNES measurements, expensive and specialized analytical instrumentation is required, while the equilibration period for DMT analysis requires days, during which the speciation in the sample may change, for example, because of ***ligand*** degradation. The Permeation Liquid Membrane (PLM) technology discussed in this chapter is attractive as it does not suffer from these shortcomings. The application of PLM to the analysis of natural waters further has the advantages of high ***selectivity***, reduced sample handling, low cost, quick analysis time (1 hour), and ease of automation (Parthasarathy et al., 1997).

PLM is based on the principles of ***liquid–liquid extraction*** and ***competitive ligand exchange*** (Slaveykova et al., 2004). Three processes are important for the manner in which PLM extracts bioavailable metals from water: liquid–liquid extraction, ***back extraction***, and ***preconcentration***. Specifically, back extraction takes advantage of the difference in solubility or partition coefficients of pollutant species in two immiscible or near-immiscible solvents (one aqueous, the other organic) to transfer or extract pollutant from one liquid solvent (aqueous, i.e., natural ***matrix***) to the other (organic). Back extraction is the process of returning the extracted pollutant from the loaded organic phase to the aqueous phase using a counter-extractant, i.e., it is the inverse process of extraction. This process is able to occur because of competitive ligand exchange, as described in the next section, and during the process of back extraction, bioavailable pollutant is preconcentrated into the back extract. Preconcentration is a term used in chemical analysis to describe the process of

DOI: 10.1201/9781003355229-16

augmenting the concentration of a pollutant to enhance the ***sensitivity*** of measurement. An appealing feature of the PLM method is that the processes of extraction, back extraction, and preconcentration of the target pollutant can be done in one step.

To date, the PLM technique has been thoroughly tested for cadmium (Cd, Bayen et al., 2006), copper (Cu, Parthasarathy et al., 1997), lead (Pb, Slaveykova et al., 2004), nickel (Ni, Bayen et al., 2007), and zinc (Zn, Gramlich et al., 2012). Three basic PLM systems can be used: bulk (BLM), supported (SLM), and emulsion (ELM) liquid membranes (Aouarram et al., 2007). In this chapter, hollow fiber PLM (HF-PLM), a type of SLM, was selected as a system for measuring bioavailable heavy metal concentration, because the small radial dimensions of the support for this type of PLM accelerate mass transfer and shorten analysis time. The geometry of the support also results in increased heavy metal analyte enrichment (Parthasarathy et al., 1997).

This chapter will introduce and provide a method for the specific use of the HF-PLM technique for measuring the bioavailability of Cd, Cu, Ni, Pb, and Zn in aqueous samples. While these metals are the most studied and are the focus of this chapter, it may be possible, however, to use or modify this method for studies on other metals. Guidance is provided for the selection and handling of materials used in the technique and how to set up and operate the experimental apparatus, as well as matters to be noted during operation. Information is also given concerning maintenance of experimental equipment. Chapters containing Terms and Definitions, guidance on avoiding contamination of samples in field and lab work, and guidance on Quality Assurance (QA)/Quality Control (QC) are also within the scope of this chapter.

13.2 PRINCIPLE

This method is based on the flux of heavy metal from a sample solution, or ***source***, to a receiving solution, or ***strip***, through a thin, permeable, hydrophobic membrane in the shape of a HF. The selected permeation membrane, the HF, is able to mimic the properties of biological membranes to some extent for the permeation and separation of heavy metal ions. Figure 13.1 graphically illustrates the basic principle of operation for determining the bioavailability of heavy metal pollutants (denoted as M^{n+} since free ion is the most common bioavailable species) using HF-PLM. As shown on the left in Figure 13.1, in HF-PLM, the permeable hydrophobic membrane is loaded with an organic solvent that is immiscible with water and contains a complexing agent, R, which is selective for the heavy metal(s) of interest and extracts metals from the sample/source that surrounds the outside of the HF-PLM. The agent R is suspended in an organic solvent comprising a ***carrier*** solution to transport heavy metals across the membrane. Once extracted into the organic phase, M^{n+} is transported across the HF-PLM and back extracted with complexing agent, L, dissolved in the aqueous strip solution located in the hollow interior of the fiber. The association constant for M^{n+}'s binding with L is greater than that of R, enabling competitive ligand exchange, i.e., L competitively replaces R. In sum, heavy metals are separated from the source by the carrier and subsequently preconcentrated in the relatively low volume of strip solution, i.e., heavy metals in a sample are separated from the source and preconcentrated into the strip solution in one step using the PLM technique.

The mechanism of extraction and back extraction is shown on the right in Figure 13.1. For this method, R has two ***components***, a crown ether (1,10-didecyl-1,10-diaza-18-crown-6, abbreviated 22DD) and a long-chain saturated fatty acid (lauric acid, abbreviated LA). The crown ether saturates the membrane surface, attracting both heavy metal ion and saturated fatty acid, thus leading to the formation of MR, a ternary metal-crown ether-fatty acid complex with sufficient stability for transport across the organic medium in the membrane. On the side of the membrane, bounded by the strip solution, back extraction occurs with complexing agent L dissolved in the aqueous strip solution. The heavy metal in the strip solution is analyzed quantitatively and is directly related to the amount of bioavailable heavy metal in the source (Wojciechowski et al., 2008; Zhang et al., 2006).

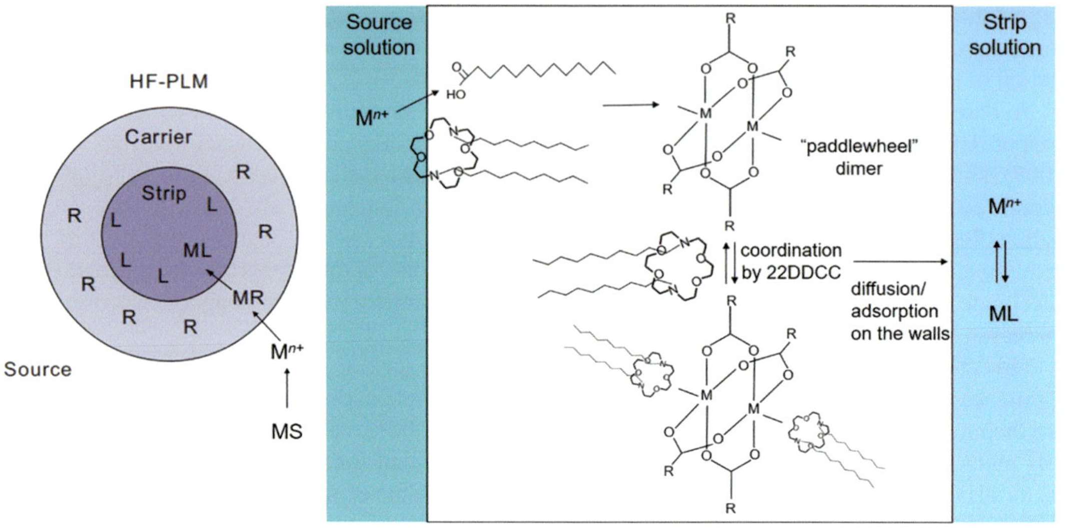

FIGURE 13.1 Schematic diagram illustrating the processes involved in the PLM technique. Left: Diagram showing process of liquid–liquid extraction of bioavailable heavy metal, M^{n+}, from the source into an organic phase carrier solution containing a complexing agent, R, followed by back extraction using competitive ligand exchange (transferring M^{n+} to ligand, L) to achieve preconcentration of the bioavailable metal into the aqueous strip solution. MR is a complex between bioavailable heavy metal M^{n+} and R on the outer surface of the HF/ permeation membrane, and ML is a complex between bioavailable heavy metal M^{n+} and L in the strip solution on the interior of the HF/permeation membrane. The charge, $n+$, associated with M^{n+} is variable, depending on the metal and valence state thereof, and hence no charges are specified for the complexes MR and ML. Right: 22DD and LA are both surface-active and interact with each other in bulk and at the fluid–fluid interfaces. When M^{n+} in the source solution passes through the membrane, it forms a M^{n+}–LA paddlewheel dimer complex that subsequently coordinates with 22DD. Then, L in the strip solution inside the HF back extracts M from the M-LA-22DD complex, and the metal bioavailability in the source is determined by measuring the concentration of metal in the strip solution.

13.3 EQUIPMENT, CONSUMABLES, AND REAGENTS

The list of equipment, consumables, and reagents given in this section is intended to be reasonably comprehensive; users should, however, read the entire method before finalizing needs and modify this indicative list accordingly.

13.3.1 Equipment and Consumables

Note: All equipment are presumed appropriately clean and ***contaminant***-free (see Chapter 3 for reference).

Sample collection/preparation-related:

Acid-washed/heavy metal clean polycarbonate bottles, 1 L, with good impact resistance – used for storage of water samples collected in the field. A minimum of 250 mL of water is needed for this method.

Fiber-reinforced tape – used to seal thermo-insulated boxes used for transport of field samples.

Field logbook – in addition to a sample collection log sheet, field personnel should record information about sample collection activities, field conditions, pH, temperature, and other unexpected events that may affect the results of experiments, which will help to assist in the subsequent analysis of the results of the experiment.

Filtration equipment, consisting of

A. a vacuum filtration device, vacuum, and ***filter membranes***;

OR

B. syringes with syringe filter cartridges.

The filter membranes associated with either option above must be 0.45 μm pore size to remove large particles and/or microorganisms from the sample. Filters composed of polypropylene are often used to avoid loss of heavy metals from samples via adhesion to the filter. In practice, the user should solicit technical specifications for products used. For option A, plastic or polytetrafluoroethylene (PTFE) is suitable for use with heavy metals, in order to avoid sample contamination, and vacuum pumps used must also be of a suitable design to avoid sample contamination. For option B, it is important to choose materials that will not contaminate the samples. Generally, high-density polyethylene and polypropylene that offer low reactivity with metals are proper materials for syringes. See Chapter 3 for more details on avoiding sample contamination.

GPS/electronic or physical devices – to ensure unambiguous location of sample ***sites*** (e.g., GPS, beacons) and depths (e.g., depth gauge).

Ice/dry ice – ice is used to ensure that the sample temperature is around 4°C during transportation (the effect on the bioavailability of heavy metals in samples is discussed below). Ice packs are preferred if they can be kept well frozen prior to use; alternatives include dry ice (subject to certain shipping requirements) or wet ice (which increases the chance of cross-contamination of samples as it melts). When using dry ice, the following points should be noted: dry ice in contact with water sublimates rapidly and generates large amounts of gas, which may result in package breakage or sample damage. Therefore, avoid direct contact between dry ice and water during transportation and ensure that thermo-insulated boxes are dry and free of moisture. During transportation, ventilation should be maintained inside the packing box to prevent carbon dioxide from accumulating. At the same time, excessive pressure or squeezing of the box should be avoided to avoid damaging the packaging material or affecting the insulation effect of dry ice. When handling dry ice, gloves, goggles, and other protective measures should be worn to avoid direct skin contact and eye injury.

Indelible pens – to record relevant data without losing fidelity in wet field environments.

Multisensor sonde – e.g., YSI or similar; used for on-site measurements during water sample collection. Calibration is required before use to ensure the ***precision*** of the instrument. Multiple sensors are desirable, including those capable of measuring temperature, pH, conductivity, and dissolved oxygen (DO).

Niskin-type water collection bottle(s) – for the collection of water samples, needs to seal well, be corrosion-resistant, and the bottle itself needs to be clean and free of contamination.

Sample identification labels – completed in indelible ink, labels are used for unique identification and shall contain (1) a unique sample identifier (also recorded in the field logbook), (2) location of the sample sampling point, (3) date of collection, and (4) type of sample collected.

Thermo-insulated boxes – clean/uncontaminated and constructed of appropriate material for the pollutants to be analyzed; must have sufficient insulation properties to keep the samples collected at a reasonably constant temperature during handling and transportation.

HF-PLM system-related:

Aluminum foil – for some photosensitive metal ions present in samples to be tested, direct sunlight should be avoided to minimize the effects of photochemical reactions. During sample storage, the container may be wrapped with aluminum foil to prevent metal ions in the sample from reacting when exposed to sunlight.

Clamp – to hold the HF-PLM coil in place. The clamp must be composed of material that will not contaminate the sample. Material made of plastic may be a good choice. Other materials that will not contaminate heavy metal samples may also be considered (refer to Chapter 3).

Heat-shrinkable polypropylene tubing – inner diameter 0.9 mm, 10 mm lengths; needs to have good sealing performance to provide a seal between the HF-PLM and PTFE tubing.

HFs – used as the support for the hydrophobic PLM. This method requires polypropylene membrane HFs having the following characteristics: inner diameter = 600 µm, outer diameter = 800 µm, pore size = 0.2 µm, porosity = 75%, Celanese Plastic. Suitable lengths for HFs are from 14 to 25 cm, with longer fibers having more resistance to pumping. The 3M Capillary Membrane MF-PP Series, Type Q3/2 fibers are most often used and may be purchased from 3M/Membrana (Wuppertal, Germany).

Hot air blower – preferably portable (e.g., hair dryer), for heat-shrinking tubing to provide a seal between the HF-PLM and PTFE tubing.

Laminar flow hood – sterile type, with HEPA filtration; for HF handling. Protects HFs from dust and airborne contaminants during various stages of handling.

Magnetic stirrer and stir bar – to stir the source solution, which allows for better transport of bioavailable heavy metals from the source solution through the HF-PLM and into the strip solution.

Nanoplast resin – to stiffen the terminal inlet and outlet tips of the HF-PLM.

Peristaltic pump and accessories – used to construct and maintain the HF-PLM and to fill the HF-PLM with strip solution, must be capable of pumping at a rate of 0.1–1 mL/min.

Plastic beakers, acrylic – size 500 mL.

PTFE tubing – inner diameter 0.8 mm, used to connect the inlet and outlet ends of the HF-PLM to the peristaltic pump.

Measurement-related:

Electronic balance – used to weigh reagents for solution preparation.

pH meter and accessories – used to measure the pH of the strip solution.

Personnel protection-related:

Lab goggles – prevents eye injuries during experiments.

Polyvinyl chloride gloves – good acid and alkali resistance and corrosion resistance, effective protection against the hazards of organic solvents.

13.3.2 Reagents

Note: All reagents must be of guaranteed reagent quality or better to ensure that the reagents used are of the highest purity to avoid contamination.

LA – a component of the carrier R (mixed with 1,10-didecyl-1,10-diaza-18-crown-6 to form R); the concentration of LA in the carrier solution is 0.1 mol/L.

Nitric acid (HNO_3) – for decontaminating laboratory and field equipment; see Chapter 3 for details.

Phenylhexane – for making the organic solvent mixture used for the solubilization of R.

Sodium hydroxide (NaOH) – to adjust the pH of the strip solution.

Toluene – for making the organic solvent mixture used for the solubilization of R.

Ultrapure water – also known as Type I water, has a resistance of 18 MΩ·cm.

1,2-Cyclohexylenedinitrilotetraacetic acid (CDTA) – the complexing agent, L, in the strip solution.

1,10-Didecyl-1,10-diaza-18-crown-6 (Kryptofix 22DD) – available from Merck, a component of the carrier R (mixed with LA to form R); the concentration of 22DD in the carrier solution is 0.1 mol/L.

3-[*N*-morpholino]propanesulfonic acid (MOPS) – a buffer for stabilizing the pH of the strip solution.

13.4 PRECAUTIONARY NOTES

When using the PLM instrumentation for quantitative analysis of bioavailable heavy metals, users should take safety precautions, including wearing personal protective equipment (e.g., gloves, goggles) to avoid injury, as well as precautions to avoid contamination. Potential sources of contamination include improperly cleaned field equipment laboratory equipment, laboratory instruments, containers that may come into contact with samples, and contaminated or insufficiently pure reagent solutions and improper procedures by laboratory personnel. Chapter 3 provides specific guidance on avoiding contamination. Specific to this chapter, all sample handling in the laboratory must be performed under clean-room or clean-room equivalent conditions, as assessed by ***blanks*** described in Chapter 4 on QA/QC.

13.5 PROCEDURE

13.5.1 Collection, Transport, and Handling of Water Samples

When collecting water samples in the field, attention should be paid to utilize properly cleaned, contaminant-free materials that are suitable for heavy metals work (Chapter 3). It is also important to avoid suspension of sediments; avoiding sediment suspension involves a judicious choice of sampling time in order to avoid resuspension caused by natural phenomena such as wind/weather.

After identifying suitable sampling locations, samples of water are collected at each sampling location from depths of 0.2 m below the water surface (for surface grab samples), or from deeper locations using Niskin-type water collection bottles. A minimum volume of 250 mL of water is needed for each sample or replicate. A Secchi disk is used to test the water transparency. A multisensor sonde is employed for on-site measurement of water temperature, pH, conductivity, and DO (Murphy et al., 2015). These field measurements of water quality are standard, though additional determinands may be added according to needs dictated by experimental objectives. Water samples are transferred from the Niskin bottle into polycarbonate containers devoid of heavy metal contamination, labeled in a sequential and clear manner for unambiguous sample identification, and transported at 4°C temperature to the laboratory within a 6-hour time frame.

Field equipment must be cleaned in between the collection of each sample according to needs dictated by the sample, e.g., for samples with little matrix, rinsing with Type I water may be sufficient. For samples with substantial organic matter, solvent rinsing or washing with detergent may be required. In all cases, care to avoid contamination must be taken (Chapter 3). To validate the absence of field contamination of samples, it is necessary to collect ***equipment blanks***. Details are in Section 13.6.2.

Upon arrival in the lab, all field samples are kept at a temperature of 4°C in a lightless environment (Zhang et al., 2020). Storage at higher temperatures may enhance bacterial growth in solution and on the container surfaces, resulting in changes in the composition of sample organic matter and attendant changes in the bioavailability of heavy metals. In our experience, the speciation of heavy metals in solution remains stable at this temperature; however, individual circumstances may differ depending on the nature of samples. In any case, due to problems with changes in sample composition at higher temperatures, storing samples at 4°C is regarded as standard best practice (Madrid and Zayas, 2007).

Water samples are typically pretreated before being subjected to HF-PLM preconcentration. This involves filtering water samples with a 0.45 μm filter membrane to remove microorganisms and nominal solids from the water, which is also useful to reduce the chances of HF-PLM fouling. Filtration must be achieved in a manner that neither removes metals from samples nor contaminates samples during filtering, and may be performed in the field, or immediately after arrival of samples in the lab. Filtering in the field increases the chances of sample contamination; however, filtering in the lab increases the chances that metal bioavailability changes during transport. Filtration should be completed within no longer than 8 hours of sample collection (Reed et al., 2023), and filtered samples should be stored in a refrigerator at 4°C. The US EPA recommends that hold times of filtered samples for biologically relevant measurements be no more than 96 hours (US EPA, 2016); however, bioavailability may begin to change from the moment samples are collected; hence, processing from sample collection to completion of HF-PLM preconcentration should be as rapid as possible.

Heavy metal bioavailability is a function of pH. In quantifying bioavailability, there is a choice between quantification at ambient sample pH (which, if different from the bioavailability ***calibration curve***, will result in a notional relative result), otherwise, samples must be buffered to a fixed pH, ideally one that approximates their original pH. The fixed pH at which samples are buffered must also be the same as the pH used for the bioavailability calibration curve (see Section 13.5.6). For different metals, the manner in which pH affects speciation is highly variable. Specifically, some buffers cause significant changes in heavy metal speciation that interfere with the measurement of bioavailability (e.g., via ***complexation*** or precipitation). Therefore, it is necessary to choose an appropriate buffer according to the specific situation. A second buffer, with a usable pH range that overlaps the first, must also be chosen to fulfill QA/QC requirements, as described in Section 13.6.1.

Buffers that are deemed acceptable, as exhibiting "no binding" or interference with heavy metal bioavailability, are as follows:

- For Cd – MES (in the pH range of 5.5–6.7), PIPES (in the pH range of 6.1–7.5), MOPSO (in the pH range of 6.2–7.6), MOPS (in the pH range of 6.5–7.9), and TEA (in the pH range of 7.3–8.3)
- For Pb – MES and MOPS (pH ranges as above)
- For Zn – MOPSO and TEA (pH ranges as above).

For Cu and Ni, also within the scope of this method (or other heavy metals for which the user may want to further develop this method for use in bioavailability studies), the suitability of buffer choice for bioavailability studies will be determined via Initial Demonstration of Proficiency (IDP, see Section 13.6.1). A list of possible buffers for Cu, Ni, and other heavy metals is as follows:

- MES in the pH range of 5.5–6.7
- PIPES in the pH range of 6.1–7.5
- MOPSO in the pH range of 6.2–7.6
- MOPS in the pH range of 6.5–7.9
- HEPES in the pH range of 6.8–8.2
- MOBS in the pH range of 6.9–8.3
- HEPPSO in the pH range of 7.1–8.5
- POPSO in the pH range of 7.2–8.5
- EPPS in the pH range of 7.3–8.7
- HEPBS in the pH range of 7.6–9.0
- CHES in the pH range of 8.6–10.0
- CAPSO in the pH range of 8.9–10.3
- CAPS in the pH range of 9.7–11.1
- CABS in the pH range of 10.0–11.4.

Each individual buffer has a pH range over which it may be used (i.e., having sufficient buffering capacity). Having determined the fixed pH to be used, it is next required to determine the buffering capacity needed. This is described in Section 13.5.6 on the preparation of standards for bioavailability calibration curves. The buffering capacity required to prepare bioavailability calibration curve standards will, in practice, meet or exceed the needs of samples. If an investigator decides to buffer samples, all samples must be buffered (QA/QC samples inclusive), and samples should be spiked in the field with the required amount of buffer ***spike***, according to sample volume collected. The preparation and amount of buffer spike to use are described further in Section 13.5.6.

13.5.2 Impregnation of Hollow Fibers to Create Permeation Liquid Membranes

The HFs need to undergo two treatment steps to prepare PLMs, and these steps are collectively referred to as ***impregnation*** here. Prior to impregnation, the carrier solution must be prepared. To do this, first prepare the organic solvent for the carrier by mixing phenylhexane and toluene at a ratio of 1:1. Next, dissolve 1,10-didecyl-1,10-diaza-18-crown-6 (hereafter 22DD) and LA (22DD and LA being collectively R hereafter) in 1:1 phenylhexane:toluene solvent such that the final concentration of each in the solvent is 0.1 mol/L. The required volume of carrier solution is determined by the number of HFs to be made; a representative volume of carrier solution needed for a batch of 20 HFs does not exceed 50 mL, though this volume may vary in practice according to user skill in preparing HF-PLMs, and more solution may always be used as a matter of convenience.

The two steps to achieve HF impregnation are as follows: (1) the HFs are rinsed with toluene three times (each time with fresh toluene) and left to air dry; to prevent contamination of the HFs during drying, they should be placed in a clean/decontaminated laminar flow hood. Note: In the next step, it is very important that the HFs be completely toluene dry or the 1:1 ratio of phenylhexane:toluene will be altered, thus compromising the integrity of the next and subsequent steps. (2) Toluene-cleaned dry HFs are impregnated with the carrier solution. To achieve impregnation, the exterior of the HFs is slowly basted with carrier solution using a pipette. The impregnation of the HF to form the HF-PLM is complete when the HF becomes transparent (HF unwetted by carrier is opaque). If any doubts exist, err on the side of caution. It is very important that the void space in the HF is filled completely with the carrier solution to form a proper PLM. If any air remains in the HF, i.e., if impregnation is incomplete, the residual air is compressible, and when strip solution is pumped through the HF-PLM, this will cause pressure differences between the inside and outside of the HF-PLM that may lead to damage of the HF and/or leaks. Either of these will render measurements with the HF-PLM incorrect.

13.5.3 Number of Hollow Fibers Needed

Multiple HFs should be prepared for serial sample analysis. Best practice is that batch sizes shall not exceed 20 samples. Batches include HF-PLM processed field samples and replicate samples. Instrument check and other QA/QC samples should also be included in each batch. The formula is as follows:

$$N = n_{\text{reps}} \times F + S + N_{QA/QC},$$

where N represents the total number of HFs needed, n_{reps} is the number of field sample replicates collected, F represents the number of field samples/sites, S is the number of standards used for constructing the bioavailability calibration curve, and $N_{\text{QA/QC}}$ represents required samples for QA/QC process (see details in Section 13.6.1).

13.5.4 Assembly of the HF-PLM System

The basic experimental configuration of the HF-PLM system, as used for determination of bioavailable heavy metals from source/samples, consists of the HF impregnated with carrier solution (i.e., an HF-PLM), PTFE tubing which is used to connect the HF-PLM to the source and the strip solutions, and a magnetic stirrer. Figure 13.2 illustrates the HF-PLM system (Parthasarathy et al., 1997). A critical aspect of assembling the HF-PLM system requires a loosely coiled polypropylene HF-PLM (the HF-PLM is manually coiled to fit the beaker into which it will be submerged) that is joined at inlet and outlet ends to short sections of PTFE tubing through the utilization of heat-shrinkable polypropylene tubing. The joins must be effected in a manner to ensure an airtight and watertight seal.

There are five main steps needed to achieve the joins: (1) stiffening approximately five mm of the terminal inlet and outlet tips of the HF-PLM by coating with resin to make them rigid. Use Nanoplast resin, and following the manufacturer's instructions for hardening the resin, apply a thin layer of resin on the surface of each end of each HF-PLM. After the application and curing of the resin, when the HF-PLM is pinched by hand, it will no longer be soft to the touch, indicating successful completion of the operation. (2) Insert one end of the stiffened HF-PLM into one end of a very short length (approximately 10 mm) of heat-shrinkable polypropylene tubing and thread it through so that the HF-PLM tip passes out of the opposite end of heat-shrinkable polypropylene tubing. (3) Insert the end of the resin-hardened HF-PLM (outer diameter 0.8 mm) that has been threaded through the short length of heat-shrinkable polypropylene tubing, approximately 2 mm, into a 15 mm length of PTFE tubing (inner diameter 0.8 mm). This is a zero-tolerance fit; thus, it is important in Step 1 to apply a very thin coat of resin, not increasing the outer diameter of the HF-PLM, while applying enough resin to harden the tip of the HF-PLM. In practice, the PTFE tubing should be soft enough to accommodate slight expansion that occurs from applying resin to the HF-PLM. (4) Adjust the edges of the heat-shrinkable polypropylene tubing to ensure that the midpoint of the heat-shrinkable polypropylene tubing approximately aligns with the join between the resin-hardened HF-PLM and the PTFE tubing. (5) Use a hot air blower (e.g., hand-held hair dryer) to apply heat to the heat-shrinkable polypropylene tubing. Shrink the tubing to create a tight

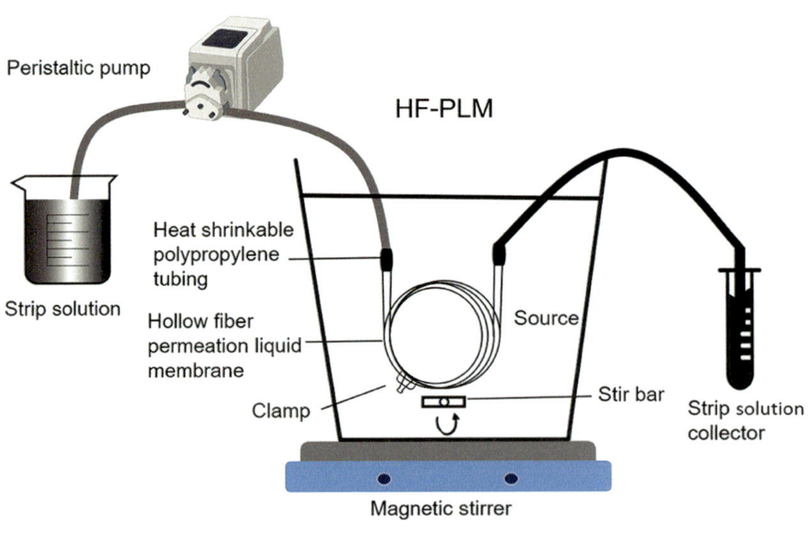

FIGURE 13.2 Schematic diagram showing the details of components required for the operation of the HF-PLM system and the configuration of the final assembled system. Note: A single-channel peristaltic pump is shown; however, a multichannel peristaltic pump is recommended; otherwise, the HF-PLM technique is not efficient. Different types of peristaltic pumps can be obtained from, e.g., GAO Tek Inc., USA.

seal, ensuring an airtight and watertight connection between the resin-hardened HF-PLM tip and the PTFE tubing. It cannot be emphasized enough that the seal between the HF-PLM and the PTFE tubing must be air and watertight; if the seal is not tight, the source solution may enter the HF-PLM directly, resulting in incorrect results.

After the resin-hardened HF-PLM-to-PTFE joins are made to the HF-PLM inlet and outlet, the final system assembly must be made to the configuration shown in Figure 13.2. Place the HF-PLM coil in a 500 mL beaker of Type I water and immerse it such that the HF-PLM coil is held vertical by means of a clamp. One end of the PTFE tubing attached to the HF-PLM becomes the HF-PLM inlet by being attached to a peristaltic pump, which pumps strip solution into the HF-PLM. The outlet end of HF-PLM is placed into a collecting tube.

13.5.5 Extraction and Preconcentration of Heavy Metals in Samples Using the HF-PLM System

After HF-PLM system assembly as described above, HF-PLMs are stored until use in a suitably sized container filled with Type I water. Immediately prior to use, each HF-PLM must be loaded with strip solution. The strip solution is 0.5 mM CDTA buffered with MOPS at a concentration of 2.5 mM, adjusted to a pH of 6.8 using 1 mol/L NaOH. For each HF-PLM, submerge in fresh Type I water in a beaker and load the strip solution container with strip solution, then add a stir bar to the beaker and stir with a magnetic stirrer at a rate of 450 rpm. (Controlling the stirring rate to around this value during all experiments with samples is important to ensure the flux of bioavailable heavy metals from the bulk of the source solution to the HF-PLM is not diffusion-limited, which in turn is important to ensure sensitivity and best precision of results.) Set the peristaltic pump to a rate of 0.48–0.6 mL/min and pass strip solution through the HF-PLM. The peristaltic pump is stopped after enough strip solution is passed through the HF-PLM to collect a 5-mL sample of strip solution at the outlet. This 5-mL sample of strip solution may be discarded or serve as a ***procedural blank*** to be analyzed as part of the QA/QC protocol (see Section 13.6.2).

After running 5 mL of strip solution through each HF-PLM, they are ready for use in preconcentrating samples. For each blank, field sample, QA/QC sample, etc., place a 250-mL portion in a beaker and immerse a HF-PLM into it. After 1 hour, when the HF-PLM will have reached equilibrium with the source (sample), the peristaltic pump is started (as before at a rate of 0.48–0.6 mL/min) and a minimum of 5 mL of the strip solution is extracted from the HF-PLM system for measurement of bioavailable heavy metals via instrumental analysis. This procedure is repeated for all types of samples (e.g., ***matrix spike*** samples, matrix spike duplicates, bioavailability calibration curve standards, and others).

Immediately after preconcentration using HF-PLM, the 5-mL sample collected must be acidified. Acidification is normally carried out by spiking a concentration of HNO_3 to reduce the pH to <2. To acidify the sample, add 1 mL of 1.9 M HNO_3 to the 5-mL sample. In practice, the user should test this spike on real samples, which vary, to ensure that it is sufficient, increasing the concentration of HNO_3 if not. Acidified samples should be stored in a refrigerator at 4°C. The US EPA recommends that hold times of acidified samples for heavy metal measurements be no more than 6 months (US EPA, 2016).

13.5.6 Preparation of Heavy Metal Standards for Bioavailability Calibration Curves

In instrumental analysis, standard solutions, which contain known concentrations of heavy metals, are used to establish calibration curves that correlate instrument signals to specific concentrations of analyte, ensuring accurate quantitative measurement of heavy metals in samples. For assured ***accuracy***, standards must be purchased commercially. Such standards for use in instrumental analysis of heavy metals are in acidic solutions to stabilize the heavy metals and, for the purposes of this method, ideally at high concentrations (e.g., 1,000 μg/g). For this method, in addition to a calibration curve for instrumental analysis, a bioavailability calibration curve is established to calculate the bioavailability of heavy metals in samples. While an instrumental standard calibration curve calculates the absolute amount of heavy metals in a sample, the bioavailability calibration curve reflects ~100% bioavailability in the absence of environmental matrix. Quantifying bioavailability in samples using such a curve reflects the influence of environmental factors on the bioavailability of heavy metals in the sample.

In order to prepare the bioavailability calibration curve, it is necessary to prepare the standards for instrumental analysis, i.e., standards in acid, to be at or near the pH range of the field samples being measured. This involves (1) diluting the stock standard solution to a series of concentrations for use in constructing a bioavailability calibration curve, which is done in a manner that reduces the acidity of the standards, and (2) buffering the resultant standards using a spike process, i.e., adding a known, high concentration, and low volume amount of a substance to a sample. The volume of a spike is no greater than 5% of the volume of solution being spiked, such that volume change associated with spiking is negligible, i.e., the concentration of substances in the sample is not altered via spiking, other than the addition of components in the spike itself (in this case, buffer).

In order to devise an appropriate buffer spike, it is necessary to consider the target spike volume of buffer to add and the final concentration of buffer desired. The buffer concentration/buffer capacity should be enough to safely effect the pH change desired and no more. A buffer system must be chosen that will keep the final pH within the intended range (see comments on buffer choices in Section 13.5.1). The following example provides illustrative guidance of the process of preparing a bioavailability calibration curve.

A user wishes to study the bioavailability of Pb and plans to analyze HF-PLM preconcentrated samples using an instrumental method based on Inductively coupled plasma–mass spectrometry (ICP–MS) that calls for constructing a calibration curve using standard solutions of 0, 0.025, 0.050, 0.100, 0.200, and 0.500 g/g Pb. For this calibration curve, the stock standard solution purchased is 1,000 g/g Pb in 2% HNO_3, i.e., approximately 2 g of nitric acid per 1,000 g of water, or ~0.32 M HNO_3. Assuming the general rule that no dilution should be more than a factor of 10, then the user will dilute the stock three times (to 100, then 10, then 1 g/g) and use the solution of 1.00 g/g Pb to make the standards for the instrumental calibration curve, and using HNO_3 for dilutions, according to instructions in the method used for instrumental analysis. For the bioavailability calibration curve, the user wishes to study the bioavailability of Pb in samples having a pH close to neutral and therefore, has elected to use MOPS as a buffer suitable for Pb. Recognizing that the bioavailability of Pb may be reduced by as much as 50% at pH 7, per Figure 13.3, the user chooses to use Pb standards of 0, 0.050, 0.100, 0.200, 0.400, and 0.800 g/g Pb for the bioavailability calibration curve.

To prepare standards for the bioavailability calibration curve, the user dilutes the stock three times (to 100, then 10, then 1 g/g), each time in water, and will use the solution of 1.00 g/g Pb to make the bioavailability calibration curve standards. Diluting in this way with water, a concentration of 0.00032 M HNO_3 (pH ~ 3.5) will be achieved for the 1.00 g/g Pb used to make bioavailability calibration curve standards. Having now diluted the acid, the user needs to buffer the 1.00 g/g Pb solution before going further. The user has purchased MOPS in sulfonic acid form (acid) and therefore, needs to add base to create a buffer. In this case, the investigator has chosen to buffer at pH 7.2, the pH equal to the pK_a of MOPS and therefore the pH at which MOPS has optimum buffering capacity. According to the Henderson–Hasselbalch equation

$$pH = pK_a - \log([HA]/[A^-])$$

such that when the concentrations [HA] (acid form of buffer) and [A] (basic form of buffer, are equal, and the ratio of $[HA]/[A^-] = 1$ and $pH = pK_a$ (if the user wanted a pH of, e.g., 7.0, the ratio of [HA]/[A] would need to be 1.58; therefore, 0.0153 moles acid form and 0.00969 moles base form of

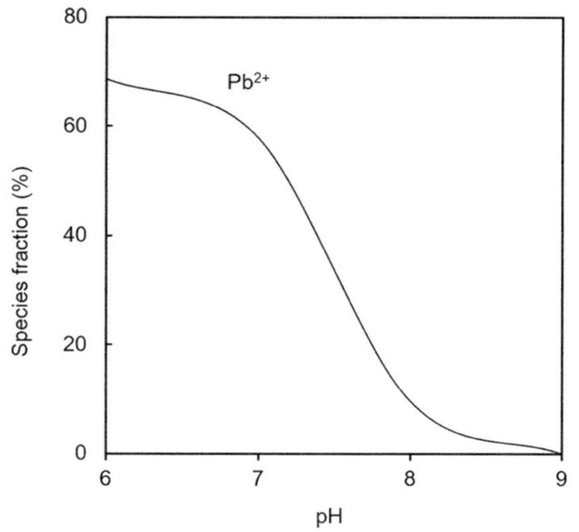

FIGURE 13.3 Percent of bioavailable Pb versus pH as Pb^{2+} in natural water. At higher pHs, bioavailability for heavy metals is often reduced due to metal insolubility/oxyhydroxide formation.

MOPS would be needed). Keeping in mind that to achieve pH stability, the minimum concentration of buffer should be on the order of 0.025 M; thus, for example a liter of 1.00 g/g Pb solution, the user needs to have 0.0125 moles of this 0.025 M in the acid form [HA] and 0.0125 moles MOPS in the base form [A] in solution. Since the user has MOPS in free-acid form, the user will need to have 0.0125 moles of a suitable strong base, NaOH being the most commonly used, to convert the needed 0.0125 moles of MOPS free acid to the base form. Further, since the 1.00 g/g Pb solution is 0.00032 M HNO_3, the user will need an additional 0.00032 moles of NaOH to neutralize the HNO_3. Finally, the user elects to use a spike volume of 2.5% of the volume of the 1 L of 1.00 g/g Pb standard solution being spiked. In total, therefore, to prepare a buffer spike for 1.0 L (~1,000 g) of 1.00 g/g Pb solution, the user will need 0.025 moles MOPS, acid form (5.23 g) and 0.00032 + 0.0125 = 0.0128 moles NaOH (0.512 g) in 25 mL solution. Checking solubilities, the user finds that these amounts of MOPS and NaOH will dissolve in 25 mL of solution. Because the user will also need buffer spike for field samples, QA/QC samples, etc., the user elects to prepare 500 mL of buffer spike solution by adding 104.6 g of MOPS free-acid and 10.2 g NaOH to a beaker and adding water till the total volume of the beaker is ~400 mL. A magnetic stir bar is added and the solution stirred until all buffer and base dissolve. The user then checks the pH and adjusts with acid (HNO_3) or base (NaOH) as necessary to achieve pH 7.2. The solution is then transferred to a 500-mL volumetric flask and diluted to 500 mL.

A 25-mL portion of this buffer spike is added to the 1 L of 1.00 g/g Pb solution, and after thorough mixing, the pH is checked to verify a pH of 7.2. This solution can now be diluted in water as needed to make the standards for the bioavailability calibration curve. The buffer spike solution may be further used to adjust the pH of other samples by adding 6.25 mL of buffer spike to each 250-mL sample.

In addition to a primary bioavailability calibration curve, as described, a secondary bioavailability calibration curve using the same pH, but a different buffer, must be prepared for QA/QC purposes (see Section 13.6.1).

13.5.7 Maintenance of HF-PLM System Components

After each HF-PLM is used, it cannot be reused as is, i.e., it must be cleaned and regenerated. To do this, first the exterior of the HF-PLMs to be regenerated are rinsed with Type I water and immersed in a fresh portion of Type I water. Strip solution is then pumped through the interior of each HF for 10 minutes at a flow rate of 0.1 mL/min to remove residual heavy metal in the membrane. Following this, Type I water is pumped through each HF for 15 minutes at a flow rate of 0.1 mL/min. The water as a medium being forced through the hollow interior of the HF-PLM serves to mechanically remove (i.e., physically scrub or abrade) excess carrier and contaminants (particulate matter, organic fouling, or other substances that may have accumulated on the membrane) from the inside surface of the HF-PLM. **Important Note**: HF-PLMs do not withstand mechanical forces corresponding to flow rates that are too high, which can cause irreversible damage to the HF-PLM. The actual limiting value will depend on the geometry and nature of the fiber, the nature of flow in the source solution, interfacial surface tension gradients, and the pressure differential along the length of the fiber due to the pumping system. A slow pumping rate, as specified here, gives the best results. After these steps, the HF-PLM is removed from the Type I water, the peristaltic pump is used to pump residual water out of the HF-PLM, and then the HF-PLM is placed in a laminar flow hood to avoid contamination by dust or other impurities while it is allowed to air dry. After air drying, re-impregnate the HF-PLM as described in Section 13.5.2. In general, a single HF can be used for analysis and then reimpregnated about 20 times according to the above steps without any breakage. Regularly replacing the HFs ensures the reliability and performance of the HF-PLM system, maintaining optimal operational conditions and minimizing the risk of failing to meet analytical Acceptance Criteria (Section 13.6.1). If the HF is damaged during use due to improper handling or lack of maintenance, the HF should be replaced immediately to avoid affecting the experiment.

13.6 VALIDITY OF RESULTS, DATA ANALYSIS, INTERPRETATION, AND REPORTING

13.6.1 Notes on QA/QC Planning Requirements

Both laboratory and field activities require a QA/QC plan to ensure the validity of measured results. Chapter 4 provides detailed guidance that is applicable to this chapter, and a full understanding of the concepts and practices described in Chapter 4 is required prior to implementing QA/QC in this method. ***Standard Methods*** for instrumental analysis listed in Chapter 5 give additional guidance pertaining to QA/QC processes. Specific QA/QC information pertinent to this chapter, and not already discussed above, is as follows:

- All analysts performing this method must undergo an IDP, both for this method, as described below, and for the instrumental method used to analyze samples prepared using HF-PLM (IDP for instrumental methods are provided in individual ***Standard Operating Procedures*** for same; this involves analysis of a Certified or Standard Reference Material and obtaining results that fall within accepted ranges for accuracy and precision).
- Routine instrumental calibration and ***calibration blank*** checks are required; guidelines for Acceptance Criteria for these are stipulated in appropriate Standard Methods used for instrumental analysis. Frequent calibration blank and standard checks are required to ensure absence of instrument drift. Drift affects both precision and accuracy. Bioavailability calibration curve construction and checks must parallel instrumental calibration and checking.
- Samples that are analyzed as having concentrations at or near the upper limit of instrument calibration must be reprocessed and reanalyzed. This involves diluting the field sample, reprocessing through the HF-PLM, and reanalysis. This process ensures that the HF-PLM preconcentration efficacy was not overloaded by high bioavailability in the original sample.
- In addition to standards necessary for instrumental analysis (described in Standard Methods for same), standards shall be prepared to construct a bioavailability calibration curve, as discussed in Section 13.5.6. Absolute concentration is assessed via instrument samples; equivalent bioavailability is assessed using the bioavailability calibration curve.

- Batch sizes for instrumental analysis shall not exceed 20 samples; samples batched include sediment and tissue digests/extracts and replicate samples; QA/QC samples are additional and should not exceed 5 per batch. Instrument check samples assess calibration stability and are not part of batching.
- Sample order in batches is to be randomized over all samples and sample types.
- One split sample is required for every two batches; this enables estimation of precision of inherent sample variability.
- Analysis of a field duplicate (for this method, processing a portion of the same field sample twice) is required for one in every ten field samples collected.
- For every batch, one randomly chosen ***equipment rinsate*** blank, OR one equipment rinsate blank per ***duplicate/replicate sample***, is required, whichever is smaller. These samples enable evaluation of whether field contamination occurred.
- For every batch, one procedural blank is required. This enables evaluation of whether contamination from the HF-PLM procedure occurred.
- One ***recovery spike*** is required per batch. This enables evaluation of whether artifactual contamination (recoveries significantly above 100%) or analyte loss (recoveries significantly below 100%) is occurring.
- One matrix spike is required per batch. Results from this indicate the short-term role of matrix in affecting bioavailability.
- One matrix spike duplicate and one recovery spike duplicate per instrument run (i.e., assumes multiple batches run sequentially), OR one matrix spike duplicate and one recovery spike duplicate every 12 hours, are required, whichever is more frequent.

13.6.1.1 Initial Demonstration of Proficiency and Handling of Standards

IDP is complicated in bioavailability studies because no Certified Reference Material or Standard Reference Material, for which "the correct result is known", exist. The concept of the bioavailability calibration curve was introduced into common use by Magrisso et al. (2009), later elaborated by, e.g. Zhang et al. (2017) and Li et al. (2022). Since then, due to the effects of even simple inorganic constituents on the bioavailability of heavy metals in solution, an operative approach to verification of "known" bioavailability is to obtain the same result (to within statistical uncertainty), when performing the same measurement twice, in different media anticipated or demonstrated to have minimal impact on bioavailability. According to this approach, IDP is accomplished when two separate bioavailability curves are constructed, based in each case on the same standard, but prepared in different buffers, as per Section 13.5.6. The bioavailability calibration curve standards, like instrumental standards, are not subject to ***matrix effects***, as occur in environmental samples, however, they are subject to limitations of permeation rate or flux imposed by the HF-PLM system and potentially to changes in speciation in result of the buffer used and/or pH. The two curves must be prepared for the same pH, which is tenable according to the buffer selections given in Section 13.5.1. For instance, for Cd, acceptable buffers are MES (in the pH range of 5.5–6.7), PIPES (in the pH range of 6.1–7.5), MOPSO (in the pH range of 6.2–7.6), MOPS (in the pH range of 6.5–7.9), and TEA (in the pH range of 7.3–8.3). MES and PIPES buffers have useful pH ranges that overlap at a pH range of 6.1–6.7, so these two buffers could be used for calibration curves at pH 6.1, for instance, because this is in the range of overlap.

For IDP, plot the instrumental response for the bioavailability of each heavy metal analyzed versus the formal concentration of heavy metal in the bioavailability calibration curve standards as determined by the instrumental calibration curve. Because the formal concentrations are "as made", and the same for both curves, no error is introduced in the x-axis. Construct one such curve for each of the two buffer systems chosen for IDP. Determine slopes and intercepts for both plots using linear regression. In the absence of buffer interference, the two curves should "agree". To evaluate this statistically, an analysis of covariance or an F-test compares the slopes and intercepts of the two regression lines to determine if they are the same to within a chosen level of confidence (p <0.05 being commonly used). If the statistical test shows that the differences between the slopes and intercepts are not significant, the calibration curves are considered to be statistically similar or "not differentiable", indicating that the two curves can be considered equivalent for practical purposes, and this suffices for IDP. If the two curves are differentiable, this indicates either an error in operator execution of the method or an issue with buffer interference. To circumvent the latter issue, it is recommended to achieve IDP for either Zn or Cd, two heavy metals whose bioavailability has been more commonly studied using PLM. Once IDP has been demonstrated by the method described herein, the bioavailability calibration curve using the primary buffer should be replotted to have an x-axis whereby bioavailability calibration curve standards are quantified using instrumental standards. In this manner, absolute bioavailable heavy metal in samples is quantified instead of formal concentrations used in IDP.

13.6.2 Data Analysis, Interpretation of QA/QC Results (Acceptance Criteria), Corrective Action, and Reporting

Data need to be analyzed, interpreted in the context of QA/QC requirements, and if needed subjected to ***Corrective Action*** prior to reporting. These include, at minimum, the following:

- Dilution factors (DFs) – Unless specified by an instrumental method, DFs are never greater than a factor of 10 for dilution of standards and samples. The basic dilution formula is $C_1V_1 = C_2V_2$. The dilution factor is given by $DF = V_2/V_1$.
- ***Calibration checks*** and metrics – Perform calibration checks according to directions specified in the method used for instrumental analysis. This will include guidance on using certified standards and ensuring regular check intervals. Metrics such as linearity, sensitivity, and precision are assessed for stability according to Acceptance Criteria given in methods of instrumental analysis. If instability occurs, Corrective Action may be taken. Only data collected under validated instrumental calibration conditions (stability of calibration) should be reported. Should the metrics associated with the instrumental calibration curve not meet requirements/Acceptance Criteria, or if calibration checks fail, as stipulated in the Standard Method chosen for instrumental analysis, it is mandatory to take Corrective Action to establish or re-establish proper calibration. Any data collected under failed calibration are not reportable.
- Precision for all instrumental duplicate/replicates – The instrumental method chosen will specify needs for analysis of duplicate or replicate instrumental analysis of processed samples. A metric for precision must be calculated and reported. Generally, the relative percent difference (RPD) or relative standard deviation (RSD) of measurements, respectively, is used to evaluate the precision for sample duplicates and replicates. The formula for RPD is as follows: $RPD = 200\% \times abs(R_1 - R_2)/(R_1 + R_2)$, where subscripts 1 and 2 represent the two duplicate samples. The formula for RSD is as follows: $RSD = stdev \times 100\%/average$. If metrics such as RPD or RSD are too high, it indicates that there is a significant amount of variability or inconsistency between the measurements of the same sample. For example, typically, an RSD value greater than 2% (acceptance criterion) for analysis of sample replicates using ICP–MS at concentration ranges in the middle of the calibration curve indicates poor precision. Poor precision for replicates of lab samples is generally indicative of an operator or instrument error, and Corrective Action involves rectifying the problem prior to analysis and reporting of results from bioavailability studies.
- Precision for ***field duplicates/replicates*** – This should be calculated using the same metrics as for instrumental duplicates/replicates above. Poor precision for field duplicates may be a measure of high environmental heterogeneity of sampled media; this is more likely to occur with soil and sediment samples, and RSDs for field duplicates of aqueous samples should normally be on the order of 10% or below as an acceptance criterion for total concentrations of heavy metal. In practice, precision for bioavailability may be considerably

poorer, a circumstance exacerbated by the complexity of HF-PLM preconcentration. To investigate issues of poor precision for field sample duplicates/replicates, compare metrics for field replicates to equivalent metrics for the bioavailability calibration curve standard in the middle of the bioavailability calibration curve. If the precision for the bioavailability calibration curve replicate standard is poor, this is indicative of a procedural problem on the part of the user and Corrective Action is required (as for comments on precision of laboratory-prepared samples, above). If, however, replicate analysis of the bioavailability calibration curve standard exhibits good precision (no higher than 15%), and analysis of blanks indicates absence of contamination, then poor precision in field replicates is likely due to field conditions, and users should check for environmental factors affecting consistency. Report RSD values and any discrepancies and provide context on how environmental heterogeneity impacts the results.

- Equipment and procedural blanks – Results for blanks must be calculated and reported in units of absolute concentration (not bioavailable) for all blank samples. The acceptance criterion for blanks is a value below the ***limit of detection*** (LOD). Blanks above the LOD are indicative of field contamination (equipment blanks) or laboratory contamination (procedural blanks). In either case, no Corrective Action is possible, and all samples must be regarded as potentially contaminated. If the contamination is effectively constant around an average value that is low, compared to field samples, then this should be reported with results and considered in interpretation of results. For highly varying contamination, in the case of field sample contamination, reporting is not possible and field work must be repeated, taking stronger measures for cleaning of field equipment and/or replacement of equipment parts responsible for contamination. If the procedural blank is above the LOD, first investigate potential contamination sources by reviewing all procedures, equipment, and reagents used. After confirming the issue, determine whether it could have resulted in contamination of samples in the lab. If so, new field samples are required and reporting is not possible. Otherwise, if field samples are deemed uncontaminated, repeat the analysis with fresh reagents and adjusted protocols to minimize contamination and ensure accurate results.
- LOD and LOQ (***limit of quantification***) – These critical parameters define the lowest concentration of heavy metal that can be reliably detected (LOD) and quantified (LOQ) using a given method of instrumental analysis. LOD is often defined as three times the standard deviation of the calibration blank divided by the slope of the calibration curve. LOQ is often defined as three times the LOD. Standard Methods for instrumental analysis commonly contain guidelines concerning what LOD and LOQ are typical. Substantial deviation from such reported guidelines may typically reflect issues with user competency in performing analysis. LOD and LOQ must be reported, as they reflect upon competence of analysis and interpretation of results (e.g., lower precision typically occurs between the LOD and LOQ).
- Split sample – A split sample involves dividing a single sample into multiple portions for parallel testing or analysis. Its function is to ensure consistency and accuracy by comparing multiple results from the same original sample. Acceptable precision is as follows: no greater than 35% RPD for soil/sediment samples; no greater than 20% RPD for other sample types. Corrective Actions include reviewing the sample handling and preparation processes to ensure consistency.
- Recovery spikes and matrix spikes – These samples and their method of preparation are discussed in more detail in Chapter 4. For both types of spikes, calculate recovery as follows: Recovery (%) = $100\% \times (C_{\text{Total}} - C_{\text{bkg}})/C_{\text{spike added}}$, where C_{Total} is the total concentration of heavy metal in the spiked sample, i.e., the concentration from the sample and the added spike; C_{bkg} is the background concentration of the analyte in the sample without the spike; $C_{\text{spike added}}$ is the known amount of metal contributed to the sample by addition of the spike. Acceptable recovery is in the range stipulated for ICP–MS analysis of aqueous samples

according to the Standard Method of analysis used. For a recovery spike, the Acceptance Criteria range is typically 90%–110%. If recovery spike values fall outside of this range, Corrective Action involves reviewing and adjusting the sample preparation and analytical procedures to identify potential sources of error. Additionally, recalibrate equipment and verify the accuracy of reagents and ***spiking solutions*** to ensure reliable recovery rates. Matrix spikes are somewhat more problematic as a result of matrix effects. In all instances, recoveries must be reported; if near 100%, this entails greater faith in the accuracy of results, and vice versa. The error for samples must be assumed to be at least the value by which spiked samples vary from 100%.

- Recovery and matrix spike duplicates – These samples should be evaluated by RPD for the duplicate samples, with results reported and explained. The recovery spike duplicate acceptance criterion is 15% RPD or less. Matrix spike duplicates may be higher if matrix effects are severe. In bioavailability studies, matrix spike results themselves may reflect the presence of environmental factors influencing bioavailability. Comments on reporting results from recovery and matrix spikes apply to their duplicates.

13.7 CONCLUSIONS AND OUTLOOK

On one hand, PLM, here HF-PLM, is considerably more laborious than some other approaches to estimating chemoavailability and even true bioavailability. On the other hand, PLM is of particular interest because of the way the liquid membrane functions in a much more bioanalogous manner than other chemoavailability approaches. Thus, PLM holds much more interest as a vehicle to assist in clarifying the environmental physicochemical processes taking place at the surface of biological membranes. As a relatively new and complicated technique, PLM has not been widely adopted. Due to the highly bioanalogous nature of the technique, however, it will be interesting in future to see if PLM can overcome the inadequacies of other chemoavailability techniques, which are unable to approximate true bioavailability in a generalizable way. In particular, HF-PLM holds promise for being able to mimic bioavailability of heavy metals to aqueous organisms, and the infinite ways in which the agent R can be varied within the liquid membrane holds great potential for future biomimetic development.

REFERENCES

Aouarram, A., Galindo-Riaño, M.D., García-Vargas, M., Stitou, M., El Yousfi, F., 2007. A permeation liquid membrane system for determination of nickel in seawater. *Talanta* 71(1), 165–170. https://doi.org/10.1016/j.talanta.2006.03.038

Bayen, S., Wilkinson, K.J., Buffle, J., 2007. The permeation liquid membrane as a sensor for free nickel in aqueous samples. *Analyst* 132(3), 262–267. https://doi.org/10.1039/B615298H

Bayen, S., Worms, I., Parthasarathy, N., Wilkinson, K., Buffle, J., 2006. Cadmium bioavailability and speciation using the permeation liquid membrane. *Anal. Chim. Acta.* 575(2), 267–273. https://doi.org/10.1016/j.aca.2006.05.096

Galceran, J., Huidobro, C., Companys, E., Alberti, G., 2007. AGNES: A technique for determining the concentration of free metal ions. The case of Zn(II) in coastal Mediterranean seawater. *Talanta* 71(4), 1795–1803. https://doi.org/10.1016/j.talanta.2006.08.027

Gramlich, A., Tandy, S., Slaveykova, V.I., Duffner, A., Schulin, R., 2012. The use of permeation liquid membranes for free zinc measurements in aqueous solution. *Environ. Chem.* 9(5), 429–437. https://doi.org/10.1071/EN12103

Kalis, E.J.J., Weng, Dousma, F., Temminghoff, E.J.M., Van Riemsdijk, W.H., 2006. Measuring free metal ion concentrations in situ in natural waters using the Donnan Membrane Technique. *Environ. Sci. Technol.* 40(3), 955–961. https://doi.org/10.1021/es051435v

Li, B., Zhang, X., Tefsen, B., Wells, M., 2022. From speciation to toxicity: Using a "Two-in-One" whole-cell bioreporter approach to assess harmful effects of Cd and Pb. *Water Res.* 217, 118384. https://doi.org/10.1016/j.watres.2022.118384

López-López, J.A., Mendiguchía, C., Pinto, J.J., Moreno, C., 2010. Liquid membranes for quantification and speciation of trace metals in natural waters. *TrAC-Trend. Anal. Chem.* 29(7), 645–653. https://doi.org/10.1016/j.trac.2010.01.007

Madrid, Y. and Zayas, Z.P., 2007. Water sampling: Traditional methods and new approaches in water sampling strategy. *TrAC-Trend. Anal. Chem.* 26(4), 293–299. https://doi.org/10.1016/j.trac.2007.01.002

Magrisso, S., Belkin, S., Erel, Y., 2009. Lead bioavailability in soil and soil components. *Water Air Soil Pollut.* 202, 315e323. https://doi.org/10.1007/s11270-009-9978-y

Murphy, K., Heery, B., Sullivan, T., Zhang, D., Paludetti, L., Lau, K.T., Diamond, D., Costa, E., O'Connor, N., Regan, F., 2015. A low-cost autonomous optical sensor for water quality monitoring. *Talanta* 132, 520–527. https://doi.org/10.1016/j.talanta.2014.09.045

Nielsen, J.S., Hrudey, S.E., Cantwell, F.F., 1984. Role of the free metal ion species in soluble nickel removal by activated sludge. *Environ. Sci. Technol.* 18(11), 883–886. https://doi.org/10.1021/es00129a013

Parthasarathy, N., Pelletier, M., Buffle, J., 1997. Hollow fiber based supported liquid membrane: A novel analytical system for trace metal analysis. *Anal. Chim. Acta.* 350(1), 183–195. https://doi.org/10.1016/S0003-2670(97)00247-X

Reed, M.H., Strope, E.K., Cremona, F., Myers, J.A., Newell, S.E., McCarthy, M.J., 2023. Effects of filtration timing and pore size on measured nutrient concentrations in environmental water samples. *Limnol. Oceanogr-Meth.* 21(1), 1–12. https://doi.org/10.1002/lom3.10529

Slaveykova, V.I., Parthasarathy, N., Buffle, J., Wilkinson, K.J., 2004. Permeation liquid membrane as a tool for monitoring bioavailable Pb in natural waters. *Sci. Total Environ.* 328(1), 55–68. https://doi.org/10.1016/j.scitotenv.2003.10.007

US EPA, 2016. Quick guide to drinking water sample collection. United States Environmental Protection Agency, CO. Washington DC.

Wojciechowski, K., Kucharek, M., Buffle, J., 2008. Mechanism of Cu(II) transport through permeation liquid membranes using azacrown ether and fatty acid as carrier. *J. Membrane Sci.* 314(1), 152–162. https://doi.org/10.1016/j.memsci.2008.01.036

Xue, H.B., Jansen, S., Prasch, A., Sigg, L., 2001. Nickel speciation and complexation kinetics in freshwater by ligand exchange and DPCSV. *Environ. Sci. Technol.* 35(3), 539–546. https://doi.org/10.1021/es0014638

Zhang, X., Li, B., Deng, J., Qin, B., Wells, M., Tefsen, B., 2020. Regional-scale investigation of dissolved organic matter and lead binding in a large impacted lake with a focus on environmental risk assessment. *Water Res.* 172, 115478. https://doi.org/10.1016/j.watres.2020.115478

Zhang, X., Qin, B., Deng, J., Wells, M., 2017. Whole-cell bioreporters and risk assessment of environmental pollution: A proof-of-concept study using lead. *Environ. Pollut.* 229, 902–910. https://doi.org/10.1016/j.envpol.2017.07.068

Zhang, Z., Buffle, J., van Leeuwen, H.P., Wojciechowski, K., 2006. Roles of metal ion complexation and membrane permeability in the metal flux through lipophilic membranes. Labile complexes at permeation liquid membranes. *Anal. Chem.* 78(16), 5693–5703. https://doi.org/10.1021/ac060299p

Section 4

Modeling

14 Introduction to Modeling Bioavailability

14.1 INTRODUCTION

As we discussed in Chapter 1, the changes in environmental factors such as pH, temperature, and ***ionic strength*** affect the solubility, reactivity, and mobility of pollutants, thereby altering pollutant ***bioavailability***. The goal of modeling bioavailability in most cases is to understand the ramifications of bioavailability, to wit, toxic effects. In most cases, the relevance of a model is assessed by understanding how such environmental factors, in the modeling context often referred to as ***toxicity-modifying factors*** (TMFs), affect bioavailability and attendant toxicity. Bioavailability-based models that are used indirectly or directly to assess toxicity are increasingly the focus for streamlining regulatory processes and the development of environmental criteria, or protective values, such as Water Quality Criteria, for instance (EC, 2000, 2013; US EPA, 2007). The changing bioavailability of pollutants in the environment poses a challenge for accurate monitoring of pollutant bioavailability (Anderson and Hillwalker, 2008). Model ***accuracy*** is evaluated by assessing the performance of the model when compared to laboratory or field observations. Whether measuring true ***bioavailability*** (harder) or ***chemoavailability*** (easier), it is difficult and/or time-consuming either way (Chen et al., 2023). Yet, if reasonable accuracy can be achieved, modeling bioavailability is an intriguing way to get around the difficulties involved in continual assessment and reassessment of bioavailability experimentally, which would otherwise be perforce necessary given environmental fluctuations in bioavailability.

The incorporation of bioavailability-based models into defensible regulatory decision-making requires robust and scientific ***model validation***. Two general approaches have been used, one, that has been referred to as ***autovalidation***, entails that the same set of experimental/observational data is used to both parameterize/calibrate a model and then validate or test for accuracy of the model's predictions. Due to the circular/non-independent nature of this process, which has also been referred to as back-calculation, while it should be a pro forma part of the model elucidation process, it is not best practice in terms of stand-alone defensibility. ***Independent validation***, whereby one observational data set is used to parameterize/calibrate a model, and a different observational data set is used to test for accuracy of the model's predictions, represents best practice in validation. This raises the question, however, of what metrics are used for accuracy? For many years, an operational standard has been the agreement of observed-versus-model predicted data to within a factor of ± 2 (Erickson et al., 1996) across a log scale variation in both. The value of a factor of 2 reflects the sometimes inherently large variability in some experimental observations. This factor of 2, however, is not a stringent level of accuracy and even admits that slopes significantly different from 1 (ideal agreement) could occur, i.e., this criterion alone as a metric would potentially enable validation of "wrong" models. Hence, various proposals are still under development for how models are validated (Garman et al., 2020), much as the thinking around what approach to modeling should be used is evolving.

Different types of pollutants have different bioavailability/TMF characteristics, and there is a general divide between strategies for modeling inorganic (heavy metals) versus organic pollutants. Arguably, in terms of standardization and use in formalized ***Environmental Risk Assessment***, modeling the environmental bioavailability of inorganic/heavy metal pollutants has progressed much more than that for organic pollutants. Models that currently receive the most focus in terms of regulatory use and potential for inorganic pollutants include multiple linear regression models and the

DOI: 10.1201/9781003355229-18

Biotic Ligand Model (BLM, Qu et al., 2013). There is less consensus and thus more variability around models for predicting the bioavailability of organic pollutants (Haws et al., 2006; Su et al., 2019). Models that are inherently mechanistic in nature are preferable in terms of defensibility and in being able to offer important insights into how environmental change impacts the manner in which pollutants behave and interact with the environments. In contrast, such models rely on accurate ***parameterization***, i.e., finding parameters associated with parametric equations in a model, which may be difficult, and opens the door to development of less explicitly mechanistic models as a matter of convenience. For many metals, the effects of abiotic factors on toxicity have been investigated and quantified, leading to the development of models ranging in complexity from empirical models (e.g., simple linear regressions) to more mechanistic speciation and toxicity models (Adams et al., 2020). The goal of future research should be to create models that are more complete and precise; standardization of practice is key to increasing recognition and regulatory use.

This chapter provides a brief and representative introduction to approaches in use for modeling the bioavailability of pollutants. It reviews the principles of different models for predicting the bioavailability of pollutants, as well as application scenarios. The topics chosen for discussion aim to give readers who are not well versed in bioavailability modeling a snapshot of the state of play in this area.

14.2 MODELING THE BIOAVAILABILITY OF ORGANIC POLLUTANTS

Because the state of practice in modeling the bioavailability of organic pollutants is less codified than for modeling the bioavailability of heavy metals, there is considerably more variability in the literature as to how to approach such modeling. The approaches adduced herein are by no means comprehensive.

14.2.1 Chemical Fate Models

Chemical fate models, also sometimes referred to as environmental multimedia models, are a class of mathematical models used to simulate and predict the behavior, transformation, and transport of chemical substances in and across different environmental compartments, such as air, water, soil, and biota. These models are important for understanding and assessing the behavior of chemical substances in environmental media and their potential impacts on ecosystems and human health (Mackay et al., 1996). The earliest chemical fate models focused on the behavior of a single chemical in a ***medium***, and as understanding of the complexity of environmental systems has increased, modern models have gradually taken into account multiphase and multicomponent effects (Wöhrnschimmel et al., 2013). Model development has been driven by increased computational power, allowing more complex environmental processes to be simulated. Such models have been used for the estimation of bioavailability and in Environmental Risk Assessment (Su et al., 2019). While these models are often associated with the fate of organic pollutants, they have been used to study inorganic pollutants as well (Diamond et al., 2000).

Some chemical fate models may be described as incorporating speciation modeling in that they rely on calculations based on fugacity (a thermodynamic term describing the effective partial pressure of a volatile or semivolatile substance that replaces mechanical partial pressure in an accurate computation of chemical equilibrium), chemical concentration, or chemical ***activity***. Chemical fate models may include mathematical representations of mass balance, chemical equilibrium, reaction kinetics, mass transfer, and substance transport. Processes at play include diffusion of substances in air, dissolution in water, ***adsorption*** in soil, and enrichment in organisms. Parameters in models are often obtained through laboratory experiments and field monitoring to ensure an accurate description of conditions in the actual environment (MacLeod, 2002). Figure 14.1 shows a schematic diagram summarizing the processes and media compartments in such models, along with some implementations from the literature.

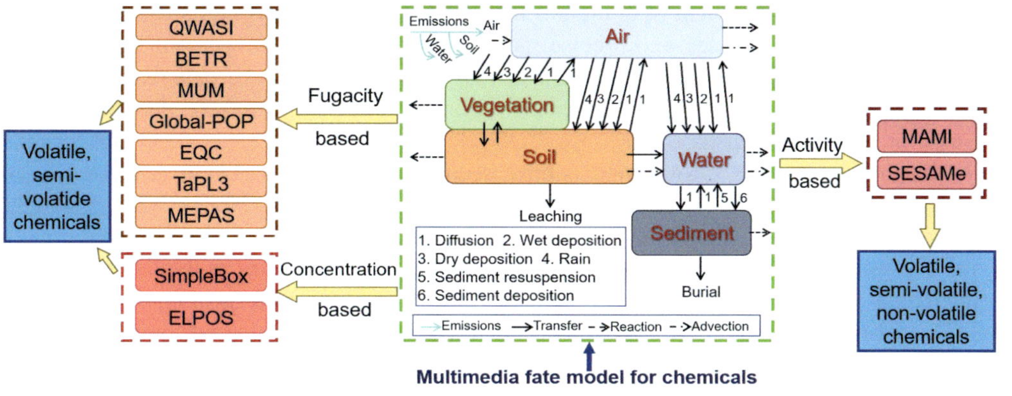

FIGURE 14.1 Schematic diagram summarizing the processes and media compartments included in chemical fate models, along with names of some implemented models from the literature: QWASI (Diamond et al., 1992), BETR (Song et al., 2016), MUM (Priemer and Diamond, 2002), Global-POP (Wania and Mackay, 1995), EQC (Mackay et al., 1996), TaPL3 (Beyer et al., 2000), MEPAS (Whelan et al., 1992), SimpleBox (Brandes et al., 1996), ELPOS (Beyer and Matthies, 2002), MAMI (Franco and Trapp, 2010), and SESAMe (Zhu et al., 2015). Models shown are not a comprehensive selection of the totality available in the literature.

The development of chemical fate modeling will continue to move toward highly integrated, multi-scale, multi-media, and multicomponent models. At the same time, the application of models will be extended to ***emerging pollutants*** and complex environmental scenarios, such as behavioral studies of nanomaterials and increasingly sophisticated chemical fate analyses in urbanized environments (Boeije et al., 1998; Datta et al., 2018).

14.2.2 Quantitative Structure–Activity Relationship Models

Quantitative Structure–Activity Relationship models, also sometimes called ***Quantitative Structure–Property Relationship*** models (QSAR and QSRP models), employ a set of descriptors or predictor variables (X_is), which characterize the structural features of a compound, and establish a mathematical relationship between these descriptors and a response variable (Y), typically some biological activity of the compound (Perkins et al., 2003). In QSAR modeling, descriptors/predictors are often numerical variables that depict characteristics of compounds and can include quantities such as, e.g., molecular weight, polarity, hydrophobicity, and alcohol–water ***distribution*** coefficient (see Figure 14.2). Early QSAR models were based on linear regression analysis, using simple molecular descriptors to predict the properties of compounds (Gharagheizi, 2009).

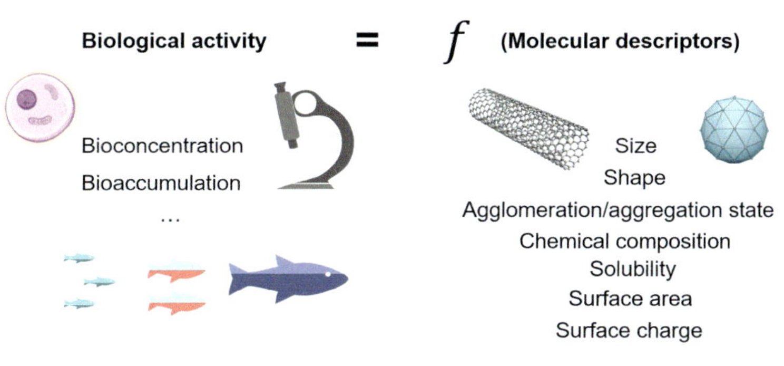

FIGURE 14.2 Schematic diagram of Quantitative Structure–Activity Relationship.

With the continuous development of computer technology and statistical methods, QSAR models have evolved into more complex and precise models, including support vector machines, Neural Networks, and random forests (Soares et al., 2022). QSAR models have extensive applications in various fields, including drug development, environmental science, and the chemical industry (Cherkasov et al., 2014; Tropsha, 2012).

In environmental science, QSAR models have been used in various ways to assess the risks of chemical substances to the environment and ecosystems (Keshavarz et al., 2021; Lavado et al., 2022). The European Centre for Ecotoxicology and Toxicology of Chemicals has disseminated guidance covering the use of QSARs endpoints including soil/sediment ***partitioning***, biodegradation, and associated bioavailability parameters of ***bioconcentration*** and ***bioaccumulation***, as well as the role of QSARs in Environmental Risk Assessment (ECETOC, 1998). More recent guidance provides information on use of QSARs to estimate K_{oc} (***soil/sediment organic carbon–water partition coefficient***) and K_{ow} (***n-octanol–water partition coefficient***) (ECETOC, 2021). These quantities are frequent correlates of bioavailability and are widely used in Environmental Risk Assessment. K_{oc} measures the tendency of a pollutant to sorb onto soil/sediment organic matter; a high K_{oc} entails low bioavailability. K_{ow} measures the distribution of a chemical between *n*-octanol and water and is used to predict the bioaccumulation potential of chemicals in organisms. Substances with high K_{ow} values tend to be more hydrophobic and ***bioaccumulate***. QSARs have also been compared to pedotransfer functions, i.e., functions that relate simple soil properties (e.g., texture, pH) to properties such as ***sorption*** of pollutants, in determining potential bioavailability for Environmental Risk Assessment purposes (Neira-Albornoz et al., 2024). While some early research on QSARs showed promise in terms of developing simple relationships between QSARs and bioavailability, their use has become increasingly focused on the ability to use QSARs to predict parameters for use in more sophisticated models (e.g., chemical fate models).

14.2.3 Machine Learning Models

In recent years, machine learning (ML) has developed into an interdisciplinary tool with applications in a number of scientific domains. The semantics of learning in this context is to generalize from experience, which entails separating the provided facts based on their various patterns, extrapolating something more general from the data, and making useful decisions in novel situations (Goyal and Bengio, 2022). ML, then, is a data-based technique that has the potential to first "learn" and later "recognize" patterns in empirical data (Jordan and Mitchell, 2015). Figure 14.3 is a schematic showing generalized schemes for how supervised (top left) and unsupervised learning (top right) occurs. In supervised learning, training data is used to develop relationships between input data, X_i (e.g., water column or soil/sediment characteristics), and desired output or "labeled" data, y' (for bioavailability models, this would be measures of bioavailability), as shown at the bottom of Figure 14.3. By iteratively minimizing an error function that relates actual output data given to calculated data, the model is "trained". After training, independent data inputs can be used to calculate outputs to verify that training was successful, i.e., the effectiveness of the trained model is tested using a separate validation data set.

Unsupervised learning occurs when the "answer" is not known. In this case, data may be clustered into different groups that are internally similar, and differentiable from other groups, based on multiple properties given as inputs. Since the goal of modeling bioavailability is prediction, one or another form of supervised learning model is most appropriate.

As an example of the use of ML for modeling bioavailability, Cipullo et al. (2019) inferred the bioavailability of petroleum hydrocarbons and heavy metals through ML. They collected bioavailability data (y' observed) from a 6-month mesocosm experiment and used soil type, pollutant amendment, initial concentration, and incubation time as inputs (X_i) to train Neural Network and Random Forest models. After training, the ***bioavailable*** pollutant concentration was predicted. The continuous output of the Neural Network was selected for toxicity estimation, and the importance

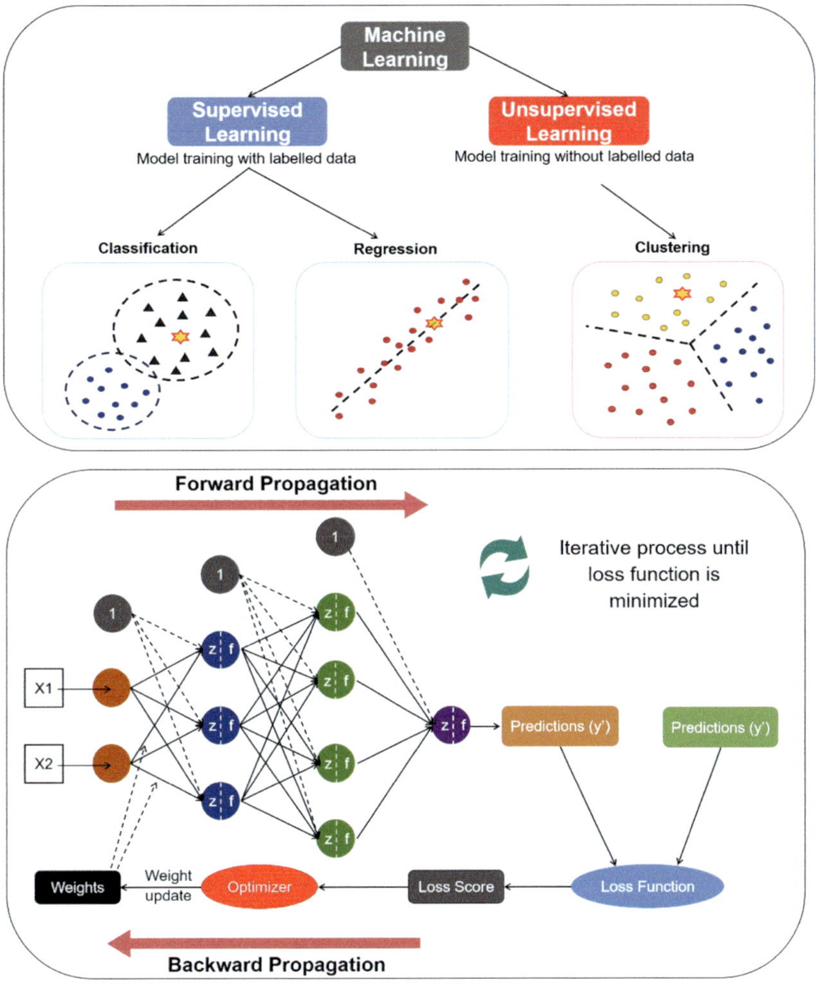

FIGURE 14.3 Schematic diagram of concepts underlying supervised and unsupervised machine learning is shown in the top part of the figure. The bottom part of the figure shows one example of how training data is used to "train" a model to be predictive. In this example, inputs X_i are ***transformed*** through a series of layers. The quantities z are linear outputs of transformations, which occur via weights (Weights in the figure, i.e., W) and bias (1 in the figure, i.e., b), according to $z = Wx + b$. For each I, a further ***activation*** function f (i.e., $f(z)$, which could be, for instance, a sigmoid or Tanh function) is applied to enable development of nonlinearity. The output of this process is a prediction of y' (brown box), which, along with an observed y' (green box), is used to compute an error or loss function, i.e., a measure of how far apart observed and predicted are. Weights and biases are adjusted, and the process repeats iteratively until agreement between observed and predicted is acceptable, at which point the model is trained. After training, an independent set of X_i and y' is processed through the model in a process to validate that the model will work effectively on data other than the training set.

of features was analyzed to understand the drivers of bioavailability. In another study on the bioavailability of polycyclic aromatic hydrocarbons (PAHs), Wu et al. (2013) collected data from an 8-month microcosm experiment. A multi-layer perceptron and radial basis function model was selected, with various soil and compost properties used as inputs to train different models. The root mean square error was used for evaluation of model parameter optimization with a subsequent tenfold cross-validation. The best model was chosen to predict the bioavailable concentration of PAHs; the importance of different variables in model construction, and in the context of temporal variation, was analyzed to understand the relationship between soil properties and pollutants.

ML is a powerful modeling technique; however, it might also be described currently as all the rage – fashionable. A concerning limitation of this approach is that, in most implementations, such models have no mechanistic basis. As such, if environmental conditions change in such a manner as to be outside the training envelope, predictive power of the model may be lost, which is problematic for bioavailability modeling since frequently the goal is to use models to inform Environmental Risk Assessment.

14.2.4 Estimating Bioavailability of Organic Pollutants in Soil/Sediments Based on Desorption Kinetics

Chemical extraction methods have become a very common approach to estimate ***chemoavailability***, i.e., availability of pollutants to a chemical proxy, and is a method of choice due to ease and convenience. For soils and sediments, chemical extraction is relatable to desorption. Parameters are obtained from experimental desorption curves, and one parameter in particular, that is, the percent pollutant rapidly desorbed, has been shown to correlate to true bioavailability, as measured by biological methods (Ortega-Calvo et al., 2013). Many different extraction techniques have been used for this kind of simple model, three common ones being Tenax porous polymer sorbent, XAD4 resin, and various cyclodextrins (Reid et al., 2000; Spasojević et al., 2015). Tenax and XAD4 are based on non-polar, hydrophobic styrene divinylbenzene copolymers (Patterson et al., 2004), and cyclodextrin molecules are a cyclic oligosaccharide with a hydrophilic shell and a toroidal-shaped non-polar cavity that can capture certain pollutants.

The operative rationale behind the use of desorption kinetics is that natural porous media such as soil/sediment contain various sorptive ***components***, and that some sorptive components bind organic pollutants so strongly as to render them effectively biounavailable in any practical sense, i.e., desorbing too slowly to exert a toxic effect. Soil/sediment desorption may be modeled as a "two-site" system with rapid and slow desorption. Alternately, some studies hold that it is unlikely that the fitting parameters of a two-site model can reasonably represent the complexity of sorptive ***sites*** in a soil/sediment, a dilemma which has been, depending on perspective, (1) resolved by the addition of more compartments with corresponding rate constants or (2) pseudoresolved through overparameterized/overfit by addition of additional parameters based on goodness-of-fit alone rather than well-elucidated mechanistic evidence of what in soil/sediments comprises the three compartments. The equation for the three-compartment model is

$$S_t / S_0 = f_{\text{rap}} \times e^{-k_{\text{rap}} \times t} + f_{\text{slow}} \times e^{-k_{\text{slow}} \times t} + f_{\text{veryslow}} \times e^{-k_{\text{veryslow}} \times t},$$

where S_t corresponds to the amount of pollutant sorbed to soil/sediment at desorption time t, S_0 is the total amount of soil/sediment associated pollutant immediately prior to desorption, and f_{rap}, f_{slow}, and f_{veryslow}, respectively, are decimal fractions cumulatively summing to 1.0 and associated with desorption rate constants k_{rap}, k_{slow}, and k_{veryslow} of rapid, slow, and very slow desorption (Spasojević et al., 2015). Values of f and k are determined by non-linear curve fitting to experimental desorption data plotted S_t/S_0 as a function of t.

A paper by Wells et al. noted that, while correlations between f_{rap} and measures of true bioavailability have often been observed, there is no systematic way to determine what f and k parameters could be used to predict bioavailability in untested soil/sediment samples, the slopes for correlations between f_{rap} and bioavailability typically are not predictive (slope $\neq$ 1.0) and are variable for different sets of samples, and finally, such correlations, when they exist, are easily disrupted if biotic factors are introduced (i.e., such correlations tend to arise and could possibly be said to be artifacts of the abiotic context of desorption experiments) (Wells et al., 2005). Thus, while more labor-saving than ML models, and arguably popular perhaps in consequence, this approach suffers from the same issues in theoretical rigor as do the ML approaches.

14.3 MODELING THE BIOAVAILABILITY OF INORGANIC POLLUTANTS – FOCUS ON HEAVY METALS

While modeling the bioavailability of heavy metals is arguably more advanced, in terms of regulatory standards and acceptance, than modeling the bioavailability of organics, the literature on modeling heavy metal bioavailability is vast. This section will concentrate on two approaches that have gained prominence: according to an extensive literature review (Garman et al., 2020), over 90% of the models in common use are either based on chemical speciation modeling (mechanistic in basis, to one extent or another) or are based on a multiple regression-type model (MRM, entirely empirical in nature), with the preponderance of the former strongly dominating. A publication on the State of the Science on Metal Bioavailability Modeling, based on a technical workshop conducted by the Society of Environmental Toxicology and Chemistry, focused on these two approaches as well (Schlekat et al., 2020).

14.3.1 Chemical Speciation Modeling

Chemical speciation modeling, based on chemical equilibria, has become a key tool for the study of environmental behavior, bioavailability, and fate of heavy metals. Chemical speciation refers to the distribution of an element among different chemical forms (chemical species) in a system, a schematic for which is given in Figure 14.4. For instance, consider two chemical reactions and their corresponding equilibria,

$$\mathrm{M_f} + \mathrm{X_{IN}} \rightleftarrows \mathrm{MX_{IN}}, \qquad K_{\mathrm{MX_{IN}}} = a_{\mathrm{MX_{IN}}} / a_{\mathrm{M_f}} \times a_{\mathrm{X_{IN}}}$$

and

$$\mathrm{M_f} + \mathrm{L} \rightleftarrows \mathrm{ML}, \qquad K_{\mathrm{ML}} = a_{\mathrm{ML}} / a_{\mathrm{M_f}} \times a_{\mathrm{L}}$$

where M_f represents the free ion of a metal (Cu^{2+} given as an example), X_{IN} is an inorganic anion that complexes or reacts with M_f, and L is an organic ***ligand*** that also complexes/reacts with M_f, to form, respectively, the reaction products MX_{IN} and ML. The chemical reactions on the left are governed by the chemical equilibria equations on the right, with the thermodynamic equilibrium

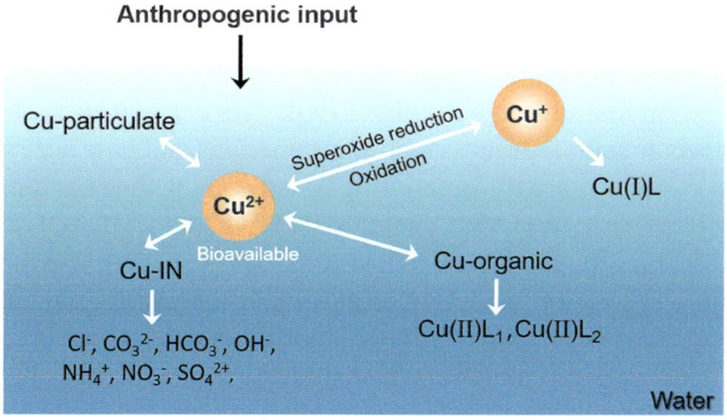

FIGURE 14.4 Some of the biogeochemical processes giving rise to different Cu species in environmental aqueous systems. Notations IN, L_i, (I), and (II) indicate, respectively, inorganic ligands, organic ligands (where $i = 1$ and 2 indicates different types of organic ligand), Cu^+, and Cu^{2+}. Examples of inorganic ligands are given in the schematic. Examples of organic ligands include amino acids and short-chain amino acids (peptides), polysaccharides, chalkophores (so-called strong Cu-binding ligands, molecules that have a high K_{ML} for binding with copper), and humic/fulvic acids derived from the decomposition of organic matter.

constants K_{MX} and K_{ML} governing the reaction with the inorganic and organic species, respectively. The quantities a represent chemical activity (related to concentration). In environmental settings, there may be a large number of such chemical reactions, leading to a number of equilibrium equations. Additional equations result from the physical requirements of mass balance (neither gained nor lost) and charge balance (charge neutrality must prevail). Using inputs such as total concentration (i.e., Cu may exist in many different forms, but the total amount can be known through chemical analysis) and temperature (which affects K values), a system of equations developed as described above is solved to reveal the distribution of species, with M_f being the primary bioavailable form of interest. Figure 14.4 summarizes some of the biogeochemical processes contributing to Cu speciation in environmental waters.

A large number of computer codes to conduct such speciation calculations/speciation modeling have been developed, beginning in the 1960s, and today these are supported with a wide range of reaction databases with values (e.g., Ks) needed for computations. Some codes are coupled with transport computations for assessment of transport and fate. A small, representative selection of examples will be mentioned here. In the early days of speciation modeling, the United States Geological Survey (USGS) led the way, starting with a publication accompanied by a digital computer code called WATCHEM in 1969 (Barnes and Clarke, 1969). Subsequent developments of WATEQ (USGS) in 1974 and MINEQL in 1976 (Massachusetts Institute of Technology) advanced the state of art considerably by incorporating more comprehensive and accurate thermodynamic data (capabilities for handling a wider range of chemical species), a more user-friendly experience, and enhanced computational programming, allowing for more powerful calculations at a time when computer memory was still limiting (Truesdell and Jones, 1974; Westall et al., 1976). MINEQL is considered by many as the seminal basis for modern speciation modeling. Another seminal development in the progression occurred with the publication of the PHREEQE code, which enabled modeling of speciation to include sophisticated handling of redox reactions, mixing of waters, addition of net irreversible reactions to solution, dissolving and precipitating phases to achieve equilibrium with the aqueous phase, and effects of changing temperature (Parkhurst et al., 1980). These two codes, MINEQL and PHREEQE, represented a very high level of sophistication for their time and were open ***source***, thus forming the basis and paving the way for the development of essentially all extant codes available, whether free or at cost, today. The program MINTEQ, later the MINTEQA2 and Visual MINTEQ programs, is one such example, having been created in the early 1980s at Battelle Pacific Northwest Laboratory (Felmy et al., 1984) through a collaborative initiative supported by the US Department of Energy and the US Environmental Protection Agency (US EPA), based on merging the mathematical framework of MINEQL with the thermodynamic database of the WATEQ3 model developed by the USGS (Ball et al., 1981). MINEQL now exists as MINEQL+ version 5.0 with a user-friendly graphical user interface.[1] PHREEQC version 3 is the latest implementation of the progression that started with PHREEQE.[2] As calculating speciation for heavy metals is complex with respect to metal–organic matter interactions and how heavy metals interact with mineral surfaces, other codes that are differentiating based on how they handle these components of modeling include ORCHESTRA, WHAM, Visual MINTEQ, and CHEAQS Next, all of which are free, except for WHAM.[3] Finally, the MARCHEMSPEC software must be mentioned as the latest important development in speciation modeling of particular interest to many marine applications.[4] The capabilities of these codes vary, along with the algorithms that run different submodels governing metal (hydr)oxide reactions and metal-aluminosilicate/phyllosilicate clay reactions (present in the water column as particulate matter), and metal-organic matter reactions.

---

[1] https://mineql.com/ers/index.html.

[2] https://water.usgs.gov/water-resources/software/PHREEQC/index.html.

[3] http://orchestra.meeussen.nl/; https://www.ceh.ac.uk/data/software-models/windermere-humic-aqueous-model-wham; https://vminteq.com/download/; https://www.cheaqs.eu/.

[4] https://marchemspec.org/wp/software/.

Ongoing issues in speciation modeling involve parameterization of some submodels, database accuracy/consistency, and kinetic effects. With regard to parameterization, for instance, ion-binding parameters governing ***complexation*** with complicated phases such as humic acids (HA) are obtained by fitting laboratory data on ion binding to model systems (e.g., isolated humic substances). In this endeavor, there is a trade-off in the number of fitting parameters (for instance, many fitting parameters needed to represent many different binding stoichiometries) and being able to derive a reasonable parameter set from multiple data sets, each with their own uncertainties and limitations of testing conditions. With respect to data sets, a vast number of parameters are needed to cover the full scale of constituents in most environmental settings. Taking equilibrium constants as one example, these are determined at specific conditions for ionic strength and temperature, and the use of these values in modeling requires adjustment to the conditions in the system being modeled. These adjustments, as well as the differences in conditions and different methods for determination, lead to uncertainty in equilibrium constants. In addition, these data sets need to be maintained and updated, and there is no central clearing house with purview over the same. Finally, on the issue of kinetics, it is well known that some environmental speciation is rate-limited/governed by kinetics (slower processes). Some of the codes mentioned herein have begun to implement capabilities to account for kinetic limitations in some subsystems, however, it is fair to say that the development of codes incorporating additional capabilities to account for kinetics is needed. Despite these issues, speciation modeling remains the most powerful and mechanistically based approach to modeling heavy metal speciation, and at the same time an approach that has a history of reasonable accuracy across a wide range of environmental conditions, accounting for the dominance of speciation-based models in modeling heavy metal bioavailability.

14.3.2 Environmentally Relevant Speciation Submodels

This section provides a cursory overview of two kinds of speciation submodel: modeling heavy metal reactions with solid phases, notably particulate matter, and heavy metal reactions with organic matter. These submodels are at once very important and simultaneously challenging due to the complexities of particulate matter and organic matter. Particulate matter includes clays, carbonates, oxides/oxyhydroxides, and biological material, either living or recently dead, being usually in the form of aggregates. While organic and inorganic particulate matter is inextricably mixed, with organics sorbing to inorganic matter, at a molecular level predominant submodels consider heavy metal interactions with inorganic surfaces and heavy metal interactions with organic matter separately.

The surface properties of carbonates, oxides/oxyhydroxides, and particularly clays have complex surface chemistries with functional groups and mineral phases that vary widely, and this heterogeneity entails a wide variety of reaction sites with different affinities for heavy metals, complicating the modeling of metal–surface interactions. Reactions of heavy metals and surfaces may involve competition for sites between different heavy metals, as well as synergism, whereby the presence of one metal enhances the adsorption of another. Models are not able to reflect the actual complexity, and hence approximations are made, such as assuming a homogeneous surface composition, assuming a uniform surface charge, and neglecting multi-layer effects (assuming a single layer of ions bound or adhered to the surface overlain by a single diffuse layer, when actually multiple layers exist). Metal–surface binding processes are greatly affected by environmental factors including pH, temperature, ionic strength, i.e., in chemical terms, leading to non-ideal behavior. It is therefore difficult to model metal–particulate matter interactions.

Modeling metal–organic matter interactions is largely made difficult by the large amount of HA and fulvic acids (FA) in many natural waters. HA and FA are the soluble fractions of humic substances, which are complex and heterogeneous mixes of polydispersed components produced by the microbial breakdown and modification of residues of plants and microorganisms in the natural environment over long periods of time (Lipczynska-Kochany, 2018). Because HA and

FA have many active functional groups, notably, multiple metal-binding carboxyl and phenolic hydroxyl groups, densely distributed in their molecular architectures, it presents a difficult problem to model reactions with heavy metals and HA/FA; there is no single chemical composition for HA or FA, and the characteristics of these molecules are highly variable in any given environment (Zhou et al., 2005). Additionally, binding of HA and FA with heavy metals includes many different types of reaction (e.g., ion exchange, complexation, ***chelation***), and, as for any other kind of reaction, these binding processes are greatly affected by pH, temperature, and ionic strength (Peng et al., 2022).

14.3.2.1 Modeling Heavy Metal Reactions with Solid Phases/Particulate Matter

There are two important classes of submodels for reactions between metals and particulate matter: (1) ion exchange models and (2) variable charge models. Regarding the former, solid particles have ion exchange sites that selectively bind and exchange ions with the surrounding aqueous solution in a manner that has been described within the framework of the Donnan model, a model which explains the distribution of ions between two solutions separated by a semipermeable membrane. The charge on the solid surface is treated as analogous to large charged anions on one side of a semipermeable membrane, charges that have no mobility with respect to the membrane. To maintain electrical neutrality, small cations (e.g., Na^+, M_f) on the other side of the membrane (surface interface) will move toward the large anionic charges, while small anions (e.g., Cl^-) will distribute to balance the charge. At equilibrium, this results in an electrostatic potential across the membrane (surface interface) known as the Donnan potential. This helps to explain the electrostatics of how ions in solution are selectively bound to surface ion exchange sites.

Ion exchange models such as the Gaines–Thomas and Vanselow have been used to model heavy metal reactions with solid phases. The Gaines–Thomas model focuses on how different cations compete for surface binding sites, assuming a predominantly negative surface charge due to the structure of most environmental solid phases. ***Selectivity*** coefficients, which are dimensionless ratios, express the preference of the ion exchange sites for particular ions. These coefficients are determined by the relative affinities of different cations for the exchange sites and are influenced by factors such as ion size, charge, and hydration energy. The equivalent fractions of charges in the solid phase, rather than their molar concentrations, enable comparison of the relative strengths of different ion exchange reactions on a common basis. Equilibrium expressions between ions in the solid phase and the aqueous solution are used to predict the distribution of ions between the two phases under different conditions (Gaines and Thomas, 1953).

The Vanselow model is similar to the Gaines–Thomas model, but, instead of selectivity coefficients, the Vanselow model uses activity ratios. This has important implications. The distribution of fixed charges on the surface, due to the ionization of functional groups or structural charges, creates a surface potential that influences the distribution of ions in the surrounding solution and affects how ions in the solution interact with the surface. The Vanselow model's use of activity ratios to express the effective concentration of ions thus accounts for the non-ideal behavior of ions in solution, providing a more complex and accurate representation of ion exchange equilibria (Miller and Benson, 1983).

Of the variable charge models, the surface chemistry assemblage model for particles was designed with submodels to represent binding of metals and across variable pHs, with different binding phases including oxides, clay minerals, carbonates, and organic matter (Lofts and Tipping, 1998). This model has shown great predictive power for some applications, however, uses a more simplistic approach to variable surface charge and empirical/simplified terms to represent electrostatic interactions in a manner that makes it less robust to ionic strength variations.

Surface complexation models (SCMs) consider the adsorption of metal ions onto mineral surfaces via formation of surface complexes, accounting for how the charge of mineral surfaces varies as a function of pH and ionic strength of the aqueous medium and associated electrostatically induced changes to the equilibrium constants governing metal-mineral surface adsorption reactions

(Goldberg, 2014; Sun and Ying, 2021; Tonkin et al., 2004). For example, in environments with higher ionic strength, the increased concentration of ions can screen surface charges, leading to changes in the adsorption behavior of metal ions. In providing a more complex treatment of surface electrostatics, SCMs offer a solid theoretical foundation for modeling adsorption processes, making them robust for predicting metal mobility and bioavailability.

An advancement of the SCM is the charge distribution multisite ion complexation model (Hiemstra and Van Riemsdijk, 1996). This model includes the competition between different metal ions for different types of adsorption sites on the mineral surface and introduces the concept of charge distribution to the model. Charge distribution contrasts with the traditional notion of point charges in SCMs, which assumes that surface sites on minerals have a uniform charge that does not vary across the surface. In contrast, charge distribution signifies that the charges on surface sites are heterogeneous along the surface and vary in location and magnitude, providing a more detailed and realistic depiction of the mineral's surface properties. This consideration is useful for accurate modeling of ion adsorption and predicting the behavior of metals in natural waters.

14.3.2.2 Modeling Heavy Metal Reactions with Organic Matter/HA and FA

While modeling heavy metal interactions with particulates is of great environmental relevance, nonetheless, strong predictive ability has been achieved for some speciation-based aqueous effects models without incorporating these reactions (*viz*., the BLM, *vide infra*). The same is not true with respect to heavy metal reactions with HA and FA, which are required for validation of such models. In the 1970s, the early models of ion binding to HA/FA were published (Mantoura et al., 1978). Since this time, development has been ongoing, with selected notable developments discussed here. The 1980s produced the Humic Ion-Binding (HIB) Model V (later followed by VI and VII, hereafter Models V, VI, and VII, implemented in the Windermere Humic Aqueous Model code or WHAM), the Stockholm Humic Model (SHM, implemented in Visual MINTEQ), and the Non-ideal Competitive Adsorption–Donnan Model (NICA–Donnan, implemented in Visual MINTEQ and other codes) (Gustafsson, 2001; Milne et al., 2001; Tipping, 1998). All of these submodels are implemented in software that has capabilities of also calculating speciation of components involved in inorganic solution chemistry, mineral reactions, and idealized clay surfaces. These metal–organic binding models, which involve different types of constitutive parameters for the electrostatic and chemical contributions of HA and FA moieties to metal ion binding, all consider two main HA/FA binding sites, those being carboxyl and phenolic functional groups.

Both Models VI and VII were developed from the basic theoretical framework of Model V, inheriting core principles and basic mathematical expressions of metal–organic matter interactions, expanding on and optimizing Model V (Tipping, 1998; Tipping et al., 2011). Model V is discrete-site binding model in which HA and FA are assumed to be rigid spheres of uniform size with ion binding groups on the surface. Binding is modified by electrostatic interactions, and counterions in the diffuse layer are described by a simple Donnan-type expression. Binding sites are represented by carboxyl and phenolic groups, each having four different binding strengths with a spread or delta. Each type of binding site (carboxyl, phenolic) is present in equal amounts, and there are half as many phenolic groups as carboxyl groups. The *K*s for proton binding are linked to *K*s for metal binding (Perdue et al., 1984). Binding takes place at both monodentate and bidentate sites. In subsequent developments, Model VI includes provisions for high-affinity tridentate sites for metals present in small amounts, which may be important when modeling the binding of trivalent metals such as Al and Fe, and the distribution factors were adjusted in Model VI (Guthrie et al., 2005). These changes allow Model VI to represent a wider range of metal binding strengths. Model VII differs from Model VI in that the description of metal–cation binding was simplified, with some parameters eliminated. The ratio of binding strengths for phenolic versus carboxylic sites for a metal is taken to be the same as that for protons. The distribution factors were entirely eliminated, and the arrangement of multidentate sites was modified. Additionally, Model VII extends and optimizes the estimation of the binding constants of metal ions and HA/FA, and the number of calculable cation species was increased to more than 40, covering, e.g., Pd^{2+}, Sn^{2+}, U^{4+}, NpO_2^{2+}, Pu^{4+}, PuO_2^{2+}, and many other radionuclide cations, enabling the calculation of equilibrium distributions of chemical species under a wider range of conditions (Stockdale et al., 2011).

Another model of metal-humic complexation is the SHM, which, instead of modeling HA/FA as hard spheres, models HA/FA as a gel, i.e., having reactions with metals at the surface of the gel, as well as within the gel volume. This approach is much more realistic. Having a surface component, the SHM shares similarities with the Models V–VII/WHAM, but it also differs in particular ways, such as how the SHM handles electrostatic effects (Gustafsson, 2001). Proton binding parameters in the SHM and WHAM V–VII are both adaptable. Similar to Models V–VII, the SHM model uses a discrete-site method, but to solve for the electrostatic contribution to cation binding, the Basic Stern Model is used as the interface model (Gustafsson, 2001).[5] The SHM enables metals to form either monodentate or bidentate complexes. In monodentate complexes, the metal attaches to a single HA/FA site, while in bidentate complexes, the metal binds to two surface sites. The SHM has a gel fraction parameter that represents the portion of HA/FA that binds with metal as gel interior (Groenenberg and Lofts, 2014). Cations bind to the surface of the gel and also within the interior ***matrix***, leading to partial neutralization of the negative charges from carboxylic acid and phenolic groups (Gustafsson and Kleja, 2005). Residual negative charge may remain within the gel fraction. Cations accumulating in the gel interior screen the negative charges on the metal binding sites, influencing the availability of binding sites. The additional screening of charge inside the gel-like structures of the humic substances is taken into account via an empirical set of equations, and an equation (Gaines–Thomas) is used to modify selectivity constants (Hiemstra and Van Riemsdijk, 1996), which helps in accurately predicting the preference of different ions for exchange positions within the gel. Unlike V–VII, the SHM does not require the computation of a diffuse layer volume and the subsequent modification of ionic concentrations in the bulk solution volume (Gustafsson, 2001). Additionally, while Models V–VII assume that the sites are different, requiring a set of additional fixed parameters, the SHM makes the simplifying assumption that ions can form mono- and bidentate complexes with the same sites. The application of the SHM as a submodel in more extensive chemical equilibrium models is also greatly facilitated by these characteristics (Gustafsson and Kleja, 2005; Gustafsson and Lumsdon, 2014).

The NICA–Donnan model has similarities with the SHM in considering organic matter as a gel within the context of electrostatic interactions between ions and the fixed charges on HA/FA (Kinniburgh et al., 1999; Marinsky, 1987). In the SHM, however, the gel-like structure does not explicitly have a distinct volume. In the Donnan approach, HA/FA is treated as an idealized electrically neutral phase with a specific volume, within which there is a uniform, average electrostatic potential called the Donnan potential. Certainly, HA and FA possess a three-dimensional structure and share many characteristics with polyelectrolyte gels, to which the Donnan model has been extensively applied (Chen and Schnitzer, 1989). The volume of the Donnan phase is connected to the swelling behavior of humic suspensions. A fundamental difference between the NICA–Donnan model and HIB V–VII and the SHM is that the NICA component of the model is based on a continuous site distribution and assumes that the distribution of the metal-binding functional groups follows statistical rules (Koopal et al., 2005). The initial version of the NICA–Donnan model assumed that metals bind only in a monodentate manner to proton-binding sites. This resulted in the release of a maximum of one proton for each metal ion bound, but it did not align well with some experimental data. In order to minimize the problematically strong positive correlation between the distribution of proton and metal cation binding sites, a new version was created that permits variable stoichiometry of metal binding to proton binding sites (ranging from monodentate to tridentate) (Kinniburgh

[5] The Basic Stern Model assumes a compact layer of ions directly adsorbed onto a charged surface, held tightly by both electrostatic forces and specific chemical binding interactions. Beyond this tightly bound layer, ions are more loosely associated with the surface in the diffuse layer. The electrostatic potential drops linearly across the Stern layer and then decreases exponentially in the diffuse layer (Westall and Hohl, 1980).

et al., 1999). To do this, a thermodynamically compatible NICA equation was redefined (Dudal and Gérard, 2004). The updated NICA model represents a notable deviation from previous models by allowing the affinity distributions for the proton and different metal ions to not only have distinct median affinities for each ion, as indicated by shifts along the affinity axis, but also to exhibit ion-specific nonideality or heterogeneity (Kinniburgh et al., 1996). The resulting model is called NICA–Donnan to highlight its consistency and needs seven parameters for each HA/FA to simulate both non-specific and specific proton binding at varying ionic strengths.

14.3.3 Multiple Regression-Type Models of Heavy Metal Bioavailability, Inferred from Toxicity

The MRM approach to bioavailability modeling is applicable to bioavailability or toxicity, however, the latter is more often the focus, although some authors refer to toxicity models as bioavailability models. Like linear regression, only with multiple variables, MRM-type models answer the question of to what extent do independent variables X_i (for i independent variables) predict Y (independent variable), in this case answered by mathematical regression techniques. Because there is no *a priori* mechanistic causation for regression, this approach is strictly empirical in nature. A recent paper (Brix et al., 2021) describes the approach for such a model, which begins with a species-specific data set for toxicity that covers a range of water chemistries; in this study, the authors used acute and chronic toxicity data sets for Cu compiled by the US EPA for BLM (discussed in more detail below) work supplemented with additional publicly available data. The authors considered three TMFs, including hardness, pH, and dissolved organic carbon, and then applied a series of regression analyses accompanied by analysis of covariance to obtain final regression equations in which hardness, pH, and dissolved organic carbon are independent variables that predict the toxicity of the different test species in their data set. In general, the empirical approach these authors took agreed with experimental data almost as well as the BLM, and for some chronic toxicity measures, better than the BLM. Less discussed was that the cumulative density functions for the empirical approach were consistently worse than for the BLM, i.e., the distribution of results for the empirical approach agreed less well with observations than did the distributions predicted via BLM. Because of the simplicity of the approach, conferring real practical advantages, it is likely that modern developments in application of such empirical approaches will breathe new life into legacy techniques of the 1980s (US EPA, 1985).

14.4 BIOTIC LIGAND MODELS

An outcome of decades of refinements in speciation modeling has been the development of the BLM for toxicity calculations. The operative idea is that there is a putative ***biotic ligand*** associated with an organism, which, when complexed with a certain concentration of any given heavy metal, results in a given toxic effect. The governing equilibrium expression is

$$M_f + BL \rightleftarrows MBL,$$

whereby it is apparent that the BL is in competition with any other species that reacts with M_f, ergo the toxicity of M_f to an organism is modulated by the number and identity of competing species present. The model is parameterized in a manner to estimate, from actual toxicity tests on a given heavy metal and organism of interest, the metal-BL Ks for binding of BL with alkaline metals, alkaline earth metals, and the heavy metal, M, of interest, as well as the fraction of MBL (of total surface BL sites) that results in the experimental toxic response variable being used (for instance, $\boldsymbol{EC_{50}}$, or the amount of metal needed to induce a given toxic effect in 50% of a population of organisms). The BL is assumed to be present at a relatively low concentration on exposure-sensitive surfaces of the organism of interest, and, for instance, if the BL is a receptor on a fish gill, the concentration of BL is not high in the scheme of the water column. In this case, the presence of the BL does not perturb the distribution of other chemical species in any given setting.

Once the needed parameters for the metal and organism of interest are determined, site-specific water quality parameters, including such parameters as pH, hardness, and dissolved organic carbon, are used as inputs to predict the toxicity of the metal to the organism. In an early study, investigators demonstrated the application of the BLM approach to accurately predict the acute toxicity of copper to the fathead minnow *Pimephales promelas* and the crustacean *Daphnia pulex* using a number of water quality data sets across waters of varying pH, hardness, and dissolved organic carbon concentrations (Santore et al., 2001). Since this early study the BLM has gained widespread acceptance and use. Since 2000, there have been more than 1,500 papers published in indexed journals. The US EPA has set the BLM as the standard for evaluating the risk of different metals in freshwater environments (US EPA, 2007, 2010). In addition, the European Commission has used the BLM as a basis for regulating environmental risks from metals (Chung et al., 2021; EC, 2011).

14.5 CONCLUSIONS AND OUTLOOK

In recent years, the use of models to simulate changes in chemical speciation and how these can be used to model the bioavailability of pollutants in the environment has received increasing attention. Some of these modeling approaches have achieved great predictive power in some settings. As yet, there is no single model that can robustly simulate changes in the bioavailability of pollutants across the full range of natural environments, due to the complexity of the natural environment. Nonetheless, the tremendous progress that has been made is an indication of the progress that will continue to come, and the flexibility of being able to use models in lieu of labor-intensive and costly laboratory investigations ensures that improved modeling of bioavailability will be ever more important in the future.

REFERENCES

Adams, W., Blust, R., Dwyer, R., Mount, D., Nordheim, E., Rodriguez, P.H., Spry, D., 2020. Bioavailability assessment of metals in freshwater environments: A historical review. *Environ. Toxicol. Chem.* 39, 48–59. https://doi.org/10.1002/etc.4558

Anderson, K.A., Hillwalker, W.E., 2008. Bioavailability. In: Jørgensen, S.E. and Fath, B.D. (eds.) *Encyclopedia of Ecology*, pp. 348–357. Academic Press, Oxford. https://doi.org/10.5860/choice.47-0032

Ball, J.W., Jenne, E.A., Cantrell, M.W., 1981. WATEQ3: A geochemical model with uranium added. US Geological Survey. 2331–1258. https://doi.org/10.3133/ofr811183

Barnes, I., Clarke, F.E., 1969. Chemical properties of ground water and their corrosion and encrustation effects on wells. USGS Professional Paper 498–D. https://doi.org/10.3133/pp498D

Beyer, A., Mackay, D., Matthies, M., Wania, F., Webster, E., 2000. Assessing long-range transport potential of persistent organic pollutants. *Environ. Sci. Technol.* 34(4), 699–703. https://doi.org/10.1021/es990207w

Beyer, A., Matthies, M., 2002. *Criteria for Atmospheric Long-Range Transport Potential and Persistence of Pesticides and Industrial Chemicals*, Erich Schmidt, Berlin.

Boeije, G., Schowanek, D., Vanrolleghem, P., 1998. Adaptation of the simple treat chemical fate model to single-sludge biological nutrient removal wastewater treatment plants. *Water Sci.* Technol. 38(1), 211–218. https://doi.org/10.1016/S0273–1223(98)00406–5

Brandes, L.J., Den, H.H., Van, M.D., 1996. SimpleBox 2.0: A nested multimedia fate model for evaluating the environmental fate of chemicals. Report 719101029, National Institute of Public Health and the Environment, Bilthoven.

Brix, K.V., Tear, L., Santore, R.C., Croteau, K., DeForest, D.K., 2021. Comparative performance of multiple linear regression and biotic ligand models for estimating the bioavailability of copper in freshwater. *Environ. Toxicol. Chem.* 40(6), 1649–1661. https://doi.org/10.1002/etc.5012

Chen, H.Y., Tian, Y.X., Cai, Y.X., Liu, Q.Y., Ma, J., Wei, Y., Yang, A.F., 2023. A 50-year systemic review of bioavailability application in soil environmental criteria and risk assessment. *Environ. Pollut.* 335, 122272. https://doi.org/10.1016/j.envpol.2023.122272

Chen, Y., Schnitzer, M. 1989. Sizes and shapes of humic substances by electron microscopy. In: Hayes, M.H.B., MacCarthy, P., Malcolm, R.L., Swift, R.S., (eds.) *Humic Substances II: In Search of Structure*, pp 621–638. Wiley & Sons, Chichester.

Cherkasov, A., Muratov, E.N., Fourches, D., Varnek, A., Baskin, I.I., Cronin, M., Dearde, J., Gramatica, P., Martin, Y.C., Todeschini, R., Consonni, V., Kuz'min, V.E., Cramer, R., Benigni, R., Yang, C., Rathman, J., Terfloth, L., Gasteige, J., Richard, A., Tropsha, A., 2014. QSAR modeling: Where have you been? where are you going to? *J. Med. Chem.* 57(12), 4977–5010. https://doi.org/10.1021/jm4004285

Chung, J., Hwang, D.S., Park, D.H., An, Y.J., Yeom, D.H., Park, T.J., Choi, J., Lee, J.H., 2021. Derivation of acute copper biotic ligand model-based predicted no-effect concentrations and acute-chronic ratio. *Sci. Total Environ.* 780, 146425. https://doi.org/10.1016/j.scitotenv.2021.146425

Cipullo, S., Snapir, B., Prpich, G., Campo, P., Coulon, F., 2019. Prediction of bioavailability and toxicity of complex chemical mixtures through machine learning models. *Chemosphere* 215, 388–395. https://doi.org/10.1016/j.chemosphere.2018.10.056

Datta, A.R., Kang, Q., Chen, B., Ye, X., 2018. Fate and transport modeling of emerging pollutants from watersheds to oceans. In: Chen, B., Zhang, B., Zhu, Z., Lee, K. (eds.) *Advances in Marine Biology*, . pp. 97–128, Cambridge, MA: Academic Press.

Diamond, M.L., Ganapathy, M., Peterson, S., Mach, C., 2000. Mercury dynamics in the Lahontan reservoir, Nevada: Application of the QWASI fugacity/aquivalence multispecies model. *Water Air Soil Pollut.* 117(1), 133–156. https://doi.org/10.1023/A:1005126704309

Diamond, M.L., Mackay, D., Welbourn, P., 1992. Models of multi-media partitioning of multispecies chemicals: The fugacity/aquivalence approach. *Chemosphere* 25(12), 1907–1921. https://doi.org/10.1016/0045-6535(92)90030-U

Dudal, Y., Gérard, F., 2004. Accounting for natural organic matter in aqueous chemical equilibrium models: A review of the theories and applications. *Earth-Sci. Rev.* 66(3), 199–216. https://doi.org/10.1016/j.earscirev.2004.01.002

European Comission (EC), 2000. Directive 2000/60/EC of the European Parliament and of the Council establishing a framework for the Community action in the field of water policy. *OJEU*, L327, 1–73.

EC, 2011. Technical guidance for deriving environmental quality standards guidance document no. 27. Common implementation strategy for the water framework directive (2000/60/EC). ISBN: 978-92-79-16228-2. https://circabc.europa.eu/sd/a/0cc3581b-5f65-4b6f-91c6-433a1e947838/TGD-EQS%20CIS-WFD%2027%20EC%202011.pdf.

EC, 2013. Directive 2013/39/EU of the European Parliament and of the Council of 12 August 2013 amending Directives 2000/60/EC and 2008/105/EC as regards priority substances in the field of water policy. *OJEU*, L226, 1–17.

ECETOC, 1998. QSARs in the assessment of the environmental fate and effects of chemicals (Technical Report No. 74). ECETOC, Brussels.

ECETOC, 2021. Persistent chemicals and water resources protection (Technical Report No. 139). ECETOC, Brussels.

Erickson, R.J., Benoit, D.A., Mattson, V.R., Nelson Jr, H.P., Leonard, E.N., 1996. The effects of water chemistry on the toxicity of copper to fathead minnows. *Environ. Toxicol. Chem.* 15, 181–193. https://doi.org/10.1002/etc.5620150217

Felmy, A.R., Reisenauer, A.E., Zachara, J.M., Gee, G.W., 1984. MININR: A geochemical computer program for inclusion in water flow models–An application study, Office of Scientific and Technical Information (OSTI). https://doi.org/10.2172/5284389

Franco, A., Trapp, S., 2010. A multimedia activity model for ionizable compounds: Validation study with 2,4-dichlorophenoxyacetic acid, aniline, and trimethoprim. *Environ. Toxicol. Chem.* 29(4), 789–799. https://doi.org/10.1002/etc.115

Gaines, G.L. Jr, Thomas, H.C., 1953. Adsorption studies on clay minerals. II. A formulation of the thermodynamics of exchange adsorption. *J. Chem. Phys.* 21, 714–718. https://doi.org/10.1063/1.1698996

Garman, E.R., Meyer, J.S., Bergeron, C.M., Blewett, T.A., Clements, W.H., Elias, M.C., Farley, K.J., Gissi, F., Ryani, A.C., 2020. Validation of bioavailability-based toxicity models for metals. *Environ. Toxicol. Chem.* 39(1), 101–117. https://doi.org/10.1002/etc.4563

Gharagheizi, F., 2009. Prediction of upper flammability limit percent of pure compounds from their molecular structures. *J. Hazard. Mater.* 167(1), 507–510. https://doi.org/10.1016/j.jhazmat.2009.01.002

Goldberg. S., 2014. Application of surface complexation models to anion adsorption by natural materials. *Environ. Toxicol. Chem.* 3310, 2172–2180. https://doi.org/10.1002/etc.2566

Goyal, A., Bengio, Y., 2022. Inductive biases for deep learning of higher-level cognition. *Proc. R. Soc. A* 478(2266), 20210068. https://doi.org/10.1098/rspa.2021.0068

Groenenberg, J.E. Lofts, S., 2014. The use of assemblage models to describe trace element partitioning, speciation, and fate: A review. *Environ. Toxicol. Chem.* 33(10), 2181–2196. https://doi.org/10.1002/etc.2642

Gustafsson, J.P., 2001. Modeling the acid–base properties and metal complexation of humic substances with the Stockholm Humic Model. *J. Colloid Interface Sci.* 244(1), 102–112. https://doi.org/10.1006/jcis.2001.7871

Gustafsson, J.P., Kleja, D.B., 2005. Modeling salt-dependent proton binding by organic soils with the NICA-Donnan and Stockholm Humic models. *Environ. Sci. Technol.* 39, 5372–5377. https://doi.org/10.1021/es0503332

Gustafsson, J.P., Lumsdon, D.G., 2014. Comment on "Citrate adsorption can decrease soluble phosphate concentration in soils: Results of theoretical modelling" by Marek Duputel, Michel Brossard, Benoit Jaillard, Davey L. Jones, Philippe Hinsinger and Frédéric Gerard (2013). *Appl. Geochem.* 46, 85–89. https://doi.org/10.1016/j.apgeochem.2014.02.006

Guthrie, J.W., Hassan, N., Salam, M., Fasfous, I., Murimboh, C., Murimboh, J., Chakrabarti, C., Grégoire, D.C., 2005. Complexation of Ni, Cu, Zn, and Cd by DOC in some metal-impacted freshwater lakes: A comparison of approaches using electrochemical determination of free-metal-ion and labile complexes and a computer speciation model, WHAM V and VI. *Anal. Chim. Acta.* 528(2), 205–218. https://doi.org/10.1016/j.aca.2004.10.003

Haws, N.W., Ball, W.P., Bouwer, E.J., 2006. Modeling and interpreting bioavailability of organic contaminant mixtures in subsurface environments. *J. Contam. Hydrol.* 82(3), 255–292. https://doi.org/10.1016/j.jconhyd.2005.10.005

Hiemstra, T., Van Riemsdijk, W.H., 1996. A surface structural approach to ion adsorption: The charge distribution (CD) model. *J. Colloid Interface Sci.* 179, 488–508. https://doi.org/10.1006/jcis.1996.0242

Jordan, M.I., Mitchell, T.M., 2015. Machine learning: Trends, perspectives, and prospects. *Science* 349(6245), 255–260. https://doi.org/10.1126/science.aaa8415

Keshavarz, M.H., Shirazi, Z., Kiani, S.P., 2021. Risk assessment of organic aromatic compounds to *Tetrahymena pyriformis* in environmental protection by a simple QSAR model. *Process Saf. Environ. Prot.* 150, 137–147. https://doi.org/10.1016/j.psep.2021.04.011

Kinniburgh, D.G., Milne, C.J., Benedetti, M.F., Pinheiro, J.P., Filius, J., Koopal, L.K., Van Riemsdijk, W.H., 1996. Metal ion binding by humic acid: Application of the NICA-Donnan model. *Environ. Sci. Technol.* 30(5), 1687–1698. https://doi.org/10.1021/es950695h

Kinniburgh, D.G., Van Riemsdijk, W.H., Koopal, L.K., Borkovec, M., Benedetti, M.F., Avena, M.J., 1999. Ion binding to natural organic matter: Competition, heterogeneity, stoichiometry and thermodynamic consistency. *Colloids Surf. Physicochem. Eng. Asp.* 151(1–2), 147–166. https://doi.org/10.1016/S0927-7757(98)00637-2

Koopal, L.K., Saito, T., Pinheiro, J.P., Riemsdijk, W.H.v., 2005. Ion binding to natural organic matter: General considerations and the NICA–Donnan model. *Colloid Surf. A.* 265, 40–54. https://doi.org/10.1016/j.colsurfa.2004.11.050

Lavado, G.J., Baderna, D., Carnesecchi, E., Toropova, A.P., Toropov, A.A., Dorne, J.L.C.M., Benfenati, E., 2022. QSAR models for soil ecotoxicity: Development and validation of models to predict reproductive toxicity of organic chemicals in the collembola *Folsomia candida*. *J. Hazard. Mater.* 423(1), 127236. https://doi.org/10.1016/j.jhazmat.2021.127236

Lipczynska-Kochany, E., 2018. Humic substances, their microbial interactions and effects on biological transformations of organic pollutants in water and soil: A review. *Chemosphere* 202, 420–437. https://doi.org/10.1016/j.chemosphere.2018.03.104

Lofts, S., Tipping, E., 1998. An assemblage model for cation binding by natural particulate matter. *Geochim. Cosmochim. Acta* 62, 2609–2625. https://doi.org/10.1016/S0016-7037(98)00183-5

Mackay, D., Di Guardo, A., Paterson, S., Cowan, C.E., 1996. Evaluating the environmental fate of a variety of types of chemicals using the EQC model. *Environ. Toxicol. Chem.* 15(9), 1627–1637. https://doi.org/10.1002/etc.5620150929

MacLeod, M.J., 2002 Contaminant fate models to support the sustainable use of chemicals in North America. Ph.D., Trent University (Canada), Canada-Ontario, CA.

Mantoura, R.F.C., Dickson, A., Riley, J.P., 1978. The complexation of metals with humic materials in natural waters. *Estuar. Coast. Mar. Sci.* 6(4), 387–408. https://doi.org/10.1016/0302–3524(78)90130–5

Marinsky, J.A., 1987. A two - phase model for the interpretation of proton and metal ion interaction with charged polyelectrolyte gels and their linear analogues. In: Stumm, W. (ed.), *Aquatic Surface Chemistry: Chemical Processes at the Particle-Water Interface* (pp.49–8) Wiley, New York.

Miller, C.W., Benson, L.V., 1983. Simulation of solute transport in a chemically reactive heterogeneous system: Model development and application. *Water Resour. Res.* 19, 381–391. https://doi.org/10.1029/WR019i002p00381

Milne, C.J., Kinniburgh, D.G., Tipping, E., 2001. Generic NICA-Donnan model parameters for proton binding by humic substances. *Environ. Sci. Technol.* 35(10), 2049–2059. https://doi.org/10.1021/es000123j

Neira-Albornoz, A., Martínez-Parga-Méndez, M., González, M., Spitz, A., 2024. Understanding requirements, limitations and applicability of QSAR and PTF models for predicting sorption of pollutants on soils: A systematic review. *Front. Environ. Sci.* 12, 1379283. https://doi.org/10.3389/fenvs.2024.1379283

Ortega-Calvo, J.J., Tejeda-Agredanoa, M.C., Jimenez-Sancheza, C., Congiua, E., Sungthonga, R., Niqui-Arroyob, J.L., Cantosa, M., 2013. Is it possible to increase bioavailability but not environmental risk of PAHs in bioremediation? *J. Hazard. Mater.* 261, 733–745. https://doi.org/10.1016/j.jhazmat.2013.03.042

Parkhurst, D.L., Plummer, L.N., Thorstenson, D.C., 1980. PHREEQE: A computer program for geochemical calculations. U.S. Geological Survey Water-Resources Investigations Report, 80–96. https://doi.org/10.3133/wri8096

Patterson, C.J., Semple, K.T., Paton, G.I., 2004. Non-exhaustive extraction techniques (NEETs) for the prediction of naphthalene mineralisation in soil. *FEMS Microbiol. Lett.* 241(2), 215–220. https://doi.org/10.1016/j.femsle.2004.10.023

Peng, X.X., Gai, S., Cheng, K., Yang, F., 2022. Roles of humic substances redox activity on environmental remediation. *J. Hazard. Mater*. 435, 129070. https://doi.org/10.1016/j.jhazmat.2022.129070

Perdue, E.M., Reuter, J.H., Parrish, R.S., 1984. A statistical model of proton binding by humus. *Geochim. Cosmochim. Acta* 48(6), 1257–1263. https://doi.org/10.1016/0016–7037(84)90060–7

Perkins, R., Fang, H., Tong, W., Welsh, W.J., 2003. Quantitative structure-activity relationship methods: Perspectives on drug discovery and toxicology. *Environ. Toxicol. Chem.* 22(8), 1666–1679. https://doi.org/10.1897/01–171

Priemer, D.A., Diamond, M.L., 2002. Application of the multimedia urban model to compare the fate of SOCs in an urban and forested watershed. *Environ. Sci. Technol*. 36(5), 1004. https://doi.org/10.1021/es001397+

Qu, R., Wang, X., Liu, Z., Yan, Z., Wang, Z., 2013. Development of a model to predict the effect of water chemistry on the acute toxicity of cadmium to *Photobacterium phosphoreum*. *J. Hazard. Mater*. 262, 288–296. https://doi.org/10.1016/j.jhazmat.2013.08.039

Reid, B.J., Stokes, J.D., Jones, K.C., Semple, K.T., 2000. Nonexhaustive cyclodextrin-based extraction technique for the evaluation of PAH bioavailability. *Environ. Sci. Technol*. 34(5), 3174–3179. https://doi.org/10.1021/es990946c

Santore, R.C., Di Toro, D.M., Paquin, P.R., Allen, H.E., Meyer, J.S., 2001. Biotic ligand model of the acute toxicity of metals. 2. Application to acute copper toxicity in freshwater fish and Daphnia. *Environ. Toxicol. Chem*. 20(10), 2397–2402. https://doi.org/10.1002/etc.5620201035

Schlekat, C., Stubblefield, W., Gallagher, K., 2020. State of the science on metal bioavailability modeling: Introduction to the outcome of a society of environmental toxicology and chemistry technical workshop. *Environ. Toxicol. Chem*. 39, 42–47. https://doi.org/10.1002/etc.4561

Soares, T.A., Nunes-Alves, A., Mazzolari, A., Ruggiu, F., Wei, G.-W., Merz, K., 2022. The (Re)-evolution of quantitative structure–activity relationship (QSAR) studies propelled by the surge of machine learning methods. *J. Chem. Inf. Model.* 62(22), 5317–5320. https://doi.org/10.1021/acs.jcim.2c01422

Song, S., Su, C., Lu, Y., Wang, T., Zhang, Y., Liu, S., 2016. Urban and rural transport of semivolatile organic compounds at regional scale: A multimedia model approach. *J. Environ. Sci.* 39, 228–241. https://doi.org/10.1016/j.jes.2015.12.005

Spasojević, J.M., Maletić, S.P., Rončević, S.D., Radnović, D.V., Čučak, D.I., Tričković, J.S., Dalmacija, B.D., 2015. Using chemical desorption of PAHs from sediment to model biodegradation during bioavailability assessment. *J. Hazard. Mater.* 283(20), 60–69. https://doi.org/10.1016/j.jhazmat.2014.09.013

Stockdale, A., Bryan, N.D., Lofts, S., 2011. Estimation of Model VII humic binding constants for Pd^{2+}, Sn^{2+}, U^{4+}, NpO_2^{2+}, Pu^{4+} and PuO_2^{2+}. *J. Environ. Monit.* 13(10), 2946–2950. https://doi.org/10.1039/C1EM10407A

Su, C., Zhang, H., Cridge, C., Liang, R., 2019. A review of multimedia transport and fate models for chemicals: Principles, features and applicability. *Sci. Total Environ*. 668, 1063–1076. https://doi.org/10.1016/j.scitotenv.2019.02.456

Sun, Y., Ying, L., 2021, Application of surface complexation modeling on adsorption of uranium at water-solid interface: A review. *Environ. Pollut*. 278, 116861. https://doi.org/10.1016/j.envpol.2021.116861

Tipping, E., 1998. Humic ion-binding model VI: An improved description of the interactions of protons and metal ions with humic substances. *Aquat. Geochem*. 4(1), 3–47. https://doi.org/10.1023/A:1009627214459

Tipping, E., Lofts, S., Sonke, J., 2011. Humic ion-binding model VII: A revised parameterisation of cation-binding by humic substances. *Environ. Chem*. 8(3), 225–235. https://doi.org/10.1071/EN11016

Tonkin, J.W., Balistrieri, L.S., Murray, J.W., 2004. Modeling sorption of divalent metal cations on hydrous manganese oxide using the diffuse double layer model. *Appl. Geochem.* 38, 81–88. https://doi.org/10.1016/S0883–2927(03)00115–X

Tropsha, A., 2012. Recent trends in statistical QSAR modeling of environmental chemical toxicity. In: Luch, A. (ed.) *Molecular, Clinical and Environmental Toxicology: Volume 3: Environmental Toxicology*, pp. 381–411. Springer, Basel. https://doi.org/10.1007/978–3–7643–8340–4–13

Truesdell, A.H., Jones, B.F., 1974. WATEQ: A computer program for calculating chemical equilibria of natural waters. *J. Res. U.S. Geol. Surv.* 2, 233–274.

US EPA, 1985. Guidelines for deriving numerical national water quality criteria for the protection of aquatic organisms and their uses. US EPA, Duluth, MN.

US EPA, 2007. Aquatic life ambient freshwater quality criteria—Copper. EPA-822-R-07-001. Washington, DC.

US EPA, 2010. Draft update of ambient water quality criteria for copper. US Environmental Protection Agency, Washington, DC.

Wania, F., Mackay, D., 1995. A global distribution model for persistent organic chemicals. *Sci. Total Environ.* 160, 211–232. https://doi.org/10.1016/0048–9697(95)04358–8

Wells, M., Wick, L.Y., Harms, H., 2005. Model polymer release system study of PAH bioaccessibility: The relationship between "rapid" release and bioaccessibility. *Environ. Sci. Technol.* 39(4), 1055–1063. https://doi.org/10.1021/es035067b

Westall, J., Hohl, H., 1980. A comparison of electrostatic models for the oxide/solution interface. Adv. *Colloid Interfac.* 12(4), 265–294. https://doi.org/10.1016/0001-8686(80)80012-1.

Westall, J.C., Zachary, J.L., Morel, F.M.M., 1976. MINEQL: A computer program for the calculation of chemical equilibrium composition of aqueous systems. Technical Note No. 18, Ralph M. Parsons Laboratory for Water Resources and Environmental Engineering, Massachusetts Institute of Technology, Boston, MA. https://hdl.handle.net/1721.1/142980

Whelan, G., Buck, J.W., Strenge, D.L., Droppo, J.G., Hooppes, B.L., Aiken, R.J., 1992. Overview of the multimedia environmental pollutant assessment system (MEPAS). *Hazard. Waste Hazard. Mater.* 9, 191–208. https://doi.org/10.1089/hwm.1992.9.191

Wöhrnschimmel, H., MacLeod, M., Hungerbuhler, K., 2013. Emissions, fate and transport of persistent organic pollutants to the Arctic in a changing global climate. *Environ. Sci. Technol.* 47(5), 2323–2330. https://doi.org/10.1021/es304646n

Wu, G., Kechavarzi, C., Li, X., Wu, S., Pollard, S.J.T., Sui, H., Coulon, F., 2013. Machine learning models for predicting PAHs bioavailability in compost amended soils. *Chem. Eng. J.* 223, 747–754. https://doi.org/10.1016/j.cej.2013.02.122

Zhou, P., Yan, H., Gu, B.H., 2005. Competitive complexation of metal ions with humic substances. *Chemosphere* 58,1327–1337. https://doi.org/10.1016/j.chemosphere.2004.10.017

Zhu, Y., Tao, S., Price, O.R., Shen, H., Jones, K.C., Sweetman, A.J., 2015. Environmental distributions of benzo[a]pyrene in China: Current and future emission reduction scenarios explored using a spatially explicit multimedia fate model. *Environ. Sci. Technol.* 49(23), 13868–13877. https://doi.org/10.1021/acs.est.5b00474

15 Speciation and Biotic Ligand Modeling

15.1 INTRODUCTION AND SCOPE

Understanding the ***bioavailability*** of heavy metal pollutants in the environment is increasingly essential for conducting effective risk assessments and making informed decisions regarding ecological management. Chemical speciation models have emerged as powerful cost- and time-effective tools for estimating the bioavailability of heavy metal pollutants. These models account for the chemical equilibria governing heavy metal pollutant ***distribution*** and also integrate environmental parameters such as pH, ***ionic strength***, and competing ions to refine predictions of bioavailability based on chemical speciation.

Speciation modeling for environmental applications is not restricted to modeling the speciation of heavy metals, although in recent years heavy metals have been a strong focus, and so modeling heavy metal speciation is the focus of this chapter. For heavy metals, a critical aspect of speciation modeling involves the accurate assessment of metal binding with dissolved organic matter (DOM), often represented as dissolved organic carbon (DOC) in models. Several approaches have been developed to refine the specification of metal–DOM interactions, ranging from empirical to mechanistic. Similarly, it is important in some environments to consider the ***sorption*** of heavy metals and organics onto mineral surfaces (present in the water column as particulates) because these interactions significantly influence both the mobility and availability of heavy metals in the environment. Minerals, e.g., iron oxyhydroxides, carbonates, and clays, due to their high surface area in particular, form in natural environments and may act as natural sorbents that bind heavy metals through processes such as ion exchange, surface ***complexation***, and van der Waals forces. Reactions between heavy metals and DOM or mineral particulates reduce the free ion concentration of heavy metals in the aqueous phase, thereby decreasing bioavailability (*NB*: The same may be said for organic ions with respect to sequestering into high-molecular weight DOM or binding to particulates).

Since the original elaboration of speciation models in the 1960s and 1970s (per Chapter 14), the late 1990s saw the development of the speciation-based ***Biotic Ligand Model*** (BLM), which includes reactions between heavy metals and biological receptors, i.e., ***biotic ligands*** (BLs), that are part of an organism's toxic response to heavy metals. BLM models have been validated without metal–surface interactions, whereas metal–organic reactions are needed for validation of this type of model, leading to an inference that metal–DOM interactions are more important than metal–mineral (particulate) interactions (Crémazy et al., 2016; Di Toro et al., 2001). Regulatory agencies worldwide have adopted the BLM for deriving environmental quality standards and assessing site-specific risks, further highlighting its utility (EC, 2010; US EPA, 2003). Figure 15.1 is a schematic diagram showing how BLM models work.

In the initial conceptualization of this chapter, to support provision of methodological examples of the application of speciation models, three widely used speciation modeling software were evaluated: Visual MINTEQ 4.0, MINEQL+, and CHEAQS Next. All three tools are free, support the addition of custom reactions, and offer capabilities for modeling metal–organic and metal–particulate (surface) interactions. Beyond these shared functionalities, the tools differ significantly in their capabilities and usability.

In terms of heavy metal–DOM (M–DOM) interactions, the tools vary in flexibility. Visual MINTEQ provides the most sophisticated options, supporting both the Stockholm Humic Model (SHM) and the Non-ideal Competitive Adsorption–Donnan (NICA–Donnan) models to describe

DOI: 10.1201/9781003355229-19

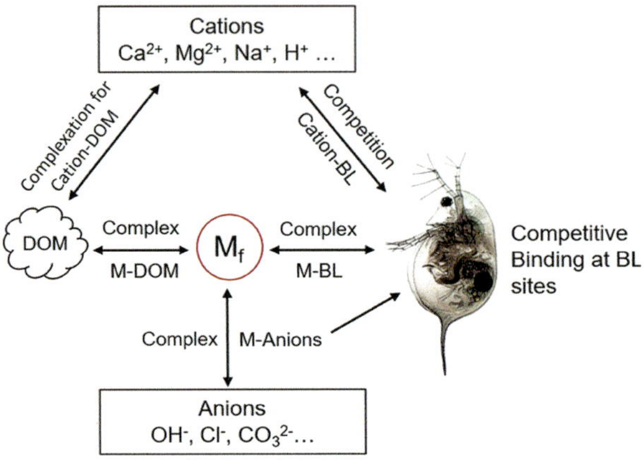

FIGURE 15.1 Schematic diagram showing the conceptual basis of speciation-based BLM modeling.

DOM reactivity, allowing users to choose, respectively, between a discrete-site gel model and a continuous-site distribution model, both of which also incorporate more advanced electrostatic interactions.[1] CHEAQS Next supports the Humic Ion-Binding V, VI, and VII models, which are robust hard-sphere discrete-site models (i.e., neglecting reactions between heavy metals in the interior volumes of DOM), while MINEQL+ offers the least flexibility for M–DOM modeling. None of these tools provide capabilities for modeling the sequestration of organic pollutants into humic or fulvic acids (HA or FA), which is known to play a part in speciation in natural environments. More specialized software, such as PHREEQC, is needed for such applications (Parkhurst and Appelo, 2013).

In terms of surface interactions, including metal–particulate and organic–particulate interactions via ion exchange or complexation, the three tools differ in their capabilities. All support surface complexation and ion exchange, but their approaches vary. To describe how ions interact with charged surfaces during surface complexation, Visual MINTEQ 4.0 has implemented the constant Constant Capacitance Model (CCM), Diffuse Layer Model (DLM), and Triple Layer Model (TLM), enabling simulation of inner- and outer-sphere complexation with electrostatic effects; CHEAQS Next and MINEQL+ offer the DLM and TLM. Visual MINTEQ supports ion exchange via the Gaines–Thomas, Vanselow, or Gapon models, with predefined ***selectivity*** coefficients available in its database, and can model precipitation and redox reactions at surfaces. CHEAQS Next uses the Gaines–Thomas model for ion exchange, with options for Vanselow and Gapon models, and also accounts for precipitation and redox reactions. MINEQL+ is more limited, supporting only the Gaines–Thomas model for ion exchange, and its surface interaction capabilities are less flexible compared to the other two tools.

Also relevant for environmental applications, CHEAQS Next and MINEQL+ offer additional capabilities for modeling kinetic processes, such as kinetics-limited ***adsorption*** and redox reactions. CHEAQS Next is particularly notable for its advanced kinetic modeling, which includes adsorption, complexation, and redox processes, making it the most versatile tool for time-dependent processes. In contrast, Visual MINTEQ is restricted to equilibrium modeling and does not support modeling of kinetic processes. Of the three software, BLM calculations are only implemented in Visual MINTEQ.

When it comes to the databases on which these software depend, Visual MINTEQ integrates the WATEQ4F thermodynamic database (USGS, 2024), whereas CHEAQS Next incorporates data from the Nuclear Energy Agency (Paris) and the International Union of Pure and Applied Chemists databases; both of these are reasonably comprehensive for environmental applications. MINEQL+ uses a proprietary database focused on inorganic species, with limited coverage of organic complexes.

[1] For further details on the theory of models discussed herein, refer to Chapter 14.

Visual MINTEQ, MINEQL+, and CHEAQS Next all support batch runs (in Visual MINTEQ via multiproblem functionality). The batch capability allows users to automate the running of multiple simulations or scenarios with varying input parameters. Analysis of outputs from these multiple input scenarios provides data for ***sensitivity*** analysis, i.e., exploration of how changes in input variables affect the output of a model. Both MINEQL+ and CHEAQS Next offer built-in functionality to analyze batch outputs and report sensitivity, whereas using Visual MINTEQ, the user must perform the sensitivity manually using outputs from batch runs. In terms of titration modeling, MINEQL+ offers more advanced tools for simulating titration experiments, whereas CHEAQS Next and Visual MINTEQ provide basic titration modeling capabilities.

From a usability perspective, all three tools are Windows-based, but macOS and Linux users can run them using emulators or virtual machines (e.g., VirtualBox for macOS or Linux). For all three, there is capability to modify parameters governing reactions. MINEQL+ provides better graphical output capabilities, such as speciation distribution plots and titration curves, while Visual MINTEQ and CHEAQS Next primarily generate detailed tabular outputs. CHEAQS Next supports more advanced graphical visualization, but this requires manual configuration, which reduces the user-friendliness of the capability. All three software will output comma-separated value files and accept Excel files for inputs. Visual MINTEQ stands out for its user-friendly interface, which makes it more accessible to beginners compared to the steeper learning curves of CHEAQS Next and MINEQL+.

Ultimately, Visual MINTEQ was selected as the most suitable software for the example problems presented in this chapter, primarily due to the superiority of its user-friendly interface. Other considerations include its comparatively advanced M–DOM modeling options and its ability to support BLM calculations, an important feature for assessing how bioavailability relates to toxicity for heavy metals in aqueous environmental systems.

This chapter will equip readers with a basic understanding of how to use Visual MINTEQ 4.0 to estimate heavy metal bioavailability, providing examples relevant to environmental applications. The four examples chosen – heavy metal speciation in water, reactions involving solids, BLM calculations, and calculations requiring alteration of Visual MINTEQ databases – were selected because they represent fundamental and widely encountered scenarios in environmental chemistry. These examples address key processes that influence heavy metal behavior: the distribution of chemical species in aqueous systems, the impact of solid-phase formation on heavy metal bioavailability, and the interaction of ***bioavailable*** heavy metals with biological receptors and how this relates to toxicity. These examples cover a range of basic scenarios so that readers can better understand the underlying theory, as well as gain some experience in application. For these examples and most other applications, speciation modeling can be performed on a personal computer, and the software is obtained by free download; accordingly there is no section for equipment in this chapter. The basic background to chemical speciation modeling is given in Chapter 14, with further elaboration herein. Chapter 2, containing Terms and Definitions, is also within the scope of this chapter.

15.2 PRINCIPLE

Visual MINTEQ is a geochemical equilibrium speciation model based on MINTEQA2, introduced in Chapter 14. As for MINTEQA2 and many other speciation software, Visual MINTEQ employs the Tableau Method (Morel, 1983), a matrix-based approach to organizing and solving chemical equilibrium equations. In the Tableau Method, it is first necessary to define system ***components*** and ***species***. Components are the basic building blocks or individual elements and compounds that make up the chemical system being studied. They represent the minimum set of independent chemical entities from which all species in the system can be derived. Components in a system could be molecular, for instance, phenol, or an element, such as Al^{3+}, or a ***ligand***, such as ethylenediaminetetraacetic acid (EDTA). Species are the different chemical forms that the components participate in forming, i.e., the actual chemical entities present at chemical equilibrium. So, for instance, OH^- can be described as a species based on a reaction involving components H^+ and H_2O, and PbEDTA is a species based on components Pb^{2+} (lead free ion) and EDTA. The distinction between ***components*** and ***species*** is not always intuitively obvious, but must be determined correctly for modeling purposes and is guided by reaction chemistry. Consider each component in terms of the reactions in which they do and can participate. For instance, given a simple aqueous system, water is always present, and hence, is a logical component. Since water is made up of H^+ and OH^-, it is necessary to add another component to describe this reaction. Although OH^- could also be chosen as a component, H^+ is more convenient as pH is a variable in many chemical problems. Dissolving a small amount of $PbCl_2$ in this system additionally adds the components Pb^{2+} and Cl^-. The key components in this simple system, then, as defined in Visual MINTEQ, are Cl^-, H^+, H_2O, and Pb^{2+}. The relevant equilibrium reactions in which these components participate are as follows:

1. Self-ionization of water:

$$H_2O \rightleftharpoons H^+ + OH^-.$$

2. Complexation of Pb^{2+} with Cl^-:

$$Pb^{2+} + Cl^- \rightleftharpoons PbCl^+,$$

$$Pb^{2+} + 2Cl^- \rightleftharpoons PbCl_2\ (aq),$$

$$Pb^{2+} + 3Cl^- \rightleftharpoons PbCl_3^-,$$

$$Pb^{2+} + 4Cl^- \rightleftharpoons PbCl_4^{2-}.$$

3. Formation of lead hydroxide species:

$$Pb^{2+} + H_2O \rightleftharpoons PbOH^+ + H^+,$$

$$Pb^{2+} + 2H_2O \rightleftharpoons Pb(OH)_2\ (aq) + 2H^+,$$

$$Pb^{2+} + 3H_2O \rightleftharpoons Pb(OH)_3^- + 3H^+,$$

$$2Pb^{2+} + H_2O \rightleftharpoons Pb_2OH^{3+} + H^+,$$

$$3Pb^{2+} + 4H_2O \rightleftharpoons Pb_3(OH)_4^{2+} + 4H^+,$$

$$4Pb^{2+} + 4H_2O \rightleftharpoons Pb_4(OH)_4^{4+} + 4H^+.$$

Using these reactions to construct a species–component Tableau for the Cl^-, H^+, H_2O, and Pb^{2+} system in Visual MINTEQ systematically represents the mass balance relationships between components and associated species. In the species–component Tableau, each row corresponds to a chemical species, such as $PbCl^+$ or $Pb(OH)_2(aq)$, while each column represents a component, i.e., Cl^-, H^+, H_2O, or Pb^{2+}. Although Visual MINTEQ does not explicitly display components as species, it is important to note that from a chemical modeling perspective, each component is also a species in the system, appearing in equilibrium calculations. Table 15.1 shows the Visual MINTEQ species–component Tableau.

Each entry in the Tableau ***matrix*** indicates the number of component atoms or molecules participating in a reaction. For example, one Pb^{2+} (component, coefficient of 1) combines with three H_2O molecules (component, coefficient of 3) to form $Pb(OH)_3^-$, releasing three protons/H^+ (component, coefficient of –3, i.e., subtracted from the H_2O) in the process. Each equilibrium reaction has an

TABLE 15.1
Species-Component Tableau for the Cl^-, H^+, H_2O, and Pb^{2+} System in Visual MINTEQ

	Cl^-	H^+	H_2O	Pb^{2+}
OH^-	0	–1	1	0
$PbCl^+$	1	0	0	1
$PbCl_2$ (aq)	2	0	0	1
$PbCl_3^-$	3	0	0	1
$PbCl_4^{2-}$	4	0	0	1
$PbOH^+$	0	–1	1	1
$Pb(OH)_2$ (aq)	0	–2	2	1
$Pb(OH)_3^-$	0	–3	3	1
Pb_2OH^{3+}	0	–1	1	2
$Pb_3(OH)_4^{2+}$	0	–4	4	3
$Pb_4(OH)_4^{4+}$	0	–4	4	4

associated equilibrium constant, K, that defines the stability of the species formed under specific conditions of temperature, and pressure. These constants are used to calculate how much of each species are present at equilibrium, by describing the extent to which reactions proceed to form products, using ***mass action equations***. These mass action equations take the form of a quotient: the numerator of the quotient is a product of the activities of the chemical reaction products, each raised to the power of their stoichiometric coefficients, and the denominator is a product of the activities of the chemical reactants, also each raised to the power of their stoichiometric coefficients. ***Activity***, a, is the effective concentration of a species in solution, related to its actual concentration, C, by an ***activity coefficient***, γ, such that $a = \gamma C$. The activity coefficient accounts for deviations from ideality due to ionic interactions and is therefore a function of ***ionic strength***.[2] At low ionic strength γ is close to 1, meaning activity and concentration are nearly equal. As ionic strength increases, inter-ionic interactions become significant, causing a to diverge from C.

As an example of an equilibrium mass action equation, consider the formation of $Pb(OH)_3^-$ in solution. The equilibrium constant, K, for this reaction is expressed in terms of activities:

$$K = \frac{a_{Pb(OH)_3^-} \times (a_{H^+})^3}{a_{Pb^{2+}} \times (a_{H_2O})^3}$$

Thermodynamically, the activity of a pure substance, in this case water, is equal to 1. Substituting $a_{H_2O} = 1$, the equation above becomes

$$K = \frac{a_{Pb(OH)_3^-} \times (a_{H^+})^3}{a_{Pb^{2+}}}$$

For each K, there is also an apparent equilibrium constant K' (also sometimes referred to as the conditional stability constant) based on concentrations instead of activities:

[2] Ionic strength is a measure of the total concentration of ions in a solution, reflecting the intensity of electrostatic interactions between charged species. It accounts for the influence of all dissolved ions, considering both their concentrations and charges. Higher ionic strength leads to greater inter-ionic interactions, which can affect chemical equilibria, solubility, and activity coefficients. In dilute solutions, ionic strength is low, and the system behaves more ideally, with activity coefficients close to 1. As ionic strength increases, deviations from ideality become significant, altering the behavior of dissolved species and their reactivity in solution.

$$K' = \frac{[\text{Pb(OH)}_3^-] \times [\text{H}^+]^3}{[\text{Pb}^{2+}]}$$

where square brackets [ ] indicate concentrations, in units of molar or molal. Substituting γC for a, the relationship between thermodynamic K and concentration-based K' equilibrium constants is given by

$$K = \frac{a_{\text{Pb(OH)}_3^-} \times (a_{\text{H}^+})^3}{a_{\text{Pb}^{2+}}} = \frac{\gamma_{\text{Pb(OH)}_3^-}[\text{Pb(OH)}_3^-] \times (\gamma_{\text{H}^+}[\text{H}^+])^3}{\gamma_{\text{Pb}^{2+}}[\text{Pb}^{2+}]} = \frac{\gamma_{\text{Pb(OH)}_3^-} \times (\gamma_{\text{H}^+})^3}{\gamma_{\text{Pb}^{2+}}} \times K'$$

which rearranges to

$$K' = \frac{\gamma_{\text{Pb}^{2+}}}{\gamma_{\text{Pb(OH)}_3^-} \times (\gamma_{\text{H}^+})^3} \times K$$

Given that, for a particular temperature and pressure, true K is constant, this equation shows how the apparent equilibrium constant, K', depends on activity coefficients, meaning that under different ionic strengths, the apparent K', i.e., concentration-based K, can vary.

Having established the Tableau, it is now possible to express as system of equations through the matrix equations

$$A = SA_{\text{Tot}} + K, \text{ and}$$

$$C = ST + K',$$

where A and C are column vectors of log species activities (a_{species}) or concentrations ([species]), S is the stoichiometric coefficient matrix (derived from the Tableau in Table 15.1), A_{Tot} and T are column vectors of log total component activities (a_{Tot}) or total component concentrations ($[\text{comp}]_{\text{Tot}}$), and K and K' and are column vectors of the logarithms of values of K and K' for the chemical reactions included in the system being modeled. The log form is used because the log product rule enables simplification of the handling of the stoichiometric coefficients (i.e., which are no longer exponentiated as they are in the mass action relationships).

In practice, most speciation programs work in activity, converting inputs in concentration into activity and then converting results of matrix calculations back into concentration for output, i.e., activity and concentration are interconvertible. The inputs required for these calculations come from several key ***sources***. First, the total concentrations of components in the system must be provided, either from measured experimental values or from predefined system conditions. Second, Ks for relevant chemical reactions are necessary to define the stability of species under given conditions. These constants are obtained from thermodynamic databases, literature, or empirical fitting procedures. Third, activity coefficients are used to correct for non-ideal behavior in solution, often calculated using models that require species-specific parameters, usually obtained from databases or literature. Lastly, while pressure does not vary greatly enough to affect thermodynamic constants in environmental applications (atmospheric pressure in the biosphere does not vary sufficiently), temperature may, and temperature is therefore required as an input.

To solve for A/C, Visual MINTEQ employs iterative numerical solvers based on mass balance and charge balance constraints. Mass balance ensures that the total concentration of each component in the system is conserved, meaning that the sum of all species containing a given component must equal its total input concentration. Charge balance acts as an additional constraint, ensuring that the sum of all positive and negative ionic charges in solution equals zero, preserving electrostatic neutrality. The Visual MINTEQ software optimizes species concentrations until a solution is found that best satisfies mass and charge balance constraints.

The BLM extends conventional speciation modeling by integrating a metal binding reaction to a biological receptor, typically represented as a BL associated with an organism. This additional reaction accounts for how heavy metal bioavailability relates to toxicity, which traditional chemical speciation models do not directly address. The key surface reaction introduced by the BLM is

$$M_f + BL \rightleftharpoons MBL,$$

where M_f represents a heavy metal free ion (e.g., Cu^{2+}), BL is the solid-phase BL, and MBL is the metal–BL complex. The BL represents a biological binding ***site***, such as a receptor on cell membranes or gill surfaces, which, when complexed with a heavy metal ion, initiates processes leading to toxic effects. Metal toxicity depends on how much of the M_f binds to the BL. Because the BL is present at very low concentrations, it does not influence overall heavy metal speciation in the system.

Free metal ions in solution, however, are not solely available to bind with the BL. They also undergo complexation with various dissolved ligands, such as hydroxide, carbonate, chloride, and DOM. These competing ligands reduce the concentration of free metal ions available to interact with the BL. Such equilibria can be described by the generic reaction,

$$M_f + L \rightleftharpoons ML,$$

where L represents competing ligands. As more of a metal binds to these ligands, the bioavailable fraction – i.e., typically defined as free metal ion concentration – decreases. Visual MINTEQ further accounts for competition between toxic heavy metals and major cations for binding sites on the BL. Because naturally occurring cations – such as Ca^{2+}, Mg^{2+}, K^+, and Na^+ – also bind to the BL, they reduce the amount of free BL binding sites available for toxic heavy metals, thereby decreasing metal toxicity. The relevant competitive binding reaction for Ca, for example, is

$$Ca_f + BL \rightleftharpoons CaBL,$$

with other reactions following this form.

Visual MINTEQ implements the BLM through two interconnected computational modes, referred to as speciation mode and toxicity mode, which collectively simulate the interactions of heavy metals with abiotic ligands and BLs to predict bioavailability and toxicity. In speciation mode, the software calculates the equilibrium distribution of metal species in aqueous systems by solving mass action equations using the Tableau Method/matrix equations. Key inputs include water chemistry parameters such as pH, temperature, concentrations of major ions (e.g., Ca^{2+}, Mg^{2+}, Cl^-), total dissolved heavy metal concentration, DOM levels (input as DOC), and optional surface interactions with mineral particulates such as iron oxyhydroxides using surface complexation models. The software has an option for inputting or calculating ionic strength and options for estimating activity coefficients as a function of ionic strength. Since BLM speciation-mode calculations have an added component, BL, the stability constants for metal–BL complexes, K_{MBL}, K_{CaBL}, K_{MgBL}, K_{NaBL}, and K_{KBL}, and in some cases others, must be included in addition to the conventional metal–ligand constants (e.g., *K*s for heavy metal binding with DOM, OH^-, CO_3^{2-}, Cl^-, and others) that are already stored in the Visual MINTEQ database. Primary outputs from speciation mode are concentrations and activities of all species in the system (e.g., MCl^+, $M(OH)_2$). The BLM speciation-mode calculation is differentiated by any other speciation calculation because of the presence of the BL, with resultant information on all BL species that form (e.g., CaBL, MBL). Speciation mode does not provide direct toxicity information.

Regarding toxicity mode calculations, the BLM links metal speciation to toxicity via one of two approaches. One approach involves quantifying the fraction, *f*, of BL sites occupied by a toxic heavy metal; *f* is expressed as the ratio

$$f = \frac{[\text{MBL}]}{[\text{BL}]_{\text{Tot}}},$$

where $[BL]_{Tot}$ is the total BL site concentration. Toxicity occurs when f exceeds a critical threshold or critical fraction, f_{crit}, which is experimentally derived for specific organisms and toxicity endpoints (US EPA, 2003). For example, Li et al. (2022) found that an f_{crit} of 0.14 was associated with a Pb EC_{50} for the ***bioreporter*** zntA. The EC_{50} (half-maximal effective concentration) refers to the concentration of a substance at which 50% of the observed biological toxic response occurs. In the case of Li et al.'s work, EC_{50} corresponds to the concentration of Pb at which 50% of bioreporter cells lost viability on ***exposure*** to Pb, as determined by a specific toxicity protocol. This means that occupancy of 14% of BL sites by Pb corresponds to this EC_{50} toxicity endpoint. In toxicity mode, using the f_{crit} approach, Visual MINTEQ calculates the total heavy metal concentration required to achieve the MBL concentration that corresponds to $f = f_{crit}$.[3] The value of f_{crit} is determined via experimental data, and based on fully mechanistic speciation expressions, as discussed hitherto.

The second approach to toxicity calculations in Visual MINTEQ uses the quantity Q, which, like f, represents the amount of heavy metal bound to the BL site at equilibrium, i.e., as affected by other competing reactions. Mathematically, however, Q is not fully mechanistic, and is instead semiempirical, being defined in part based on equilibrium expressions that include both competing metal–BL binding affinity (e.g., CaBL) and in part on fitting functions from experimental data. As with the f approach, toxicity is predicted to occur when the system exceeds a certain threshold value. In the Q-based approach, this threshold is denoted Q^*, which is conceptually analogous to f_{crit}, and is determined experimentally for specific organisms and endpoints. Also, as for the f_{crit} approach, in the Q^* approach Visual MINTEQ calculates the total metal concentration required to achieve the MBL concentration that corresponds to Q^*.

The inputs for BLM toxicity-mode calculations are the same as for speciation-mode calculations, save for one additional parameter being required, i.e., toxicity-mode calculations additionally require a value for either f_{crit} or Q^*.

15.3 VISUAL MINTEQ, INTRODUCTION TO USE, PROCEDURE, AND EXAMPLES

This section provides detailed instructions for using Visual MINTEQ 4.0 to perform equilibrium calculations to model heavy metal speciation in water, reactions involving solids, and interactions with BLs. Instructions are given for how to configure the software, input parameters, and validate results, as well as providing examples of how to make custom alterations to Visual MINTEQ database files. The examples of problems presented in this chapter begin with a very simple example aimed at users who are not already somewhat conversant with speciation modeling and segue to more sophisticated problems. The Visual MINTEQ User Guide (Gustafsson, 2024), which comes with the software, assumes a greater level of user knowledge than the simplest examples given here; however, the User Guide covers a much wider range of applications. Some of the examples given here, particularly as relates to the BLM, provide information beyond that given in the User Guide.

15.3.1 Visual MINTEQ Installation and Orientation

Download the Windows Installer Package file from OneDrive or Google Drive using the web links on the Visual MINTEQ webpage (https://vminteq.com/download/; click "link to user guide (pdf)" on this same webpage to download the current User Guide). Please note that there are two versions

[3] NB: While thermodynamic calculations are in terms of activity, the conversion between activity and concentration is effected by activity coefficients, and the toxicity endpoints calculated by Visual MINTEQ are in terms of total heavy metal concentration associated with such endpoints, since this is consistent with regulatory reporting standards.

of the Visual MINTEQ download file: the 64-bit version (×64) should work on most computers. There is, however, also a 32-bit version available for download. Installing the wrong version will produce an error when attempting to run the software.

After the download is complete, double-click on the installation zip file (Vminteq40_setup.zip). Then double-click on the icon setup to open the installer wizard. The first screen you will see is the Welcome to the Visual MINTEQ Setup Wizard. Click "Next" to proceed. You will then be prompted to choose the installation type, where you can select either "Only for me (your user name)" to install for a single user or "Everybody (all users)" to install for all users on the computer. After selecting the appropriate option, click "Next". The End-User License Agreement will then be displayed; read the terms and conditions, then select "I accept the terms in the License Agreement", and then click "Next" to continue to the "allow installer to make changes to your computer" dialog. Once the setup is completed and the installation location is confirmed, Visual MINTEQ will be installed and will launch automatically. There may be a prompt for the folder to use for user-editable files, in which case select one of the options. Make sure that the user has full file permissions for the folder where user-editable files are located. After the initial install, the main menu will appear in its default mode, with no components added, as shown in Figure 15.2. After closing Visual MINTEQ, you can reopen it by double-clicking the new Visual MINTEQ icon on the desktop.

The user interface of Visual MINTEQ is organized around a series of menu-driven functionalities that enable multifaceted control over geochemical/biotic modeling parameters and processes. The "File" menu item on the main menu bar leads to a drop-down menu that allows a user to load a Visual MINTEQ input file or to save an existing file for reuse later. The "Parameters" menu item leads to a drop-down menu that allows a user to specify ***alkalinity***, ***redox potential*** (as Eh or pE, electron activity), add/edit BLM (discussed in Examples 3 and 4 below), alter various default settings, and show initial charge balance. The "Solid phases and excluded species" menu item leads to a drop-down menu that allows a user to specify infinite, finite, possible solid phases (discussed in Example 2 below) or specify excluded species. The "Adsorption" menu item leads to a drop-down menu that allows a user to edit surface complexation reactions (maximum five reactions) or view adsorption isotherms. The "Gases" menu item allows a user to set fixed equilibrium pressures for

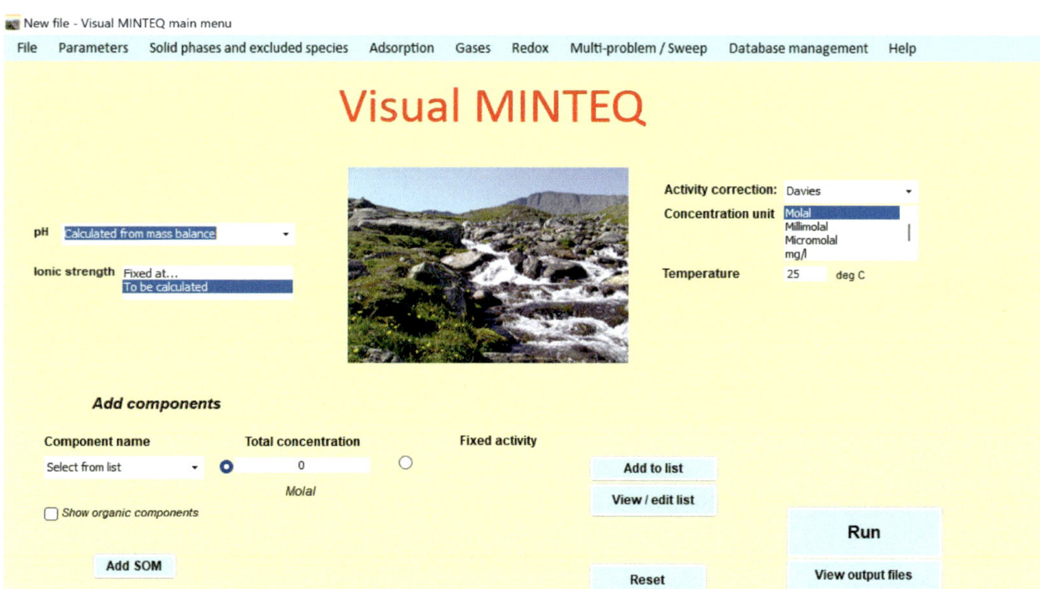

FIGURE 15.2 The Visual MINTEQ main menu in its default mode as it appears after installation.

gases. Users can either specify a fixed CO_2 partial pressure, with a default atmospheric CO_2 pressure of 0.00041 atm, which can be modified, or select a gas from the database and specify its partial pressure. The "Redox" menu item allows a user to add fixed-ratio redox couples by selecting from a predefined list or entering custom values. Users can define the log *K*s (equilibrium constant) and ΔH_r (enthalpy change) for the redox reaction in kJ/mol. The "Multi-Problem Menu" allows users to perform advanced simulations by varying specific parameters across multiple calculations. Users can choose from several options, including sweeping one parameter, titration/mixing with a defined titrant, simulating evaporation/concentration, or generating multiple problems in a single run. The "Database Management" menu item leads to a drop-down menu that allows a user to select the component, thermodynamic, and solids databases, and to review and alter component information (in practice, functionally limited to altering some parameters for estimating activity coefficients). New components cannot be added to Visual MINTEQ without manually altering the component database file, discussed further below. Four of the drop-down choices for this menu item allow the user to add new aqueous, redox, solid, and gas species. The remaining choices allow users to alter SHM gel parameters and add or edit temperature-dependent equilibrium constants for selected species. The "Help" menu leads to the Visual MINTEQ User Guide and version information. For the examples discussed herein, the majority of operations are conducted through buttons, text boxes, and dialogs on the Visual MINTEQ main workspace below the main menu bar.

Before starting to run problems in Visual MINTEQ, it is useful to gain familiarity with the main workspace, which contains options that must be reviewed when creating a problem and is divided into inputs such as pH, ionic strength, activity/concentration information, and temperature in the top half of the window and component input dialogs on the bottom half of the window.

At the top left of the main menu space, Visual MINTEQ provides three choices for pH: (1) "Fixed at…," a fixed input, i.e., using a measured or known pH, e.g., the pH might be 6.5 for natural freshwater; (2) "Calculated from mass balance"; or (3) "Calc from mass & charge balance". Choose the "Fixed at…" option when the pH value is known. After choosing this option, enter the known pH value in the text box that appears to the immediate right. Using a measured pH is indispensable to ***accuracy*** when the actual pH is known. As a result of choosing a fixed pH, the H^+ activity will be fixed and added to the list of fixed species in Visual MINTEQ (per above, each component also being a species). The other two options will calculate a pH. It is desirable to calculate the pH in certain situations, for instance, if the actual pH is unknown for some reason, for modeling a theoretical composition scenario (pH unknown) or for scenario exploration where the pH is known, but the user is interested in seeing how speciation would change with pH. In some cases, a calculated pH may be used for verification, i.e., if the actual measured pH is consistent with the overall mass/charge balances in the system, then the calculated pH should agree very well with the measured pH, and conversely a discrepancy between measured and calculated pH could indicate problems with other inputs, usually component inputs. When pH is calculated from mass balance, Visual MINTEQ solves the mass action equations in a manner to best reconcile the calculated total component concentrations with inputs, i.e., mass balance (for all components including H^+), but no explicit consideration of charge balance is made. In contrast, the charge balance option additionally requires that the sum of all positive and negative charges in the system equals zero, providing a more physically realistic pH estimate, since in real physical systems charge balance must exist in the system overall. Nominally, results from the two approaches to calculating pH should be the same. Differences in the two calculations may be an indicator of inaccurately specified components/concentrations or unaccounted species and needs to be investigated.

In Visual MINTEQ, ionic strength may be "Fixed at…" a chosen value or "To be calculated". Ionic strength, I, is calculated using the formula: $\frac{1}{2}\sum_i C_i Z_i^2$, where C_i is the molar concentration of ion i and Z_i is its charge. The default choice of "To be calculated" is always useful. If the actual ionic strength is not known, it is the only practicable choice, and if the ionic strength is known, verifying that the calculated result agrees with the known value serves as a check to ensure that inputs are well specified.

In the upper left of the Visual MINTEQ window, a choice for activity correction must be made. In ideal solutions, activity and concentration are nearly equal, but in real solutions, activity is different than concentration – for ions, for instance, due to inter-ionic forces. Using activities instead of concentrations in mass action equations ensures accurate predictions by accounting for the actual amount of a species or component participating in reactions, which is greatly influenced by ionic strength. This avoids the non-ideal behavior that would otherwise arise if concentrations are used, as concentration does not accurately represent the amount of species participating in reactions as ionic strength increases. Most applied environmental problems rely on measured concentrations for model inputs; hence, speciation software must have a way to estimate activity from concentration using activity coefficients. Visual MINTEQ provides three methods for estimating activity coefficients for converting concentration to activity: (1) The Davies equation calculates activity coefficients using an empirical expression that is a function of ionic strength and ion charge; this approach to calculating activity coefficients is suitable for low-to-moderate ionic strength systems (≤0.5 M). (2) The extended Debye–Hückel equation (listed as Debye–Hückel in Visual MINTEQ) calculates activity coefficients using an empirical expression that is a function of ionic strength and ion charge, but also includes a term adjusting for ionic size; the extended Debye–Hückel equation is considered effective for ionic strengths up to 0.1 M. (3) The Specific Ion-Interaction Theory (SIT) equation calculates activity coefficients according to an empirical expression that is a function of ion charge and ionic strength and a series of additional terms that describe interactions between each specific ion in solution with every other ion in solution; this approach for calculating activity coefficients is designed for high-ionic strength solutions (≥0.5 M). In the absence of other information, the user chooses the method for calculating activity coefficients that is appropriate to the ionic strength of the solution modeled. The Davies equation is the default and is generally suitable. While Visual MINTEQ has the SIT option for calculating activity coefficients at high ionic strengths, the inadequacies of the SIT for very high ionic strength conditions such as seawater are well known. For seawater, users should consider using the software MARCHEMSPEC[4]; this software calculates activities coefficients using Pitzer equations, which are known to result in the best accuracy for very high ionic strengths.

Below the input for activity correction on the Visual MINTEQ main menu is the input for concentration unit. The choices are molal (moles per kilogram or mol/kg), millimolal, micromolal, mg/L, µg/L, and log molal. The choice of unit affects how input data are interpreted and converted for input into equilibrium calculations. Molal, millimolal, and micromolal are molality-based units, directly applicable in equilibrium calculations, whereas mg/L and µg/L are mass-to-volume-based units, that are convenient because they are often used in Environmental Monitoring. When one of these mass-to-volume-based units is selected, Visual MINTEQ automatically converts them into molal concentrations using molecular weights and solution density before performing calculations. Log molal is sometimes used for modeling scenarios with orders of magnitude differences in input concentrations, and in some areas of environmental science, the use of log concentration units is standard practice. Although users can input concentrations in different units, calculations within Visual MINTEQ are ultimately performed in molal, i.e., Visual MINTEQ converts all inputs to molal internally. Molal concentrations are independent of temperature and solvent density changes, making them more suitable for precise thermodynamic calculations. Upon completion of calculations, Visual MINTEQ converts units of molal to units of molar (mol/L), and all speciation calculation results are presented in molar. If concentration units are available for input in mol/L, it is necessary to convert to molal, though, in practice, for many situations, the error in using molar instead of molal for inputs is generally negligible.

After selecting concentration units, the next input below is temperature. The default temperature in Visual MINTEQ is 25°C, however, users may change this value within a much broader range (0°C–99.99°C). While the software accepts temperature inputs in this broad range, it is optimized

[4] Download MARCHEMSPEC at https://marchemspec.org/wp/software/.

for environmental applications typically within 0°C–40°C. Use of input temperatures beyond this 0°C–40°C range may reduce accuracy of results due to the lack of availability of appropriate thermodynamic data correction to reflect such conditions.

15.3.2 Example 1: Calculation of the Speciation of Cadmium Nitrate in Water

In this example, speciation is calculated for a sample of lake water that is spiked with cadmium nitrate ($Cd(NO_3)_2$) to investigate how Cd speciation is influenced by natural water chemistry, particularly the interactions with major cations, anions, and DOC, which govern Cd bioavailability and potential toxicity. The water composition used in this study is based on laboratory measurements and represents a sample used by Li et al. (2023). The measured parameters for the lake water include 3.72 mg/L DOC, major cations (46.2 mg/L Na^+, 4.34 mg/L K^+, 8.81 mg/L Mg^{2+}, and 21.6 mg/L Ca^{2+}), and major anions including 54.4 mg/L Cl^- and 62.4 mg/L SO_4^{2-}. Additionally, nutrient species are present as 0.09 mg/L NH_4^+ (ammonium), 0.049 mg/L NO_3^- (nitrate), 0.001 mg/L NO_2^- (nitrite), and 0.004 mg/L PO_4^{3-} (phosphate). Results for two levels of $Cd(NO_3)_2$ (0.01 and 1.0 mg/L, i.e., 10 ppb and 1 ppm) ***spike*** will be calculated to simulate low and high levels of contamination. Concentrations of Cd were chosen to reflect potential Cd levels in environments affected by anthropogenic inputs, such as industrial discharge (Rantetampang and Mallongi, 2013). The use of a nitrate spike is quite artificial. In contaminated waters collected from the field and subject to chemical analysis, the appropriate level of counterion is present. For this example and Example 2 below, it is necessary to preserve charge balance for the calculation, and doing so with nitrate is convenient because of its tendency not to interact with other solution components (~95+% remaining as NO_3^-). Two different pH levels (pH = 8 and pH = 5.5) will also be used to discuss the differences for metal speciation as a function of pH.

Procedure:

1. The first run in Visual MINTEQ will assume lake water spiked with Cd to a final concentration of 1.0 mg/L Cd_{Tot} at a pH = 8. To set this up, launch Visual MINTEQ and set "pH fixed at…" to 8.0; this reflects typical alkaline freshwater conditions that are sometimes present in mildly eutrophic waters with algal buildups/blooms and concomitant other sources of pollution. Next, choose "Ionic strength" to be calculated so that Visual MINTEQ can determine I based on the actual concentrations of all dissolved ions, ensuring consistency with the system's overall composition. On the right-hand side of the window, for activity correction, leave the choice of activity correction as the default (the Davies equation). Set the concentration initially for mg/L, since most of the inputs have this unit and leave the temperature at the default, 25°C, for this example.
2. Input total concentrations for components present in the lake water sample, as follows:
   - Under the "Add components" drop-down at the bottom left of the main working space, select the "Component name" "Na+1", then input a "Total concentration" of 46.2 mg/L in the text box. Notice that next to the text box for total concentration, there is a button that can be toggled to "Fixed activity". Fixing activity is only appropriate when activity is actually known (e.g., such as pH, which is an activity-based measurement), or when the user wishes to simulate scenarios wherein species have a fixed activity. In this example, the activity of H^+ is fixed via fixing pH, and no other activities will be fixed. Next, click the button "Add to list" to the right of the "Fixed activity" option. The Visual MINTEQ window at this point is as shown in Figure 15.3.
   - Repeat this process for the remainder of the ionic components, "K+1", "Mg+2", "Ca+2", "Cl–1", "SO4–2", "NH4+", "NO3–", "NO2–", and "PO4–3".
   - For this example, M–DOC interactions will be handled using the SHM. To enter DOC, under "Add components" select the component name "DOC (SHM)", input the total concentration of 3.72 mg/L, and click the button "Add to list". Generally, the

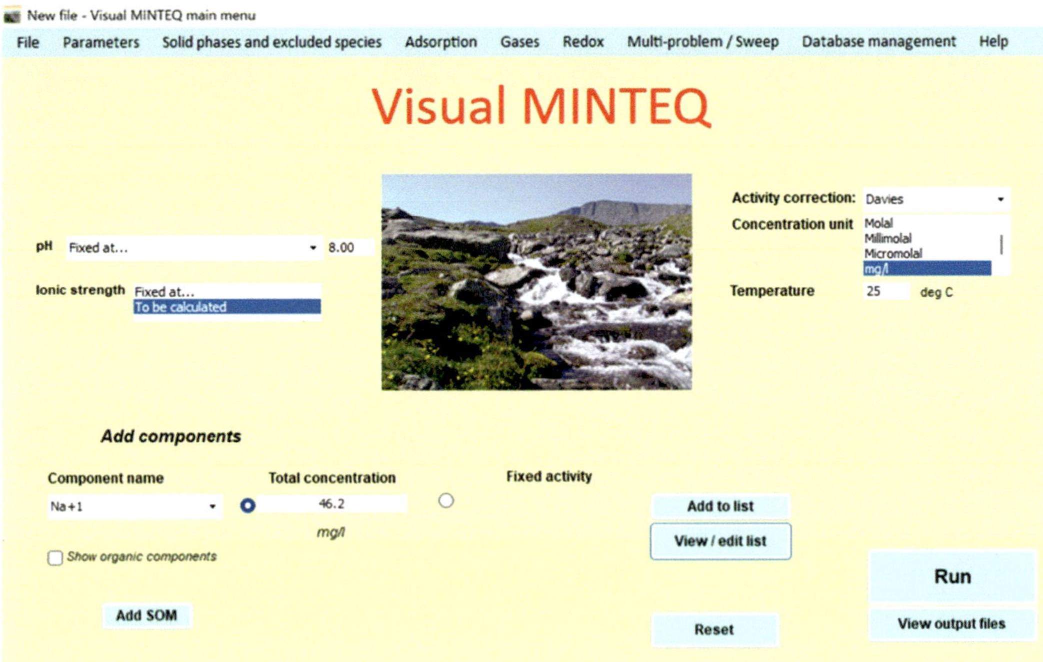

FIGURE 15.3 The Visual MINTEQ main window after initial setup and addition of the first component of 46.2 mg/L Na+.

SHM is well suited for natural waters with moderate DOC levels and provides a well-validated approach for modeling metal–DOM interactions. For both the SHM and the NICA–Donnan models in Visual MINTEQ, the parameters governing the distribution of HA and FA in DOC, binding strength parameters, and some details of how electrostatics are dealt with, may be altered if the user has more information (see, for instance, Zhang et al., 2020a). For this simple example, the Visual MINTEQ default ratio of HA to FA is edited. HA has a higher average molecular weight than FA and is more refractory and harder to break down. In result, HA can be more environmentally persistent. For an agricultural lake with both forests and farmland within its catchment, HA typically constitutes 25%–50%of DOC, with 33% being a reasonable estimate for the purposes of illustration. To modify the HA/FA ratio in Visual MINTEQ, users adjust the fraction of HA and FA within the DOC settings. After having entered "DOC (SHM)" as a component, two additional dialogs for editing organic matter parameters will appear on the bottom left of the Visual MINTEQ main working space. One is the button "Edit DOC parameters", and the other is a drop-down menu to the left of this button. Make sure that "SHM" is chosen in the drop-down menu, and then click on the "Edit DOC parameters" button to the right of this. A new window, the SHM menu, will open. In this window, at the top left, there is a text box labeled "% of active DOM that is FA". Enter 67 in this text box (i.e., 67% FA, 33% HA), then click on the button "Save and back to main menu" to the bottom left.

3. In the main menu, by clicking "View/edit list" in the bottom middle part of the screen, the "List of components" window will open showing entries for the total concentrations of components in the natural lake water sample (see Figure 15.4). In this window, the concentration can be changed or specific components may be deleted entirely. By convention in Visual MINTEQ, any components entered with fixed activity will show up with a total concentration of zero in this window. The components at the bottom of the

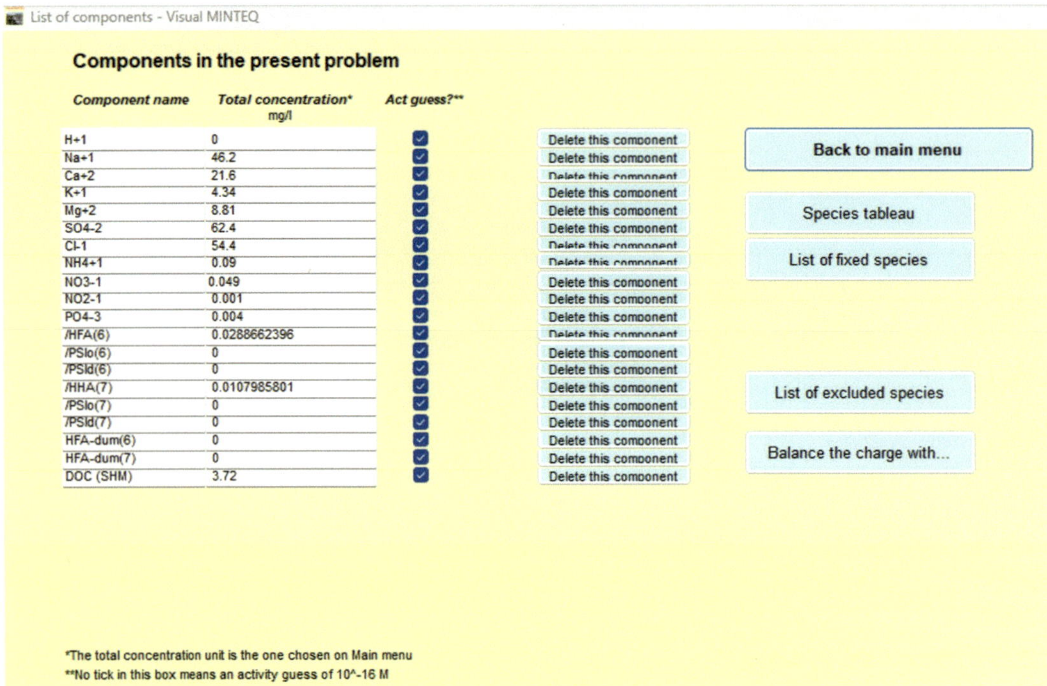

FIGURE 15.4 Visual MINTEQ list of components window summarizing natural lake water total component concentrations that were entered for Example 1; this window opens when "View/edit list" button on the main menu of Visual MINTEQ is selected. Note: The concentration of H^+ is not shown due to this being a fixed activity species.

list in Figure 15.4 represent DOC species. Below the inorganic species entries, the species listed – HFA(6), PSIo(6), PSId(6), and HFA-dum(6), HHA(7), PSIo(7), PSId(7), and HFA-dum(7) – represent different fractions or functional groups of DOM as modeled in the SHM framework. HFA(6) corresponds to FA, and HFA-dum(6) is a dummy species required to account for charge variations in the FA gel phase of the SHM model, while PSIo(6) and PSId(6) represent potentials associated with the Basic Stern Model incorporated in SHM (as detailed in Chapter 14). The numeral 6 is assigned to FA. When both HA and FA are present, as in this example, these species are divided into "6" for FA-related terms and "7" for HA-related terms.

In addition to viewing and modifying component concentrations, the list of components window provides several other functionalities. The "Act guess?" column allows users to enable or disable default initial guesses of activity for individual components. Normally the guess option is ticked (default), enabling Visual MINTEQ to set the initial guess; no tick in this box sets the initial activity guess to 10^{-16} M, which helps Visual MINTEQ to converge, or find a solution, when activities of components are very low. The "Delete this component" buttons next to each entry enable users to remove specific components from the model if needed. On the right side, the "Species tableau" button opens a new window that provides the Tableau for the possible chemical species formed from the entered components. Note that HFA-related species do not appear in the species Tableau, because Visual MINTEQ models HAs and FAs as binding sites in the SHM sub-model. Components are also not shown in the Visual MINTEQ species Tableau, though both HA, FA, and components are shown in various speciated forms in the output/results after a model is run. Clicking on the "List of Fixed Species" button in the list of components window opens a new window that provides users with a review of species (components) that were input as having a fixed activity rather than letting the model calculate an equilibrium concentration. In this example, the fixed species window shows that H^+ is set to a log activity of –8, which corresponds to a pH of 8. This setting ensures that the pH remains constant during calculations, preventing changes due to system interactions. Fixing species like H^+ is useful when modeling controlled pH conditions, such as in laboratory experiments or specific environmental scenarios where buffering effects dominate. Other species, such as redox couples or gaseous species, can also be fixed, and activities changed here if needed; likewise, fixed activity species may be deleted in fixed species window. Clicking on the "List of excluded species" button opens a window where users review and modify any species excluded from calculations. The reasons for potentially excluding species are mentioned further in Example 2 below. Note, for this example, the Stern potential for HA and FA show up as excluded species. While these are needed for the SHM calculation, they are not actual chemical species and are excluded from the calculation as such. The "Balance the charge with…" button at the bottom left of the list of components window provides options for charge balancing, ensuring electroneutrality by allowing Visual MINTEQ to adjust the concentration of selected components, such as H^+, OH^-, Na^+, or Cl^-. When results from calculations reflect charge imbalance, this is an indication that the component inputs may be misspecified or some system component has been neglected. The charge balancing feature helps refine the input data when it can be safely assumed that the component source of charge imbalance is known. After checking that the component list appears as expected based on inputs, click on "Back to main menu". Click the "File" on the top left and select save the input file as "LakeWater" for later use (see Example 2).

4. Add the 1.0 mg/L Cd^{2+} in the form of $Cd(NO_3)_2$. Select the component name "Cd+2", input a total concentration of 1.0 mg/L and click the button "Add to list". Then click "View/edit list" and change NO_3^- concentration from the original (0.049 mg/L) to 1.153 mg/L NO_3^- (0.049 mg/L from lake water + 1.104 additional NO_3^- contributed from the $Cd(NO_3)_2$ spike).
5. After entering the $Cd(NO_3)_2$ components, click "View/edit list" again. The spike components are added, as shown in Figure 15.5. Click on "Back to main menu".

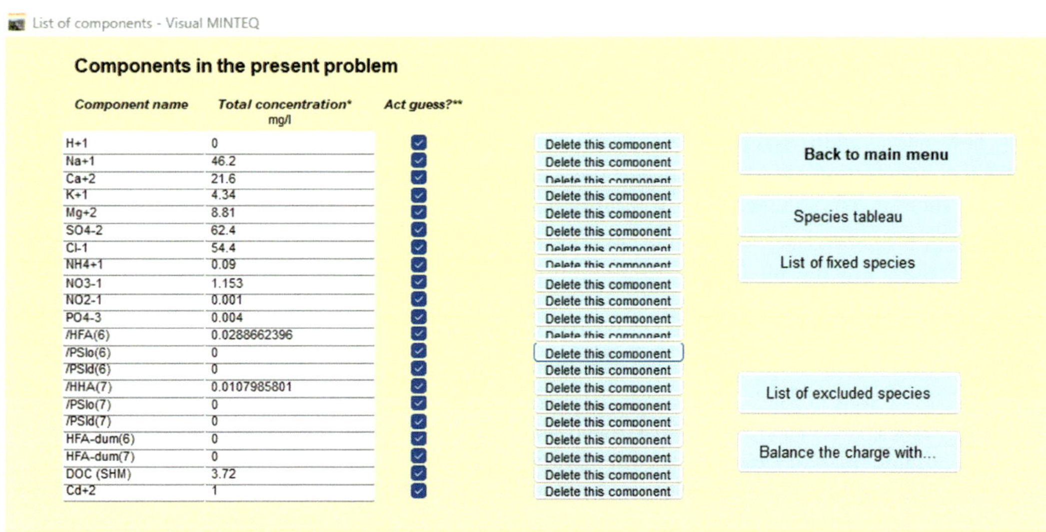

FIGURE 15.5 Window summarizing component total concentrations after spiking the lake water sample with $Cd(NO_3)_2$ for a total Cd concentration of 1.0 mg/L.

6. The problem is now ready; click on the "Run" button in the main menu working space to perform calculations. A black command prompt window may briefly appear, and after Visual MINTEQ is finished executing the calculation, the window "Visual MINTEQ – Output" will automatically appear, as shown in Figure 15.6. This window displays the output summary from the Visual MINTEQ calculation. The pH of the system is shown as 8.000 at the top left, as was set as a fixed input value. The ionic strength is 4.70×10^{-3} M, which falls within the range that would be expected for a freshwater lake. In the top middle of the output window, the number of iterations indicates how many computational cycles were needed for the solver to converge to a stable solution. The sum of cations in the system is 3.81×10^{-3} eq/kg, and the sum of anions is 2.75×10^{-3} eq/kg, revealing an imbalance between positively and negatively charged species. The charge difference is 16.16%, which suggests a significant deviation from charge neutrality. While this is significant, it is not necessarily unusual for such an example. A number of minor ions are present in natural waters and have not been included in this calculation, as well as lower molecular weight organic ions that may not be captured by the SHM, which models DOC as HA and FA. Additionally, normal uncertainties in the laboratory analysis of the lake water components entered can be a source of charge imbalance. At the top right of the output window, the "See calculated adsorption parameters" button allows users to view the Stern potentials and charge densities used in the SHM for modeling adsorption to FA (surface 6) and HA (surface 7).[5] The "Gases" button opens a window that provides the result for the partial pressure of ammonia gas (NH_3) in solution, which is 4.28×10^{-9} atm. Partial pressure represents the equilibrium pressure exerted by NH_3 (g) (in the gas phase), indicating its volatility under the given water chemistry conditions. This low value suggests that NH_3 is mostly retained in the aqueous phase.

To obtain an immediate impression of overall species distribution, it is usually helpful to look at relative results. Start reviewing these by clicking the "View species distribution" button on the lower left of the Visual MINTEQ – Output page. This will open the Visual MINTEQ – Species table window showing the relative calculated distribution among dissolved and adsorbed species (Figure 15.7).

For convenience in discussing the selected species of interest, the species table in Figure 15.7 has been exported using the "Print to Excel" button and partial data are formatted in Table 15.2. Also included in Table 15.2 for comparison are results from two additional runs. For the first additional run, the Cd^{2+} and NO_3^- concentrations were changed to reflect spiking to a total Cd concentration of 0.01 mg/L (with 0.060 mg/L NO_3^-, 0.049 mg/L from lake water, and 0.011 mg/L from the $Cd(NO_3)_2$ spike). The second additional run explores the impact of pH on Cd speciation and bioavailability (at a total Cd concentration of 0.01 mg/L), as Cd is known to be highly sensitive to pH variations. Lower pH levels, such as 5.5, enhance Cd^{2+} solubility and bioavailability, which is particularly relevant in the context of acid rain, a growing environmental concern that exacerbates heavy metal mobilization in natural waters. The partial data shown in Figures 15.6 and 15.7 and Table 15.2 should be sufficient for the reader to determine if Example 1 results have been duplicated. The remainder of this discussion assumes that results have been duplicated and that the reader has access to full Visual MINTEQ outputs, in order to follow along with the discussion here (as with Examples 2–4 below).

The Cd speciation in Table 15.2 illustrates the influence of total Cd concentration, DOM in the form of FA and HA and pH on metal bioavailability. Free-ion Cd^{2+}, for many organisms the primary bioavailable form of Cd, is present at significantly different levels across the three conditions. At Cd 1.0 mg/L and pH 8, 69.4% of the total Cd remains as free-ion Cd^{2+}, indicating high relative bioavailability. However, when the total Cd concentration is reduced to 0.01 mg/L at the same pH, free-ion Cd^{2+} drops to 16.3%, demonstrating that at lower Cd concentrations, competitive

[5] See Gustaffson (2001) for full details.

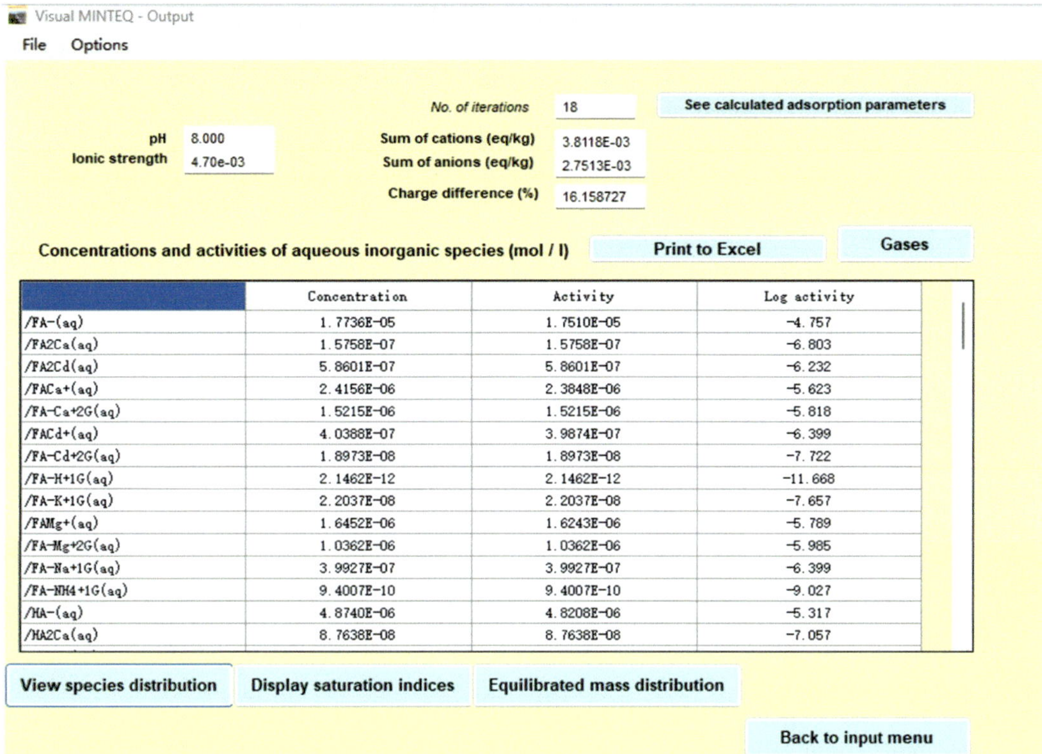

	Concentration	Activity	Log activity
/FA-(aq)	1.7736E-05	1.7510E-05	-4.757
/FA2Ca(aq)	1.5758E-07	1.5758E-07	-6.803
/FA2Cd(aq)	5.8601E-07	5.8601E-07	-6.232
/FACa+(aq)	2.4156E-06	2.3848E-06	-5.623
/FA-Ca+2G(aq)	1.5215E-06	1.5215E-06	-5.818
/FACd+(aq)	4.0388E-07	3.9874E-07	-6.399
/FA-Cd+2G(aq)	1.8973E-08	1.8973E-08	-7.722
/FA-H+1G(aq)	2.1462E-12	2.1462E-12	-11.668
/FA-K+1G(aq)	2.2037E-08	2.2037E-08	-7.657
/FAMg+(aq)	1.6452E-06	1.6243E-06	-5.789
/FA-Mg+2G(aq)	1.0362E-06	1.0362E-06	-5.985
/FA-Na+1G(aq)	3.9927E-07	3.9927E-07	-6.399
/FA-NH4+1G(aq)	9.4007E-10	9.4007E-10	-9.027
/HA-(aq)	4.8740E-06	4.8208E-06	-5.317
/HA2Ca(aq)	8.7638E-08	8.7638E-08	-7.057

FIGURE 15.6 Speciation results for 1.0 mg/L Cd^{2+} spiked as the form of $Cd(NO_3)_2$ into a natural water sample at a pH of 8.0; units are in mol/L.

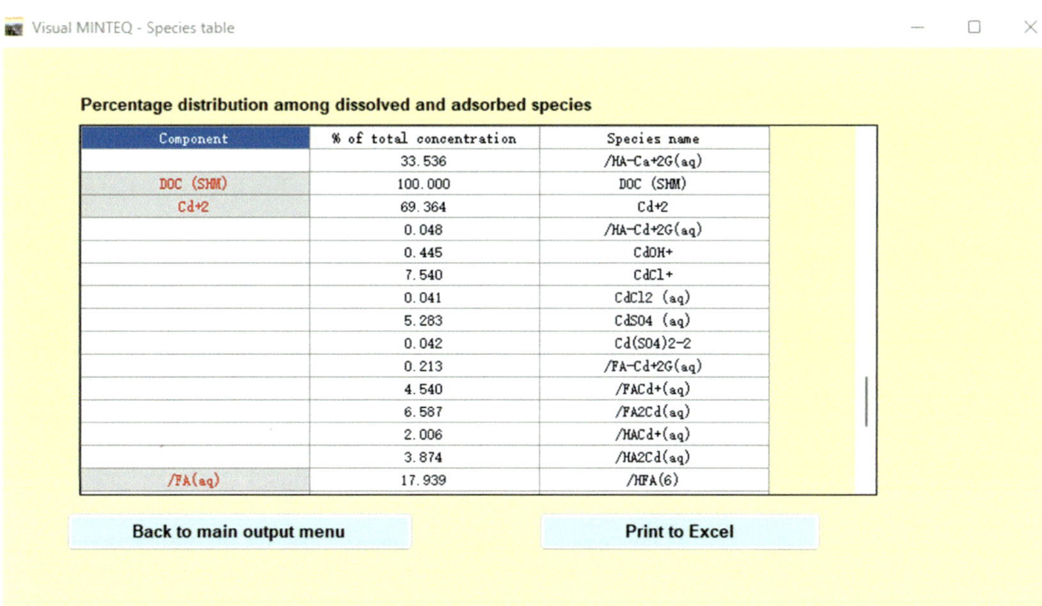

Component	% of total concentration	Species name
	33.536	/HA-Ca+2G(aq)
DOC (SHM)	100.000	DOC (SHM)
Cd+2	69.364	Cd+2
	0.048	/HA-Cd+2G(aq)
	0.445	CdOH+
	7.540	CdCl+
	0.041	CdCl2 (aq)
	5.283	CdSO4 (aq)
	0.042	Cd(SO4)2-2
	0.213	/FA-Cd+2G(aq)
	4.540	/FACd+(aq)
	6.587	/FA2Cd(aq)
	2.006	/HACd+(aq)
	3.874	/HA2Cd(aq)
/FA(aq)	17.939	/HFA(6)

FIGURE 15.7 Visual MINTEQ species table window showing the relative speciation distribution results for $Cd(NO_3)_2$ spiked into a natural lake water sample (pH fixed at 8.0) to a total Cd concentration of 1.0 mg/L.

complexation with stronger ligands, particularly the DOM constituents HA/FA, becomes more dominant as the ligand-to-metal ratio increases. Weaker inorganic complexes such as $CdSO_4$ and $CdCl^+$ decrease in proportion, in result of the differential binding by DOM. In contrast, maintaining total Cd at 0.01 mg/L, nonetheless at pH 5.5, free-ion Cd^{2+} increases again to 69.8%, as expected, given the known effect of pH on metal speciation and bioavailability. At lower pH, proton competition significantly reduces heavy metal complexation with organic ligands HA/FA, leading to increased free Cd^{2+}.

The SHM allows for monodentate and bidentate binding of Cd to HA and FA. A monodentate complex refers to a coordination structure where the metal/Cd ion binds to the DOM via a single functional group or binding site. In contrast, a bidentate complex involves the metal/Cd ion binding simultaneously to two distinct functional groups on the same DOM molecule, resulting in a more stable chelated structure. These two binding modes differ in their affinity, stability, and prevalence under different environmental conditions such as pH and total heavy metal concentration. For instance, at pH 8.0, the bidentate complex $/FA_2Cd(aq)$ increases from 6.6% at 1.0 mg/L Cd to 48.9% at 0.01 mg/L Cd (relative to total Cd concentration), while the corresponding HA complex, $/HA_2Cd(aq)$, increases from 3.9% to 28.8%. At a pH of 8.0, the proportion of total FA and HA monodentate binding is ~40% and ~33% at 1.0 mg/L Cd, whereas at 0.01 mg/L Cd, this proportion is ~4% and ~3%, respectively. At high Cd loadings, the number of available DOM functional groups per Cd^{2+} becomes limiting, and the relative proportion of monodentate binding increases. At pH 5.5 and low Cd concentration, both monodentate and bidentate complexation are suppressed due to proton competition for ligand sites. Monodentate species ($/FACd^+$ and $/HACd^+$), however, exceed bidentate ($/FA_2Cd$ and $/HA_2Cd$) in relative proportion. The strong protonation of DOM under acidic conditions hinders the simultaneous deprotonation of two adjacent binding sites necessary for bidentate coordination. Indeed, the relative abundance of the protonated species HFA(6) and HFA(7) compared to /FA-(aq) and /HA-(aq) increases substantially at pH 5.5 compared to pH 8.0, as seen in Table 15.2. These conditions favor simpler, monodentate interactions, as only a limited number of binding sites remain available for metal ***chelation***.

Comparing the absolute concentration data (i.e., in molar) for HA/FA aqueous to HA/FA gel phase, only ~20% of the organic matter binding sites for HA/FA are in the gel phase. Most of the gel-phase sites are not bound to metal cations – the HFA-dum(6) and HFA-dum(7) species account for 41.5%–45.3% for FA and 38.4%–43.4% for HA, across the three calculation scenarios. At maximum, ~0.5% of Cd is complexed with gel-phase binding sites. This large disparity between the Cd bound to aqueous versus gel HA/FA is in large part explained by competitive exclusion, where the comparatively high environmental concentrations of Ca^{2+} and Mg^{2+}, combined with how these cations interact with the gel, enable them to outcompete Cd^{2+} for gel-phase binding sites. Relatively, a total of 56% and 53% of HA and FA gel sites, respectively, are bound to Ca^{2+} and Mg^{2+}.

At all pHs, the major cations (Ca^{2+}, Mg^{2+}, K^+, Na^+) and anions (Cl^-, SO_4^{2-}, NO_3^-, NO_2^-) for the Example 1 calculation remain predominantly in their free ionic forms, with more than 90% of their total concentration existing as free ions, with the notable exception of phosphate (PO_4^{3-}), which participates in buffering (for instance, at pH 8, 69.32% is dibasic HPO_4^{2-} with monobasic $H_2PO_4^-$ being only 8.76%, which shifts to ~96% $H_2PO_4^-$ at pH 5.5). The dibasic form of PO_4^{3-} also binds with the alkaline earth elements (8.66% $CaHPO_4(aq)$ and 8.16% $MgHPO_4(aq)$ at pH 8), reflecting the strong interactions of phosphate with divalent cations and its role in forming mineral complexes.

In natural systems, the ability to cope with Cd at ambient versus polluted levels depends largely on the complexation capacity of DOM. At lower Cd concentrations (0.01 mg/L) and a pH of 8, most Cd is complexed with DOM. At elevated Cd concentrations (1.0 mg/L), the proportion of bioavailable Cd^{2+} concentration increases dramatically as the available binding sites on DOM become saturated. Similarly, under acidic conditions (e.g., pH 5.5), Cd speciation shifts toward the free Cd^{2+} ion due to proton competition for DOM binding sites. The results from this example illustrate how DOM is an important mediator of Cd fate in natural waters, however, its ability to mitigate Cd

TABLE 15.2
Visual MINTEQ Relative Speciation Distribution Results for Three Different Cd Spikes (Cd^{2+} = 1.0 mg/L at pH 8, Cd^{2+} = 0.01 mg/L at pH 8 and pH 5.5) into a Natural Lake Water Sample

		% of Total Concentration		
		Cd 1.0 mg/L	Cd 0.01 mg/L	Cd 0.01 mg/L
Component	Species Name	pH 8	pH 8	pH 5.5
Cd+2	Cd+2	69.4	16.3	69.8
	CdCl+	7.5	1.8	7.6
	/FA2Cd(aq)	6.6	48.9	4.5
	CdSO4(aq)	5.3	1.2	5.3
	/FACd+(aq)	4.5	1.9	6.9
	/HA2Cd(aq)	3.9	28.8	2.2
	/HACd+(aq)	2.0	0.89	3.4
	CdOH+	0.45	0.10	<0.10
	/FA-Cd+2G(aq)	0.21	0.52	0.20
	/HA-Cd+2G(aq)	0.05	0.012	0.045
	Cd(SO4)2 – 2	0.048	<0.10	0.042
	CdCl2(aq)	0.041	<0.10	0.042
/FA(aq)	/FA(ag)	61.4	62.8	55.9
	/HFA(6)	17.9	19.6	33.5
	/FACa+(aq)	8.4	9.4	6.3
	/FAMg+(aq)	5.7	6.4	4.3
	/FA2Cd(aq)	4.1	0.30	0.027
	/FACd+(aq)	1.4	<0.10	0.021
	/FA2Ca(aq)	1.1	1.5	<0.10
HFA-dum(6)	HFA-dum(6)	41.7	41.5	45.3
	/FA-Ca+2G(aq)	31.8	32.1	30.1
	/FA-Mg+2G(aq)	21.7	21.9	20.5
	/FA-Na+1G(aq)	4.2	4.2	3.9
	/FA-Cd+2G(aq)	0.40	<0.10	<0.10
	/FA-K+1G(aq)	0.23	0.23	0.22
/HA(aq)	/HA(ag)	45.1	46.8	38.9
	/HFA(7)	26.0	28.6	46.3
	/HACa+(aq)	11.4	13.1	8.7
	/HAMg+(aq)	7.8	8.9	6.0
	/HA2Cd(aq)	6.4	0.48	0.036
	/HACd+(aq)	1.7	<0.10	0.028
	/HA2Ca(aq)	1.6	2.2	<0.10
HFA-dum(7)	HFA-dum(7)	38.6	38.4	43.4
	/HA-Ca+2G(aq)	33.5	33.9	31.1
	/HA-Mg+2G(aq)	22.8	23.1	21.2
	/HA-Na+1G(aq)	4.4	4.4	4.1
	/HA-Cd+2G(aq)	0.42	<0.10	<0.10
	/HA-K+1G(aq)	0.24	0.25	0.22
	/HA-NH4+1G(aq)	0.010	0.010	0.010

bioavailability declines significantly at higher pollution (Cd) levels, and similar comments pertain to acid rain. These two pollution scenarios lead to increased bioavailability and toxicity risks for aquatic organisms.

There are two more features on the Visual MINTEQ – Output window to take note of – these are the "Display saturation indices" and "Equilibrated mass distribution" buttons (see bottom buttons in Figure 15.6). In Visual MINTEQ, clicking on the "Display saturation indices" button opens a new window, as shown in Figure 15.8, with information indicating whether minerals are likely to precipitate or dissolve based on their saturation state, while clicking on the "Equilibrated mass distribution" button opens a new window that provides the final equilibrium distribution of chemical species among dissolved, precipitated, and adsorbed phases. Understanding information in the "Display saturation indices" window is important because it helps determine whether solids/minerals would precipitate under the given environmental conditions reflecting Visual MINTEQ inputs. For this example, the results in Figure 15.8 show that all saturation indices all below zero, indicating undersaturation, but there will be cases in which the "saturation indices" will above zero, indicating oversaturation; Example 2 deals with one such case. The "Equilibrated mass distribution" window provides insight into how different chemical species are partitioned among dissolved, sorbed, and precipitated phases, the distribution between which strongly influences bioavailability and potential environmental impacts, thus providing information that assists in understanding potential exposure risks to aquatic organisms. In engineered systems, these outputs can help design treatment strategies, such as pH adjustment to promote precipitation or DOM enhancement to facilitate complexation, and reduce free ion concentrations.

15.3.3 Example 2: Calculations Involving Solids

In Example 1, we explored the saturation indices window and discussed how all saturation indices were negative, indicating undersaturation. However, this is not always the case – under certain conditions, the ***saturation index*** (SI) may be positive, in which case solid-phase formation should occur. In this example, we examine an environmentally relevant scenario involving lead (Pb) precipitation

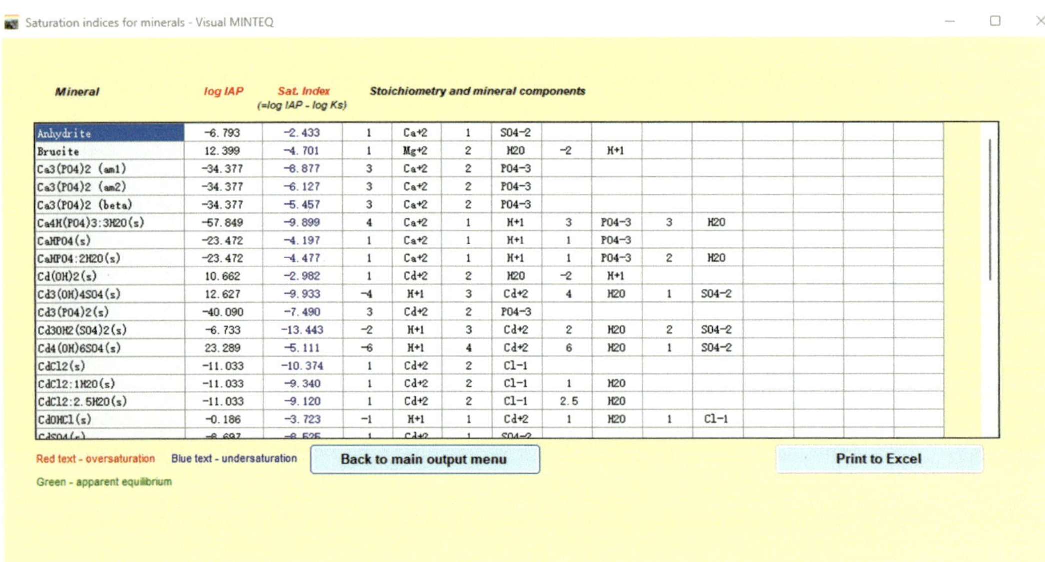

Mineral	log IAP	Sat. Index (=log IAP - log Ks)	Stoichiometry and mineral components							
Anhydrite	-6.793	-2.433	1	Ca+2	1	SO4-2				
Brucite	12.399	-4.701	1	Mg+2	2	H2O	-2	H+1		
Ca3(PO4)2 (am1)	-34.377	-8.877	3	Ca+2	2	PO4-3				
Ca3(PO4)2 (am2)	-34.377	-6.127	3	Ca+2	2	PO4-3				
Ca3(PO4)2 (beta)	-34.377	-5.457	3	Ca+2	2	PO4-3				
Ca4H(PO4)3:3H2O(s)	-57.849	-9.899	4	Ca+2	1	H+1	3	PO4-3	3	H2O
CaHPO4(s)	-23.472	-4.197	1	Ca+2	1	H+1	1	PO4-3		
CaHPO4:2H2O(s)	-23.472	-4.477	1	Ca+2	1	H+1	1	PO4-3	2	H2O
Cd(OH)2(s)	10.662	-2.982	1	Cd+2	2	H2O	-2	H+1		
Cd3(OH)4SO4(s)	12.627	-9.933	-4	H+1	3	Cd+2	4	H2O	1	SO4-2
Cd3(PO4)2(s)	-40.090	-7.490	3	Cd+2	2	PO4-3				
Cd3OH2(SO4)2(s)	-6.733	-13.443	-2	H+1	3	Cd+2	2	H2O	2	SO4-2
Cd4(OH)6SO4(s)	23.289	-5.111	-6	H+1	4	Cd+2	6	H2O	1	SO4-2
CdCl2(s)	-11.033	-10.374	1	Cd+2	2	Cl-1				
CdCl2:1H2O(s)	-11.033	-9.340	1	Cd+2	2	Cl-1	1	H2O		
CdCl2:2.5H2O(s)	-11.033	-9.120	1	Cd+2	2	Cl-1	2.5	H2O		
CdOHCl(s)	-0.186	-3.723	-1	H+1	1	Cd+2	1	H2O	1	Cl-1
CdSO4(s)	-8.697	-8.525	1	Cd+2	1	SO4-2				

FIGURE 15.8 The Visual MINTEQ saturation indices window showing the saturation indices for minerals from the results for $Cd(NO_3)_2$ spiked into a natural lake water sample (pH fixed at 8.0) to a total Cd concentration of 1.0 mg/L.

in a system influenced by phosphate. This example will illustrate how Visual MINTEQ calculates solid-phase formation.

Some heavy metals are more prone than others to form oxyhydroxide species at alkaline pHs, and, for example, Pb forms a number of highly insoluble species in the presence Cl^- and/or PO_4^{3-}. Therefore, the same type of calculation as shown in Example 1 for Pb requires consideration of solid-phase formation. Two scenarios are considered in this example, both involving a lake water sample of the composition described in Example 1, only with a high concentration of phosphate (1 mg/L PO_4^{3-}, mimicking a fertilizer runoff event) and levels of Pb of either 0.1 mg/L (100 ppb) or 250 mg/L (250 ppm). For the two Pb concentrations chosen, 0.1 mg/L reflects moderate contamination levels found in impacted natural waters, whereas 250 mg/L represents an extreme pollution case, such as an accidental spill or industrial discharge (Schneider et al., 2019). A pH of 6.5 is selected as many surface waters fall within a pH range of 6 to 8, and 6.5 provides a relevant midpoint where both dissolved and solid-phase Pb species coexist. Pb solubility is highly pH-dependent, and keeping the pH at 6.5 is useful to specifically focus on the role of phosphate in lead precipitation.

For calculations involving solids, there are four opinions in the "Solid phases and excluded species menu" on the main menu bar in Visual MINTEQ: "Specify infinite solid phase", "Specify finite solid phase", "Specify possible solid phase", and "Specify excluded species". The infinite solid phase choice represents a scenario in which an exceedingly large quantity of the solid is assumed to be present – this condition simulates a system where the solid phase is present in such an abundance that it would never completely dissolve. In the case of a finite solid phase, a specific amount of solid exists or is introduced into the system. For example, if exactly 1 g of a solid is added to 1 L of solution, the total quantity of the solid phase is constrained; it may dissolve completely if it is highly soluble and a small amount is present, for instance. Alternately, this option admits the possibility that reactions in the system may result in additional precipitation of the solid. The possible solid phase option assumes that a given sample contains some solute in solution, which might reach saturation at equilibrium. This option is particularly relevant for evaluating the formation of solid precipitates under such circumstances. The option to exclude solids explicitly prevents the formation of a specific solid phase, either due to prior knowledge that the species in question will not form under the given conditions, or for the purpose of conducting calculations under the assumption that the solid phase does not exist in the system.

Procedure:

1. Define settings:
   - Launch Visual MINTEQ and on the main menu and click "File" on the top left and select "Open input file". This example uses the same lake water sample as reported above in Example 1. Open the saved file "LakeWater".
   - Set the pH fixed at 6.5.
   - Ensure the ionic strength option is set to be calculated.
   - Change the concentration unit to mg/L.
2. As for Example 1, spike 250 mg/L Pb, in the form of $Pb(NO_3)_2$ (in order to maintain charge balance) into the natural lake water, and add phosphate.
   - Select the component name "Pb+2", input a total concentration of 250 mg/L, and click the button "Add to list".
   - Select the component name "NO–3", input a total concentration of 149.69 mg/L (The initial NO_3^- concentration is 0.049 mg/L, and adding 250 mg/L Pb in the form of $Pb(NO_3)_2$ introduces 149.54 mg/L NO_3^-; the updated total NO_3^- concentration is 149.69 mg/L), and click the button "Add to list".
   - Click "View/edit list" and change the "PO4–3", input to a total concentration of 1.0 mg/L.
   - While still in the List of components window, verify the total concentrations of input components (see Figure 15.9), then click "Back to main menu".
3. Run the calculation and check saturation indices.

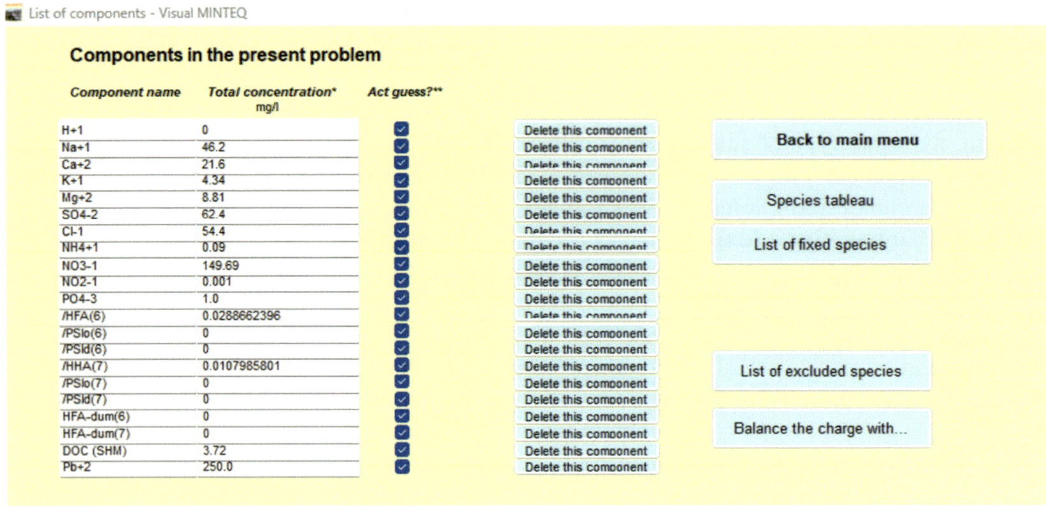

FIGURE 15.9 Visual MINTEQ list of components window showing the total concentrations of input components for Example 2 with Pb^{2+} at 250 mg/L.

- In the Visual MINTEQ main menu, click on the "Run" button to perform calculations. After clicking the "Run" button, results of the speciation calculation will automatically appear, as shown in Figure 15.10.
- Click "Display saturation indices". The saturation indices for different solid phases that might form will be as shown in Figure 15.11. This saturation indices window from Visual MINTEQ shows the log IAP (***ion activity product***) and SI[6] for various species, indicating their potential to precipitate. The red text highlights species that are oversaturated and thus likely to precipitate under the given conditions. For this example, 12 species are oversaturated, all of which contain Pb.
- Click "Print to Excel" to export all SIs, For this example here, the oversaturated species are Anglesite ($PbSO_4$), Chloropyromorphite ($Pb_5(PO_4)_3Cl$, two forms), Hydroxylpyromorphite ($Pb_5(PO_4)_3OH$), Lanarkite ($Pb_2(SO_4)O$, misspelled as Larnakite in Visual MINTEQ), $Pb(OH)_2(s)$, $Pb_2(OH)_3Cl(s)$, $Pb_3(PO_4)_2(s)$, $Pb_3O_2SO_4(s)$, $Pb_4(OH)_6SO_4(s)$, $Pb_4O_3SO_4(s)$, and $PbHPO_4(s)$. In Visual MINTEQ, the appearance of two forms of chloropyromorphite – "chloropyromorphite (c)" and "chloropyromorphite (soil)" – reflects differences in the value of K_{sp}, dependent on the degree of crystallinity. The "(soil)" form, with a lower K_{sp} ($\log K_{sp} = -80.4$) than the crystalline form "(c)" ($\log K_{sp} = -79.6$), is less soluble under standard conditions. The "chloropyromorphite (soil)" species forms in a chemically complex milieu and has slightly different properties than the crystalline form. For theoretical or lab-scale simulations, the crystalline

[6] The ion activity product (IAP) is the product of the activities of ions that are in solution when a given solid dissolves, raised to their respective stoichiometric coefficients. The IAP has the same mathematical form as the ***solubility product constant***, K_{sp}, which is an equilibrium constant that specifically determines the activities of ions in solution, *at equilibrium*, when a given solid dissolves. IAP reflects the current state of a solution with respect to activities of ions in the potential presence of a solid, such that if IAP $< K_{sp}$, the solution is undersaturated, and more solid can dissolve; if IAP $= K_{sp}$, the solution is at equilibrium, and no more solid will dissolve; and if IAP $> K_{sp}$, the solution is supersaturated, and precipitation should occur. The saturation index is related to IAP, as SI = log (IAP/K_{sp}), such that if SI <0, the solution is undersaturated; if SI = 0, the solution is saturated and in equilibrium with the solid phase; and if SI >0, the solution is supersaturated, and precipitation of the solute should occur. The SI is convenient as no knowledge of K_{sp} is required to understand the status of solution with respect to being undersaturated, at equilibrium, or oversaturated.

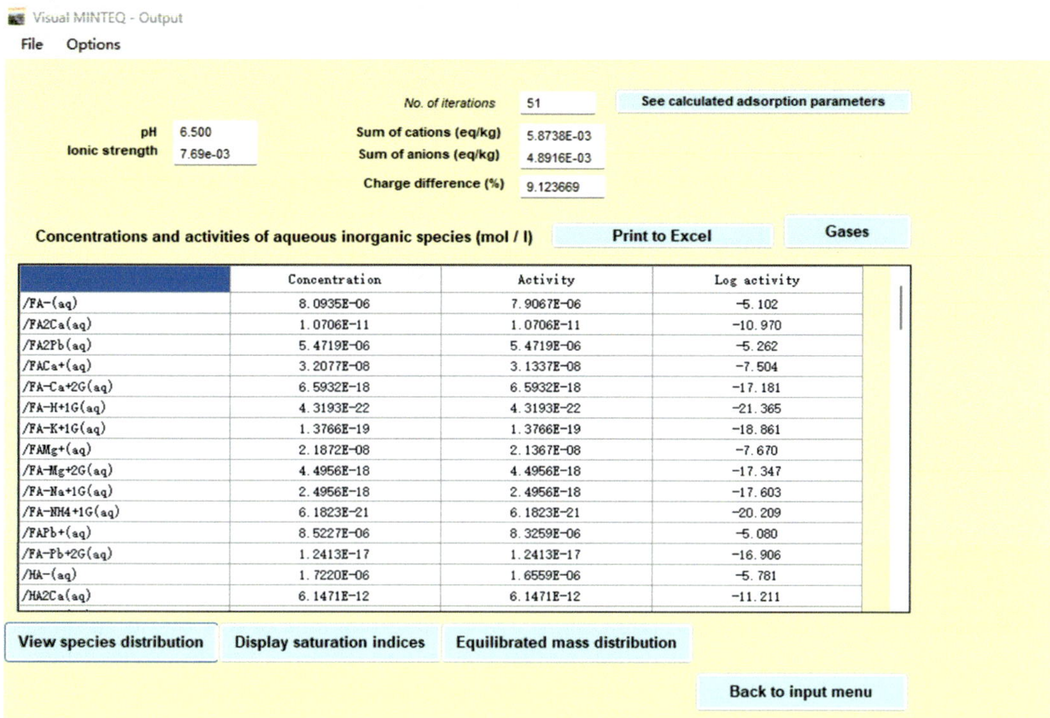

Visual MINTEQ - Output

File Options

pH 6.500
Ionic strength 7.69e-03

No. of iterations 51
Sum of cations (eq/kg) 5.8738E-03
Sum of anions (eq/kg) 4.8916E-03
Charge difference (%) 9.123669

See calculated adsorption parameters

Concentrations and activities of aqueous inorganic species (mol / l)

Print to Excel Gases

	Concentration	Activity	Log activity
/FA-(aq)	8.0935E-06	7.9067E-06	-5.102
/FA2Ca(aq)	1.0706E-11	1.0706E-11	-10.970
/FA2Pb(aq)	5.4719E-06	5.4719E-06	-5.262
/FACa+(aq)	3.2077E-08	3.1337E-08	-7.504
/FA-Ca+2G(aq)	6.5932E-18	6.5932E-18	-17.181
/FA-H+1G(aq)	4.3193E-22	4.3193E-22	-21.365
/FA-K+1G(aq)	1.3766E-19	1.3766E-19	-18.861
/FAMg+(aq)	2.1872E-08	2.1367E-08	-7.670
/FA-Mg+2G(aq)	4.4956E-18	4.4956E-18	-17.347
/FA-Na+1G(aq)	2.4956E-18	2.4956E-18	-17.603
/FA-NH4+1G(aq)	6.1823E-21	6.1823E-21	-20.209
/FAPb+(aq)	8.5227E-06	8.3259E-06	-5.080
/FA-Pb+2G(aq)	1.2413E-17	1.2413E-17	-16.906
/HA-(aq)	1.7220E-06	1.6559E-06	-5.781
/HA2Ca(aq)	6.1471E-12	6.1471E-12	-11.211

View species distribution Display saturation indices Equilibrated mass distribution

Back to input menu

FIGURE 15.10 Speciation results for 250 mg/L Pb in natural lake water with a high concentration of phosphate (1.0 mg/L PO_4^{3-}); calculated results are in units of mol/L.

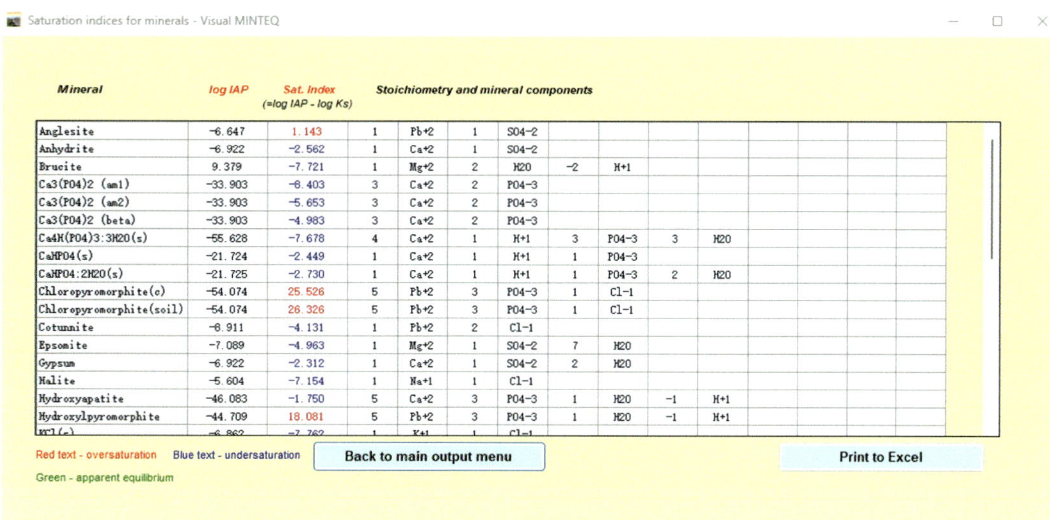

Saturation indices for minerals - Visual MINTEQ

Mineral	log IAP	Sat. Index (=log IAP - log Ks)	Stoichiometry and mineral components									
Anglesite	-6.647	1.143	1	Pb+2	1	SO4-2						
Anhydrite	-6.922	-2.562	1	Ca+2	1	SO4-2						
Brucite	9.379	-7.721	1	Mg+2	2	H2O	-2	H+1				
Ca3(PO4)2 (am1)	-33.903	-8.403	3	Ca+2	2	PO4-3						
Ca3(PO4)2 (am2)	-33.903	-5.653	3	Ca+2	2	PO4-3						
Ca3(PO4)2 (beta)	-33.903	-4.983	3	Ca+2	2	PO4-3						
Ca4H(PO4)3:3H2O(s)	-55.628	-7.678	4	Ca+2	1	H+1	3	PO4-3	3	H2O		
CaHPO4(s)	-21.724	-2.449	1	Ca+2	1	H+1	1	PO4-3				
CaHPO4:2H2O(s)	-21.725	-2.730	1	Ca+2	1	H+1	1	PO4-3	2	H2O		
Chloropyromorphite(c)	-54.074	25.526	5	Pb+2	3	PO4-3	1	Cl-1				
Chloropyromorphite(soil)	-54.074	26.326	5	Pb+2	3	PO4-3	1	Cl-1				
Cotunnite	-6.911	-4.131	1	Pb+2	2	Cl-1						
Epsomite	-7.089	-4.963	1	Mg+2	1	SO4-2	7	H2O				
Gypsum	-6.922	-2.312	1	Ca+2	1	SO4-2	2	H2O				
Halite	-5.604	-7.154	1	Na+1	1	Cl-1						
Hydroxyapatite	-46.083	-1.750	5	Ca+2	3	PO4-3	1	H2O	-1	H+1		
Hydroxylpyromorphite	-44.709	18.081	5	Pb+2	3	PO4-3	1	H2O	-1	H+1		
KCl(s)	-6.862	-7.762	1	K+1	1	Cl-1						

Red text - oversaturation Blue text - undersaturation Back to main output menu Print to Excel

Green - apparent equilibrium

FIGURE 15.11 Saturation indices for minerals window for a 250 mg/L Pb spike into a natural lake water sample with a high concentration of phosphate (1.0 mg/L PO_4^{3-}).

form is suitable, while in natural systems, such as the Example 2 conditions, the "soil" form is more appropriate, as it better represents *in situ* conditions.

4. Define solid phases:
   - Go to the "Solid phases and excluded species" menu on the main menu bar of Visual MINTEQ. Choose "Specify possible solid phases" from the drop-down menu. Selecting this option allows the oversaturated species to form precipitates. Select the option "Limit list to those species that can form based on the components that have been defined", as shown in Figure 15.12, to prevent the inclusion of irrelevant species in the list of choices. Select "Anglesite" and click "Add"; the parameters governing the formation of this mineral will show in this window, including the logK_{sp} (showing as *K*s in the Visual MINTEQ window for specification of possible species). The stoichiometry of the solid is indicated below, showing that the solid consists of one Pb^{2+} and one SO_4^{2-}.
   - Continue to add all the oversaturated species listed above to the list of possible solid phases, then click on the button "List of possible species" at the bottom of the possible species window. A new window will open, which should look like the one in Figure 15.13.
5. Results: Click "Back to main menu" and in the main Visual MINTEQ menu, and use the "Run" button to perform calculations. View the output, including ionic strength, species concentrations, and activities (Figure 15.14). In the saturation indices window, three species will now have a SI of zero, indicating that precipitates formed for these species up to the point of equilibrium between the solid and aqueous phase. Click on the "View species distribution" button to view the relative speciation (inset of Figure 15.14). At equilibrium, Pb^{2+} accounts for 57% of the total aqueous concentration of Pb. Over 30% of the remaining aqueous Pb is bound to DOM, indicating that DOM serves as a significant sink modulating the bioavailability of aqueous Pb that has not precipitated.

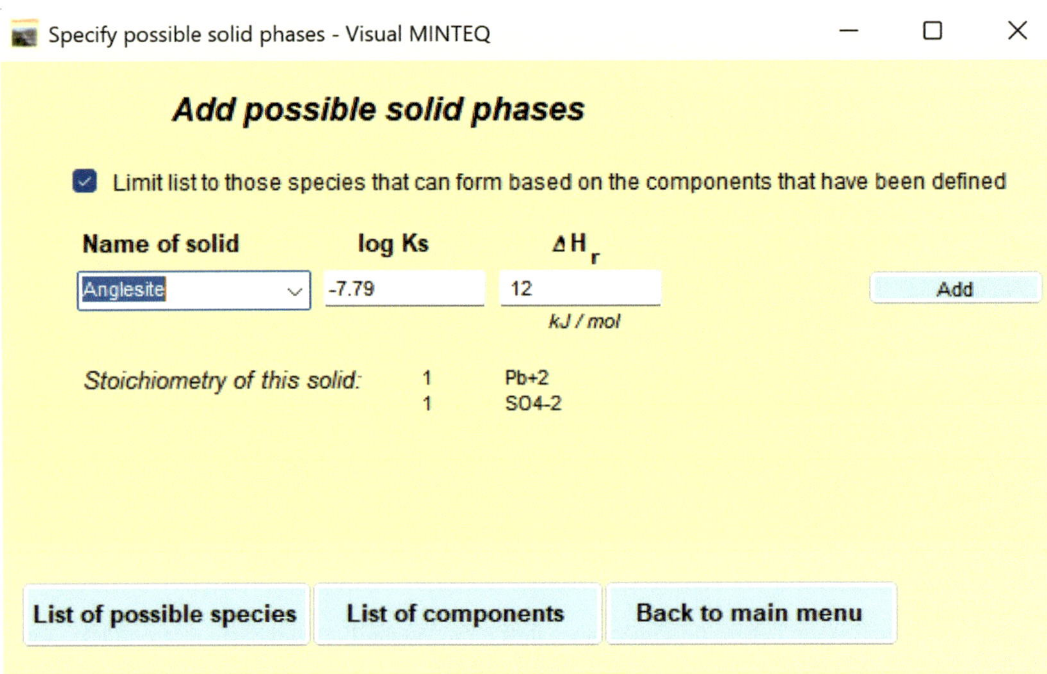

FIGURE 15.12 The specify possible solid phases window in Visual MINTEQ showing the addition of Anglesite.

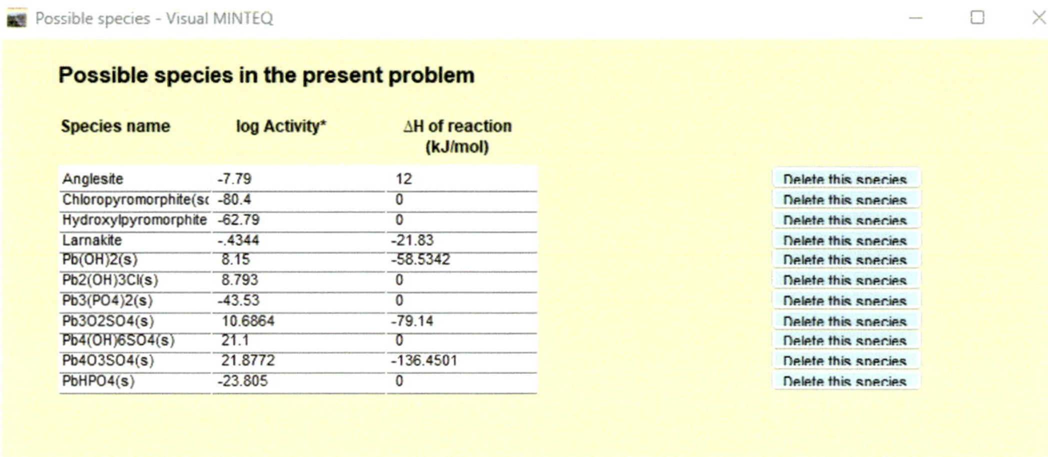

FIGURE 15.13 The list of possible solid species, based on saturation index, in Example 2.

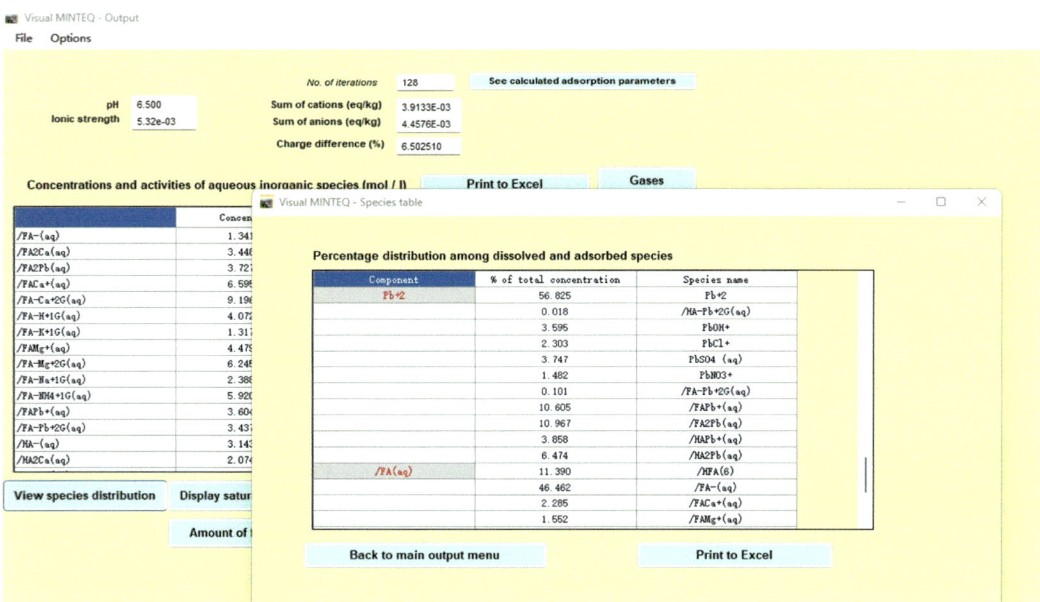

FIGURE 15.14 Speciation results for 250 mg/L Pb in natural lake water with a high concentration of phosphate (1.0 mg/L PO_4^{3-}); for this calculation, possible solid phases are specified, as shown in Figure 15.13.

Next, in the Output window, view the mass distribution by clicking "Equilibrated mass distribution"; in the new window that opens, we see that 97% of the Pb in the system formed precipitates (Figure 15.15), i.e., the 57% of Pb present as Pb^{2+} is comprised of the remaining 3% of Pb that remained aqueous and did not precipitate.

Click "Back to main output window" and then view the amount of finite solids by clicking the "Amount of finite solids" button (as shown in the bottom of the Output window, Figure 15.14). A new window will open, as shown in Figure 15.16. Based on the Visual MINTEQ simulation results, of the twelve species that were supersaturated, three Pb-containing solid phases, chloropyromorphite ($Pb_5(PO_4)_3Cl$), Lanarkite ($Pb_2(SO_4)O$), and lead hydroxide ($Pb(OH)_2$), are predicted to precipitate under the given environmental conditions. In comparison to chloropyromorphite ($K_{sp} = -80.4$), the other phosphate-containing solids, hydroxylpyromorphite ($Pb_5(PO_4)_3OH$), $Pb_3(PO_4)_2(s)$, and

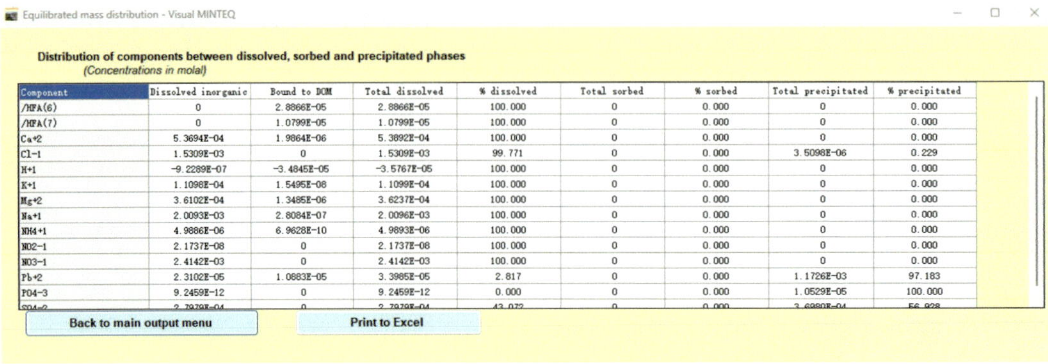

Component	Dissolved inorganic	Bound to DOM	Total dissolved	% dissolved	Total sorbed	% sorbed	Total precipitated	% precipitated
/HFA(6)	0	2.8866E-05	2.8866E-05	100.000	0	0.000	0	0.000
/HFA(7)	0	1.0799E-05	1.0799E-05	100.000	0	0.000	0	0.000
Ca+2	5.3694E-04	1.9864E-06	5.3892E-04	100.000	0	0.000	0	0.000
Cl-1	1.5309E-03	0	1.5309E-03	99.771	0	0.000	3.5098E-06	0.229
H+1	-9.2289E-07	-3.4845E-05	-3.5767E-05	100.000	0	0.000	0	0.000
K+1	1.1098E-04	1.5495E-08	1.1099E-04	100.000	0	0.000	0	0.000
Mg+2	3.6102E-04	1.3485E-06	3.6237E-04	100.000	0	0.000	0	0.000
Na+1	2.0093E-03	2.8084E-07	2.0096E-03	100.000	0	0.000	0	0.000
NH4+1	4.9886E-06	6.9628E-10	4.9893E-06	100.000	0	0.000	0	0.000
NO2-1	2.1737E-08	0	2.1737E-08	100.000	0	0.000	0	0.000
NO3-1	2.4142E-03	0	2.4142E-03	100.000	0	0.000	0	0.000
Pb+2	2.3102E-05	1.0883E-05	3.3985E-05	2.817	0	0.000	1.1726E-03	97.183
PO4-3	9.2459E-12	0	9.2459E-12	0.000	0	0.000	1.0529E-05	100.000

FIGURE 15.15 Mass distribution results for 250 mg/L Pb in natural lake water with a high concentration of phosphate (1.0 mg/L PO_4^{3-}) when solid-phase formation is enabled.

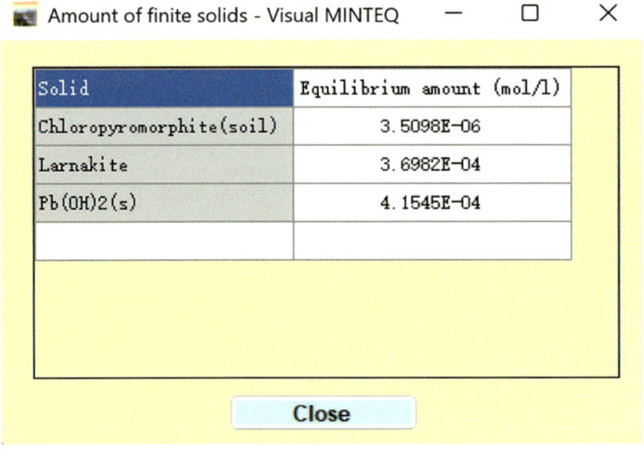

Solid	Equilibrium amount (mol/l)
Chloropyromorphite(soil)	3.5098E-06
Larnakite	3.6982E-04
Pb(OH)2(s)	4.1545E-04

FIGURE 15.16 Visual MINTEQ window showing the amount of finite solids forming after possible solid phases have been defined for the case of spiking 250 mg/L Pb into a natural lake water sample having a high concentration of phosphate (1.0 mg/L PO_4^{3-}).

$PbHPO_4$(s), have K_{sp}s, respectively, of –62.79, –43.53, and –23.81; although these are also all highly insoluble, they are relatively more soluble than chloropyromorphite, and after precipitation of chloropyromorphite, the SIs of these other phosphate-containing lead species are below zero; hence, these do not precipitate. The same principle determines why, of the sulfate-containing Pb species, Lanarkite ($Pb_2(SO_4)O$, K_{sp} = –0.434) precipitates before $Pb_3O_2SO_4$(s) (K_{sp} = 10.6864), $Pb_4(OH)_6SO_4$(s) (K_{sp} = 21.1), and $Pb_4O_3SO_4$(s) (K_{sp} = 21.8772), however, stoichiometry plays a role as well – while the K_{sp} for Anglesite is = –7.79, i.e., indicating overall less solubility than Lanarkite, there are two Pb2+ per SO_4^{2-}– in Lanarkite and only one in Anglesite, and the high Pb in the system causes the IAP/SI of Lanarkite to exceed that of Anglesite. A similar pattern is seen with hydroxide-containing Pb species. In all cases, phosphate-, sulfate-, and hydroxide-containing Pb species, precipitation of the first results in undersaturation of the remaining possible solid phases.

Next, to investigate the bioavailability of Pb under varying concentrations in natural aquatic systems (with identical phosphate levels of 1.0 mg/L PO_4^{3-}), prepare a new calculation, using 0.1 mg/L Pb. The initial NO_3^- concentration is 0.049 mg/L, and adding 0.1 mg/L Pb in the form of $Pb(NO_3)_2$ introduces 0.060 mg/L NO_3^-; the updated total NO_3^- concentration is 0.109 mg/L. Follow the previous procedure of running the calculation without solids, then checking saturation indices, and then defining possible solid phases. Run this second calculation. For discussion here, selected results from both calculations are arranged in Table 15.3; these include relative amounts aqueous Pb species and total Pb species. Additionally, Table 15.3 has information on HA and FA speciation.

The Visual MINTEQ simulations for Example 2 demonstrate that at 250 mg/L Pb, Pb speciation is dominated by precipitation processes, with $Pb(OH)_2$(s) and $Pb_2(SO_4)O$/Lanarkite accounting

TABLE 15.3
Results Showing Pb and DOC Species for 250 and 0.1 mg/L Pb Spikes into a Natural Lake Water Sample with a High Concentration of Phosphate (1.0 mg/L PO_4^{3-})[a]

	250 mg/L Pb			0.1 mg/L Pb		
Species Name	Concentration	% of Total Aqueous Concentration	% of Total Concentration	Concentration	% of Total Aqueous Concentration	% of Total Concentration
			Pb			
Pb+2	19	57	1.6	4.4*	2.5	0.92
/FA2Pb (aq)	3.7	11	0.31	106*	59	22
/FAPb+ (aq)	3.6	11	0.30	4.7*	2.6	0.98
/HA2Pb (aq)	2.2	6.5	0.18	60*	33	12
/HAPb+ (aq)	1.3	3.9	0.11	2.3*	1.3	0.48
PbSO4 (aq)	1.3	3.7	0.11	0.70*	0.40	0.15
PbOH+	1.2	3.6	0.10	0.28*	0.16	0.059
PbCl+	0.78	2.3	0.065	0.18*	0.10	0.038
PbNO3+	0.50	1.5	0.042	0.000085*	<0.01	0.000018
/FA-Pb+2G (aq)	0.034	0.10	0.0028	0.013*	<0.01	0.0027
/HA-Pb+2G (aq)	0.0061	0.018	0.0005	0.0028*	<0.01	0.00058
Pb(OH)2(s)	415	–	34.4	–	–	–
$Pb_2(SO_4)O$(s)/ Lanarkite	370	–	61.3	–	–	–
$Pb_5(PO_4)_3Cl$(s)	3.5	–	1.5	61*	–	63.1
HA (aq) – 91% of total HA				**HA (aq) – 88% of total HA**		
/HA2Pb (aq)	2.2	41	–	0.06	1.1	–
/HA (aq)	3.1	29	–	4.7	43	–
/HFA(7)	1.5	14	–	3.8	35	–
/HAPb+ (aq)	1.3	12	–	0.0023	0.021	–
/HACa+ (aq)	0.24	2.3	–	1.3	12	–
/HAMg+ (aq)	0.17	1.5	–	0.90	8.3	–
HA (gel) – 9% of total HA				**HA (gel) – 12% of total HA**		
HFA-dum(7)	0.51	46	–	0.78	41	–
/HA-Ca+2G (aq)	0.16	29	–	0.31	33	–
/HA-Mg+2G (aq)	0.11	20	–	0.21	22	–
/HA-Na+1G (aq)	0.042	3.8	–	0.082	4.3	–
/HA-Pb+2G (aq)	0.0061	1.1	–	0.0000028	0.24	–
FA (aq) – 84% of total FA				**FA (aq) – 81% of total FA**		
/FA (aq)	13.4	46	–	17	60	–
/FA2Pb (aq)	3.7	26	–	0.11	0.73	–

(Continued)

TABLE 15.3 (*Continued*)
Results Showing Pb and DOC Species for 250 and 0.1 mg/L Pb Spikes into a Natural Lake Water Sample with a High Concentration of Phosphate (1.0 mg/L PO_4^{3-})[a]

	250 mg/L Pb			0.1 mg/L Pb		
Species Name	**Concentration**	**% of Total Aqueous Concentration**	**% of Total Concentration**	**Concentration**	**% of Total Aqueous Concentration**	**% of Total Concentration**
			Pb			
/FAPb+ (aq)	3.6	12	–	0.0047	0.016	–
/HFA(6)	3.3	11	–	7.0	24	–
/FACa+ (aq)	0.66	2.3	–	2.6	8.9	–
/FAMg+ (aq)	0.45	1.6	–	1.7	6.0	–
FA (gel) – 16% of total FA				**FA (gel) – 19% of total FA**		
HFA-dum(6)	2.9	46	–	4.0	43	–
/FA-Ca+2G (aq)	0.92	29	–	1.5	31	–
/FA-Mg+2G (aq)	0.62	20	–	1.0	21	–
/FA-Na+1G (aq)	0.24	3.8	–	0.38	4.1	–
/FA-Pb+2G (aq)	0.034	1.1	–	0.000013	<0.01	–
/FA-K+1G (aq)	0.013	0.21	–	0.021	0.23	–

[a] Units are in M unless indicated by an *, in which case units are in nM; results for each component add to a total of 100%, to within round-off error.

for 34% and 61% of the total Pb concentration, respectively. The aqueous phase still retains a substantial portion of Pb, and complexation with HA and FA plays a significant role in modulating bioavailability. For 250 mg/L Pb, the PbHA and PbFA species cumulatively account for ~32% of aqueous Pb species, demonstrating the importance of organic complexation in regulating Pb mobility. In contrast, gel-phase complexes (HA-gel and FA-gel) contribute marginally, together representing <1% of total Pb. In terms of toxicity, however, free Pb^{2+} is 19 μM/3.9 mg/L in absolute terms, indicating that despite the precipitation of most of the Pb, an exposure issue may nonetheless be present. For instance, the United States Environmental Protection Agency's ***Criterion Maximum Concentration***, i.e., the highest concentration of a pollutant in water that is not expected to cause acute toxicity to aquatic life, is around ~12 mg/L (US EPA, 2024) for Pb at the hardness used in this example – since the chronic metric is 26 times lower, chronic toxicity effects would be expected to pertain from 3.9 mg/L bioavailable Pb.

As compared to 250 mg/L Pb, when Pb concentration drops to 0.1 mg/L, the oversaturation calculated by Visual MINTEQ is only for $Pb_5(PO_4)_3Cl$. The total amount of Pb precipitated is still high at Pb = 0.1 mg/L, with 63% precipitated as $Pb_5(PO_4)_3Cl$; the lower Pb concentration drives less precipitation of sulfate- and hydroxide-containing Pb species. In contrast to 250 mg/L Pb, at 0.1 mg/L Pb, the higher ligand-to-Pb ratio favors M–DOM complexation. Dissolved Pb^{2+} constitutes only 0.92% of the total Pb, and the majority of aqueous Pb is bound to dissolved organic ligands; FA_2Pb(aq) and HA_2Pb(aq) represent 59% and 33% of aqueous Pb, respectively (closely mimicking the proportion of DOM, i.e., FA versus HA), with PbHA and PbFA species cumulatively accounting for ~96% of aqueous Pb species. As with Example 1, this highlights the dominant role of DOM in controlling Pb bioavailability at low concentrations. While the results in this example represent relatively high levels of Pb contamination, they serve to highlight the sensitivity of Pb to precipitation, which is not uncommon in highly contaminated environments.

15.3.4 Example 3: BLM Calculations Using the Default BLM Parameters in Visual MINTEQ

This example demonstrates how to use Visual MINTEQ for BLM calculations in both speciation and toxicity mode. An example using published BLM parameters for Cu included in the Visual MINTEQ database will be used. The US EPA (2007) guidelines focus on Cu as the primary heavy metal to illustrate metal toxicity prediction using the BLM, and the Cu-*Daphnia magna* example used here from Visual MINTEQ is the earliest Cu-BLM implementation included in the software, based on BLM parameters published by de Schamphelaere and Janssen (2002), as later refined by de Schamphelaere et al. (2002). *D. magna*, commonly known as a water flea, is a planktonic freshwater crustacean that is widely used as ***test organism*** in aquatic toxicology due to its ecological relevance and sensitivity to pollutants. The toxicity endpoint applied by de Schamphelaere and Janssen was an acute toxicity test measuring the Cu concentration that causes immobility in 50% of juvenile *D. magna* (<24 hours old) after a Cu exposure period of 48 hours. This example demonstrates how to apply the Cu-BLM in Visual MINTEQ, considering competition with major cations (Ca^{2+} is used as our example here) and how this affects Cu speciation and toxicity.

It is also possible for Visual MINTEQ users to input new components as well as BLM parameters that are not yet available in the database. The process of manually adding a component and new complexation reaction is shown in this example; inputting new BLM parameters will be described in Example 4 below. Before introducing the BLM toxicity mode in Visual MINTEQ, the BLM speciation-mode calculations are run.

To assess how increasing Ca concentration influences Cu speciation, three different Ca concentrations will be evaluated: 0.25, 0.50, and 1.0 mM, all having the same ambient total Cu of 40 nM. The input water chemistry conditions are based on de Schamphelaere and Janssen (2002), which is based on synthetic water samples that do not contain organic matter. In each scenario, Cl^- concentrations also vary to maintain charge balance. For the Ca^{2+} = 0.25 mM condition, Cl^- is 0.58 mM. For Ca^{2+} = 0.50 mM, Cl^- increases to 1.08 mM. In the Ca^{2+} = 1.0 mM condition, Cl^- reaches 2.08 mM (*NB*: The concentration of Cl^- reflects its presence as a counterion to Ca^{2+}, in addition to a small amount of Cl^- that is also in the ***medium*** from other constituents). Other components are kept constant: Mg^{2+} at 0.25 mM, Na^+ at 0.077 mM, K^+ at 0.078 mM, CO_3^{2-} at 91.3 μM (0.0913 mM), and SO_4^{2-} at 0.25 mM. The pH was set to be fixed at 6.90. The choice of 40 nM (0.00004 mM) for ambient Cu concentration is lower from the levels reported by de Schamphelaere et al. (2002); while this concentration is lower than what these and other authors typically observe in natural waters, the experimental conditions are inherently artificial due to the exclusion of DOM, which would otherwise alter the speciation and bioavailability of Cu substantively. In the absence of DOM, a lower Cu concentration is better for the purposes of illustration.

Procedure:

1. Define settings:
   - Launch Visual MINTEQ and access the main menu.
   - Set the pH to be fixed at 6.90.
   - Leave the choice of activity correction as the default (the Davies equation).
   - Set the temperature to 15°C, the prescribed temperature for BLM models (US EPA, 2002).
   - Set the concentration for millimolal.
2. Define components and input total concentrations
   - Add Ca^{2+}, Mg^{2+}, Na^+, K^+, Cl^-, CO_3^{2-}, SO_4^{2-}, and Cu^{2+} as components and with concentrations as reported above. For accurate concentration data, first convert molar units to molal – these differ according to volume versus weight of the solution at 15°C, i.e., by the amount by which the density of the solution at 15°C differs from 1.0. For dilute solutions, as in the example here, this difference amounts to less than 0.1%, in which

case, for convenience it is easier to skip the conversion and use molar concentration units as approximately equal to molal. For the Ca = 0.25 mM set, the input components list should appear as in Figure 15.17.

3. Specify a BLM calculation
   - Navigate to the Parameters category on the main menu bar of Visual MINTEQ, and in the drop-down menu click the "Add/Edit Biotic Ligand Model"; The BLM window will open. Under the "Select metal" dialog, choose Cu^{2+} from the drop-down menu, then, to the right a new dialog, "Select organism/endpoint/reference", will appear. Under this dialog, select "*Daphnia Magna*/LC_{50}/de Schamphelaere and Janssen (2002)".[7] Since the needed BLM parameters are already in the Visual MINTEQ database, there is no further action needed; the appropriate parameters are entered automatically. The f_{crit} of 0.47 is shown in the Biotic Ligand Model window, next to which there are two input choices for BLM calculations in speciation or toxicity mode. *NB*: The user has the option to "View/edit BLM parameter file", which will open a new window from which all BLM parameters for this organism and endpoint can be viewed.
   - Choose "Speciation" model (see Figure 15.18).
   - In the Biotic Ligand Model window, click "Save this problem and Back to the main menu", then view the component list in the problem after the Cu-BLM has been added (shown in Figure 15.19). Compared to the components in Figure 15.18, before the BLM was specified, now the BLsite-1 component appears. Click "Back to main menu", then on the main menu bar, choose "File" → "Save input file", and name the saved file "Cu_Daphnia_BLM.vda". This file will be used again later in this example.
4. Results: Use the "Run" button to perform calculations. View the output, including ionic strength, species concentrations, and activities (as shown in Figure 15.20). The charge balance difference for this calculation is 0.18%. Such a low discrepancy between positive and negative charges reflects that all major ionic species contributing to charge have been properly accounted for, and no significant charge imbalance, i.e., as can be achieved with a synthetic solution whose components are all well known. The calculated f value is now given in the results window, and at $f = 0.6053$, this is higher than the f_{crit} of 0.47, i.e., above the toxicity threshold. The free Cu^{2+} is 28.9 nM. To a great extent, this is also a consequence of the synthetic solution, since no organic matter is present to complex with Cu^{2+} and decrease its bioavailability.
   - View the speciation distribution window (Figure 15.21). For the BL site, 13.4% of the BL site binds with Ca, and 56.8% is speciated as BL-Cu, with two other species, BL-CuOH and BL-$CuCO_3^-$, making up a total of 60.53% of the BL site bound to Cu, corresponding to the f value of 0.6053.

Repeat the steps above and run BLM speciation-mode calculations for Ca = 0.50 and Ca = 1.0 mM. Results show that free Cu^{2+} concentration is 28.9, 29.1, and 29.5 nM for total Ca concentrations of 0.25, 0.5, and 1.0 mM, respectively. While these free Cu^{2+} concentrations remain relatively stable, the distribution of Cu and Ca bound to the BL changes markedly with increasing Ca. Specifically, the fraction of Cu bound as BL-Cu decreases from 56.8% to 40.4%, while BL-Ca increases from 13.4% to 37.3%. Ca exhibits a clear competitive effect; consequently, the calculated f drops from 0.6053 to 0.4301. Notably, at 1.0 mM Ca, f has fallen below the f_{crit} = 0.47 threshold value for *D. magna* toxicity, reflecting a substantial reduction in Cu toxicity risk under the protective effects of Ca. If 3.72 mg/L DOM is added to the synthetic water matrix (as in Examples 1 and 2) for

7 The parameters used in this study are not from the original publication listed in the Visual MINTEQ database, but rather from a refined data set reported in a subsequent study (see de Schamphelaere et al., 2002). Notably, although the endpoint is labeled as LC_{50} in Visual MINTEQ, the original source, and the later study reporting the refined data set, actually report EC_{50} values for immobilization of *D. magna*.

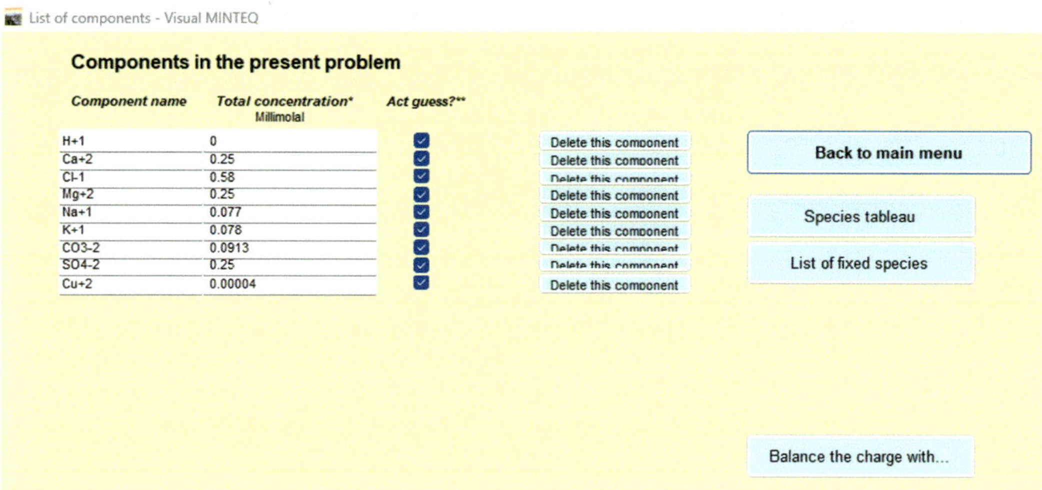

FIGURE 15.17 Visual MINTEQ input components for the Ca = 0.25 mM Cu-BLM scenario with 40 nM total ambient Cu.

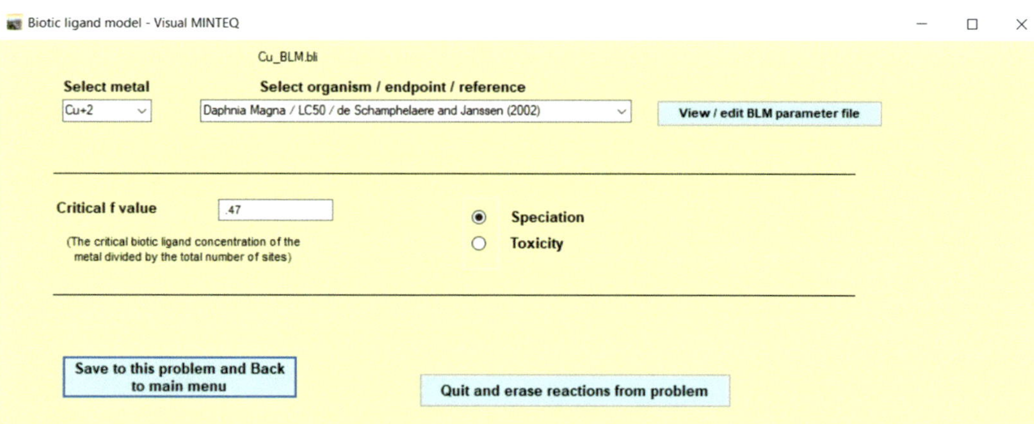

FIGURE 15.18 Biotic Ligand Model window in Visual MINTEQ with the Cu-BLM for *D. magna* from de Schamphelaere and Janssen (2002) specified for a speciation mode calculation.

Ca = 0.25 mM with 40 nM Cu, the f value drops markedly from 0.61 to 1.67×10^{-4}, and the free Cu^{2+} concentration decreases to ~0.003 nM (data not shown). Hence, while Ca has an important competitive effect, as for Cd and Pb, DOM is very effective at reducing Cu bioavailability under environmentally relevant conditions.

To investigate the effect of Ca on Cu toxicity to *Daphnia magna* using the BLM module of Visual MINTEQ, the same calculation performed in speciation mode may be repeated in toxicity mode. The input parameters will be the same as for the speciation mode examples as reported above.

Procedure:

1. Define settings:
   - Either launch Visual MINTEQ and access the main menu, or, if Visual MINTEQ is running, click "Reset" on the bottom right of the main menu workspace.

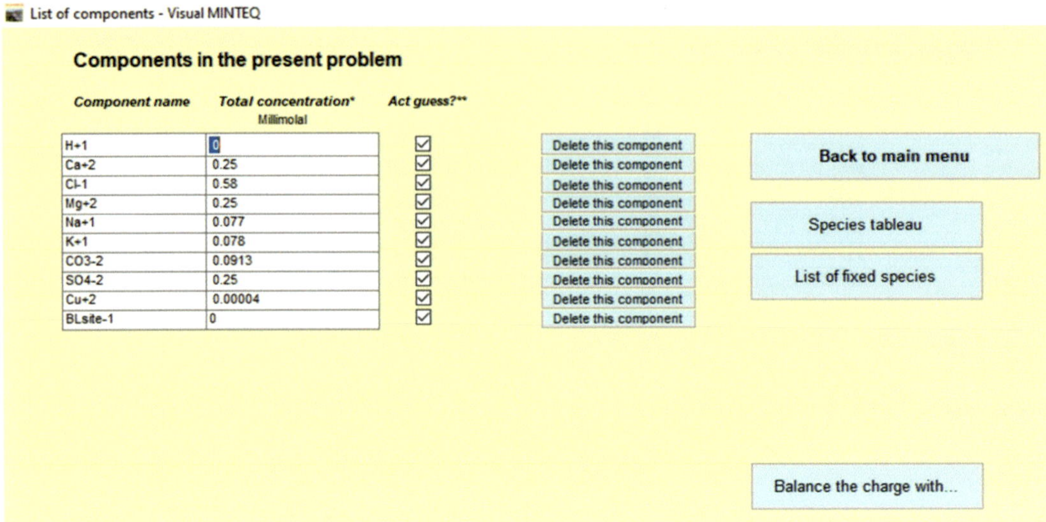

FIGURE 15.19 Input components for the Ca = 0.25 mM scenario after the Cu-BLM speciation-mode calculation was specified. The BL, component name BLsite-1, now appears in the component list.

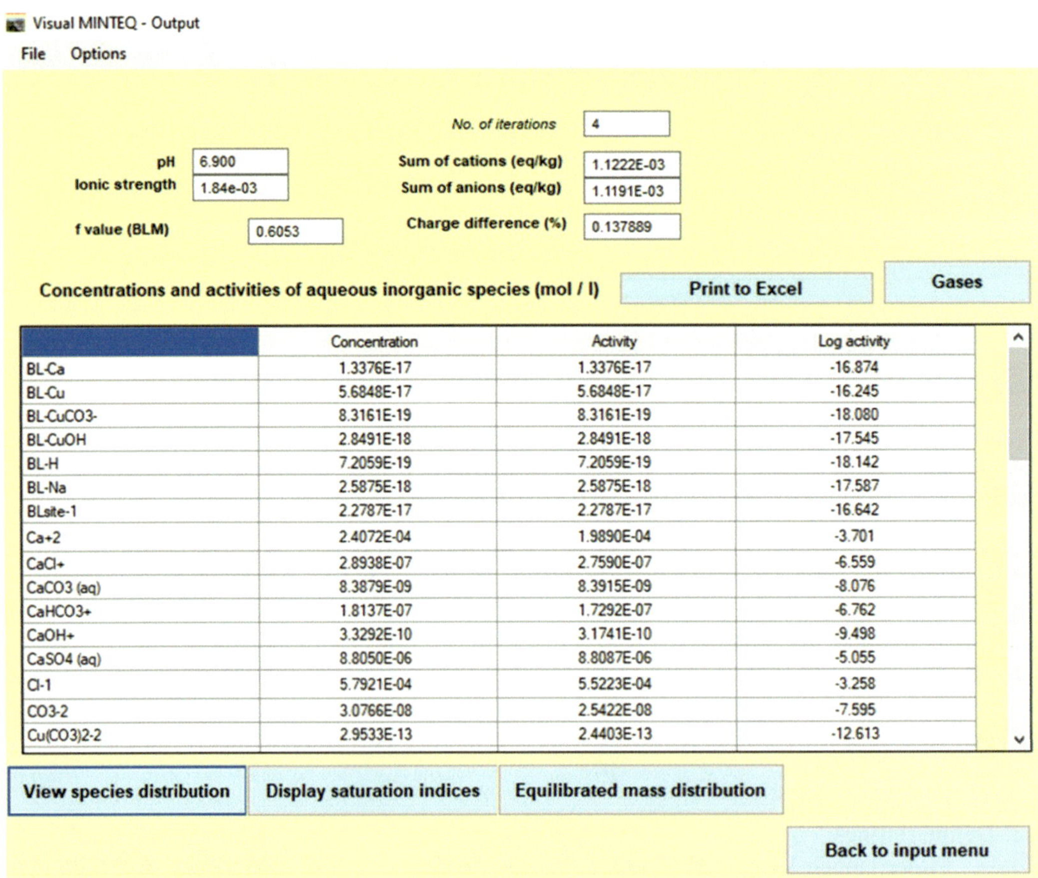

FIGURE 15.20 Visual MINTEQ output window showing results for the Cu-BLM speciation-mode calculation with Ca = 0.25 mM and 40 nM Cu.

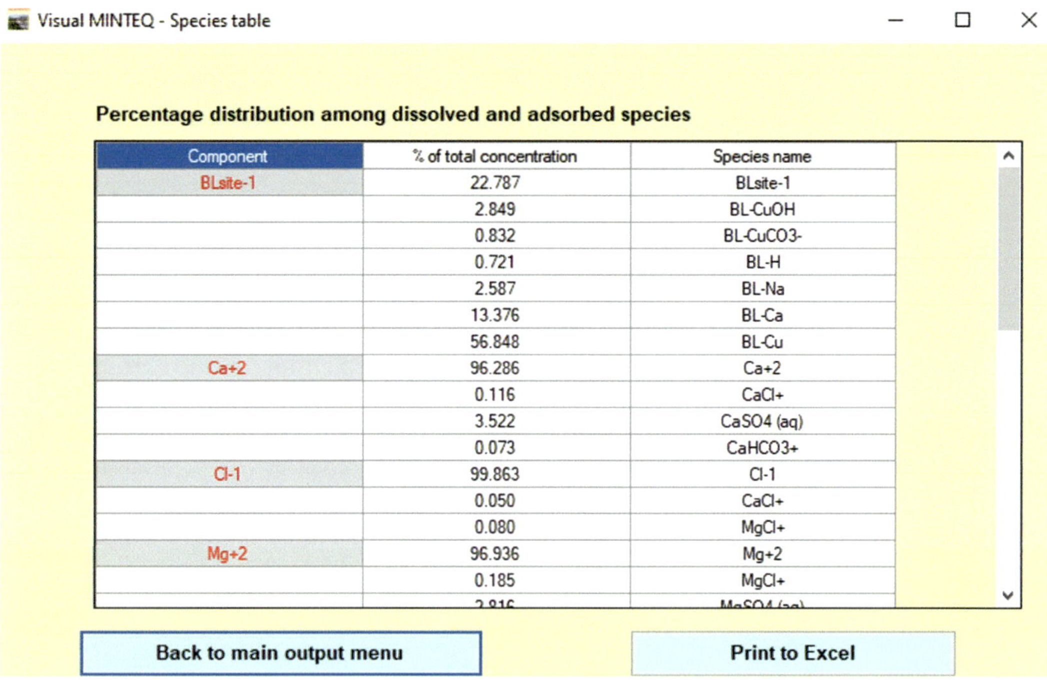

Component	% of total concentration	Species name
BLsite-1	22.787	BLsite-1
	2.849	BL-CuOH
	0.832	BL-CuCO3-
	0.721	BL-H
	2.587	BL-Na
	13.376	BL-Ca
	56.848	BL-Cu
Ca+2	96.286	Ca+2
	0.116	CaCl+
	3.522	CaSO4 (aq)
	0.073	CaHCO3+
Cl-1	99.863	Cl-1
	0.050	CaCl+
	0.080	MgCl+
Mg+2	96.936	Mg+2
	0.185	MgCl+

FIGURE 15.21 Relative speciation distribution results for the Cu-BLM speciation-mode calculation with Ca = 0.25 mM and 40 nM Cu.

2. Open the file
   - Open the file "Cu_Daphnia_BLM.vda", and click on "View/edit list" on the main menu to view the list of components window and verify that Ca^{2+}, Mg^{2+}, Na^{+}, K^{+}, Cl^{-}, CO_3^{2-}, and SO_4^{2-} components are entered properly and with concentration as given for the speciation calculation.
3. Add BLM
   - Navigate to the "Parameters" selection on the main menu bar and from the drop-down, click the "Add/Edit Biotic Ligand Model"; as before, select metal for Cu^{2+} and select "*Daphnia Magna*/LC_{50}/de Schamphelaere and Janssen (2002)".
   - In the middle of the BLM window, choose "Toxicity" (mode) and press "Save to this problem and Back to main menu".
   - From the main menu, view the components in the present problem again to verify that the BLsite-1 components now appear.
4. Results: Use the "Run" button to perform calculations and view the output.

Repeat steps above and run the toxicity mode calculation for Ca = 0.50 and Ca = 1.0 mM.

Table 15.4 shows the speciation results for Cu under the three toxicity mode calculations. For each calculation, the f value shown in the results window is 0.47, indicating that Visual MINTEQ varies the total amount of Cu in the system until this f_{crit} value occurs. The total Cu concentration for each run, i.e., the sum of all Cu-species, is given at the bottom of Table 15.4. These represent the total Cu concentration at which f_{crit} occurs; hence, the EC_{50}s for Cu with Ca at 0.25, 0.5, and 1.0 mM are 23.1, 31.2, and 47.0 nM, respectively. As for the BLM speciation calculation, this trend reflects increasing competition between Ca^{2+} and Cu^{2+} for binding to the BL.

To complete this example, we illustrate the process of manually adding a new component and its associated complexation reaction into the Visual MINTEQ database, using the example of chalkophores (Chlk) – a class of ligands that are present to varying degree in freshwater environments and

TABLE 15.4
Speciation Distribution Results for the Cu-BLM Toxicity-Mode Calculations with Ca = 0.25, 0.5 and 1.0 mM Ca. (all concentrations of Cu species are in nM)

	Ca 0.25	Ca 0.5	Ca 1.0
Cu(CO3)2−2	1.71E−04	2.29E−04	3.40E−04
Cu(OH)2 (aq)	5.12E−02	6.74E−02	9.78E−02
Cu(OH)3−	1.66E−05	2.21E−05	3.24E−05
Cu(OH)4−2	1.24E−11	1.68E−11	2.57E−11
Cu+2	1.67E+01	2.27E+01	3.47E+01
Cu2(OH)2+2	4.65E−04	8.34E−04	1.85E−03
Cu2OH+3	4.52E−07	8.44E−07	2.00E−06
Cu3(OH)4+2	2.05E−08	4.86E−08	1.56E−07
CuCl+	1.59E−02	3.91E−02	1.09E−01
CuCl2 (aq)	2.31E−06	1.04E−05	5.43E−05
CuCl3−	1.25E−11	1.04E−10	1.05E−09
CuCl4−2 1	3.99E−17	6.31E−16	1.25E−14
CuCO3 (aq)	2.06E+00	2.70E+00	3.86E+00
CuHCO3+	6.22E−02	8.21E−02	1.19E−01
CuHSO4+	7.69E−08	9.62E−08	1.28E−07
CuOH+	3.65E+00	4.86E+00	7.14E+00
CuSO4 (aq)	6.10E−01	7.57E−01	9.93E−01
Total Cu (nM), i.e., EC_{50} in nM total Cu	23.1	31.2	47.0
EC_{50} (μ/L total Cu)	1.47	1.98	2.99
EC_{50} (mg/L total Cu)	1,470	1,980	2,990

exhibit strong binding affinity for Cu, thereby influencing Cu bioavailability and toxicity in aquatic environments (Li et al., 2023). Since Chlk is not included in the default component or thermodynamic database of Visual MINTEQ, we demonstrate how to incorporate both the Chlk species and the Chlk-Cu complexation reaction, and further explore the implications of Chlk presence on Cu speciation and ecological risk.

When adding a new component to the Visual MINTEQ database, it is necessary first to understand the structure and format of the component database file. In Visual MINTEQ version 4.0, this file is normally located at: C:\Program Files (x86)\Vminteq40\comp_2008.vdb). Open this component file with a simple text editor (e.g., Word Pad) and verify that the contents are as shown in Figure 15.22. Note that each line in this file corresponds to a specific component and consists of six fields separated by commas.

As an illustrative example, consider the entry for the 23rd line of the file: "150, Ca+2, 2, 6, 0.165, 40.08". The structure for this entry is as follows:

- The number 150 is a unique internal identification number for the component (not used directly in calculations but needed for system indexing);
- Ca+2 is the name of the component, including its charge notation;
- 2 is the ionic charge of the component (+2 in this case for Ca^{2+});
- 6 is the Debye–Hückel size parameter, used to calculate the activity coefficient of the ion in solution;
- 0.165 is the Debye–Hückel empirical correction parameter, a dimensionless constant that further refines activity coefficient estimation under higher ionic strength conditions.
- 40.08 is the molar mass of the component, in grams per mole (g/mol).

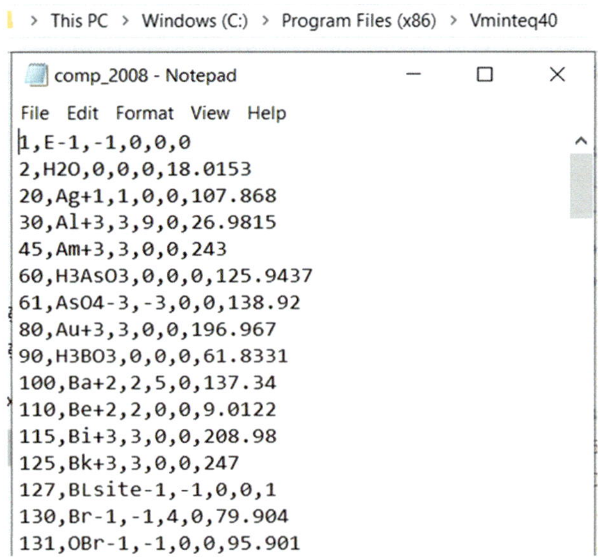


```
1,E-1,-1,0,0,0
2,H2O,0,0,0,18.0153
20,Ag+1,1,0,0,107.868
30,Al+3,3,9,0,26.9815
45,Am+3,3,0,0,243
60,H3AsO3,0,0,0,125.9437
61,AsO4-3,-3,0,0,138.92
80,Au+3,3,0,0,196.967
90,H3BO3,0,0,0,61.8331
100,Ba+2,2,5,0,137.34
110,Be+2,2,0,0,9.0122
115,Bi+3,3,0,0,208.98
125,Bk+3,3,0,0,247
127,BLsite-1,-1,0,0,1
130,Br-1,-1,4,0,79.904
131,OBr-1,-1,0,0,95.901
```

FIGURE 15.22 Component information for each component in the Visual MINTEQ database file comp_2008.vdb.

Understanding the meaning and function of these fields is needed when defining new components, such as organic chelators (e.g., Chlk), to ensure consistency and correct calculation of chemical speciation within the MINTEQ model framework.

Once a new component is added, reactions involving the added component can be added either by adding a new reaction to the Visual MINTEQ thermodynamic database or by editing the thermodynamic database file (thermo.vdb) from within Visual MINTEQ. The former is preferable, in which case it is necessary first to understand the structure and format of the thermodynamic database file. This file is located in the same folder as the component file comp_2008.vdb. In the thermo.vdb file, species such as aqueous metal–ligand compounds are defined using a three-line format. The first line contains key thermodynamic and structural information, while the second line specifies the stoichiometric composition of the species in terms of its basic components, and the third line provides the molar mass of the component and references. For example, consider the following entry:

"2317320,CuSO4 (aq),8.7,2.36,0,0,0,0,0,0,2
1,231,1,732
159.609,NIST 46.7,NIST 46.7".

This entry is for the reaction that corresponds to $Cu^{2+} + SO_4^{2-} \rightleftharpoons CuSO_4$ (aq), which plays an important role in modeling Cu speciation in aqueous environments. The number 2317320 is a unique species ID – the first three digits are the component ID for copper, the second three digits are the component ID for sulfate, and the seventh digit is designated to ensure uniqueness among species composed of the same elemental components. For instance, both $CuOH^+$ and $Cu(OH)_2$ consist of Cu and OH groups. To differentiate such species, the identifier assigns sequential terminal digits – e.g., 0 for the first occurrence ($CuOH^+$), 1 for the second ($Cu(OH)_2$), and so forth. As no more than nine distinct species typically exist for any given combination of components, this numbering scheme is sufficient to maintain identifier uniqueness. The second field is the name of the species, in this case $CuSO_4$ (aq). The value 8.7 is the enthalpy of reaction (ΔH, in kJ/mol), used for temperature-dependent corrections of database *K*s (adjusting database *K*s for changes in temperature), and the value 2.36 represents the log of the equilibrium constant (log*K*) for the formation reaction at 25°C. The next six entries are for the maximum and minimum log*K* of the reaction, species charge, the Debye–Hückel *a* and *b* parameters, and an alkalinity factor. The species charge is 0 for $CuSO_4$, and all other entries are zero because data are unavailable, or in the case of alkalinity factor because there are only relevant to certain types of reactions (*viz*., involving carbonate). The last digit in the first line is the number of components; in the case of $CuSO_4$ there are two, which is also an instruction as to how many fields of component information are read on line 2.

The second line gives the stoichiometry and component ID for the first component, Cu (1,231), followed by the stoichiometry and component ID for the second component, SO_4^{2-} (1,732). The third line for the species $CuSO_4$(aq) is comprised of the molar mass of the species (159.609 g/mol), followed by the next two text entries, which give the source(s), respectively, of the stability constant (K) and enthalpy of reaction, in this case both referenced from NIST (National Institute of Standards and Technology) 46.7, indicating that these thermodynamic values are derived from a reliable and standardized data source.

Procedure:

1. Adding Chlk as a new component in Visual MINTEQ:
   - Copy the original component database file comp_2008.vdb in the Visual MINTEQ folder to a new file and rename. For instance, the file might be copied and renamed "comp_2008_wChlk.vdb". This ensures that if a mistake is made editing the new file, users may revert to the original component file.
   - Open the file "comp_2008_wChlk.vdb" (use a simple text editor), find the line beginning with the number 951, and create a new line after this. Add "952,Chlk–2,–2,0,0,500" to the new line and save the file. When adding the new line, ensure that there are no spaces between commas. Note: In the Visual MINTEQ components file, the entries up to ~900 are used for inorganic constituents. The ID number 952 was arbitrarily used here for Chlk, because that number has not yet been assigned in the database and because Chlk is an organic component; whatever ID is chosen, users should make sure the new ID number is entered in the correct number sequence and is not used for other components in the component database file. For this entry, the charge of Chlk is assumed to be –2, with an assumed molar mass on the order of 500 amu. Debye–Hückel size parameters are not known for Chlk and are thus entered as zero. Casual examination of the database file reveals that many other components also have entries of zero for these parameters. After modification, the file should appear as shown in Figure 15.23.
2. Adding the formation of CuChlk as a new reaction in Visual MINTEQ:
   - Copy the thermo.vdb database file in the Visual MINTEQ folder and rename it (e.g., "thermo_wChlk.vdb").
   - Add the three lines:
     "2319520,CuChlk,0,13.5,0,0,0,0,0,0,2
     1,231,1,952
     563.546, Cha and Abruna (1990), n.a."
     into thermo_wChlk.vdb and save (Figure 15.24). In the first line, 2319520 is the species ID (composed of Cu, ID 231, Chlk, ID 952, and 0), and CuChlk is the designated name of the neutral aqueous complex. The next entry is zero since the enthalpy of reaction is not known, followed by a log$K$ value of 13.5. According to Cha and Abruna (1990), the log$K$ for the reagent used in chalkophore assays developed by Zhang et al. (2020b) is 13.2 – since the assay is based on ***competitive ligand exchange***, therefore, Chlk must have a log$K$ of greater than 13.2 (in some cases much greater); thus, a value of 13.5 used here is conservative. The remaining entries save the last are all zero for the same reasons as given for $CuSO_4$ above. The third line gives the molar mass of CuChlk along with the relevant reference for the log$K$.
3. Select the new component and thermodynamic databases in Visual MINTEQ:

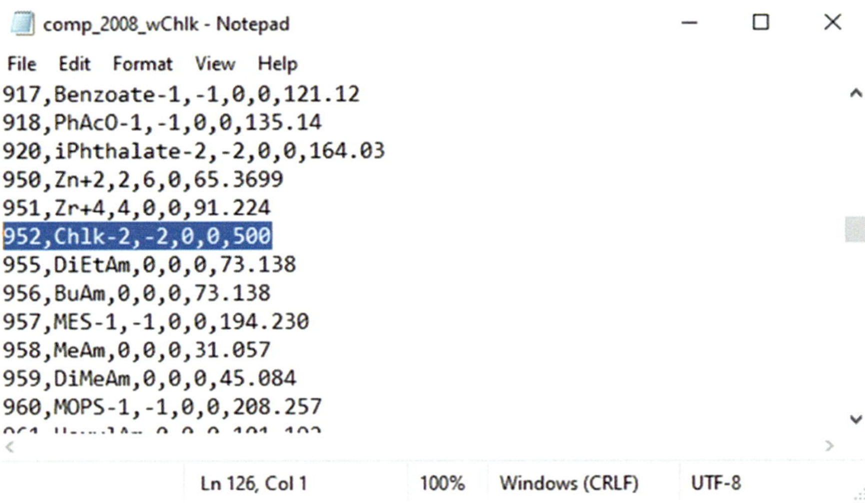

FIGURE 15.23 Adding Chlk to the component file ("comp_2008_wChlk.vdb") in the Visual MINTEQ database. The entry is placed within the number range used for organic components, with a component ID being a previously unused number, and between/near complexing species iPhthalate and Diethylamine (as named in the Visual MINTEQ database). The entry reflects that the new component (Chlk) has a component ID (952), charge (–2), and molar mass (500 amu).

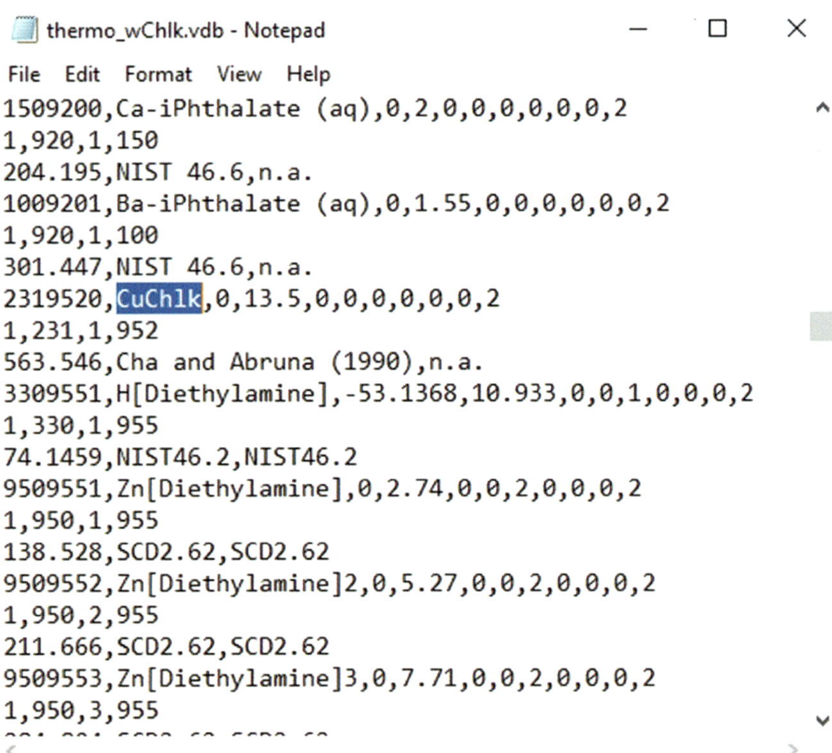

FIGURE 15.24 Adding the CuChlk complexation reaction to the thermodynamic database (file thermo_wChlk.vdb) in Visual MINTEQ. The entry is placed between the blocks of reaction data for the organic complex species having the nearest component IDs above and below that chosen for the component Chlk (nearest species iPhthalate and Diethylamine); in practice, it should not matter where the new species data is inserted, as long as it does not interrupt the three-line structure of entries for other species.

- Open Visual MINTEQ and on the main menu bar, navigate to "Database management;" in the drop-down, click "Database selection" and a new database selection window will open. In the database selection window, click "change" to select the new component and thermodynamic database files created above (comp_2008_wChlk and thermo_wChlk, Figure 15.25). For the solids database, since the reaction of Cu and Chlk will not involve any solid phase, leave the setting for the original solid database from Visual MINTEQ.

4. Run a scenario with 0.25 mM Ca, 40 nM total Cu, and 2 µM Chlk:
   - Open the file "Cu_Daphnia_BLM.vda" as saved above; on the bottom left of the Visual MINTEQ main menu working space, click the tick box labeled "Show organic components". Choose the component Chlk-2 and input a total concentration of 2 µM (0.002 mM); the concentration chosen here is from (Li et al., 2023), who reported freshwater Chlk ranging from 1 to 7 µeq EDTA/L (converting to µM for the 1:1 Cu:Chlk reaction); the list of components will be as shown in Figure 15.26.
   - Run the calculation; the results are shown in Figure 15.27.

For this level of Ca, in the absence of Chlk, the concentration of free Cu^{2+} is 28.9 nM and *f* is 0.61. Upon the addition of Chlk, due to its strong binding affinity with Cu, the free Cu^{2+} concentration drastically decreases to ~10^{-6} nM or less (highly variable due to difficulty of getting repeatable convergence at such a low Cu^{2+} concentration), and the corresponding *f* value is ~10^{-7}–10^{-8}, i.e., Chlk substantially eliminates the bioavailability and consequent toxicity of Cu to *D. magna*. This result is not necessarily representative of Cu availability in freshwaters since the Chlk concentration determined by Li et al. (2023) includes both extracellular and intracellular pools of Cu-binding ligands. Consequently, the actual fraction of Chlk available to reduce Cu bioavailability in the water column remains unclear. While extracellular Chlk may participate in Cu complexation and thereby mitigate toxicity in the dissolved phase, intracellular Chlk likely functions in micronutrient acquisition and homeostasis, serving a fundamentally different physiological purpose, as discussed in more detail in Li et al. (2023).

FIGURE 15.25 Database selection for new files adding Chlk as a component and CuChlk as a species.

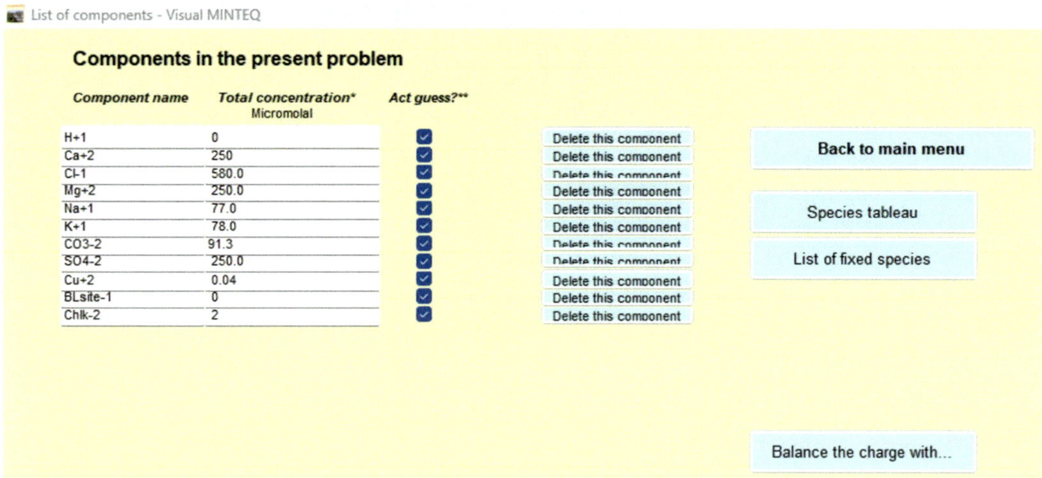

FIGURE 15.26 Components for the Example 3 problem with Ca = 0.25 mM, now with 2 μM Chlk added.

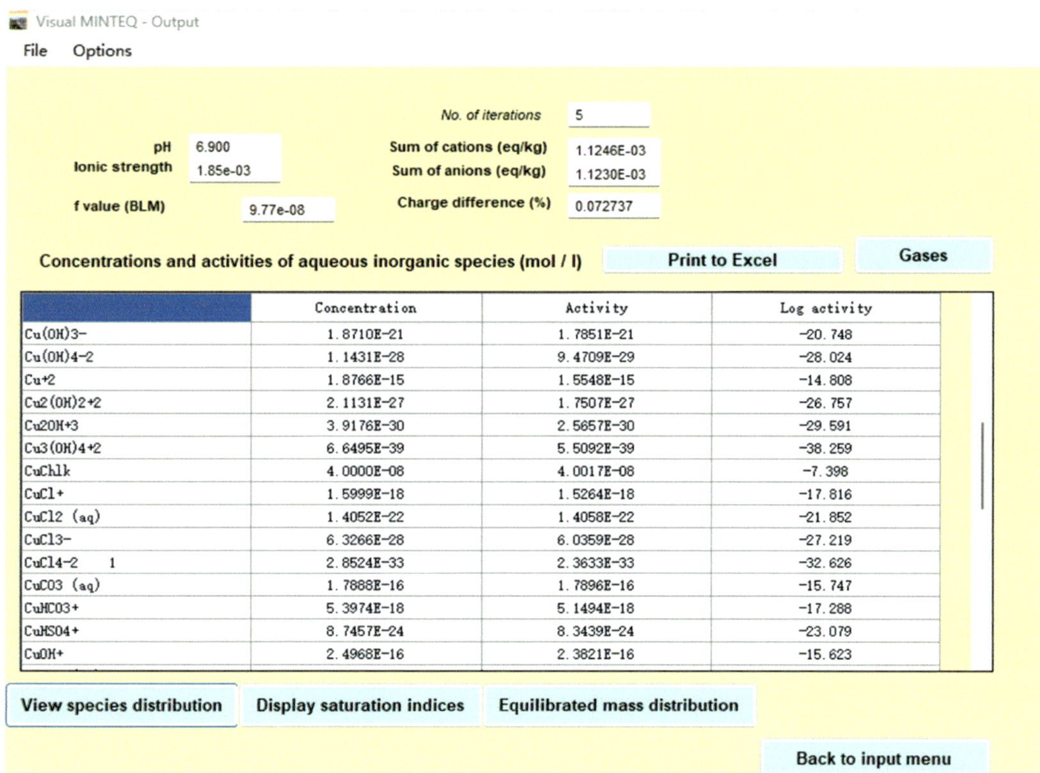

FIGURE 15.27 Results for Cu-BLM speciation (Ca = 0.25 mM with 40 nM Cu) after the addition of 2 μM Chlk.

15.3.5 Example 4: The Process of Manually Adding BLM Parameters

This example demonstrates the process of manually adding BLM parameters for a metal not included in the Visual MINTEQ database (steps of which are not provided in the Visual MINTEQ User Guide). Currently, the Visual MINTEQ database contains only four metals with BLM parameters for Cu, Ni, Pb, and Zn. This example will illustrate the addition of the toxic metal Cd into the Visual MINTEQ BLM database, using log$K$ values for Cd interactions with the zntA bioreporter BL, as reported by Li et al. (2022). This approach establishes a framework for extending BLM applications to other metals unsupported in the database, enhancing ***Environmental Risk Assessments*** for heavy metals.

The values of parameters to be added in this example are as follows: $\log K_{\text{CaBL}} = 3.67$, $\log K_{\text{MgBL}} = 3.82$, $\log K_{\text{NaBL}} = 2.48$, $\log K_{\text{KBL}} = 2.72$, $\log K_{\text{CdBL}} = 6.40$, with $f_{\text{crit}} = 0.50$. A new database file named Cd_BLM.bli must be created. The .bli files in the Visual MINTEQ database contain information for metal–BL reactions. To understand the structure of the .bli files, first open the file Cu_BLM.bli (in the same folder as the comp_2008.vdb and thermo.vdb database files); it should appear as shown in Figure 15.28. The first line has three entries. The first identifies the organism (*Pimephales promelas*), the second is the toxicity metric used (LC_{50}, a measure of acute toxicity that indicates the concentration of an aqueous pollutant that is fatal to 50% of a test population of *P. promelas* within a specified time frame), and the final value (5) indicates how many additional organisms with the same BL binding parameters will be included in inputs. The second line cites the source of the BLM parameters (Carbonaro and Di Toro 2006; Di Toro et al., 2001). The third

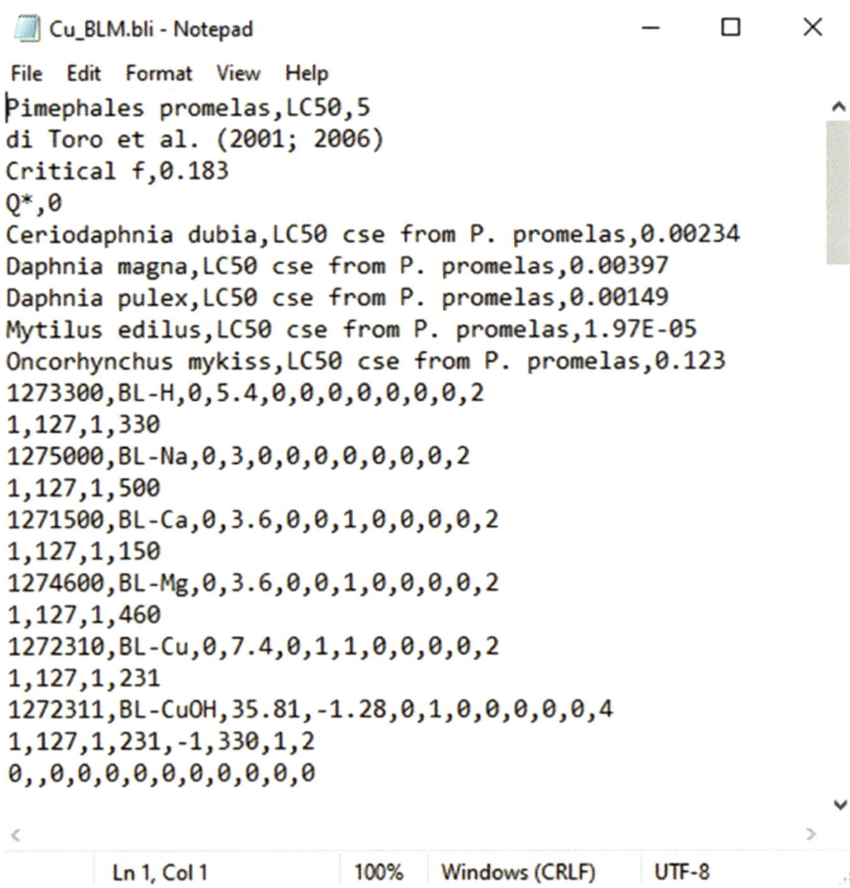

FIGURE 15.28 First block of data in the Cu–BLM.bli database file with information needed to conduct BLM calculations involving for Cu.

line defines the value of f_{crit} for calculations using the *f* approach, or alternately a dummy value of *f* may be used. The fourth line specifies Q^*, as either 0 or an experimentally determined Q^* value. If $Q^* = 0$, this indicates that a calculation based on f_{crit} will be performed, as is the case for the *P. promelas* Cu-BLM. Alternately (scrolling down further in the Cu_BLM.bli file) if Q is nonzero, as for *Chlorella vulgaris*, this indicates that a calculation based on Q^* will be performed, and a dummy value of f_{crit} is used (typically either 0.5 or –99). If the last entry in the first row, *n*, is nonzero, after the fourth line in this file, there will be an additional *n* rows of entries, each of which will have the format organism, endpoint, f_{crit}, or Q^* value (i.e., lines 5–9 in the file Cu_BLM.bli contain f_{crit} values associated with LC_{50}s for *Ceriodaphnia dubia*, *D. magna*, *Daphnia pulex*, *Mytilus edulis*, and *Oncorhynchus mykiss*), and all of which calculations are run with the Cu binding parameters for *P. promelas*. After the *n* rows of entries with f_{crit} or Q^* values for additional species, there follow $2 \times n_{species}$ lines to represent information for the different chemical species that arise from reactions with an organism's BL component. For *P. promelas*, there are six species: BL-H, BL-Na, BL-Ca, BL-Mg, BL-Cu, and BL-CuOH. Taking the information for BL-Ca as an example, the two lines describing this species in the Cu_BLM.bli file are

"1271500,BL-Ca,0,**3.6**,0,**0,1**,0,0,0,0,**2**
1,127,1,150",

where the first number in the first line, 1271500, is the species ID number, as described in Example 3. The second entry in the first line is the species name. Of the remaining ten numbers on the first line, most are placeholders and normally require no modification; the second, fourth, fifth, and tenth numbers after the species name in the first line for each species, indicated in large bold font in the example above, do require specific inputs: The second number of 10 after the species name is the log of the equilibrium constant associated with the reaction that forms the species. For BL-Ca, for example, this is $\log K_{CaBL}$, which has a value of 3.6. The fourth number after the species name indicates whether the species contributes to toxicity in an organism. A value of 0 denotes no contribution to toxicity, while a value of 1 indicates a contribution to toxicity. In Figure 15.28, cations including Ca, Mg, Na, and K do not contribute to the toxicity when binding with BL, so the entries for BL-Ca, BL-Mg, BL-Na, and BL-K contribution to toxicity are 0. However, Cd contributes to the toxicity of BL, and thus, this entry for BL-Cd is 1. The fifth number is the overall charge of the species; in the Visual MINTEQ database, the BL is assigned a charge of –1, and hence, for BL-Ca, where Ca has a charge of +2, the overall charge becomes +1, entered as 1 (and, for instance, for BL-Na, where Na has a charge of +1, the overall charge is neutral and entered as 0). The final number in the first line for each species specifies how many components are to be read from the second line, which is two for BL-Ca; thus, in the second line there are two values of 1, followed by a component ID, which together specify that both one BL and one Ca component are present in the species. After all species are entered, the line "0,,0,0,0,0,0,0,0,0,0,0" must be entered as a formatting instruction (for reading the first line in each species). After this formatting line, there must be a ***blank*** line. Next, there can be additional blocks of data for different BLM models (or not). The final lines in the file must be the end-of-file formatting indicator lines:

"xxx,yyy,0
0,0".

Procedure:

1. Make alterations to Visual MINTEQ BLM database files:
   - In the Visual MINTEQ folder, first find the document named "BLMmetals.txt". Open this using a simple text editor, and the file contents should appear as shown in the top panel of Figure 15.29; only the four metals Cu+2, Ni+2, Pb+2, and Zn+2 for BLM

calculations are shown, followed by an "xxx" for end of file. This file will need to be modified, and it is not possible to use a different file name for use in BLM calculations. It is prudent to make a copy of this file, in case of the need to recover the file later. The copy could be named, for instance, "BLMmetals_original.txt".

- Add "Cd+2" at the end of the file, as shown in the bottom panel of Figure 15.29; this will allow the Visual MINTEQ database to provide the opinion of "Cd^{2+}" for adding BLM parameters. Save changes and close the file.
- For the next steps, the component ID number for the heavy metal that will be added, and any other components involved in reactions with the BL, will be needed. This information is found in the primary Visual MINTEQ component database "comp_2008.vdb", in the same folder as the BLMmetals.txt file. Scrolling down in this file, the line "160,Cd+2,2,0,0,112.3994"
  reveals that the component ID for Cd^{2+} is 160. The components Ca^{2+} (component ID 150), Mg^{2+} (component ID 460), Na^{+} (component ID 500), K^{+} (component ID 410), and BLsite-1 (component ID 127) will also be needed.
- Next, create a .bli file, "Cd_BLM.bli", in the Visual MINTEQ folder for Cd. Since no native file extension of this type exists, the two choices for doing this include (1) opening one of the other .bli files (for Cu, Ni, Pb, and Zn), deleting contents, and inserting information for Cd, or (2) writing a new simple text file with the information for Cd, saving the file, then renaming the file to have the file extension .bli. To add information to perform Cd BLM calculations for the bioreporter zntA using parameters from Li et al. (2022), the Cd_BLM.bli file should be populated to be identical to that shown in Figure 15.30. For the Cd BLM example given here, the organism is zntA, the endpoint is the EC_{50} (cell viability), there are no additional organisms defined with the same BL

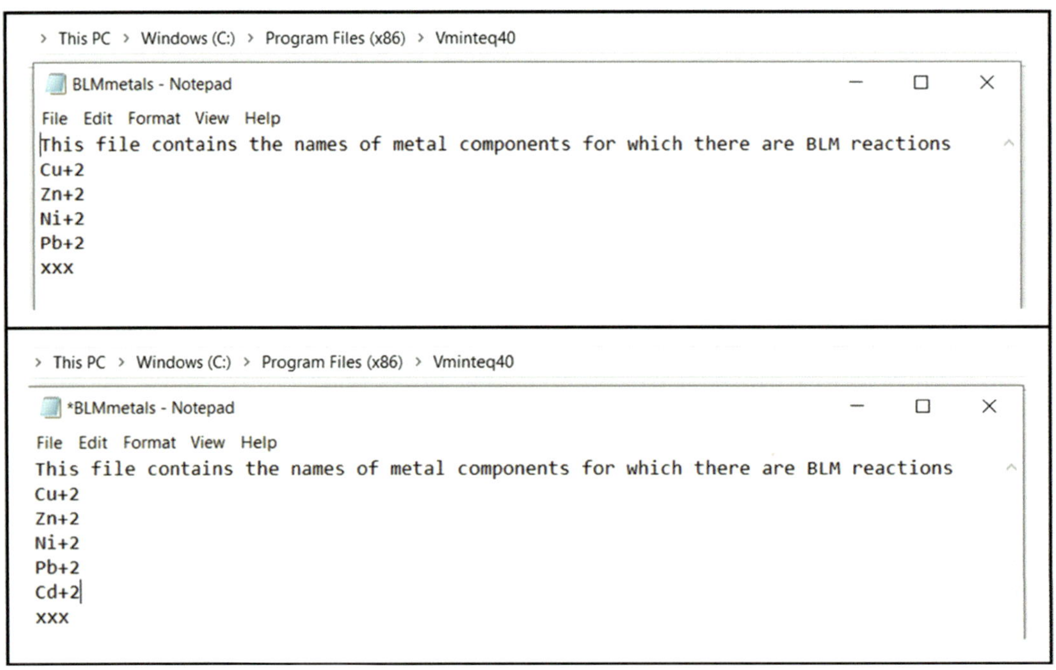

FIGURE 15.29 Adding Cd^{2+} to the BLMmetals.txt file in the Visual MINTEQ database.

```
zntA,EC50,5
Li et al. (2022)
Critical f,0.5
Q*,0
1271500,BL-Ca,0,3.67,0,0,1,0,0,0,0,2
1,127,1,150
1274600,BL-Mg,0,3.82,0,0,1,0,0,0,0,2
1,127,1,460
1275000,BL-Na,0,2.48,0,0,0,0,0,0,0,2
1,127,1,500
1274100,BL-K,0,2.72,0,0,0,0,0,0,0,2
1,127,1,410
1271600,BL-Cd,0,6.40,0,1,1,0,0,0,0,2
1,127,1,160
0,,0,0,0,0,0,0,0,0,0,0

xxx,yyy,0
0,0
```

FIGURE 15.30 Cd_BLM.bli file for the bioreporter zntA using BLM parameters as reported from Li et al. (2022).

binding parameters (0), the reference is Li et al., and the f_{crit} method is used (as signified by Q^* being 0), with $f_{crit} = 0.5$ (coincidentally – not a dummy value). The log*K*s for binding of BL with Ca, Mg, Na, K, and Cd are 3.67, 3.82, 2.48, 2.72, and 6.40, respectively (all 1:1 binding with BL). Species charges and toxicities are as explained for the Cu_BLM.bli file above. Save changes to this Cd .bli file.

Once this file is established, users may expand its scope by appending additional Cd-BLM entries (e.g., BLM f_{crit}/Q^* for other organisms/toxicity endpoints that have the same BL binding parameters as zntA or BLM information for other organisms that have different BLM parameters) either by direct manual editing of the .bli file or through VM's graphical interface (via the "View/edit BLM parameter file" button in the BLM window).

- After adding Cd2+ to the BLMmetals.txt file and creating the Cd_BLM.bli file, launch Visual MINTEQ (important – if the software is open when files are created/edited, it must be closed and relaunched), navigate to "Parameters" on the main menu bar, and chose "Add/Edit Biotic Ligand Model" from the drop-down. In the BLM window that opens, under "Select metal", choose "Cd2+", and under "Select organism/endpoint/reference", choose "zntA/EC50/Li et al. (2022)", as shown in Figure 15.31. If these options do not appear, there is either a problem with editing of the files, or with the path (wrong files in wrong place for Visual MINTEQ to locate them).
- Press "View/edit BLM parameter file"; a window will appear as shown in Figure 15.32. Check and confirm the species name, log*K* value for the reaction forming the species, if the species contributes to toxicity or not, the species charge, number of components forming the species, and reaction stoichiometry for species formation.
- It is now possible to run Visual MINTEQ with BLM speciation- or toxicity-mode calculations for the bioreporter zntA for Cd. To conduct test runs, the user might, for example, use the file LakeWater and vary Cd concentrations to assess the effects of varying different water quality conditions on Cd speciation and toxicity. Refer to Li et al. (2022) for comparative data.

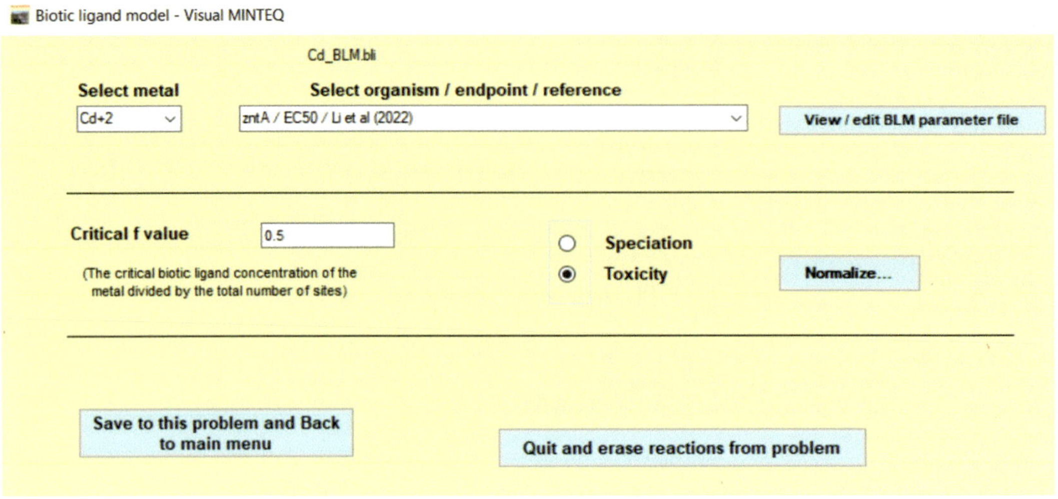

FIGURE 15.31 Using the new BLM database capability to set up a calculation for Cd toxicity to the bioreporter zntA (Li et al., 2022).

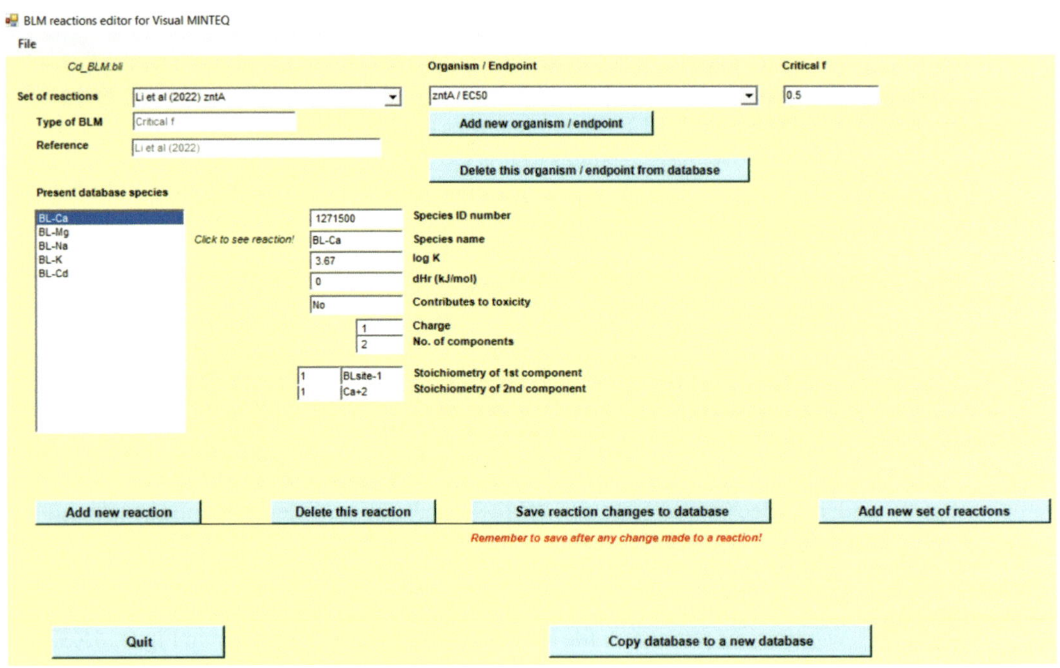

FIGURE 15.32 The Visual MINTEQ BLM reactions editor for Cd with the organism zntA and an EC_{50} endpoint, based on data as reported by Li et al. (2022).

15.4 CONCLUSIONS AND OUTLOOK

Chemical speciation models and the speciation-based BLM provide valuable tools for understanding and assessing heavy metal bioavailability in environmental systems. By having the ability to integrate metal complexation, precipitation, ion exchange, sorption reactions, and BL interactions, these models enhance our ability to predict heavy metal pollutant fate and toxicity under varying conditions. This chapter systematically demonstrates the application of Visual MINTEQ 4.0 in speciation and speciation-based BLM modeling through examples, including step-by-step procedures, illustrating how to model heavy metal speciation in aqueous systems, including solid-phase formation, toxicity estimation, and basic database file manipulations. Methodologically, the strength of this approach illustrated lies in having a user-friendly interface, which simplifies complex geochemical calculations – such as configuring inputs for pH, ionic strength, and DOM parameters – and having multiple approaches for interpreting outputs, such as species distributions, saturation indices, and toxicity endpoints, all without requiring advanced computational expertise. The flexibility of the software is highlighted through its integration of default parameters with user-customizable features, such as adjusting the HA to FA ratio in DOM or adding novel BLM parameters for metals such as Cd, enabling tailored modeling for site-specific environmental conditions.

In speciation modeling, challenges remain in capturing kinetic effects, and improving approaches to modeling heterogeneous reaction sites and complex multi-metal interactions. Future efforts should focus on refining model ***parameterization***, expanding thermodynamic databases, and incorporating further capacity to model kinetic effects. Continued advancements in speciation modeling will enhance Environmental Risk Assessment and regulatory decision-making, ensuring a more accurate and mechanistic understanding of heavy metal impacts to natural systems.

ACKNOWLEDGMENT

We wish to acknowledge Boling Li, who was the second author contributing to the content of this chapter. While circumstances have prevented Professor Li's formal listing as an author, we fully recognize and appreciate his intellectual and creative contributions to the authorship of this work.

REFERENCES

Carbonaro, R.F., Di Toro, D.M., 2006. Estimating metal ligand binding section. Center for the Study of Metals in the Environment, Final Report. U.S. Environmental Protection Agency. Washington, DC.

Cha, S.K., Abruna, H.D., 1990. Determination of copper at electrodes modified with ligands of varying coordination strength: A preamble to speciation studies. *Anal. Chem.* 62(3), 274–278. https://doi.org/10.1021/ac00202a010

Crémazy, A., Wood, C.M., Smith, D.S., Ferreira, M.S., Johannsson, O.E., Giacomin, M., Val, A.L., 2016. Investigating copper toxicity in the tropical fish cardinal tetra (*Paracheirodon axelrodi*) in natural Amazonian waters: Measurements, modeling, and reality. *Aquat. Toxicol.* 180, 353–363. https://doi.org/10.1016/j.aquatox.2016.10.011

de Schamphelaere, K.A.C., Heijerick, D.G., Janssen, C.R., 2002. Refinement and field validation of a biotic ligand model predicting acute copper toxicity to *Daphnia magna*. *Comp. Biochem. Physiol.* 133(1), 243–258. https://doi.org/10.1016/S1532-0456(02)00087-X

de Schamphelaere, K.A.C., Janssen, C.R., 2002. A biotic ligand model predicting copper toxicity for *Daphnia magna*: The effects of calcium, magnesium, sodium, potassium, and pH. *Environ. Sci. Technol.* 36(1), 48–54. https://doi.org/10.1021/es000253s

Di Toro, D.M., Allen, H.E., Bergman, H.L., Meyer, J.S., Paquin, P.R., Santore, R.C., 2001. Biotic ligand model of the acute toxicity of metals. 1. Technical basis. *Environ. Toxicol. Chem.* 20(10), 2383–2396. https://doi.org/10.1002/etc.5620201034

EC (European Commission), 2010. *Nickel and Its Compounds*. Environmental Quality Standards Sheet. Danish Environmental Protection Agency, Copenhagen, Denmark.

Gustafsson, J. P. 2001. Modeling the acid–base properties and metal complexation of humic substances with the Stockholm Humic Model. *J. Colloid Interf. Sci.* 244(1), 102–112. https://doi.org/10.1006/jcis.2001.7871

Gustafsson, J.P., 2024. Visual MINTEQ 4.0 user guide. https://vminteq.com/download/

Li, B., Zhang, X., Tefsen, B., Wells, M., 2022. From speciation to toxicity: Using a "Two-in-One" whole-cell bioreporter approach to assess harmful effects of Cd and Pb. *Water Res.* 217, 118384. https://doi.org/10.1016/j.watres.2022.118384

Li, B., Zhang, X., Wu, G., Qin, B., Tefsen, B., Wells, M., 2023. Toxins from harmful algal blooms: How copper and iron render chalkophore a predictor of microcystin production. *Water Res.* 244, 120490. https://doi.org/10.1016/j.watres.2023.120490

Morel, F.M.M., 1983. *Principles of Aquatic Chemistry*. John Wiley, New York.

Parkhurst, D.L., Appelo, C.A.J., 2013. Description of input and examples for PHREEQC version 3—a computer program for speciation, batch-reaction, one-dimensional transport, and inverse geochemical calculations. U.S. Geological Survey Techniques and Methods, Book 6, Chapter A43. https://pubs.usgs.gov/tm/06/a43/

Rantetampang, A.L., Mallongi, A., 2013. Assessment of cadmium through aquatic biota consumption from Sentani Lake in Papua. *IJERT* 2(6), 2492–2500.

Schneider, L., Mariani, M., Saunders, K.M., Maher, W.A., Harrison, J.J., Fletcher, M.S., Zawadzki, A., Heijnis, H., Haberle, S.G., 2019. How significant is atmospheric metal contamination from mining activity adjacent to the Tasmanian Wilderness World Heritage Area? A spatial analysis of metal concentrations using air trajectories models. *Sci. Total Environ.* 656, 250–256. https://doi.org/10.1016/j.scitotenv.2018.11.241

US EPA, 2002. *Edition of the Drinking Water Standards and Health Advisories*. US Environmental Protection Agency Washington, DC.

US EPA, 2003. *The Biotic Ligand Model: Technical Support Document for Its Application of the Evaluation of Water Quality Criteria for Copper*. US Environmental Protection Agency, Washington, DC.

US EPA, 2007. Update of ambient water quality criteria for copper. EPA-822-F-07-001. US Environmental Protection Agency, Washington, DC.

US EPA, 2024. *National Recommended Water Quality Criteria – Aquatic Life Criteria Table*. US Environmental Protection Agency, Washington, DC.

USGS, 2024. WATEQ4F database update (USGS Open-File Report). U.S. Geological Survey. Reston, VA. https://doi.org/10.3133/ofr91183.

Zhang, X., Li, B., Deng, J., Qin, B., Wells, M., Tefsen, B., 2020a. Regional-scale investigation of dissolved organic matter and lead binding in a large impacted lake with a focus on environmental risk assessment. *Water Res.* 172, 115478. https://doi.org/10.1016/j.watres.2020.115478

Zhang, X., Li, B., Deng, J., Qin, B., Wells, M., Tefsen, B., 2020b. Quantitative high-throughput approach to chalkophore screening in freshwaters. *Sci. Total Environ.* 735, 139476. https://doi.org/10.1016/j.scitotenv.2020.139476

Section 5

Case Studies on the Assessment of Bioavailability

16 Case Study
Evaluating Organic Diffusive Gradients in Thin-Film Devices as Proxies to Measure Bioavailability of Methamphetamine in Zebrafish

16.1 INTRODUCTION – DGT, o-DGT, AND o-DGT CASE STUDY

Diffusive Gradients in Thin-Films (DGT) is a technique that was developed in the 1990s and has since become extremely popular, primarily in two contexts: (1) as a way to detect pollutants below their analytical detection limits in result of the ability of DGT devices to act as passive samplers and concentrate environmental pollutants over a given deployment time and (2) as a ***chemoavailability*** proxy to ***bioavailability*** measurements. DGT has been used in many matrices, notably aqueous, soils, and sediments. The principle behind the function of DGT is discussed in detail in Chapter 10.

The DGT technique has been often used to study heavy metal pollutants, being particularly applicable to species with good solubility in water (Zhu et al., 2023). Particular focus has been on whether or not DGTs can be effective proxies of bioavailability. Unquestionably, for metals, DGTs often correlate with bioavailability for representative examples of both small fish/invertebrates and plants (e.g., Han et al., 2019; Li et al., 2019a; Pelcová et al., 2017; Philipps et al., 2018; Yin et al., 2014). DGT measurements plotted versus bioavailability, however, often yield graphs with slopes that deviate drastically from a slope of 1.0 that would signify that DGT can predict bioavailability. In other words, DGT results may correlate with bioavailability, but, according to the distinction between chemoavailability and bioavailability that we make here, they are not bioavailability. In some cases, DGT results are not even a good correlate of bioavailability. As examples, in a study focusing on copper (Cu), Philipps et al. (2018) found that DGT-measured Cu concentrations were linearly correlated to fathead minnow-accumulated Cu, but not to lampmussel-accumulated Cu. These authors found that DGT as a ***passive sampling*** technique correlates with Cu concentrations in fathead minnow in waters with low natural organic matter concentrations, pointing out that larger concentrations might interfere unduly with the measurement. Yabuki et al. (2014) have also noted that use of DGT may be problematic in cases where the ***ionic strength*** of the water is low, or where significant organic matter is present. Unquestionably, the DGT technique, even if it is not effective at measuring true bioavailability, is of great use in Environmental Monitoring of heavy metal pollutants due to the ability to function as a passive sampler and concentrate pollutants.

In theory, the DGT technique should be suitable for use with a wider variety of pollutants since the gel layer that captures pollutant analytes can be chemically tailored. It has only been relatively recently, however, since the 2010s, that serious work has begun on using the DGT technique to study organic pollutants as a modified technique referred to as o-DGT (organic-DGT, Guibal et al., 2019; Ji et al., 2022; Wang et al., 2019). While development in this direction has been relatively recent, progress has been substantial. The o-DGT technique has now been used to study a variety of organic pollutants, including antibiotics (Chen et al., 2012; Li et al., 2022; Xie et al., 2018), pesticides/herbicides (Guibal et al., 2017; Li et al., 2019b; Taylor et al., 2020), pharmaceuticals and personal care products (Cao et al., 2022; Chen et al., 2017), endocrine disrupting compounds (Chen

DOI: 10.1201/9781003355229-21

et al., 2018), illicit drugs (Guo et al., 2017; Zhang et al., 2018), and flame retardants (Zou et al., 2018). Most pollutant o-DGT studies are aimed at aqueous matrices, however, progress begins to be made on soils as well (Chen et al., 2015; Ren et al., 2020, Weng et al., 2019). While these studies and others have focused on demonstrating the use of o-DGT in Environmental Monitoring, there is a distinct paucity of studies so far that examine the ability of o-DGTs to predict the bioavailability of organic pollutants, particularly with respect to aquatic biota.

Given this gap, i.e., a relative lack of o-DGT studies on bioavailability, one study[1] made a pivotal advance and has therefore been chosen as a case study for consideration here. Yin et al. (2019, hereafter "Yin et al.") used o-DGT passive samplers to measure the concentration of methamphetamine and its metabolite amphetamine from aqueous solution, in an experimental design wherein Zebrafish were simultaneously exposed under the same experimental conditions. This study is significant as a case study because it is pivotal towards trying to understand the possible connection between o-DGT results and bioavailability (as reflected by ***bioconcentration***) and also as it involves a very important and ***emerging pollutant*** of interest. One confounding factor with environmental bioavailability of organic pollutants is that, unlike metals, organics may degrade. That Yin et al. included amphetamine in their study is a useful aspect in considering this case study. Employing Zebrafish is also another interesting feature, as this organism's attributes are well understood in the context of ***biomonitoring*** (Kelly et al., 2021).

Methamphetamine use is recognized as a global problem and of great public health concern (Kohno et al., 2020). Methamphetamine is highly addictive (Brecht and Herbeck, 2014; Galbraith, 2015) and an illicit drug (specifically, subject to international control, INCB, 2022) that is used in modern times effectively exclusively as a recreational drug. Human health effects include neurotoxicity, elevated mortality, increased incidence of HIV and hepatitis C infection, poor mental health (suicidality, psychosis, depression, and violence), increased risk of adverse cardiovascular events (Farrell et al., 2019; Galbraith, 2015; Kohno et al., 2020), and cognitive and behavioral problems in persons exposed to methamphetamine prenatally (LaGasse et al., 2012). It is estimated that over 60% of individuals treated for methamphetamine addiction will relapse into use within one year (Brecht and Herbeck, 2014), a circumstance ensuring an ongoing global problem. Drug testing for methamphetamine and its degradation by-product amphetamine is often performed using urine (Kraemer and Maurer, 1998), and thus, as the global problem develops, methamphetamine and amphetamine are increasingly prevalent in environmental matrices. Studies have shown that methamphetamine is a persistent pollutant in soils (Pal et al., 2011) and it is one of the most common illicit drugs in surface waters (Bartelt-Hunt et al., 2009; Castrignano et al., 2018; Li et al., 2016; Loganathan et al., 2009; Mastroianni et al., 2016; Wang et al., 2018; Zuccato et al., 2008), with detection frequencies being higher than 90% in some locales (e.g., Wang et al., 2016). Concentrations in surface water have been reported in the range of 0.1–917 ng/L (Jiang et al., 2015 and Supporting Information; Wang et al., 2016).

The human health exigency of methamphetamine is well understood. Comparatively little research has been done on the ecotoxicity of methamphetamine, however, a number of studies have demonstrated that, unsurprisingly, it is quite ecotoxic (Santos et al., 2020; Wang et al., 2020a, 2020b), with one study having found that methamphetamine induced developmental defects in fish, resulting in a significant decrease in reproductive capacity and a concomitant high teratogenic risk level – Wang et al. (2020a) found evidence of transgenerational toxicity and ecosystem-level consequences for fish exposed to methamphetamine at environmentally relevant concentrations. Due to the risk to ecosystems, moreover, human food ***sources*** such as fish, it is important to understand the bioavailability of methamphetamine.

In their work, Yin et al.'s stated objectives were (1) to characterize the relationship between methamphetamine and amphetamine measured by o-DGT and in aqueous phase samples as measured by ultraperformance liquid chromatography–tandem mass spectrometry (UPLC–MS/MS), (2)

[1] Entitled "Biomimetic Accumulation of Methamphetamine and Its Metabolite Amphetamine by Diffusive Gradients in Thin-Films to Estimate Their Bioavailability in Zebrafish".

to measure ***bioaccumulation*** (actually, bioconcentration) and biodegradation of methamphetamine in Zebrafish, (3) to investigate whether o-DGT-measured concentrations of methamphetamine or amphetamine could be used to predict bioavailability of these in Zebrafish. Here we will review their work as a case study, with a critical analysis of the work, particularly from the standpoint of bioavailability measurement.

16.2 SUMMARY OF YIN ET AL.'S EXPERIMENTAL APPROACH AND RESULTS

16.2.1 Experimental Approach and Methods

The general experimental approach was factorial and is graphically depicted in Figure 16.1. Zebrafish and o-DGT devices were exposed to methamphetamine in aqueous solution together, i.e., under identical conditions, and during the course of the ***exposure*** solutions, o-DGT devices, and Zebrafish were simultaneously sampled for chemical analysis of methamphetamine and amphetamine. Factors varied were concentration (four levels, a ***blank*** and three methamphetamine concentrations: nominally 0.01, 1.0, and 100 ng/L) and time (six levels, sampling occurring at 1, 2, 4, 7, 10, and 14 days). The rationale behind the choice of concentrations was blank (required), environmentally relevant low (0.01 ng/L), "low" (1 ng/L), and "high" (100 ng/L).

Experimental materials used for the work are summarized as follows:

- Ultrapure deionized water (Milli-Q water purification system)
- Charcoal for cleaning up tap water
- 5-L glass beakers
- Oasis MCX (60 mg, 3 cc) solid-phase extraction cartridges (Waters, Milford, MA, USA)
- Methamphetamine, amphetamine, and their deuterated ***internal standards***, and high-purity acetonitrile, formaldehyde, hydrochloric acid, 5+% ammonia, and formic acid (Sigma-Aldrich, St. Louis, MO, USA)

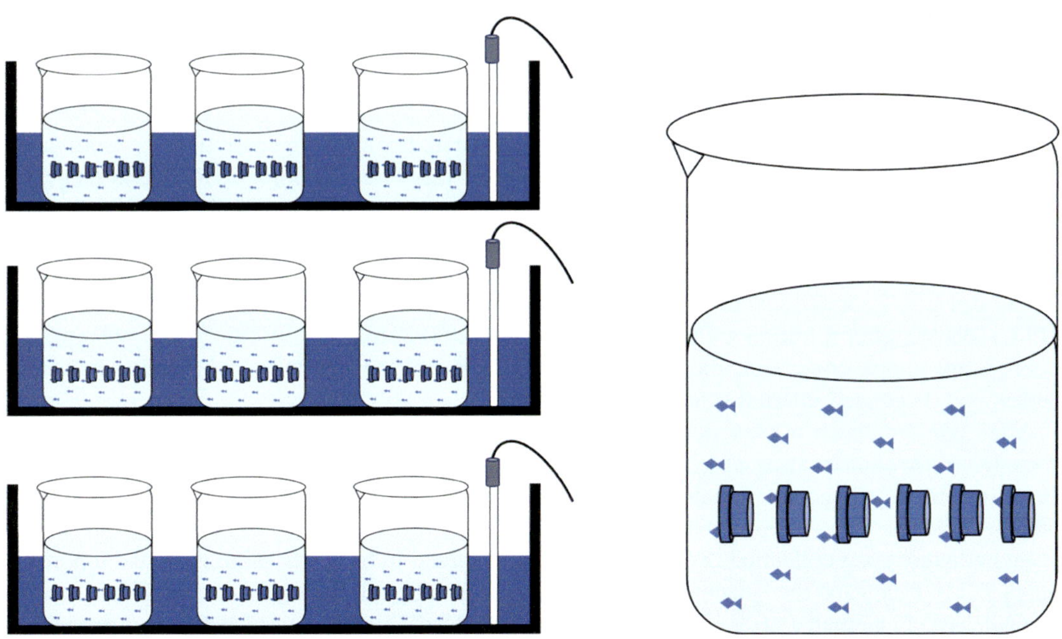

FIGURE 16.1 Graphical depiction of the factorial experimental approach to methamphetamine and amphetamine exposure experiments utilizing Zebrafish and o-DGT devices. Factors varied were concentration (four levels, a blank and three methamphetamine concentrations: nominally 0.01, 1.0, and 100 ng/L), and time (six levels, sampling occurring at 1, 2, 4, 7, 10, and 14 days).

- o-DGT devices (DGT Research Ltd., UK); the work used o-DGTs consisting of an XAD 18 agarose-***binding gel*** (0.05 cm thickness), an agarose ***diffusive gel*** (0.075 cm thickness), and a polyether-sulfone ***filter membrane*** (0.014 cm in thickness), with an exposure window on the o-DGT device of 3.14 cm^2
- Zebrafish ***test organisms*** (raised in flow-through aquaria) and *Daphnia pulex* feed
- Tricaine methyl sulfonate (Sigma-Aldrich Corp)
- UPLC–MS/MS system consisting of a triple quadrupole mass spectrometer (Xevo TQ–S micro IVD system, Waters, Manchester, UK) equipped with electrospray ionization. UPLC equipped with a Waters C18 column (2.1 mm × 150 mm, 1.7 μm particle size, Milford, MA)
- 0.22 μm nylon needle filter
- N_2 for solvent evaporation
- Glass microfiber filters (GF/C™)
- Water bath for temperature control
- –20°C freezer.

The mode of operation of the "bespoke" o-DGT devices used was the subject of previous study (Guo et al., 2017), including assessment of the ***sorption*** of methamphetamine and amphetamine to o-DGT device surfaces relative to the gels, sorption capacity of the devices, sorption kinetics, efficiency of methamphetamine and amphetamine elution, and determination of diffusion coefficients. After establishing that the o-DGT devices function suitably (i.e., sorption and elution), it is particularly important to know sorption capacities to ensure that the o-DGT device will not saturate with analyte (methamphetamine or amphetamine) at the maximum exposure concentrations and times for any given experiment. Therefore, prior to commencing exposure experiments with o-DGT and Zebrafish, Yin et al. confirmed that the o-DGT devices that they used had excess sorption capacity under their planned experimental conditions.

Juvenile Zebrafish used in experiments were hatched from roe and raised in a flow-through aquatic system to an age of two months, followed by ***acclimation*** in the exposure setting for 10 days in 5-L glass beakers containing 4 L of charcoal-filtered and aerated tap water, 24 fish per beaker. During acclimation and the exposure experiments, the temperature was kept at 26°C ± 0.5°C with a light/dark cycle of 14/10 hours. Zebrafish were fed twice daily with newly hatched *D. pulex* at a rate of 1% of the total Zebrafish weight.

After acclimation, exposure experiments were conducted. For each concentration exposure there were three replicates, i.e., three beakers. In each beaker there were enough o-DGTs and fish (two individual juvenile Zebrafish per sample). Exposure solutions were renewed daily during the exposure period to achieve a quasi-steady level of methamphetamine and amphetamine exposure. At each sampling time, exposure water, o-DGT devices, and fish were sampled. All samples (water, o-DGT devices, Zebrafish) were processed on the day of sampling and analyzed immediately by UPLC–MS/MS after filtration with a nylon needle filter of 0.22 μm.

For water sample collection, 900 μL was collected before changing the exposure solution in each beaker, and deuterated internal standards were added.

After sampling each o-DGT device, the XAD binding gel was removed and placed in 10 mL of methanol for elution, after which deuterated internal standards were added. After 24 hours, the eluent was collected and dried under a weak N_2 stream, then reconstituted to 1 mL with 10% acetonitrile in water.

Immediately upon collection, Zebrafish were anesthetized in 0.03% tricaine, weighed (approximately 0.25 g each), and then placed in a 15-mL centrifuge tube and homogenized with 10 mL of 1% formaldehyde in methanol after adding the deuterated internal standards. Each sample was subject to ultrasound-assisted extracted for 20 minutes, followed by freezing at –20°C for 30 minutes to remove lipids before centrifugation at 4°C. After the first extraction, a second identical extraction was performed and the extracts combined. Extracts were diluted to 200 mL with Milli-Q water, filtered through glass microfiber (GF/C™) filters to remove solid particles, and then acidified to a pH of 2 with concentrated HCl. The solution was loaded onto Oasis MCX cartridges that were preconditioned with consecutive exposures to 5 mL of methanol and 5 mL of Milli-Q water. The cartridges were washed with 10 mL of Milli-Q water and then dried under a vacuum for 30 minutes. Analytes were eluted from the cartridges with 5% $NH_3 \cdot H_2O$ in methanol. The eluates were concentrated under a weak N_2 stream in a water bath at 40°C to near dryness and reconstituted to 1 mL with 10% acetonitrile in water.

For analysis by UPLC–MS/MS, a triple quadrupole mass spectrometer equipped with an electrospray ionization source was used. The UPLC column was kept at an operating temperature of 40°C. The mobile phases were 0.1% formic acid in water (polar, A) and acetonitrile (nonpolar, B) with a flow rate of 0.45 mL/min. A mobile-phase gradient was used, starting at 98% A and 2% B for 0.5 minutes, then ramping up to 50% B by 2.5 minutes, switching to 98% B by 2.6 minutes, staying at this mixture until 4 minutes, then returning to 98% A and 2% B at 4.2 minutes. After a return to the 98% A and 2% B conditions, the column was given 1.3 minutes to re-equilibrate before injecting another sample, i.e., the total run time per sample was 5.5 minutes. Sample ionization was conducted with the electrospray ionization chamber set to 450°C, with desolvation gas flow at 1,000 mL/min. The sample injection volume was 5 µL. Positive ionization was used with a cone voltage of 22 V and 14 V for methamphetamine and amphetamine, respectively. Mass-selective detection employed multiple-reaction monitoring mode. The theoretical mass, quantifying ion, and qualifying ion m/z ("/" collision energy in eV) were 149.24, 91.00/16, and 119.05/10 and 135.21, 91.00/17 eV, and 119.10/9 eV for methamphetamine and amphetamine, respectively. Methamphetamine and amphetamine were quantified against standards, and working standard mixtures were prepared weekly in acetonitrile and were stored at –20°C. Concentration results for water, o-DGT extracts, and extracts of Zebrafish homogenate concentrations were corrected for dilution factors, as applicable, and reported as ng/L (C_{sol}, concentration of methamphetamine and amphetamine in exposure water), ng/o-DGT device, and ng/g, respectively. The concentration of methamphetamine or amphetamine in exposure water can also be inferred from the mass on o-DGTs if enough parameters are known. The needed information includes m (the mass of methamphetamine or amphetamine accumulated in the o-DGT binding layer over the exposure time, in ng), Δg (the thickness of the o-DGT agarose diffusive gel + thickness of the polyether-sulfone filter membrane, in cm), D (the coefficient of diffusion of methamphetamine or amphetamine, in cm^2/s), A (the area of the DGT's exposure window, in cm^2), and t (the exposure time, in seconds). Using this information, the o-DGT-inferred concentration, $C_{o\text{-DGT}}$, of methamphetamine or amphetamine in exposure water is calculated as

$$C_{o\text{-DGT}} = \frac{m\Delta g}{DAt}.$$

16.2.2 Results

The key results of Yin et al. are shown in Figures 16.2 and 16.3. Figure 16.2 primarily relates to Quality Assurance and Quality Control (QA/QC) for methamphetamine. Data in Figure 16.3 are those on which the authors based their primary conclusions. Based on the findings in Figure 16.3, in particular that o-DGT methamphetamine concentrations correlate with Zebrafish bioconcentration, Yin et al. "*conclude that the* [o-]*DGT technique represents a useful surrogate to estimate the bioavailability of METH* [methamphetamine]". In the remainder of this case study we will conduct a critical analysis of this case study, evaluating both the data and the conclusions of the study.

16.3 CRITICAL ANALYSIS OF o-DGT CASE STUDY

16.3.1 Experimental Design

Yin et al.'s o-DGT case study incorporated many positive aspects of best practice in research and experimental design. The authors succinctly analyze the gap in the state of knowledge (lack of

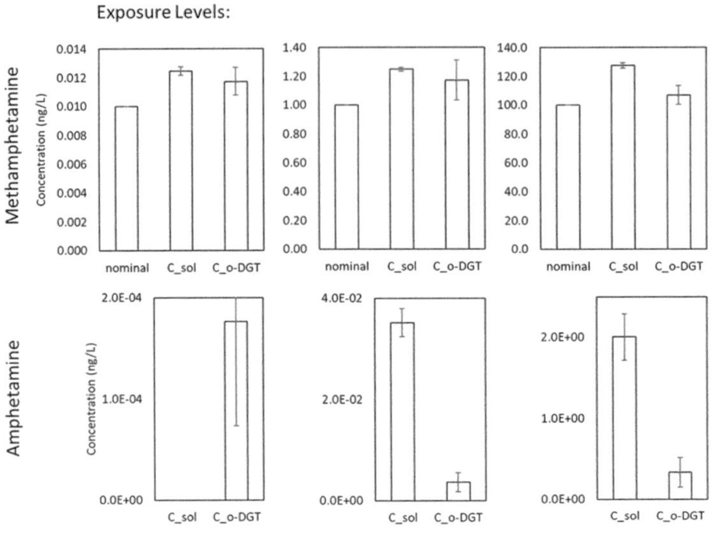

FIGURE 16.2 Data showing concentrations of methamphetamine (top row) for three exposure levels (0.01, 1.0, and 100. ng/L) nominally dosed, measured in solution at each sampling time before replacement of exposure water (C_{sol}, labeled C_sol in the graphs), and as calculated for the solution based on o-DGT parameters ($C_{o\text{-}DGT}$, labeled C_o-DGT in the graphs). There are no error bars for nominal concentration since this represents the concentration as prepared and dosed. Bottom row shows data for amphetamine; no detectable amphetamine was present in "pre-experiments" on o-DGTs in beakers without Zebrafish. Since amphetamine is a degradation by-product, there is no nominal dosage. For the lowest methamphetamine dosage level, C_{sol} amphetamine was below limits of detection. All results calculated using data from Yin et al. (2019) and reported here as averages over all exposure times.

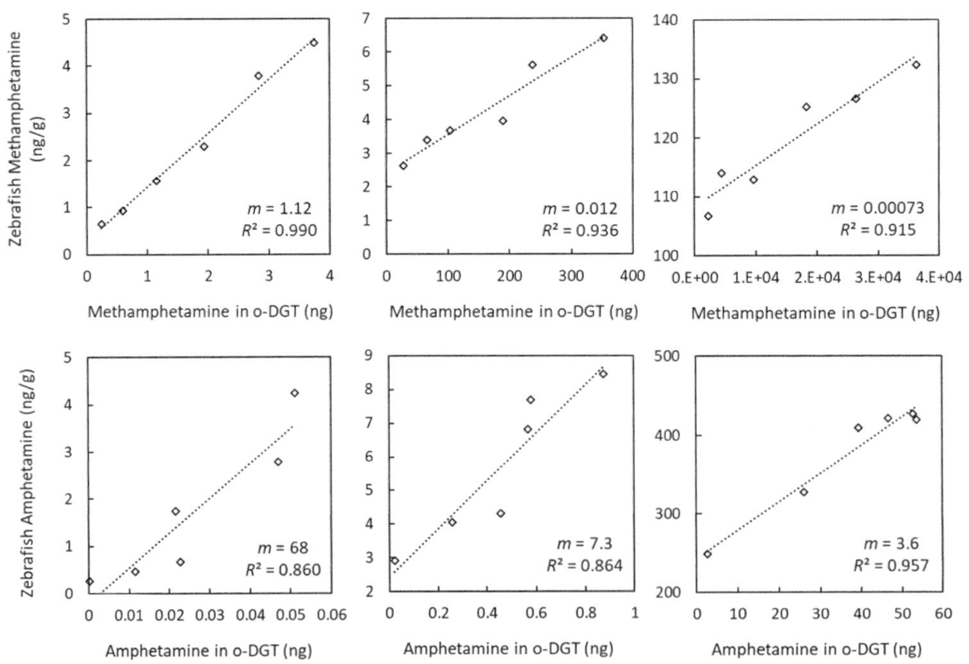

FIGURE 16.3 Data showing correlations between Zebrafish bioconcentrated and o-DGT methamphetamine/amphetamine. (Adapted from data in Yin et al., 2019.)

studies on the ability of o-DGT to serve as a proxy for bioavailability). The study employed a simple and well-known factorial design, and the chemical analysis was conducted by a method known to be highly suitable for both the media analyzed (extracts of water, o-DGT devices, and Zebrafish tissue homogenates) and analytes (methamphetamine and amphetamine). The authors further noted that their approach followed guidelines of the Organization for Economic Co-operation and Development for testing of chemicals. One flaw in the experimental design was the sampling frequency, described further below.

16.3.2 Quality Assurance and Quality Control

On one hand, the approach that the case study took to QA/QC was lacking. On the other hand, it could be credibly asserted that the case study attended to QA/QC to a greater degree than many research publications do. Overall, the lack of a suitable level of QA/QC and QA/QC reporting for research studies where procedures are known and readily available is regrettable, as QA/QC is vital to evidence-based assessment of results and whether claims based on results are valid. Because so many journals now provide the avenue of publishing Supplementary Material/Supporting Information, there is more scope than ever for authors to conveniently provide complete QA/QC information.

The overarching omission in QA/QC reporting for the case study considered here is lack of demonstration of laboratory capability, i.e., whereby it is demonstrated that accurate analysis can be and has been performed for a sample of similar ***matrix*** and for the same analyte. Often this involves analysis of a Certified or Standard Reference Material, or a "blind" sample provided by an external lab with suitable expertise (i.e., the external lab provides a sample that they have analyzed and in which analysis they have confidence and the testing laboratory's task is to demonstrate that they can obtain a correct result without knowing what the sample's concentration is). Additional QA/QC details omitted in the case study include information on calibration assessment, procedures to avoid contamination, assessment of ***matrix spikes*** to test interferences, contamination assessment and other QA/QC controls, and importantly, potential loss of analyte that might have occurred in extraction (no information on surrogate ***spikes***). It is best practice to report method detection limits and the ***limit of quantification***, which were also not reported by Yin et al. While the authors were careful to note that their approach followed guidelines for testing of chemicals, they did not mention use of a ***Standard Method*** of analysis for the sample preparation or analysis using UPLC–MS/MS. Standard Methods incorporate appropriate guidelines for QA/QC; hence, use of such a method would have avoided omissions in the case study.

Lacking full QA/QC, it is perhaps not surprising that some irregularities occur in the data. The authors state that "*the deviations of measured METH* [methamphetamine] *concentrations were within 20% of the nominal concentrations*", and "*There was no significant difference (p >0.05) in the METH concentrations between grab water and* [o-]*DGTs throughout the entire 2 weeks, confirming the accuracy of the* [o-]*DGT method*". From reanalysis of the data provided in tables by Yin et al. (Supporting Information) and Yin et al.'s data digitized for this case study (accurate to that provided in tables to within 0.4%, on average, in all cases less than 1%), these statements are inaccurate. The concentration of methamphetamine in exposure solutions was, on average, 26% higher than the C_{sol} measured, not "*within 20% of the nominal concentrations*". Unlike the assertion above, reanalysis of the data shows that there was a significant difference between methamphetamine C_{sol} and $C_{o\text{-}DGT}$ for the experiments as a whole and for two of the three exposure levels individually (criterion of $p <0.05$; for the entire experimental data set, $p <0.0045$). In other words, it is not possible that the data are "*confirming the accuracy of the* [o-]*DGT method*" (if by accuracy it is meant that $C_{o\text{-}DGT}$ is representative of C_{sol}) since the data instead confirm the inaccuracy of $C_{o\text{-}DGT}$ as a reflection of C_{sol}. While the factorial design of the experiment is constructive, the lack of controls is problematic. In the case of comparing methamphetamine C_{sol} and $C_{o\text{-}DGT}$, the latter is 17% lower than the former for the 100 ng/L methamphetamine exposure level ($p <0.00058$).

It would be useful to confirm whether the relationship between C_{sol} and $C_{o\text{-}DGT}$ in the absence of Zebrafish would be the same as with Zebrafish – this control measurement was not made. Similarly, no control measurement was made to confirm that the o-DGT devices had no effect on Zebrafish bioconcentration.

In some respects, sampling frequency was inadequate, which is something that might be avoided with robust QC Planning. While amphetamine is known to be a metabolic byproduct of methamphetamine, the concentration of amphetamine in exposure water might foreseeably change considerably during the time course of the experiment. The experimental plan called for changing the exposure water daily, and water was sampled before changing, i.e., when amphetamine concentrations would be highest. The case study speculated that values of C_{sol} were so much higher than $C_{o\text{-}DGT}$ for amphetamine because exposure water was sampled only when amphetamine was likely to be highest. Given that for one exposure level C_{sol} was ~50 times higher than $C_{o\text{-}DGT}$, this explanation seems unsatisfactory. The matter would not be a subject of speculation if (1) controls had been performed and (2) water samples had been taken more frequently. Indeed, while the authors' inclusion of the metabolite amphetamine was a strong point of the study, because of the inadequacy of sampling frequency, the overall quality of the amphetamine data set was compromised.

In reanalyzing the digitized data, the original regression statistics of slope and intercept reported were confirmed to within 3%. The R^2 "goodness of fit" was not reproducible from the original reported results. For instance, Figure 16.4 shows two regression lines for the correlation between Zebrafish bioconcentration and o-DGT for amphetamine under a nominal methamphetamine exposure of 1.0 ng/L. The solid-black regression line is that for the entire set of six regressors ($R^2 = 0.86$). The dotted regression line is the regression line published, which is reproducible only if using the five best-linear-fit regressors, i.e., the original published data showed six regressors, with a regression line based on the five best, producing a considerably higher R^2 ($R^2 = 0.95$). In another example (for the correlation between Zebrafish bioconcentration and o-DGT for amphetamine under a nominal methamphetamine exposure of 0.01 ng/L), the error in R^2 reporting was even larger, and in the other direction. Yin et al. reported $R^2 = 0.62$, whereas the R^2 based on a high-fidelity digitization of their data is 0.86. Another useful feature of using a Standard Method, when one exists, is that such methods also typically contain guidance on reporting, which are intended to guide users to "flag" data quality issues and may help to reduce reporting errors of the type described here.

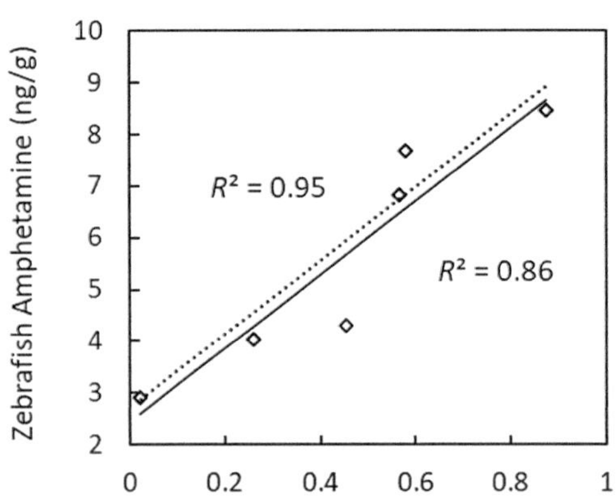

FIGURE 16.4 Graph showing a line based on six regressors and having an $R^2 = 0.86$ (solid black line). For this data, Yin et al. reported an $R^2 = 0.95$ (dashed line), which is only obtained by deleting one regressor (low point between 0.4 and 0.6 on the *x*-axis).

16.3.3 Conclusions of Case Study

One of the main messages of study considered here was that "*We therefore conclude that the* [o-] *DGT technique represents a useful surrogate to estimate the bioavailability of METH* [methamphetamine]". This conclusion was primarily based on the observed correlations between Zebrafish bioconcentrated and o-DGT methamphetamine/amphetamine. In the sense that correlation in linear regression represents statistical quantification of the associations between a predictor variable (abscissa, here o-DGT concentration of methamphetamine and amphetamine) and an outcome variable (ordinate, here Zebrafish bioconcentration), the data argue for the opposite conclusion that o-DGT as used by Yin et al. is not able to function as a surrogate to estimate methamphetamine bioavailability to Zebrafish, as illustrated in Figure 16.5. The top panel shows all the data for the three methamphetamine exposure levels together, from which there are clearly major discontinuities in the correlation for different exposure levels. The remaining graphs illustrate why such discontinuity develops. The discontinuities in question render the approach unsuitable for "real world" environmental applications, because in field conditions, methamphetamine/amphetamine exposure levels will vary; hence, this loss of overall linearity results in lack of predictive capability.

The bottom two graphs in Figure 16.5 demonstrate that the behavior of the o-DGT is fundamentally different than that of Zebrafish bioconcentration. In developing their rationale for the use of o-DGT as a bioavailability surrogate, Yin et al. point out that the o-DGT diffusive layer behind which methamphetamine concentration occurs via a binding resin is similar to the uptake of

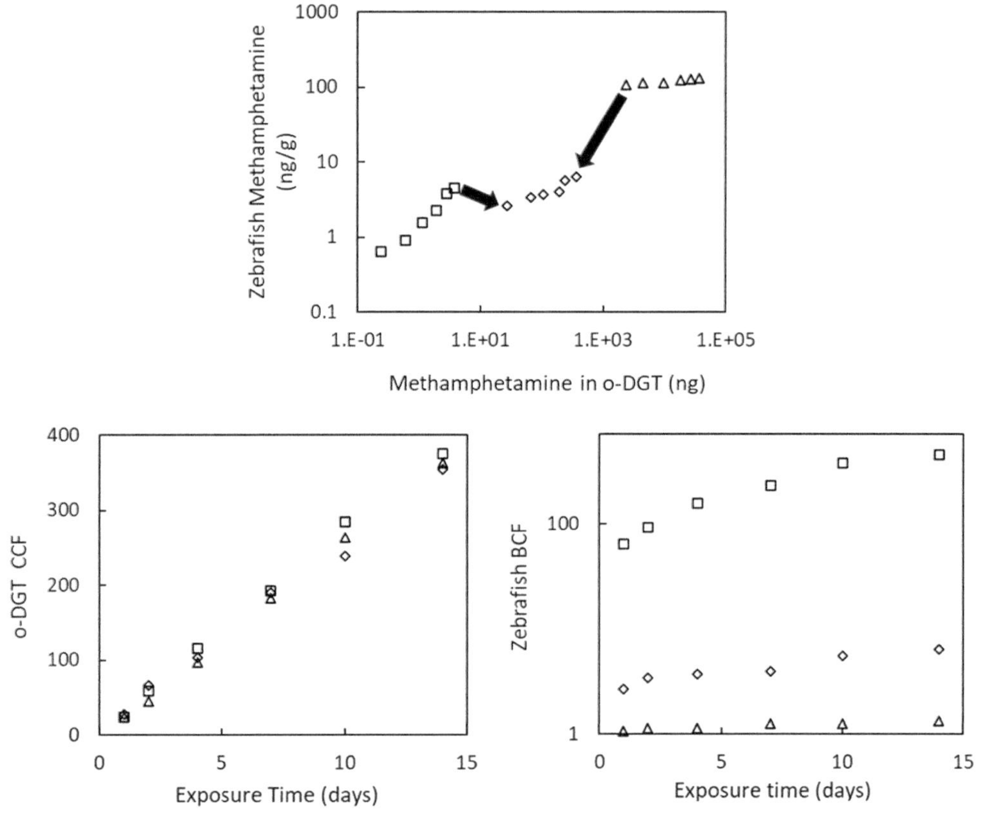

FIGURE 16.5 Top panel: data for the three methamphetamine exposure levels (0.01, 1.0, and 100 ng/g) shown together, from which there are clearly major discontinuities in the correlation for different exposure levels (black arrows). Bottom panel: graphs of CCF for o-DGT and BCF for Zebrafish over time. The differing behaviors of the CCFs and BCFs are the underlying cause of the discontinuities in the top panel.

methamphetamine by Zebrafish, as pollutants in water are accumulated mainly by passive diffusion through gills and skin. While the basis for this rationale is valid, as the graphs in Figure 16.5 show, the behavior of the chemoavailability concentration factors (CCFs) of the o-DGTs is very different from the actual bioconcentration factors (BCFs) exhibited by Zebrafish. Both exhibit linearity at all methamphetamine exposure levels, however, the CCFs for the o-DGT are essentially invariant – the regression lines for all three superimpose, whereas BCFs show a saturation behavior as the exposure level increases. At the highest methamphetamine exposure level, the slope of the BCF versus time is close to zero and the BCF values are near 1, i.e., very little bioconcentration is occurring at this high level of methamphetamine exposure. Yin et al. discuss these differing behaviors in their paper, noting that Zebrafish exhibit lower uptake and increase efflux rates as concentrations of toxic organics increases. This is a biological protective mechanism, and in the case of methamphetamine metabolism to amphetamine comes into play as well. Yin et al. further note that for some other organic pollutants, bioconcentration in Zebrafish ceases in as little as seven days, and that moreover, for some aquatic organisms, bioconcentration is not linearly correlated with the concentrations in the exposure environment. All of these points explain why chemoavailability, in this case as measured by o-DGT, may often not represent a useful surrogate to estimate the bioconcentration or bioaccumulation.[2]

Reviewing the three objectives of the work, the results failed to characterize the relationship between methamphetamine and amphetamine measured by o-DGT and in aqueous-phase samples (i.e., $C_{o\text{-}DGT}$ and C_{sol}). In the case of methamphetamine, the problem was a lack of suitable controls to help understand the potential origin of differences in $C_{o\text{-}DGT}$ and C_{sol}. In the case of amphetamine, the problem was insufficient sampling frequency to understand changes in C_{sol} for amphetamine and how these would be reflected in $C_{o\text{-}DGT}$. The study did successfully measure bioconcentration of methamphetamine in Zebrafish. Results for bioconcentration and biodegradation of methamphetamine in Zebrafish were less convincing due to uncertainties in amphetamine C_{sol}. The study failed to demonstrate that o-DGT could predict bioavailability (i.e., bioconcentration in the case of the case study) of methamphetamine and amphetamine in Zebrafish. Though the majority of the study's primary objectives were not met, it is still a very important study, both as an early work in the overall effort to develop o-DGT for practical environmental use in understanding the bioavailability of organic pollutants and due to the implications of the work, as discussed in the next section.

16.4 IMPLICATIONS OF o-DGT CASE STUDY AND FUTURE OUTLOOK

At the outset of the Yin et al.'s paper under consideration as a case study here, the authors remark that "*Traditional extraction methods (soil solution and solvent extraction) are simple to use and conventionally employed to assess pesticide chemical form* [e.g. speciation] *and bioavailability in soils. However, whilst convenient for regulatory testing, it has been suggested that these approaches may be too crude or are poor predictors of bioavailability, due to their arbitrary original development to detect 'total' concentration using exhaustive extraction*". Yet, this same statement could be similarly true with respect to the alternative of o-DGT. In other words, the authors' comment above could be restated, with reference to o-DGT, as "*the o-DGT approach is too crude and is a poor predictor of bioavailability, due to its arbitrary original development to detect chemoavailable concentration*". For two of three exposure levels in the Yin et al.'s study, like extractions, o-DGT *overestimated*

[2] It should be noted that for this case study, the authors use the term bioaccumulation, when bioconcentration would be more appropriate. Bioconcentration has been defined as "*the intake and retention of a substance in an organism entirely by respiration from water in aquatic ecosystems or from air in terrestrial ones*". Bioaccumulation is "*the intake of a chemical and its concentration in the organism by all possible means, including contact, respiration and ingestion*" (Alexander, 1999). Bioavailability in turn, is an ineluctable prerequisite to bioconcentration and bioaccumulation, and in this sense monitoring for bioconcentration or bioaccumulation is a measure of bioavailability. It would likely be inaccurate to say that 100 ng/L methamphetamine at the high exposure rate was biounavailable to Zebrafish; hence, it is useful to clarify that this case study is about predicting BCFs rather than predicting bioavailability per se.

Zebrafish methamphetamine levels (by factors of *~80 and ~1,500*). Moreover, the only exposure level at which the o-DGT was a reasonable proxy for Zebrafish bioconcentration was the exposure level that might be described as less relevant for such studies (0.01 ng/g), i.e., high levels of pollutant are the focus for practical environmental work in impact and risk assessment. Contrary to Yin et al.'s assertions, because of the differential behavior of CCF versus BCF as a function of exposure level, o-DGT is not a suitable surrogate for methamphetamine uptake and concentration by Zebrafish.

While the points above may not seem intuitively encouraging vis-à-vis the development of o-DGT for studies such as Yin et al. envisage, and despite some inadequacies of QA/QC in this initial study, nonetheless the results strongly argue for expanded potential to use o-DGT as a chemoavailability proxy for bioavailability in future. Yin et al.'s results show strong and consistent relationships with respect to how o-DGT devices and Zebrafish interact with methamphetamine, and to some extent secondarily a similar potential for amphetamine, though the data is not as sound. An important implication of these relationships is that the behaviors of both methamphetamine and amphetamine may be parameterized. Indeed, in obtaining parameters to calculate $C_{o\text{-}DGT}$ and demonstrating that $C_{o\text{-}DGT}$, while differing from C_{sol} in some cases, nonetheless approximates C_{sol}, it can be said that the behavior of methamphetamine has to some extent already been parameterized. Based on Yin et al.'s results, there is no impediment in practice to parameterizing the interactions of Zebrafish with respect to methamphetamine using o-DGT results as a predictor. Though not an equilibrium process, this could be achieved in a ***Biotic Ligand Model*** like framework (which can include kinetic expressions for some processes). To be generalizable, environmental factors such as ionic strength (solution composition), the effect of natural organic matter, and temperature would need to be incorporated into such a model for both o-DGTs and Zebrafish. Such information can be gathered in laboratory studies, for subsequent validation in field conditions (field validation being particularly important given some issues with DGTs in field conditions, Wang et al., 2020c).

Yin et al.'s study was a first and an important step toward relating o-DGT chemoavailability to bioavailability. It is likely unreasonable to expect that chemovailability techniques such as o-DGT ever can fully represent bioavailability, however, to take matters further, it is very important to recognize the limitations of chemoavailability techniques, and to strive for modeling formalisms that systematically relate chemoavailability measurements to bioavailability in a truly validatable predictive framework. The severe drawback of most bioavailability methods is time, with attendant expense, and the typically higher inherent variability of biological measurements. Despite the effort required to implement such modeling formalisms and validation frameworks, in the long run, investment in time and research now will enable collection of more robust data using cheaper approaches, such as o-DGT, in future. Another advantage of such formalisms is the development of standardized approaches, i.e., Standard Methods, which typically also include prescribed QA/QC needs. One particularly attractive feature of o-DGT is that devices with different chemical properties and thicknesses of diffusive gel can be designed to meet specific environmental testing needs (i.e., better mimic bioavailability of specific organisms, ensure low detection limits, etc.). While DGT and emerging o-DGT techniques, as used in contemporary environmental studies, are largely still "too crude" and "a poor predictor of bioavailability", they hold great potential to be time-saving and reasonably adequate predictors of bioavailability in future.

REFERENCES

Alexander, D.E., 1999. Bioaccumulation, bioconcentration, biomagnification. In: *Environmental Geology. Encyclopedia of Earth Science*. Springer, Dordrecht, https://doi.org/10.1007/1-4020-4494-1_31

Bartelt-Hunt, S.L., Snow, D.D., Damon, T., Shockley, J., Hoagland, K., 2009. The occurrence of illicit and therapeutic pharmaceuticals in wastewater effluent and surface waters in Nebraska. *Environ. Pollut.* 157, 786–791. https://doi.org/10.1016/j.envpol.2008.11.025

Brecht, M.L., Herbeck, D., 2014. Time to relapse following treatment for methamphetamine use: A long-term perspective on patterns and predictors. *Drug Alcohol Depend.* 139, 18–25. https://doi.org/10.1016/j.drugalcdep.2014.02.702.

Cao, H.M., Bu, Q.W., Li, Q.S., Gao, X.H., Xie, H.J., Gong, W.W., Wang, X.X., Yang, L., Tang, J.F., 2022. Development and applications of diffusive gradients in thin films for monitoring pharmaceuticals in surface waters. *Environ. Pollut.* 311, 119979. https://doi.org/10.1016/j.envpol.2022.119979

Castrignano, E., Yang, Z., Bade, R., Baz-Lomba, J.A., Castiglioni, S., Causanilles, A., Covaci, A., Gracia-Lor, E., Hernandez, F., Kinyua, J., McCall, A.-K., van Nuijs, A.L.N., Ort, C., Plósz, B.G., Ramin, P., Rousis, N.I., Ryu, Y., Thomas, K.V., de Voogt, P., Zuccato, E., Kasprzyk-Hordern, B., 2018. Enantiomeric profiling of chiral illicit drugs in a pan-European study. *Water Res.* 130, 151–160. https://doi.org/10.1016/j.watres.2017.11.051

Chen, C.E., Chen, W., Ying, G.G., Jones, K.C., Zhang, H., 2015. *In Situ* measurement of solution concentrations and fluxes of sulfonamides and trimethoprim antibiotics in soils using o-DGT. *Talanta* 132, 902–908. https://doi.org/10.1016/j.talanta.2014.08.048

Chen, C.E., Zhang, H., Jones, K.C., 2012. A novel passive water sampler for in situ sampling of antibiotics. *J. Environ. Monitor.* 14, 1523–1530, https://doi.org/10.1039/C2EM30091E

Chen, W., Li, Y.Y., Chen, C.E., Sweetman, A.J., Zhang, H., Jones, K.C., 2017. DGT passive sampling for quantitative in situ measurements of compounds from household and personal care products in waters. *Environ. Sci. Technol.* 51, 13274–13281. https://doi.org/10.1021/acs.est.7b03940

Chen, W., Pan, S.H., Cheng, H., Sweetman, A.J., Zhang, H., Jones, K.C., 2018. Diffusive gradients in thin-films (DGT) for in situ sampling of selected endocrine disrupting chemicals (EDCs) in waters. *Water Res.* 137, 211–219. https://doi.org/10.1016/j.watres.2018.03.029

Farrell, M., Martin, N.K., Stockings, E., Bórquez, A., Cepeda, J.A., Degenhardt, L., Ali, R., Tran, L.T., Rehm, J., Torrens, M., Shoptaw, S., McKetin, R., 2019. Responding to global stimulant use: Challenges and opportunities. *Lancet* 394, 1652–1667. https://doi.org/10.1016/S0140-6736(19)32230-5

Galbraith, N., 2015. The methamphetamine problem. *BJPsych Bull.* 39, 218–220. https://doi.org/10.1192/pb.bp.115.050930

Guibal, R., Buzier, R., Charriau, A., Lissalde, S., Guibaud, G., 2017. Passive sampling of anionic pesticides using the diffusive gradients in thin films technique (DGT). *Anal. Chim. Acta* 966, 1–10. https://doi.org/10.1016/j.aca.2017.02.007

Guibal, R., Buzier, R., Lissalde, S., Guibaud, G., 2019. Adaptation of diffusive gradients in thin films technique to sample organic pollutants in the environment: An overview of o-DGT passive samplers. *Sci. Total Environ.* 693, 133537. https://doi.org/10.1016/j.scitotenv.2019.07.343

Guo, C.S., Zhang, T.T., Hou, S., Lv, J.P., Zhang, Y., Wu, F.C., Hua, Z.D., Meng, W., Zhang, H., Xu, J., 2017. Investigation and application of a new passive sampling technique for in situ monitoring of illicit drugs in waste waters and rivers. *Environ. Sci. Technol.* 51, 9101–9108. https://doi.org/10.1021/acs.est.7b00731

Han, L., Zhao, X., Jin, J., Gao, B., Yang, Y., Sun, K., Li, F., 2019. Plant- and manure-derived hydrochar and pyrochar for alleviating the bioavailability of Cd in soils. *Sci. Total Environ.* 678, 543– 550. https://doi.org/10.1016/j.scitotenv.2019.05.039

International Narcotics Control Board (INCB), 2022. The Report of the International Narcotics Control Board for 2022 (E/INCB/2022/1), United Nations/INCB, Vienna.

Ji, X.W., Challis, J.K., Brinkmann, M., 2022. A critical review of diffusive gradients in thin films technique for measuring organic pollutants: Potential limitations, application to solid phases, and combination with bioassays. *Chemosphere*. 287, 132352. 618, 284–290. https://doi.org/10.1016/j.chemosphere.2021.132352

Jiang, J., Lee, C., Fang, M., Tu, B., Liang, Y., 2015. Impacts of emerging contaminants on surrounding aquatic environment from a youth festival. *Environ. Sci. Technol.* 49, 792–799. https://doi.org/10.1021/es503944e

Kelly, J.R., Shelton, S.G., Daniel, D.K., Bhat, A., Mondal, R., Nipple, F., Amro, H., Bower, M.E., Isaac, G., McHaney, G., Martins, E.P., Shelton, D.S., 2021. Wild Zebrafish sentinels: Biological monitoring of site differences using behavior and morphology. *Toxics* 9, 165. https://doi.org/10.3390/toxics9070165

Kohno, M., Beste, C., Pilhatsch, M., 2020. Editorial: The global methamphetamine problem: Approaches to elucidate the neurobiology, epidemiology, and therapeutic effectiveness. *Front. Psychiatry* 11. https://doi.org/10.3389/fpsyt.2020.00850

Kraemer, T., Maurer, H.H., 1998. Determination of amphetamine, methamphetamine and amphetamine-derived designer drugs or medicaments in blood and urine. *J. Chromatogr. B.* 713, 163–187. https://doi.org/10.1016/S0378-4347(97)00515-X

LaGasse, L.L., Derauf, C., Smith, L.M., Newman, E., Shah, R., Neal, C., Arria, A., Huestis, M.A., DellaGrotta, S., Lin, H., Dansereau, L.M., Lester, B.M., 2012. Prenatal methamphetamine exposure and childhood behavior problems at 3 and 5 years of age. *Pediatrics* 129, 681–8. https://doi.org/10.1542/peds.2011-2209

Li, K., Du, P., Xu, Z., Gao, T., Li, X., 2016. Occurrence of illicit drugs in surface waters in China. *Environ. Pollut.* 213, 395–402. https://doi.org/10.1016/j.envpol.2016.02.036

Li, X.H., Meng, G.J., Chang, Z.B., Lian, X.X., Ma, J.H., Guo, R.C., Wang, Y.L., 2022. Development of organic-diffusive gradients in thin films technique for measuring freely dissolved concentrations of tetracyclines using a commercial SPE packing. *Ecotox. Environ. Saf.* 234, 113359. https://doi.org/10.1016/j.ecoenv.2022.113359

Li, Y., Rothwell, S., Cheng, H., Jones, K.C., Zhang, H., 2019a. Bioavailability and metabolism in a soil-crop system compared using DGT and conventional extraction techniques. *Environ. Int.* 130, 104924. https://doi.org/10.1016/j.envint.2019.104924

Li, Y.Y., Chen, C.E.L., Chen, W., Chen, J.W., Cai, X.Y., Jones, K.C., Zhang, H., 2019b. Development of a passive sampling technique for measuring pesticides in waters and soils. *J. Agr. Food Chem.* 67, 6397–6406. https://doi.org/10.1021/acs.jafc.9b00040

Loganathan, B., Phillips, M., Mowery, H., Jones-Lepp, T.L., 2009. Contamination profiles and mass loadings of macrolide antibiotics and illicit drugs from a small urban wastewater treatment plant. *Chemosphere* 75, 70–77. https://doi.org/10.1016/j.chemosphere.2008.11.047

Mastroianni, N., Bleda, M.J., Lopez de Alda, M., Barceló, D., 2016. Occurrence of drugs of abuse in surface water from four Spanish river basins: Spatial and temporal variations and environmental risk assessment. *J. Hazard. Mater.* 316, 134–142. https://doi.org/10.1016/j.jhazmat.2016.05.025

Pal, R., Megharaj, M., Kirkbride, K.P., Heinrich, T., Naidu, R., 2011. Biotic and abiotic degradation of illicit drugs, their precursor, and by-products in soil. *Chemosphere* 85, 1002–1009. https://doi.org/10.1016/j.chemosphere.2011.06.102

Pelcová, P., Vičarová, P., Ridošková, A., Doćekalová, H., Kopp, R., Mareš, J., Poštulková, E., 2017. Prediction of mercury bioavailability to common carp (*Cyprinus carpio L.*) using the diffusive gradient in thin film technique. *Chemosphere* 187, 181–187. https://doi.org/10.1016/j.chemosphere.2017.08.097

Philipps, R.R., Xu, X., Mills, G.L., Bringolf, R.B., 2018. Impact of natural organic matter and increased water hardness on DGT prediction of copper bioaccumulation by Yellow Lampmussel (*Lampsilis cariosa*) and Fathead Minnow (*Pimephales promelas*). *Environ. Pollut.* 241, 451–458. https://doi.org/10/1015j.envpol.2018.05.059

Ren, S.Y., Wang, Y., Cui, Y., Wang, Y., Wang, X.C., Chen, J.W., Tan, F., 2020. Desorption kinetics of tetracyclines in soils assessed by diffusive gradients in thin films. *Environ. Pollut.* 256, 113394. https://doi.org/10.1016/j.envpol.2019.113394

Santos, M.E.S, Grabicová, K., Steinbach, C., Schmidt-Posthaus, H., Šálková, E., Kolářová, J., Staňová, A.V., Grabic, R., Randák, T., 2020. Environmental concentration of methamphetamine induces pathological changes in brown trout (*Salmo trutta fario*). *Chemosphere* 254, 126882. https://doi.org/10.1016/j.chemosphere.2020.126882

Taylor, A.C., Fones, G.R., Mills, G.A., 2020. Trends in the use of passive sampling for monitoring polar pesticides in water. *Trends Environ. Anal.* 27, e00096. https://doi.org/10.1016/j.teac.2020.e00096

Wang, D.G., Zheng, Q., Wang, X.P., Du, J., Tian, C.G., Wang, Z., Ge, L.K., 2016. Illicit drugs and their metabolites in 36 rivers that drain into the Bohai Sea and North Yellow Sea, North China. *Environ. Sci. Pollut. Res.* 23, 16495–16503. https://doi.org/10.1007/s11356-016-6824-9.

Wang, R., Zou, Y., Luo, J., Jones, K.C., Zhang, H., 2019. Investigating potential limitations of current diffusive gradients in thin films (DGT) samplers for measuring organic chemicals. *Anal. Chem.* 91, 12835–12843. https://doi.org/10.1021/acs.analchem.9b02571.

Wang, R.M., Jones, K.C., Zhang, H., 2020c. Monitoring organic pollutants in waters using the diffusive gradients in the thin films technique: Investigations on the effects of biofouling and degradation. *Environ. Sci. Technol.* 54, 7961–7969. https://doi.org/10.1021/acs.est.0c00224

Wang, Z., Han, S., Cai, M., Du, P., Zhang, Z., Li, X., 2020b. Environmental behavior of methamphetamine and ketamine in aquatic ecosystem: Degradation, bioaccumulation, distribution, and associated shift in toxicity and bacterial community. *Water Res.* 174, 115585. https://doi.org/10.1016/j.watres.2020.115585

Wang, Z., Mao, K., Du, W., Cai, M., Zhang, Z., Li, X., 2020a. Diluted concentrations of methamphetamine in surface water induce behavior disorder, transgenerational toxicity, and ecosystem-level consequences of fish. *Water Res.* 184, 116164. https://doi.org/10.1016/j.watres.2020.116164

Wang, Z., Xu, Z., Li, X., 2018. Biodegradation of methamphetamine and ketamine in aquatic ecosystem and associated shift in bacterial community. *J. Hazard Mater.* 359, 356–364. https://doi.org/10.1016/j.jhazmat.2018.07.039

Weng, Z., Rose, M.T., Tavakkoli, E., Van Zwieten, L., Styles, G., Bennett, W., Lombi, E., 2019. Assessing plant-available glyphosate in contrasting soils by diffusive gradient in thin-films technique (DGT). *Sci. Total Environ.* 646, 735–744. https://doi.org/10.1016/j.scitotenv.2018.07.221

Xie, H.J., Chen, J.W., Chen, Q.N., Chen, C.E.L., Du, J., Tan, F., Zhou, C.Z., 2018. Development and evaluation of diffusive gradients in thin films technique for measuring antibiotics in seawater. *Sci. Total Environ.* 618, 1605–1612. https://doi.org/10.1016/j.scitotenv.2017.09.330

Yabuki, L.N., Colaço, C.D., Menegário, A.A., Domingos, R.N., Kiang, C.H., Pascoaloto, D., 2014. Evaluation of diffusive gradients in thin films technique (DGT) for measuring Al, Cd, Co, Cu, Mn, Ni, and Zn in Amazonian rivers. *Environ. Monit. Assess.* 186, 961–969. https://doi.org/10.1007/s10661-013-3430-x

Yin, H., Cai, Y., Duan, H., Gao, J., Fan, C., 2014. Use of DGT and conventional methods to predict sediment metal bioavailability to a field inhabitant freshwater snail (*Bellamya aeruginosa*) from Chinese eutrophic lakes. *J. Hazard Mater.* 264, 184–194. https://doi.org/10.1016/j.jhazmat.2013.11.030

Yin, X., Guo, C., Lv, J., Hou, S., Zhang, Y., Jin, X., Teng, Y., Xu, J., 2019. Biomimetic accumulation of methamphetamine and its metabolite amphetamine by diffusive gradients in thin films to estimate their bioavailability in Zebrafish. *Environ. Sci. Technol. Lett.* 6, 708–713. https://doi.org/10.1021/acs.estlett.9b00669

Zhang, Y., Zhang, T.T., Guo, C.S., Hou, S., Hua, Z.D., Lv, J.P., Zhang, Y., Xu, J., 2018. Development and application of the diffusive gradients in thin films technique for simultaneous measurement of methcathinone and ephedrine in surface river water. *Sci. Total Environ.* 618, 284–290. https://doi.org/10.1016/j.scitotenv.2017.11.068

Zhu, Q., Ji, J., Tang, X., Wang, C., Sun, H., 2023. Bioavailability assessment of heavy metals and organic pollutants in water and soil using DGT: A review. *Appl. Sci.* 13, 9760. https://doi.org/10.3390/app13179760

Zou, Y.T., Fang, Z., Li, Y., Wang, R., Zhang, H., Jones, K.C., Cui, X., Shi, X., Yin, D., Li, C., Liu, Z., Ma, L.Q., Luo, J., 2018. Novel method for in situ monitoring of organophosphorus flame retardants in waters. *Anal. Chem.* 90, 10016–10023. https://doi.org/10.1021/acs.analchem.8b02480

Zuccato, E., Castiglioni, S., Bagnati, R., Chiabrando, C., Grassi, P., Fanelli, R., 2008. Illicit drugs, a novel group of environmental contaminants. *Water Res.* 42, 961–968. https://doi.org/0.1016/j.watres.2007.09.010

17 Case Study
Assessing A Two-in-One Whole-Cell Bioreporter Approach for Bioavailability and Toxicity of Cadmium and Lead

17.1 INTRODUCTION

Bioreporters, sometimes called whole-cell biosensors, are living microbial cells that have been genetically engineered to produce a measurable signal in response to a specific chemical or physical agent in their environment. The manner in which these modified organisms achieve this depends on a promoter:***reporter gene*** fusion. The ***promoter***, a DNA sequence that regulates the expression of a gene in response to an environmental stimulus, is chosen such that it will have a response to the inducing agent (for environmental studies, a target pollutant or class of pollutants). As a result of being fused to a ***reporter gene***, ***activation*** of the promoter also causes expression of the reporter gene, leading to the production of ***reporter molecules*** that provide a "report" (usually bioluminescence or fluorescence), i.e., the presence of a signal indicates that the bioreporter has sensed ***bioavailable*** pollutant in its environment (Belkin, 2003; van der Meer and Belkin, 2010). Bioreporters have an important role in Environmental Monitoring, providing real-time and cost-effective tools for assessing pollutants in complex ecosystems. Because traditional methods of chemical analysis often require extensive sample preparation and costly instrumentation, the ability of bioreporters to offer direct, real-time data on pollutants makes them particularly valuable for monitoring ***contaminants*** in dynamic environments, such as water bodies, where pollutant concentrations can vary rapidly (Zhang et al., 2017). This rapid response is particularly useful for tracking pollutants, including both organics and heavy metals, when concentrations might fluctuate due to factors such as industrial runoff, weather events, or human ***activity*** (Liu et al., 2022; Zhang et al., 2020a).

Two main categories of bioreporters exist: ***lights-on bioreporters*** and ***lights-off bioreporters*** (Belkin, 2003; Xu et al., 2014). Each type is suited to specific monitoring needs. Lights-off bioreporters are primarily used to measure toxicity. Such organisms emit a continuous signal under non-toxic conditions. When the concentration of a toxic substance exceeds a critical threshold, signal is suppressed, indicating toxic response (Abbas et al., 2018; An et al., 2012). This approach is useful for detecting whether some substance or mixture of substances in an environment has reached a toxic threshold, but provides limited insight into what pollutants are present. Lights-off bioreporters can give a single-point measurement of toxicity, such as the $\boldsymbol{EC_{50}}$ (the effective concentration at which 50% of ***test organisms*** are adversely affected) (Xu et al., 2014). Despite their utility in signaling toxicity, these bioreporters are inadequate for providing detailed, continuous data on pollutant behavior in the environment (Xu et al., 2014). Lights-on bioreporters, on the other hand, produce a signal in response to the chemical speciation of bioavailable pollutants. The intensity of the signal increases with the concentration of bioavailable pollutant, allowing users to monitor the pollutant availability for biological uptake, i.e., true ***bioavailability*** (Magrisso et al., 2009; Zhang et al., 2017). Lights-on reporters are typically responsive to a specific pollutant or class of pollutants

DOI: 10.1201/9781003355229-22

(Zhang et al., 2017). Even though often quite resistant to toxicity of target pollutants to which they respond, the signal from lights-on bioreporters typically diminishes at some high concentration (Li et al., 2022; Zhang et al., 2024).

Despite the strengths of both lights-on and lights-off bioreporters, a gap remains to be bridged between speciation/bioavailability and toxicity (Kessler et al., 2012; Wells, 2012; Zhang et al., 2017). In response to this gap, one study by Li et al. (2022, hereafter "Li et al.")[1] demonstrated an approach wherein a lights-on bioreporter was used to measure both speciation and toxicity of cadmium (Cd) and lead (Pb). Both Cd and Pb are globally recognized as hazardous contaminants with severe consequences for human health and the environment (Alengebawy et al., 2021; Pure Earth/Green Cross, 2015). Cd is classified as a Group 1 carcinogen by the International Agency for Research on Cancer, and chronic ***exposure*** to Cd has been linked to numerous health issues, including kidney disease, lung cancer, and osteoporosis (Khan et al., 2021; Song et al., 2015). Cd persists in the environment, particularly in water bodies and soils, and poses a long-term threat to ecosystems due to its ***bioaccumulation*** in the food chain (Pure Earth/Green Cross, 2015). Pb, similarly, is another major public health concern, particularly in urban and industrialized areas where Pb from older paints, pipes, and gasoline has accumulated over decades (Grandjean and Landrigan, 2014). Lead is especially harmful to children, leading to cognitive impairments, developmental delays, and behavioral problems (Grandjean and Bellange, 2017). While the World Health Organization has established that there is no safe level of blood Pb (World Health Organization, 2021), a study by Pure Earth and United Nations International Children's Emergency Fund (UNICEF) estimates that around one in three children (approximately 800 million globally) have blood Pb levels at or above 5 μg/dL, a blood Pb level that causes irreversible neurological and cognitive impairment in children (Pure Earth and UNICEF, 2020). The economic cost of Pb-induced cognitive deficits alone is estimated at over 1.15 trillion US dollars annually (Grandjean and Bellange, 2017).

In developing a two-in-one speciation/toxicity approach to detection of bioavailable Cd and Pb, Li et al. also integrated bioreporter response with speciation modeling validation (bioavailability) and the ***Biotic Ligand Model*** (BLM, toxicity, see Chapter 14), which can predict the bioavailability and consequent toxicity of heavy metals in aquatic systems based on water chemistry information. This approach offers a significant improvement for the use of bioreporter technology, conveying the benefits of both lights-on and lights-off bioreporters (Zhang et al., 2020b). While the BLM has enjoyed great success in regulatory settings based on its ability, once validated, to predict toxicity for a given toxicity metric and species, validation of a BLM can be extremely time-consuming as it involves the time and labor of raising organisms and subsequent performance of toxicity tests. In contrast, because bioreporters are fast to self-replicate/easy to grow, it is much less work to conduct such validation with these organisms. Further, Li et al. showed how, because bioreporters have a robust response to a range of Cd and Pb concentrations, bioreporter response was able to predict toxicity endpoints from below the 1st and up to the 50th percentile range of ***Genus Mean Acute Values*** (GMAVs) from data available for a large range of aquatic species. An individual GMAV is the geometric mean of EC_{50} values for a genus and are critical metrics for determining toxicity thresholds in aquatic species. GMAVs for sensitive species contribute to development of regulatory criteria, such as the ***Criterion Maximum Concentration*** (CMC), which is the highest concentration of a pollutant in ambient water that an aquatic community can tolerate for brief durations (acute basis) without resulting in any significant effect. The CMC is set below the GMAV for the most sensitive species to provide a safety margin. This ability of the bioreporter to predict toxicity to key sensitive aquatic organisms makes the approach particularly valuable for monitoring priority pollutants in a wide range of aquatic environments.

Given the pervasive nature of Cd and Pb contamination, the need for sensitive, real-time monitoring approaches that can assess both the bioavailability and toxicity of these metals is important for effective environmental management. Li et al.'s approach provides a two-in-one solution that

[1] Entitled "From speciation to toxicity: Using a "Two-in-One" whole-cell bioreporter approach to assess harmful effects of Cd and Pb".

addresses this need, offering real-time data across a concentration continuum that can improve ***Environmental Risk Assessments*** and provide a cost-effective, scalable solution for monitoring priority pollutants.

17.2 SUMMARY OF LI ET AL.'S EXPERIMENTAL APPROACH AND RESULTS

17.2.1 Experimental Approach and Methods

The general experimental approach employed by Li et al. is graphically depicted in Figure 17.1. *Escherichia coli* PHL268 strain zntA (hereafter zntA), sensitive to Cd and Pb, served as the lights-on bioreporter for both metal speciation and toxicity measurement, enabling the detection and response to bioavailable Cd and Pb in aqueous environments. The bioreporter was exposed to solutions with varying Cd and Pb concentrations to assess both bioavailability and toxicity. The bioavailability of Cd or Pb was first measured though bioluminescence produced by zntA lights-on response when in contact with bioavailable Cd or Pb below toxic levels. The toxicity of Cd or Pb was measured by bioluminescence inhibition, and, subsequently using a LIVE/DEAD® BacLight™ Bacterial Viability Kit (L13152, Thermo Fisher). In the LIVE/DEAD assay, dye reagents are added to the bioreporter assay and the live and dead cells are differentiated, respectively, by SYTO 9 and propidium iodide (PI) fluorescent dye staining; the LIVE/DEAD fluorescence emission ratio is then used to calculate percent live versus dead cells (nominal working definition).

Experimental materials used for the work are summarized as follows:

- Type I water
- 100 mL bacterial culture flasks

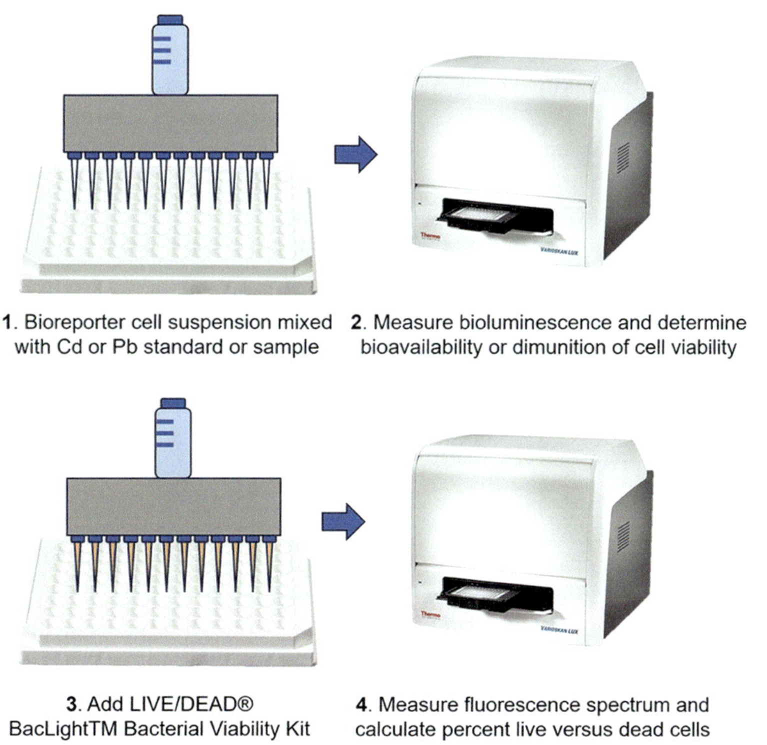

FIGURE 17.1 Graphical depiction of the experimental approach for two-in-one whole-cell bioreporter approach for bioavailability and toxicity of Cd and Pb.

- Autoclave for sterilizing glassware and growth media to prevent contamination during experiments
- 96-well white microtiter plate (Nunc, Denmark)
- Bioreporter strain: zntA, expressing *zntA* reporter gene, which encodes a metal efflux pump that enables the reduction of intracellular toxic metals
- Lysogeny Broth (LB) ***medium*** amended with 40 μg/mL ampicillin
- Optimized minimal medium: 6.06 g of 3-[*N*-morpholino]propanesulfonic acid (MOPS), 2 g of sodium gluconate, 0.23 g of β-glycero-phosphate, 4.68 g of NaCl, 1.07 g of NH_4Cl, 0.43 g of Na_2SO_4, 0.2 g of $MgCl_2{\cdot}6H_2O$, and 0.03 g of $CaCl_2{\cdot}2H_2O$, dissolved in 1,000 mL of distilled water and adjusted to pH 7.0 ± 0.1
- pH meter
- Single-channel and 12-channel pipette (Thermo Scientific, USA)
- Incubator (Thermo Scientific, USA) for maintaining a controlled temperature environment for bacterial culture
- −80°C freezer
- Cd and Pb standard solutions (in 0.002% HNO_3)
- Ca, Mg, K, and Na (major inorganic constituents) stock solutions
- Humic acid (HA, selected to represent dissolved organic matter in aqueous system) solution
- Natural water samples from Lake Tai in China
- LIVE/DEAD® BacLightTM Bacterial Viability Kit (L13152, Thermo Fisher Scientific)
- Inductively coupled plasma–mass spectrometry (ICP–MS, NexION300, PerkinElmer, USA)
- Luminometer (Varioskan LUX, Thermo Fisher Scientific, USA).

The zntA strain was stored on LB agar plates amended with 40 μg/mL ampicillin at 4°C. Overnight cultures of zntA were prepared by growing the cells at 30°C, shaking at 90 rpm for 16 hours. Subsequently, the overnight culture was diluted 100-fold in fresh LB medium without ampicillin and incubated at 26°C with shaking at 200 rpm until reaching an optical density at 600 nm (OD_{600}) of 0.10. Once the desired OD_{600} was achieved, cells were harvested by centrifugation at 13,200×*g*, and the supernatant was discarded. The resulting cell pellet was resuspended in optimized minimal medium to an OD_{600} of 0.20.

For the bioluminescence measurements, 50 μL of either a Cd or Pb standard solution or sample solution was transferred into each well of a 96-well white microtiter plate. Subsequently, 50 μL of $OD_{600} = 0.2$ bioreporter cell suspension was added. Twelve different concentrations of Cd or Pb standards with three duplicates were added to the plate. The metal standards were prepared following the method outlined by Zhang et al. (2017) (in low acid in order to be buffered to appropriate pH by culture medium) and analyzed using ICP–MS (NexION300, PerkinElmer, USA). The plates were then incubated at 26°C, with shaking at 180 rpm, while bioluminescence measurements were recorded every 10 minutes for a total duration of 3 hours using a luminometer (Varioskan LUX, Thermo Fisher Scientific, USA).

To perform the LIVE/DEAD assay, a working solution of the LIVE/DEAD staining reagent was made by mixing solubilized SYTO 9 and PI dyes from the kit to achieve a final concentration of 12 μM SYTO 9 and 60 μM PI. To perform LIVE/DEAD experiments, 100 μL of the prepared LIVE/DEAD working solution was added to each well of the microplate 1:1 with cell suspensions to stain live and dead cells, and then mixed thoroughly by pipette. After 15 minutes incubation in the dark at room temperature, the fluorescence emission at 490–700 nm (emission maxima, λmax, for SYTO 9 and PI being 500 and 635 nm, respectively) was then measured according to the L13152 kit protocol, with excitation at 470 nm. The LIVE/DEAD fluorescence emission ratio was calculated from spectra and the relative amount of live and dead cells determined using a calibration plot, according to the L13152 kit protocol.

The BLM needs to be ***parameterized*** by determining specific binding constants and toxicity thresholds for the target organism, as illustrated and discussed in Chapter 15. For ***parameterization*** of the BLM, it is necessary to first characterize the individual effect of cations Ca, Mg, K, and Na (major inorganic constituents) on Cd and Pb toxic response. The Cd and Pb toxicity tests for BLM parameterization contained four sets of experiments with varying major cation concentrations. Each set was composed of a series of media having different concentrations of one target cation, with the remaining major cation and other ***component*** concentrations being fixed (US EPA, 2002). For each specific target cation concentration in each set, toxicity bioassays across a range of Cd or Pb concentrations were conducted, allowing the calculation of the EC_{50} (Sebaugh, 2011) and the subsequent systematic analysis of the effects of inorganic cations on EC_{50}. Parameters were extracted from experimental data based on standard protocols. For validation of BLM in speciation- and toxicity-mode calculations, three types of validation experiments were conducted, two using lab-prepared samples to mimic the environmental variability of cations and/or humic acids (HA), and the third type of validation experiment used real complex ***matrix*** environmental field samples.

17.2.2 Results

The key results of Li et al. are shown in Figures 17.2–17.4. Figures 17.2 and 17.3 present the speciation and toxicity validation results, respectively, for Cd and Pb, measured using the two-in-one bioreporter (lights-on, x-axis) and predicted by calculation (speciation, BLM, y-axis). The results in Figure 17.2 reflect a bioreporter response to bioavailable Cd or Pb in lights-on mode, as measured by chemical ***activity***.[2] The results in Figure 17.3 reflect a bioreporter response to bioavailable Cd or Pb, at toxic levels, via bioluminescence inhibition (lights-off) and LIVE/DEAD. The standard by

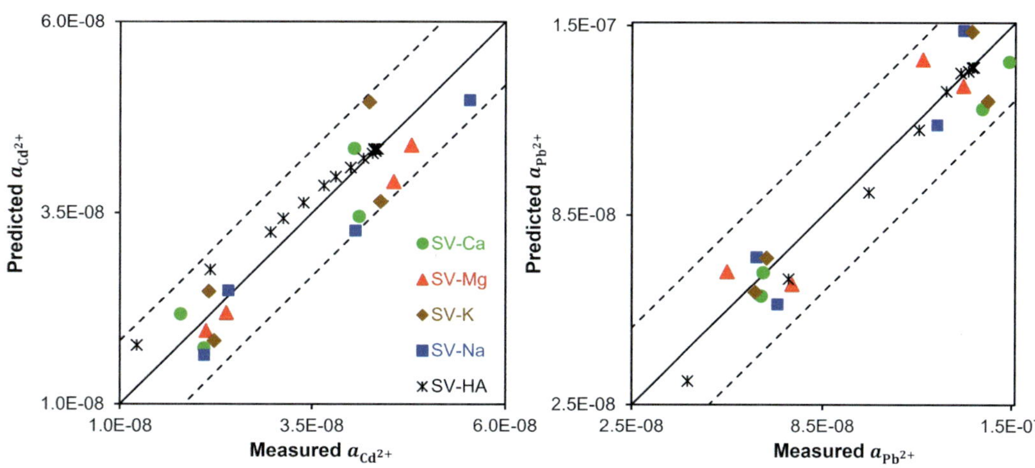

FIGURE 17.2 Validation of bioreporter speciation measurements (x-axis) as compared to model predicted (y-axis) for Cd (left-panels) and Pb (right-panels), where a indicates chemical activity. For both panels, the solid line indicates an exact match between measured and predicted activities, and the dashed lines are plus or minus a factor of twice the pooled average relative percent difference between measured and predicted over the range of Cd and Pb response; this amounts to much better agreement than is required for validation according to US EPA guidance, which is a factor of 2. Legend indicates different samples used for validation: SV indicates speciation validation; Cations Ca, Mg, K, and Na indicate samples with randomly varying major cation concentrations; HA indicates a series of samples with varying humic acid (dissolved organic matter) concentrations.

[2] In chemistry, activity is a measure of the "effective" concentration of a species in a mixture, taking into account interactions with other species that affect its behavior, such as ions in a solution.

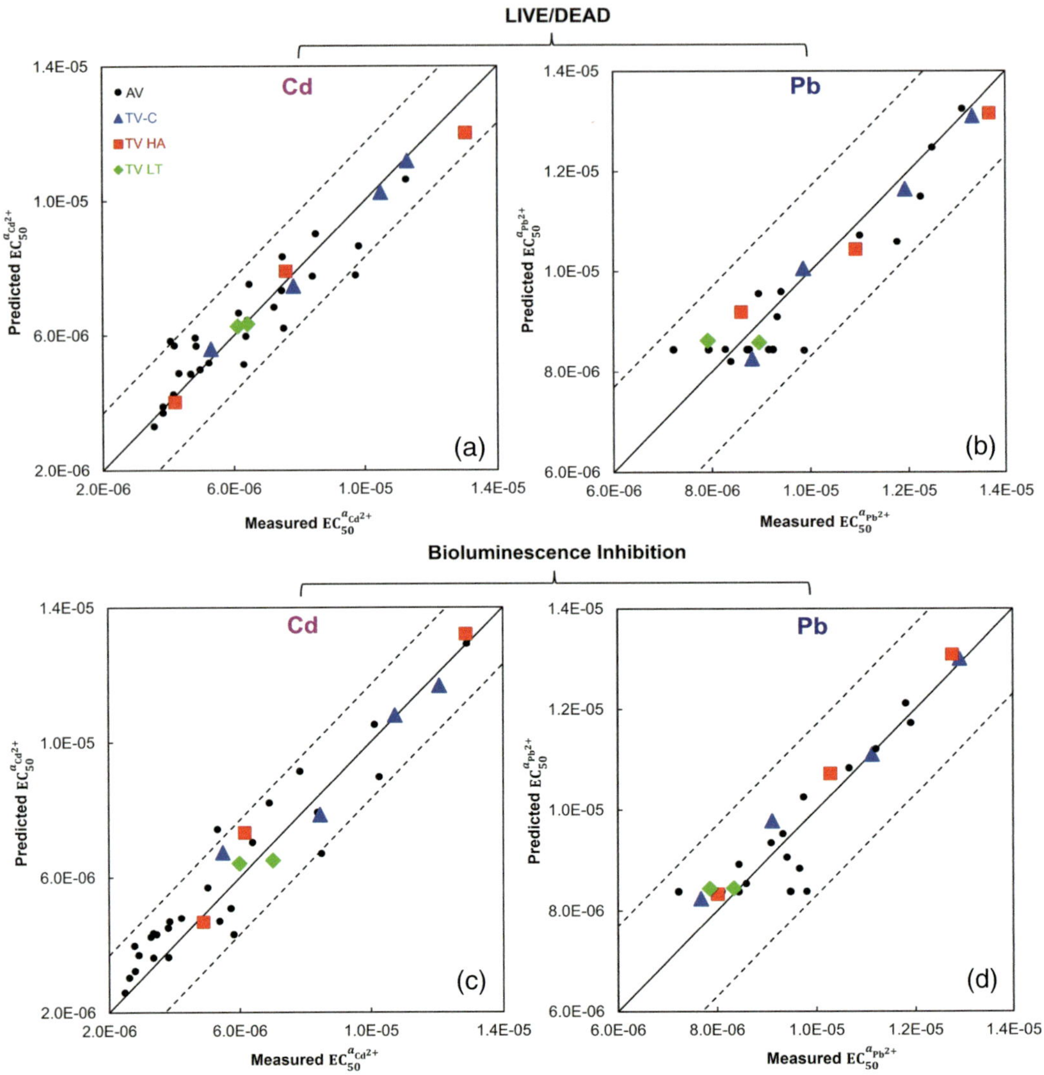

FIGURE 17.3 Validation of bioreporter and LIVE/DEAD toxicity measurements (*x*-axis) as compared to model predicted (*y*-axis) for Cd (left-panels) and Pb (right-panels) according to EC_{50}s. Solid and dashed lines are as for Figure 17.2. Bottom row shows results from bioluminescence inhibition of the bioreporter and the top row shows results using the LIVE/DEAD assay. Legend indicates different samples used for validation: AV indicates ***autovalidation*** data (refer to Chapter 14); TV indicates results from independent toxicity validation experiments; C indicates samples with randomly varying major cation concentrations; HA indicates samples with varying humic acid concentrations; LT indicates freshwater samples from Lake Tai, China.

which Li et al. validated the two-in-one functionality of the bioreporter approach was well below the validation guideline from US EPA (2003).

In Figure 17.4, the BLM-predicted EC_{50} values for Cd and Pb (*y*-axis, predicted using zntA-parameterized BLM) are plotted versus the ranked ***sensitivity*** (*x*-axis) GMAVs for various genera of aquatic organisms based on data from Water Quality Criteria (WQC) guidance documents (US EPA, 1984 for Pb and US EPA, 2016 for Cd). WQC is a regulatory framework that defines acceptable concentrations of pollutants in water to protect aquatic life and ecosystems and hence, gathers GMAVs for calculation of CMCs. In this figure, the most sensitive species used to derive the CMC (*Salvelinus* for Cd and *Gammarus* for Pb) are plotted alongside the least sensitive

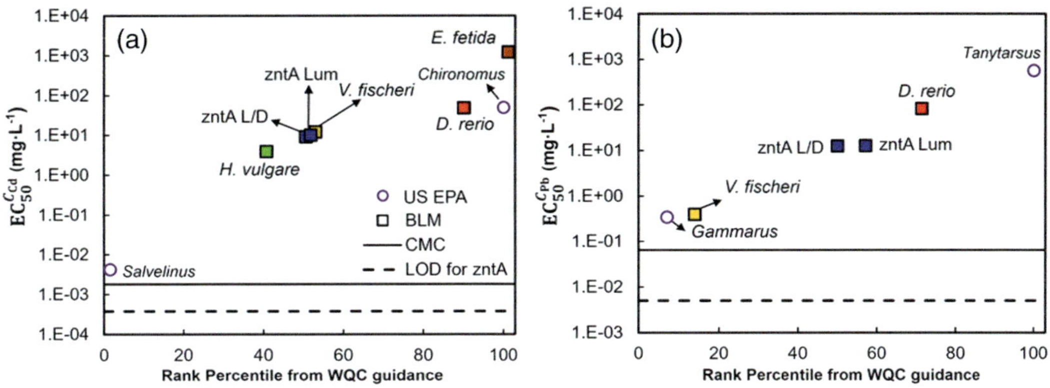

FIGURE 17.4 EC_{50} for Cd (a) and Pb (b) versus rank percentile GMAVs from WQC guidance documentation. Open circles show data for the most sensitive genera used in formulating the CMC (*Salvelinus* for Cd, *Gammarus* for Pb) and for the 100th percentile according to rank (*Chironomus* for Cd, *Tanytarsus* for Pb). For comparison, EC_{50}s of species having validated BLMs are shown in squares. The EC_{50} for *E. fetida* is above the 100th percentile from WQC guidance documentation. The CMCs (solid lines) and zntA LOD (dashed lines) for Cd and Pb are given for reference. All data reflect a hardness of 100 mg/L. The labels L/D and Lum in the figures indicate BLMs for zntA based on LIVE/DEAD and bioluminescence inhibition assays, respectively. Values of EC_{50} on the *y*-axis are given in Cd and Pb concentrations instead of Cd and Pb activity, since concentration is used in regulatory frameworks.

species (*Chironomus* for Cd and *Tanytarsus* for Pb), per data from WQC guidance. These points illustrate the range of sensitivity across genera. For comparison, EC_{50}s for various organisms for which BLMs have been validated are shown (square symbols). The EC_{50}s for the BLMs span a range from very sensitive (*Vibrio fischeri/Aliivibrio fischeri* for Pb) to very insensitive (*Eisenia fetida* for Cd). Because the EC_{50} is a single-point measurement, all of the organisms for which BLMs exist are not particularly useful in the scheme of environmental protection because their EC_{50}s are above the CMC. The exception is zntA. While the zntA EC_{50}s for both Cd and Pb are about mid-range (50th percentile) of the GMAVs, because of the two-in-one functionality of Li et al.'s approach, the zntA bioreporter can report EC_{50}s of other organisms. The EC_{50} range that zntA can report spans the zntA ***limit of detection*** (LOD) for both Cd and Pb, which is notably below the CMC for both, to higher values up to the EC_{50} of zntA, corresponding to the 50th and 55th percentile sensitivity for Cd and Pb. This is a broad and useful detection range, bridging the gap between toxicity assessment and bioavailability monitoring for environmental samples. This means that zntA can measure, and with the validated BLM, predict Cd and Pb EC_{50}s for any genus for which the GMAV falls in the range of 0 to ~50th of GMAVs, i.e., the most sensitive half of all species for which data have been reported in setting the WQC guidance values.

17.3 CRITICAL ANALYSIS OF "TWO-IN-ONE" BIOREPORTER CASE STUDY

17.3.1 Experimental Design and Modeling

The experimental design implemented in Li et al. is innovative, offering a framework for addressing the gap between bioavailability and toxicity measurements, yet grounded in established methodologies. Specifically, the study leverages the BLM, a widely used and rigorously validated approach that has been applied for over two decades, to predict heavy metal toxicity in aquatic organisms by accounting for metal speciation, competition with other ions, and organism-specific sensitivities. The BLM was first formalized in the late 1990s (Di Toro et al., 2001) and has since been integrated into regulatory frameworks, such as the US EPA's guidance for WQC (US EPA, 2003). Importantly, the BLM's validation process continues to evolve and mature, as described by Garman et al. (2019), who emphasize robust model calibration and ***accuracy*** testing across diverse water chemistry conditions to ensure the BLM's reliability for regulatory applications.

Li et al.'s approach incorporates a wide range of dissolved organic matter and major inorganic cation concentrations for these critical parameters known to influence metal bioavailability and toxicity. These factors, explicitly identified as ***toxicity-modifying factors*** in bioavailability-based models, enhance the ecological relevance of the findings by simulating realistic environmental conditions. Garman et al. (2019) highlighted that incorporating such parameters is essential for ensuring that models perform well under diverse and geographically relevant scenarios. Furthermore, the study's experimental validation largely aligns with recommendations from Garman et al. (2019), which stress the importance of testing model predictions against observed data. Figures 17.2 and 17.3 demonstrate that predicted EC_{50} values closely parallel the 1:1 line, with data points distributed above and below the line within a factor of 2, consistent with US EPA guidelines and Garman's validation standards.

While Li et al.'s validation results are generally convincing, as a first demonstration study, there are limitations. The results for Cd speciation modeling in Figure 17.2 indicate some features that Garman et al. highlight for scrutiny. For both Na and Mg, the average predicted values are low compared to measured values. Similarly, at the highest HA concentrations tested, agreement is excellent, however, there appears to be a small but systematic positive bias (predicted values higher than measured) at lower concentrations of HA. The study's experimental conditions, though realistic, did not represent a sufficiently varied envelope of environmental conditions. The organic matter and cation composition of samples used for validation may not have provided sufficient complexity to fully challenge the bioreporter's response envelope or the limitations of current speciation models. In more realistic environmental conditions, with a greater diversity of organic compounds and higher variability in ionic composition, bioreporter and/or model performance might differ. As highlighted by Garman et al. (2019) on validation of bioavailability models, experiments should span orders of magnitude in parameter variation to ensure model reliability across diverse scenarios. For the validation experiments performed in Figure 17.2, the low and high concentrations used for validation of Ca, Mg, K, Na, and HA were 0.01/11 mM, 0.005/50 mM, 0.025/25 mM, 14/80 mM, and 0.598 ppb/19.6 ppm, respectively. These concentrations span approximately 2, 4, 3, <1, and ~1+ orders of magnitude for Ca, Mg, K, Na, and HA, respectively. For Na, using a larger range was not practically possible. Li et al. did not use a higher concentration of Na because the bioreporter zntA is not adapted for higher ***ionic strengths***. On the lower end, reducing Na concentration further was not feasible because the experimental medium for the bioreporter required a minimum Na level to maintain physiological conditions necessary for the bioreporter's functionality. For HA, the use of the concentration range used, 0.598–19.6 ppm adequately covers the range of reported dissolved organic carbon concentrations over a decades-long time span in Li et al.'s large study area (2.86–11.83 ppm, Wang et al., 2020), ensuring that experimental conditions encompassed the natural variability observed in the field.

The aforementioned limitations notwithstanding, Li et al.'s study tested cation variation across 46 sets[3] for Cd and 37 sets for Pb, and dissolved organic matter (HA) variation was tested across 19 sets. In contrast, only two sets of experiments were conducted using real environmental samples for each metal. Compared to previous works, such as An et al. (2012), who tested 12 cation sets, and de Schamphelaere and Janssen (2002), who tested 30 cation sets, Li et al. explored a wider range of cation variations and organic matter concentrations. While inadequate from the perfect ideal, this nonetheless represents an improvement in testing scope.

This example of the number of sets tested reflects how some of the limitations of Li et al.'s work are shared with the state of practice in general. That Li et al. did not test under "tougher" conditions,

[3] A set refers to the manner in which major cation concentrations are varied; each set has a unique combination of cation concentrations, and across the totality of sets each major cation concentration will vary across orders of magnitude. Similar comments pertain to organic matter (HA – humic acid) testing.

such as higher ionic strengths or more extreme pH levels—conditions that could offer more insight into the bioreporter's functionality under stress—is a characteristic in common with many other studies. Also, during parameterization of the BLM, while Ca and Mg consistently showed protective effects, i.e., zntA EC_{50} plotted versus cation concentration was linear with a positive slope (indicating less exposure as the presence of these cations increased), the same was not observed for K and Na in the case of Pb. For zntA exposure to Pb under increasing concentrations of K and Na, Li et al. observed a non-linear response, which for some ranges of K and Na concentrations corresponded to low increases in Pb toxicity, i.e., an anti-protective effect. Li et al.'s results align with prior findings of others (e.g., An et al., 2012; Qu et al., 2013), however, no clear mechanistic explanation for this effect exists, and, as discussed by Li et al., previously proposed theories by other investigators to explain this observation are problematic (Qu et al., 2013). In result, Li et al. excluded K and Na from Pb BLM parameterization, consistent with prior work of others (e.g. de Schamphelaere and Janssen, 2002). While this approach avoids overfitting, it reflects a broader research gap in understanding alkali metal interactions with heavy metal bioavailability.

17.3.2 Quality Assurance and Quality Control

Many research papers do not implement or discuss Quality Assurance and Quality Control (QA/QC). The study by Li et al. implemented QA/QC to ensure the accuracy, reliability, and reproducibility of results from instrumental analysis, as well as following standard guidelines for both speciation and toxicity assessments. For instrumental analysis, ICP–MS was employed to ensure accuracy of the concentrations of Cd and Pb in solutions used for dosing the bioreporter, in laboratory-based model validation samples, and in samples of natural freshwater used in model validation. QA/QC for ICP–MS analysis was performed according to US EPA guidelines (US EPA, 2014), including analysis of a Certified Reference Material (GBW10020 from Beijing Wanjiashouhua Biological Technology Co., Ltd, recovery 104% and 102% for Cd and Pb, respectively).

The true strength of the QA/QC measures for BLM modeling is inherent to the US EPA (2003) standard guidelines, such that each data point in validation results should have replicates and validation should utilize data from independent experiments and independent calculations. For Li et al.'s work, every experiment was replicated at least three times under consistent conditions to ensure reproducibility, with an average relative standard deviation lower than 8%. A combination of autovalidation[4] and independent validation was used, while some studies still rely on autovalidation (An et al., 2012; de Schamphelaere and Janssen, 2002). Moreover, all results for predicted versus measured speciation and toxicity values remained within predefined error tolerance thresholds of less than 20% relative percent difference, reflecting a high level of confidence in execution of work. A unique aspect of Li et al.'s results was the speciation validation, given that most BLM papers focus on toxicity only.

17.3.3 Analysis of Detection Characteristics

The bioreporter used in Li et al.'s study shows remarkable sensitivity at low concentrations, detecting metals below the CMC for both Cd and Pb. This feature is particularly relevant in providing an early indication of contamination, especially in cases where heavy metal concentrations approach but do not exceed regulatory thresholds (toxicity). Combined with the ability of the zntA bioreporter for detection of Cd and Pb across a continuous spectrum of speciation to the relatively robust zntA toxicity endpoints provides a wide range of measurement capacity. Detecting concentrations below

[4] Autovalidation refers to a validation process in which the same dataset used to parameterize or calibrate a model is also used to assess its accuracy. While autovalidation is a standard step in model development, it is not considered a best practice for standalone validation. Instead, independent validation using a separate dataset is recommended to ensure the model's accuracy and generalizability (Chapter 14).

the CMC is important, as it ensures that the bioreporter will always be able to provide a margin of safety in detection capability with reference to regulatory criteria established to protect the most sensitive species. Because the construction of lights-on bioreporters is such that the regulatory system governed by the promoter itself conveys a resistance to the target pollutant to which the bioreporter responds, it seems more likely than not that the approach of Li et al., with respect to Environmental Monitoring of Cd and Pb bioavailability *and* toxicity, will be extensible to a wide range of pollutants for which there are lights-on bioreporters. As the work by Li et al. is a first study exploring the two-in-one bioreporter approach, it remains to be demonstrated to what extent their detection range could be broadened or how effectively the system might be adapted for other contexts.

While Li et al.'s approach is generally demonstrably sound/extensible, with very promising results, a significant drawback applicable specifically to Li et al.'s study is that the bioreporter zntA is selective, not specific. It is possible to validate its use on samples that contain Cd or Pb, but in polluted samples that might contain both, validation would be more complex, perhaps impossible. While the approach is groundbreaking, due to being selective, the bioreporter used by Li et al. does not have strong potential for practical application in water quality management based on quantitative metrics at this time. Bioreporter technology continues to advance, and some recent developments include the engineering of reporters that are specific to individual heavy metal pollutants. Future advancement of Li et al.'s work could focus on using a reporter with more specificity to individual heavy metals. This would enable real progress toward practical use and provide a more comprehensive toolkit for Environmental Risk Assessments, while ensuring that the bioreporter approach remains aligned with ecological and regulatory priorities.

17.4 CONCLUSIONS AND IMPLICATIONS

The two-in-one bioreporter approach developed by Li et al. is an interesting advancement in the field of Environmental Monitoring, offering a way forward to simultaneously evaluate the bioavailability and toxicity of hazardous pollutants such as Cd and Pb, and across the most environmentally relevant concentration ranges. Such an integrated framework enhances our ability to understand the dynamic behavior of pollutants in aquatic environments, which is essential for informed environmental risk management. Because Li et al.'s approach leverages the inherent nature of the bioengineering behind lights-on bioreporters, along with very well-accepted methodologies in experimentation, modeling, and validation, the results are arguably extensible to other bioreporters, providing a foundation for cost-efficient Environmental Risk Assessment of aquatic ecotoxicity across a wide spectrum of priority toxic pollutants. Despite these promising features, challenges remain in translating into fully real-world aquatic environments, which present complex matrices having fluctuating physicochemical characteristics (e.g., pH, organic matter composition, and temperature) that could influence performance in such an approach. Although the zntA bioreporter employed by Li et al. demonstrates the remarkable ability to continuously measure from speciation to toxicity, it would be interesting to develop such an approach further to deal with more pollutants simultaneously. The small size of bioreporters are readily amenable to the use of bioreporter panels, which could address such a scenario. Future work should focus on developing an advanced bioreporter-based system, similar to Li et al.'s zntA approach, but capable of detecting and differentiating between multiple bioavailable pollutants in mixed samples. Such an advancement would represent a major breakthrough in Environmental Monitoring, addressing the complexity of real-world pollution scenarios and enhancing the agility of risk assessments.

ACKNOWLEDGMENT

We wish to acknowledge Boling Li, who was first author crafting the major content of this chapter. While circumstances have prevented Professor Li's formal listing as an author, we fully recognize and appreciate his intellectual and creative contributions to the authorship of this work.

REFERENCES

Abbas, M., Adil, M., Ehtisham-ul-Haque, S., Munir, B., Yameen, M., Ghaffar, A., Shar, G.A., Tahir, M.A., Iqbal, M., 2018. *Vibrio fischeri* bioluminescence inhibition assay for ecotoxicity assessment: A review. *Sci. Total Environ.* 626, 1295–1309. https://doi.org/10.1016/j.scitotenv.2018.01.066

Alengebawy, A., Abdelkhalek, S.T., Qureshi, S.R., Wang, M.Q., 2021. Heavy metals and pesticides toxicity in agricultural soil and plants: Ecological risks and human health implications. *Toxics*, 9, 42. https://doi.org/10.3390/toxics9030042

An, J., Jeong, S., Moon, H.S., Jho, E.H., Nam, K., 2012. Prediction of Cd and Pb toxicity to *Vibrio fischeri* using biotic ligand-based models in soil. *J. Hazard. Mater.* 203, 69–76. https://doi.org/10.1016/j.jhazmat.2011.11.085

Belkin, S., 2003. Microbial whole-cell sensing systems of environmental pollutants. *Curr. Opin. Microbiol.* 6, 206–212. https://doi.org/10.1016/S1369-5274(03)00059-6

de Schamphelaere, K.A.C., Janssen, C.R., 2002. A biotic ligand model predicting copper toxicity for *Daphnia magna*: The effects of calcium, magnesium, sodium, potassium, and pH. *Environ. Sci. Technol.* 36, 48–54. https://doi.org/10.1021/es000253s

Di Toro, D.M., Allen, H.E., Bergman, H.L., Meyer, J.S., Paquin, P.R., Santore, R.C., 2001. Biotic ligand model of the acute toxicity of metals. 1. Technical basis. *Environ. Toxicol. Chem.* 20, 2383–2396. https://doi.org/10.1002/etc.5620201034

Garman, E.R., Meyer, J.S., Bergeron, C.M., Blewett, T.A., Clements, W.H., Elias, M.C., Farley, K.J., Gissi, F., Ryani, A.C., 2019. Validation of bioavailability-based toxicity models for metals. *Environ. Toxicol. Chem.* 39, 101–117. https://doi.org/10.1002/etc.4563

Grandjean, P., Bellange, M., 2017. Calculation of the disease burden associated with environmental chemical exposures: Application of toxicological information in health economic estimation. *Environ. Health*, 16, 1–13. https://doi.org/10.1186/s12940-017-0340-3

Grandjean, P., Landrigan, P.J., 2014. Neurobehavioural effects of developmental toxicity. *Lancet Neurol.* 13, 330–338. https://doi.org/10.1016/S1474-4422(13)70278-3

Kessler, N., Schauer, J.J., Yagur-Kroll, S., Melamed, S., Tirosh, O., Belkin, S., Erel, Y., 2012. A bacterial bioreporter panel to assay the cytotoxicity of atmospheric particulate matter. *Atmos. Environ.* 63, 94–101. https://doi.org/10.1016/j.atmosenv.2012.09.048

Khan, S., Naushad, M., Lima, E.C., Zhang, S., Shaheen, S.M., Rinklebe, J., 2021. Global soil pollution by toxic elements: Current status and future perspectives on the risk assessment and remediation strategies–A review. *J. Hazard. Mater.* 417, 126039. https://doi.org/10.1016/j.jhazmat.2021.126039

Li, B., Zhang, X., Tefsen, B., Wells, M., 2022. From speciation to toxicity: Using a "Two-in-One" whole-cell bioreporter approach to assess harmful effects of Cd and Pb. *Water Res.* 217, 118384. https://doi.org/10.1016/j.watres.2022.118384

Liu, C., Yu, H., Zhang, B., Liu, S., Liu, C.G., Li, F., Song, H., 2022. Engineering whole-cell microbial biosensors: Design principles and applications in monitoring and treatment of heavy metals and organic pollutants. *Biotechnol. Adv.* 108019. https://doi.org/10.1016/j.biotechadv.2022.108019

Magrisso, S., Belkin, S., Erel, Y., 2009. Lead bioavailability in soil and soil components. *Water Air Soil Poll.* 202, 315–323. https://doi.org/10.1007/s11270-009-9978-y

Pure Earth (Blacksmith Institute) and Green Cross, 2015. *World's Worst Pollution Problems: Top Six Toxic Threats*. Pure Earth and Green Cross, New York, Zurich.

Pure Earth and UNICEF (United Nations International Children's Emergency Fund), 2020. *The Toxic Truth: Children's Exposure to lead pollution Undermines a Generation of Future Potential*. Pure Earth and UNICEF, New York.

Qu, R., Wang, X., Liu, Z., Yan, Z., Wang, Z., 2013. Development of a model to predict the effect of water chemistry on the acute toxicity of cadmium to *Photobacterium phosphoreum*. *J. Hazard. Mater.* 262, 288–296. https://doi.org/10.1016/j.jhazmat.2013.08.039

Sebaugh, J.L., 2011. Guidelines for accurate EC_{50}/IC_{50} estimation. *Pharm Stat.*, 10, 128–134. https://doi.org/10.1002/pst.426

Song, W.E., Chen, S.B., Liu, J.F., Chen, L., Song, N.N., Ning, L., Bin, L., 2015. Variation of Cd concentration in various rice cultivars and derivation of cadmium toxicity thresholds for paddy soil by species-sensitivity distribution. *J. Integr. Agr.* 14, 1845–1854. https://doi.org/10.1016/S2095-3119(14)60926-6

US EPA, 1984. Ambient water quality criteria for lead. EPA-440/5-84-027. US Environmental Protection Agency, Washington, DC.

US EPA, 2002. Review of metals action plan: An EPA science advisory board report, EPA-SAB-EC-LTR-03-001. US EPA, Washington, DC.

US EPA, 2003. The biotic ligand model: Technical support document for its application of the evaluation of water quality criteria for copper. EPA-822-R-03-027. US EPA, Washington, DC.
US EPA, 2014. Method 6020B: Inductively Coupled Plasma - Mass Spectrometry. US EPA, Washington, DC.
US EPA, 2016. Ambient water quality criteria for cadmium. EPA-820-R-16-002. US Environmental Protection Agency, Washington, DC.
van der Meer, J.R., Belkin, S., 2010. Where microbiology meets microengineering: Design and applications of reporter bacteria. *Nat. Rev. Microbiol.* 8, 511–522. https://doi.org/10.1038/nrmicro2392
Wang, S., Wang, W., Chen, J., Zhang, B., Zhao, L., Jiang, X., 2020. Characteristics of dissolved organic matter and its role in lake eutrophication at the early stage of algal blooms—A case study of Lake Taihu, China. *Water* 12, 2278. https://doi.org/10.3390/w12082278
Wells, M., 2012. Polycyclic aromatic hydrocarbon (PAH) sensitive bacterial biosensors in environmental health. In: Hunter, R.J., Preedy, V.R. (eds.) *Biosensors and Environment Health.* CRC Environmental Health Series, pp. 170–182, Enfield, NH.
World Health Organization, 2021. Guideline for clinical management of exposure to lead. World Health Organization, Geneva.
Xu, T., Perry, N., Chuahan, A., Sayler, G., Ripp, S., 2014. Microbial indicators for monitoring pollution and bioremediation. In: Das, S., (ed.) *Microbial Biodegradation and Bioremediation,* pp. 115–136. Elsevier.
Zhang, X., Li, B., Deng, J., Qin, B., Wells, M., Tefsen, B., 2020a. Advances in freshwater risk assessment: Improved accuracy of dissolved organic matter-metal speciation prediction and rapid biological validation. *Ecotox. Environ. Saf.* 202, 110848. https://doi.org/10.1016/j.ecoenv.2020.110848
Zhang, X., Li, B., Deng, J., Qin, B., Wells, M., Tefsen, B., 2020b. Regional-scale investigation of dissolved organic matter and lead binding in a large impacted lake with a focus on environmental risk assessment. *Water Res.* 172, 115478. https://doi.org/10.1016/j.watres.2020.115478
Zhang, X., Qin, B., Deng, J., Wells, M., 2017. Whole-cell bioreporters and risk assessment of environmental pollution: A proof-of-concept study using lead. *Environ. Pollut.* 229, 902–910. https://doi.org/10.1016/j.envpol.2017.07.068
Zhang, X., Zhu, Y., Elçin, E., He, L., Li, B., Jiang, M., Yang, X., Yan, X-P., Zhao, X., Wang, Z., Wang, F., Shaheen, S.M., Rinklebe, J., Wells, M., 2024. Whole-cell bioreporter application for rapid evaluation of hazardous metal bioavailability and toxicity in bioprocess. *J. Hazard. Mater.* 461, 132556. https://doi.org/10.1016/j.jhazmat.2023.132556

Index

Note: **Bold** page numbers refer to tables; *italic* page numbers refer to figures and page numbers followed by 'n' refer to notes.

D

I

N

O

P

Q

R

S